Lecture Notes in Computer Science 5852

Commenced Publication in 1973
Founding and Former Series Editors:
Gerhard Goos, Juris Hartmanis, and Jan van Leeuwen

Petra Wiederhold   Reneta P. Barneva (Eds.)

Combinatorial Image Analysis

13th International Workshop, IWCIA 2009
Playa del Carmen, Mexico, November 24-27, 2009
Proceedings

 Springer

Volume Editors

Petra Wiederhold
CINVESTAV- IPN
Centro de Investigación y de Estudios Avanzados
Department of Automatic Control
Av. I.P.N. 2508, Col. San Pedro
Zacatenco, 07000 D.F., Mexico
E-mail: biene@ctrl.cinvestav.mx

Reneta P. Barneva
Department of Computer Science
State University of New York at Fredonia
Fredonia, NY 14063, USA
E-mail: barneva@cs.fredonia.edu

Library of Congress Control Number: 2009938059

CR Subject Classification (1998): I.4, I.5, I.3.5, F.2.2, G.2.1, G.1.6

LNCS Sublibrary: SL 6:
Image Processing, Computer Vision, Pattern Recognition, and Graphics

ISSN   0302-9743
ISBN-10   3-642-10208-5 Springer Berlin Heidelberg New York
ISBN-13   978-3-642-10208-0 Springer Berlin Heidelberg New York

springer.com

© Springer-Verlag Berlin Heidelberg 2009
Printed in Germany

Typesetting: Camera-ready by author, data conversion by Scientific Publishing Services, Chennai, India
Printed on acid-free paper   SPIN: 12785728   06/3180   5 4 3 2 1 0

Preface

The articles included in this volume were presented at the 13th International Workshop on Combinatorial Image Analysis, IWCIA 2009, held at Playa del Carmen, Yucatan Peninsula, Mexico, November 24-27, 2009. The 12 previous meetings were held in Paris (France) 1991, Ube (Japan) 1992, Washington DC (USA) 1994, Lyon (France) 1995, Hiroshima (Japan) 1997, Madras (India) 1999, Caen (France) 2000, Philadelphia (USA) 2001, Palermo (Italy) 2003, Auckland (New Zealand) 2004, Berlin (Germany) 2006, and Buffalo (USA) 2008.

Image analysis is a scientific discipline which provides theoretical foundations and methods for solving problems appearing in a range of areas as diverse as biology, medicine, physics, astronomy, geography, chemistry, robotics, and industrial manufacturing. It deals with algorithms and methods aimed at extracting meaningful information from images. The processing is done through computer systems, and the focus is, therefore, on images presented in digital form. Unlike traditional approaches, which are based on continuous models requiring float arithmetic computations and rounding, "combinatorial" approaches to image analysis (also named "discrete" or "digital" approaches) are based on studying the combinatorial properties of the digital images. They provide models and algorithms, which are generally more efficient and accurate than those based on continuous models. Some recent combinatorial approaches aim at constructing self-contained digital topology and geometry, which might be of interest and importance not only for image analysis, but also as a distinct theoretical discipline.

Following the call for papers, IWCIA 2009 received 70 submissions. After a rigorous review process, 32 were accepted for inclusion in this volume. The Reviewing and Program Committee consisted of 86 scientists from 29 countries and five continents. The whole submission and review process was carried out with the help of the EasyChair conference system. Paper review assignments were largely done matching paper topics to topics of expertise submitted by the reviewers. The selection of papers was very rigorous: each paper was sent to at least four members of the Program Committee for a double-blind review.

The volume starts with the invited paper "Digital Circularity and Its Applications" of Partha Bhowmick, Sahadev Bera, and Bhargab B. Bhattacharya. It summarizes recent research results about properties of digital circles, circular arcs, discs, and surfaces of revolution, which are helpful for their characterization and construction, as well as their detection, segmentation, and reconstruction. In particular, the authors present an efficient and easy to implement algorithm for segmentation and parametrization of circular arcs extracted from graphical regions of digital documents.

The contributed papers included in this volume are grouped into seven sections. The first two sections include nine papers related to digital geometry. The first one focuses on curves, straightness, and convexity, while the second one is devoted to 3D transformations, 3D discrete rotation and angles, medial axis, as well as to metrics and distance transforms on tilings. The third section contains four works on segmentation considering various problems and approaches to their solution, including probabilistic image modeling, straight segment detection in gray-scale images, noise detection along contours, application of Image Foresting Transform, as well as two papers on thinning in 2D and 3D. The next section is devoted to specific algorithms for image representation, processing, analysis, reconstruction and recognition, and their applications in stereo vision, color image compression, and vision-based driver assistance. The fifth section includes papres on digital tomography. The next section considers various image models and related approaches based on combinatorics, arithmetics, algebra, mathematical morphology, topology and grammars. The last section of this book focuses on digital topology and its applications to image modeling and analysis. It studies general and combinatorial topology stuctures such as abstract simplicial and cell complexes, homology groups, and Alexandroff spaces from an image modeling point of view, as well as conjectured properties of Jordan curves in a digital setting.

We would like to express our deepest gratitude to the Center of Research and Advanced Studies CINVESTAV-IPN at Mexico City for their support, to CONACYT (Consejo Nacional de Ciencia y Tecnología, Mexico) for their generous sponsorship through Grant IWCIA09/CONACYT/NSF, and to NSF (National Science Foundation, USA) for their encouragement.

We would also like to thank to everyone who contributed to the success of IWCIA 2009. First of all, the Chairs are indebted to IWCIA's Steering Committee for endorsing the candidacy of Mexico for the 13th edition of the Workshop. Our most sincere thanks go to the members of the Reviewing and Program Committee who worked very hard to ensure the timely review of all submitted manuscripts, as well as to our Co-chair Mariano J.J. Rivera, from CIMAT (Centro de Investigación en Matemáticas) Guanajuato, Mexico, for his valuable help in the analysis of the reviews and in the decision process about paper acceptance.

We wish to express our special thanks to the invited speakers for their remarkable talks and overall contribution to the workshop program:

- Bhargab B. Bhattacharya from the Indian Statistical Institute at Kolkata, India
- Reinhard Klette from the University of Auckland, New Zealand
- Konrad Polthier from "Freie Universität Berlin," Germany
- Jorge Urrutia from "Universidad Nacional Autónoma de México" (UNAM), Mexico

The workshop was organized locally by members from CINVESTAV-IPN Mexico City. We are grateful to our hosts at the Hotel "Iberostar Quetzal" at Playa del Carmen, for providing us with all comforts in this beautiful setting. We also thank the organizers for the interesting cultural events realized, and hope that participants enjoyed their visit to the Mayan ruins at Chichen Itza and Tulum, the Mexican dances and the regional culinary specialities.

We thank all authors for making it so easy to document the state of the art with interesting articles and we hope that you, the reader, will find these contributions interesting. We wish to thank the participants and all who made this workshop an enjoyable and fruitful scientific event. We hope that the attendees benefited from the scientific program and were inspired with new ideas. Finally, we express our gratitude to the Springer editorial team, in particular to Alfred Hofmann, Anna Kramer, Christine Reiss, and Peter Strasser for the efficient and kind cooperation in the timely production of this book.

November 2009   Petra Wiederhold
                                                             Reneta P. Barneva

Organization

The 13th International Workshop on Combinatorial Image Analysis, IWCIA 2009, was held in Playa del Carmen, Quintana Roo State, Mexico, November 24-27, 2009.

General Chair

Petra Wiederhold   CINVESTAV-IPN, Mexico

Steering Committee

Valentin E. Brimkov   SUNY Buffalo State College, USA
Gabor T. Herman   CUNY Graduate Center, USA
Petra Wiederhold   CINVESTAV-IPN, Mexico

Invited Speakers

Bhargab B. Bhattacharya   Indian Statistical Institute, India
Reinhard Klette   University of Auckland, New Zealand
Konrad Polthier   Freie Universität Berlin, Germany
Jorge Urrutia   UNAM, Mexico

Reviewing and Program Committee

Md. Atiqur Rahman Ahad   Kyushu Institute of Technology, Japan,
                                   and University of Dhaka, Bangladesh
Lyuba Alboul   Sheffield Hallam University, UK
Eric Andres   University of Poitiers, France
Akira Asano   Hiroshima University, Japan
Tetsuo Asano   JAIST, Japan
Péter Bálazs   University of Szeged, Hungary
Jacky Baltes   University of Manitoba, Canada
Reneta P. Barneva   SUNY Fredonia, USA
George Bebis   University of Nevada, Reno, USA
Gilles Bertrand   ESIEE, France
Arindam Biswas   Bengal Engin. and Science University, India
Valentin E. Brimkov   SUNY Buffalo State College, USA
Alfred M. Bruckstein   Technion, I.I.T., Israel
Chia-Yen Chen   National University of Kaohsiung, Taiwan
David Coeurjolly   LIRIS-CNRS, University of Lyon, France
Salvador Correa Ruíz   CIMAT, Guanajuato, Mexico

Michel Couprie	ESIEE, France
Marco Cristani	University of Verona, Italy
Guillaume Damiand	LIRIS-CNRS, Université de Lyon, France
Isabelle Debled-Renneson	Henri Poincaré Univ. (Nancy I), LORIA, France
Leila De Floriani	University of Genova, Italy
Eduardo Destefanis	Univ. Tecnol. Nacional Córdoba, Argentina
Antoine Deza	McMaster University, Canada
Ulrich Eckhardt	Universität Hamburg, Germany
Fabien Feschet	LAIC, IUT, France
Boris Flach	Dresden University of Technology, Germany
Chiou-Shann Fuh	National Taiwan University, Taiwan
Jürgen Gall	Max-Plank-Institut Saarbrücken, Germany
Edel Garcia Reyes	CENATAV, Cuba
Edgar Garduño Angeles	IIMAS, UNAM, Mexico
Concettina Guerra	Università di Padova, Italy
Martin Held	Universität Salzburg, Austria
Gabor T. Herman	CUNY Graduate Center, USA
Atsushi Imiya	IMIT, Chiba University, Japan
Ramakrishna Kakarala	Nanyang Technological University, Singapore
Kamen Kanev	Shizuoka University, Japan
Yukiko Kenmochi	CNRS, Univ. of Marne-de-Vallée, France
Christer O. Kiselman	Uppsala University, Sweden
Gisela Klette	University of Auckland, New Zealand
Ullrich Köthe	Universität Hamburg, Germany
T. Yung Kong	CUNY Queens College, USA
Kostadin Koroutchev	Universidad Autonoma de Madrid, Spain
Andreas Koschan	University of Tennessee, USA
Walter G. Kropatsch	Vienna University of Technology, Austria
Norbert Krüger	University of Southern Denmark, Denmark
Fajie Li	Huaqiao University at Xiamen City, China
Jerome Z. Liang	SUNY Stony Brook, USA
Pascal Lienhardt	University of Poitiers, France
Shih-Schon Lin	University of Pennsylvania, USA
Joakim Lindblad	Swedish Univ. of Agricult. Sciences, Sweden
Hongbing Lu	Fourth Military Medical University, China
Rémy Malgouyres	LAIC, IUT, Université d'Auvergne, France
Erik Melin	Uppsala University, Sweden
Christian Mercat	Université de Montpellier 2, France
Benedek Nagy	University of Debrecen, Hungary
Akira Nakamura	Hiroshima University, Japan
Kálmán Palágyi	University of Szeged, Hungary
João Paulo Papa	State University of Campinas, Brazil
Christian Perwass	Institut für Informatik, Germany

Hemerson Pistori	Dom Bosco Catholic University, Brazil
Ioannis Pitas	University of Thessaloniki, Greece
Ralf Reulke	Humboldt Universität Berlin, Germany
Mariano J.J. Rivera	CIMAT, Mexico
Jos Roerdink	University of Groningen, The Netherlands
Christian Ronse	Université Louis Pasteur, France
Bodo Rosenhahn	University of Hannover, Germany
Arun Ross	West Virginia University, USA
Jose Ruiz Shulcloper	CENATAV, Cuba
Henrik Schulz	Forschungszentrum Dresden-Rossendorf, Germany
Isabelle Sivignon	LIRIS-CNRS, University of Lyon, France
Wladyslaw Skarbek	Warsaw University of Technology, Poland
Gerald Sommer	Christian-Albrechts-Universität Kiel, Germany
Alberto Soria	CINVESTAV-IPN, Mexico
Peer Stelldinger	Universität Hamburg, Germany
Robin Strand	Uppsala University, Sweden
Peter Sturm	INRIA Grenoble, France
K. G. Subramanian	Universiti Sains, Malaysia
Akihiro Sugimoto	National Institute of Informatics, Japan
João Manuel R.S. Tavares	University of Porto, Portugal
Mohamed Tajine	Université Louis Pasteur, France
Iván R. Terol-Villalobos	CIDETEQ, Mexico
Peter Veelaert	Hogeschool Gent, Belgium
Young Woon Woo	Dong-Eui University Busan, Korea
Yasushi Yagi	Osaka University, Japan
Gengsheng Lawrence Zeng	University of Utah, USA

Organizing Committee

Reneta P. Barneva, Co-chair	SUNY at Fredonia, USA
Mariano J.J. Rivera, Co-chair	CIMAT, Guanajuato, Mexico
Tonatiuh Matos	CINVESTAV-IPN, Mexico
Maria de la Luz Rodríguez	CINVESTAV-IPN, Mexico
Martha P. Rubio-Gonzalez	CINVESTAV-IPN, Mexico
Mario Villafuerte	CINVESTAV-IPN, Mexico
Petra Wiederhold	CINVESTAV-IPN, Mexico

Table of Contents

Segmentation, Thinning and Skeletonization

Image Representation, Processing, Analysis, Reconstruction and Recognition - Algorithms and Applications

Digital Tomography

Image Models Based on Geometry, Combinatorics, Arithmetics, Algebra, Mathematical Morphology, Topology and Grammars

Digital Topology and its Applications to Image Modeling and Analysis

Digital Circularity and Its Applications*

Partha Bhowmick[1], Sahadev Bera[2], and Bhargab B. Bhattacharya[2]

[1] Department of Computer Science and Engineering
Indian Institute of Technology, Kharagpur - 721302, India
bhowmick@gmail.com
[2] Advanced Computing and Microelectronics Unit
Indian Statistical Institute, Kolkata - 700108, India
sahadev_r@isical.ac.in, bhargab@isical.ac.in

Abstract. This paper presents a brief survey of some recently discovered properties of digital circles, digital discs, and digital surfaces of revolution, which are related with many theoretical challenges in the digital space and have diverse applications in image processing, pattern recognition, and computer graphics. These digital-geometric properties are helpful in today's context of digital revolution to eliminate the inherent discrepancies and limitations of real circles in solving various problems on characterization, construction, reconstruction, and segmentation of digital circles/discs. As a very contemporary problem, we have discussed why and how the notion of digital circularity leads to an efficient and algorithm-friendly interpretation for segmentation and parametrization of circular arcs from graphic regions in a digital document. Another problem is generating a digital surface of revolution, which needs detecting and fixing the absentee voxels in the 3D digital space in order to ensure the irreducibility and continuity of the voxel-surface of revolution. The pattern of absentee voxels in such a surface corresponds to that of absentee pixels in a cover of digital disc with concentric digital circles. Experimental results demonstrate the impact and effectiveness of digital circularity on all these state-of-the-art practices.

Keywords: Computer graphics, digital circle, digital geometry, image analysis, segmentation.

1   Introduction

Characterization, construction, reconstruction, and segmentation of digital circles have many theoretical challenges and practical applications, and have gained much significance in the last few years with the evolution of new paradigms, such as digital calculus [58], digital geometry [47], theory of words and numbers in the digital perspective [8,48,56], etc., These properties are useful not only to generate digital circles, but also to characterize and recognize digital circles and circular arcs from a given set of digital curves and curve segments. In fact, the

---

* Plenary talk by Bhargab B. Bhattacharya at IWCIA 2009.

P. Wiederhold and R.P. Barneva (Eds.): IWCIA 2009, LNCS 5852, pp. 1–15, 2009.
© Springer-Verlag Berlin Heidelberg 2009

latter problem on characterization and recognition is of higher relevance today with the advent of digitization in general, and vectorization in particular [21,25,66,67,36].

Further, apart from the circle generation algorithms, since the properties, parameterization, characterization, and recognition of digital circles and circular arcs constitute a very engrossing area of research, several other works on digital circles and related problems have also appeared from time to time, some of which are as follows:

- *polygonal approximation of digital circles*: [7], [40];
- *characterization of digital circles*: [9], [27], [35], [57], [70];
- *detection/segmentation of circular arcs/objects in digital images*: [14], [15], [16], [18], [23], [38], [46], [51], [58], [62], [63];
- *parameterization of circular arcs*: [13], [24], [69], [74];
- *anti-aliasing solutions for digital circles*: [28]; etc.

It may be mentioned here that, if S is an arbitrary digital curve segment, then each individual run of S is always a run of one or more digital circles. In other words, given any positive integer λ, there always exists a range of digital circles with their radii lying in $[r', r'']$, such that each of these circles (in each of the eight octants, and hence in Octant 1) contains some run(s) having length λ. Thus, trivially, each run of S is always a digital circular arc. The problem becomes non-trivial when two successive runs of S are considered, since we have to decide whether there exists a common digital circle having these two runs in succession. The problem complexity eventually increases when more run-lengths are considered from S.

1.1   Definitions and Preliminaries

Several definitions of digital circles may be seen in the literature [2,3,8,30,60,70]. If we consider the radius $r \in \mathbb{Z}^+$ and the center $c = o(0,0)$, then the first octant of the corresponding digital circle may be defined as follows:

$$\mathcal{C}_1^{\mathbb{Z}}(o, r) = \left\{ (i, j) \in \mathbb{Z}^2 \mid 0 \leqslant i \leqslant j \leqslant r \wedge |j - \sqrt{r^2 - i^2}| < \frac{1}{2} \right\}. \tag{1}$$

Hence, using the 8-octant symmetry of a digital circle [8], the complete circle is given by

$$\mathcal{C}^{\mathbb{Z}}(o, r) = \left\{ (i, j) \mid \{|i|, |j|\} \in \mathcal{C}_1^{\mathbb{Z}}(o, r) \right\}. \tag{2}$$

For an arbitrary center $p(i_p, j_p) \in \mathbb{Z}^2$, the corresponding digital circle is

$$\mathcal{C}^{\mathbb{Z}}(p, r) = \left\{ (i + i_p, j + j_p) \mid (i, j) \in \mathcal{C}^{\mathbb{Z}}(o, r) \right\}. \tag{3}$$

2   Constructing a Digital Circle Using Number Theory

There exists a rich collection of algorithms for construction of digital circles, which were developed mostly in the early period of scan-conversion technique

[4,11,17,20,26,39,50,61,65]. Subsequently, with the advent of newer techniques, several other algorithms were proposed in later periods [10,12,41,54,68,71,72,73]. All these algorithms are essentially based on appropriate digitization of 1st order and 2nd order derivatives (differences), which are predominantly useful to analyze and solve problems involving curves and curve segments in the Euclidean/real plane.

Recently, certain interesting properties of digital circles related with the distribution of perfect squares (square numbers) in integer intervals, have been established [8]. Based on these number-theoretic properties, the problem of constructing a digital circle or a circular arc maps to the new domain of number theory. Given an integer radius and an integer center, the corresponding digital circle can be constructed efficiently by using number-theoretic properties only. These properties enrich the understanding of digital circles from a perspective that is different from those derived earlier by using digital calculus and other methods. A brief review of these number-theoretic properties [8] is presented in this section.

If $p(i, j)$ be a point that lies in $\mathcal{C}_1^{\mathbb{Z}}(o, r)$, then it can be shown that

$$r^2 - j^2 - j \leqslant i^2 < r^2 - j^2 + j. \tag{4}$$

From Eqn. 4, we get the pattern of grid points constituting the first octant of $\mathcal{C}^{\mathbb{Z}}(o, r)$. Since the first grid point in the first octant is always $(0, r)$ (considering the clockwise enumeration), the topmost grid points $(j = r)$ satisfy $0 \leqslant i^2 < r^2 - r^2 + r = r$, or, $0 \leqslant i^2 \leqslant r - 1$. Thus, in Octant 1, the grid points with ordinate r have the squares of their abscissae in the (closed) interval $I_0 := [0, r - 1]$. Similarly, the integer interval containing the squares of abscissae of all the Octant 1 grid points with ordinate $r - 1$, can be shown to be $I_1 = [r, 3r - 3]$. In general, we have the following lemma.

Lemma 1. *[8] The interval $I_k = [(2k - 1)r - k(k - 1), (2k + 1)r - k(k + 1) - 1]$ contains the squares of abscissae of the grid points of $\mathcal{C}^{\mathbb{Z}, I}(o, r)$, whose ordinates are $r - k$, for $k \geqslant 1$.*

From Lemma 1, for $k \geqslant 1$, the length of I_k is $l_k = 2r - 2k$, that of I_{k+1} is $l_{k+1} = 2r - 2(k + 1) = l_k - 2$, whence we have Lemma 2, and subsequently Theorem 1.

Lemma 2. *[8] The lengths of the intervals containing the squares of equiordinate abscissae of the grid points in $\mathcal{C}^{\mathbb{Z}, I}(o, r)$ decrease constantly by 2, starting from I_1.*

Theorem 1. *[8] The squares of abscissae of grid points, lying on $\mathcal{C}_1^{\mathbb{Z}}(o, r)$ and having ordinate $r - k$, lie in the interval $[u_k, v_k := u_k + l_k - 1]$, where*

$$u_k = \begin{cases} u_{k-1} + l_{k-1} & \text{if } k \geqslant 1 \\ 0 & \text{if } k = 0 \end{cases} \tag{5}$$

$$l_k = \begin{cases} l_{k-1} - 2 & \text{if } k \geqslant 2 \\ 2r - 2 & \text{if } k = 1 \\ r & \text{if } k = 0 \end{cases} \tag{6}$$

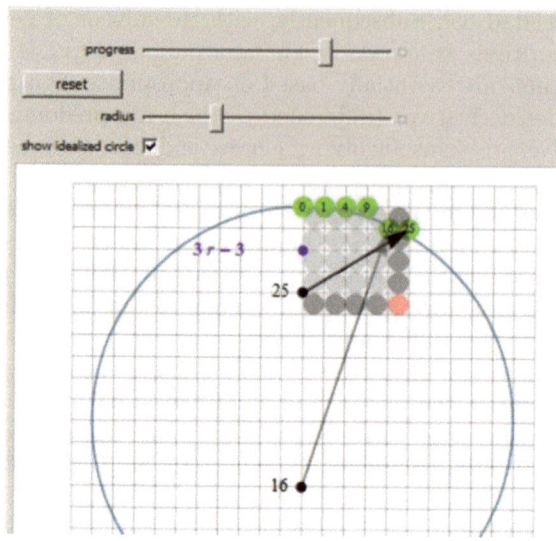

Fig. 1. An implementation of DCR in Wolfram Mathematica [43]

An algorithm DCS (Fig. 1) for construction of digital circles has been designed by Bhowmick and Bhattacharya in [8] based on (numbers of) square numbers in the intervals $I_k (k \geqslant 0)$, using Theorem 1. The algorithm DCS is further improved to another algorithm, DCR, and finally culminated to a hybrid algorithm, DCH, using certain run length properties of a digital circle, as captured in the following theorems.

Theorem 2. [8] *The run length of grid points of $C^{\mathbb{Z}}(o, r)$ with ordinate $j - 1$ never exceeds one more than the run length of its grid points with ordinate j.*

Theorem 3. [8] *If λ_j be the run length of grid points of $C^{\mathbb{Z},I}(o, r)$ with ordinate j, then the run length of grid points with ordinate $j - 1$ for $j \leqslant r - 1$ and $r \geqslant 2$, is given by*

$$\lambda_{j-1} \geqslant \left\lfloor \frac{\lambda_j - 1}{2} \right\rfloor - 1.$$

Combining Theorem 2 and Theorem 3, therefore, we obtain Eqn. 7, which can be used to derive the the horizontal run of grid points with ordinate $j - 1$, from the previous run with ordinate j, for $j \leqslant r$.

$$\left\lfloor \frac{\lambda_j - 1}{2} \right\rfloor - 1 \leqslant \lambda_{j-1} \leqslant \lambda_j + 1 \tag{7}$$

To find the value of λ_{j-1}, a binary search is applied in $[\lfloor (\lambda_j - 3)/2 \rfloor, \lambda_j + 1]$, using a Look-Up-Table (of size N), implemented in the form of a 1-dimensional array that contains the square of each integer $n = 0, 1, 2, \ldots, N$, where N^2 is the largest square not exceeding the maximum value R of radius r. A visual demonstration of Algorithm DCR may be found in Wolfram Mathematica [43].

3   Digital Circularity and Arc Segmentation

Most of the earlier works on the reconstruction/segmentation problem of digital discs/circles did not focus on the inherent digital-geometric properties of digital circles [22,23,35,51,58]. The *circularity measure* [35] is based on an approximate Euclidean measure and fails to provide the exact information even when the concerned object is exactly circular in $\mathbb{Z}^2$ [14,44]. The concepts of *digital convexity* and *digital polygonality* were proposed by Kim [44] and with certain improvements [45], which, however cannot determine the circularity of an arbitrary fragment of a digital circle/disc. The problem of *arc separability* was linked to recognize digital disks, and an algorithm based on Voronoi diagrams in $\mathbb{R}^2$ was proposed by Fisk [29] to find the convex solution-space. Based on a similar approach, an algorithm to recognize digital circles from digital curve segments was presented by Kovalevsky [49]. Sauer [64] proposed a computational-geometric technique was proposed to achieve linear time complexity for the circle recognition algorithm, using the *minimum covering circle* algorithm [55], It was shown that linear programming can be used to design a linear-time algorithm for detecting arcs of digital circles [19]. Characterization of digital circular arcs was related with curvature estimation [70]. Hough transform, a standard practice to segment circular arcs [34,53], was used in several works in improved forms [16,31,42,46]. A geometric characterization was done by Chattopadhyay et al. [14] with an objective of *domain reconstruction* for the full circle from the domains of individual quarter circles. Coeurjolly et al. [18], computed the *arc center domain* of two point sets from the generalized Voronoi cell using the Euclidean metric. A detailed comparison of the above works has been presented Pal and Bhowmick [59].

A very recent technique based on a novel number-theoretic analysis has been reported [59]. Its first step owes to the following fact: Since the *squares of abscissae* of the grid points constituting the $k(\geqslant 0)$th run (in Octant 1) of $\mathcal{C}^{\mathbb{Z}}(o, r)$ lie in the interval $I_k := [\max\{0, (2k-1)r - k(k-1)\}, (2k+1)r - k(k+1) - 1]$ (Lemma 1), there is no change in the corresponding run-length with a change in r as long as I_k "slides on the integer axis" without any change in the squares contained by it. Hence, there exists a set of finitely many circles with identical length of their $k(\geqslant 0)$th runs. With increasing k, the number of squares in I_k usually decreases and occasionally increases by at most unity, provided the radius r remains fixed (Theorem 2). For some $r' > r$ and some $k' > k \geqslant 0$, the interval $I_{k'}$ corresponding to r' may contain a larger number of squares compared to I_k corresponding to r, which opens up many new analytical possibilities [59]. As shown there, it may happen that a digital curve segment S does not belong to any digital circle, or belongs to a particular digital circle, or belongs to a range of digital circles of consecutive integer radii.

The notion of *radii nesting* may be used to successively analyze the runs of digital points that constitute a digital curve segment, S [59]. An interval or a set of intervals of *conflicting radii* plays a crucial role during the analysis, and subsequently the *rate of convergence* of the radius interval depends on the pattern of runs that constitute S. Two algorithms, namely *DCT* and *DCG*, have been proposed, and a simple-yet-effective solution to expedite them using an *infimum circle* and a *supremum circle*, has been developed. It has been also shown how a proper combination of DCT and DCG can be used for segmentation of an arbitrary digital curve segment into a sequence of circular arcs. Integer arithmetic has been used for computing successive radius intervals to test the property of *radii nesting* can be used to decide whether or not a digital curve segment S is digitally circular. The algorithm DCT is very efficient when the circularity testing has to be done on S with the correspondence of its first run with one of the extremum runs of the underlying digital circle/circular arc.

As the algorithm DCT is based on run-length analysis and not on the constituent digital points/pixels of S, the time complexity of DCT is effectively linear on the number of runs, which is one of its major features. The concept of *conflicting radii* has also been discussed to show how the rate of convergence of the effective radius interval depends on the run-length pattern of a digital curve segment. Demonstrations and elaborations detailed out in this paper clearly reveal the challenge of getting a digital-geometric solution to digital circularity.

The general-case solution that needs to compute a *region of radii*, namely $\mathcal{R}_{ak}$, instead of an *interval of radii* as in the case of DCT, is more complex, since it involves the intersection of four parabolic arcs in the real plane, and subsequently taking their appropriate integer rounding. The algorithm DCG solves the decision problem of digital circularity for the general case, which is also extendible for the circular segmentation, as shown there. To expedite the algorithm DCG, especially when the runs in S correspond to those of a digital circle with arbitrary run-positions, a constant-time procedure to bound the circle of solution by an *infimum circle* and a *supremum circle* has been also provided. Experimental results on the related set of digital curves exhibit the potential and elegance of the two algorithms, especially when a curve segment actually consists of a sequence of well-defined digitally circular arcs.

In a recent work [6], the authors have proposed another algorithm for detection of digital circles and circular arcs using *chord property* and *sagitta property*. If S is a circular arc, then its sagitta is the line segment drawn perpendicular to the chord pq, which connects the midpoint m of pq with S. The sagitta property is the following: if the perpendicular to pq at m intersects S at s, then the radius of the circle whose arc is S, is given by

$$r = \frac{d^2(m, s)}{8d(p, q)} + \frac{d(p, q)}{2}$$

where, $d(p, q)$ denotes the Euclidean distance between the points p and q, and $d(m, s)$ is that between m and s.

Since a digital image may contain thick curves, we use thinning as a preprocessing before applying our algorithm. The subsequent steps are (i) finding the intersection points among the digital curve segments and endpoints for open digital curves, and storing them in a list $\mathcal{P}$; (ii) for each point $p_i \in \mathcal{P}$, the corresponding segment(s) incident at p_i is/are extracted and stored in a list of segments, L; (iii) verifying the circularity using a variant of chord property; (iv) parameter estimation using the sagitta property, with necessary updates in

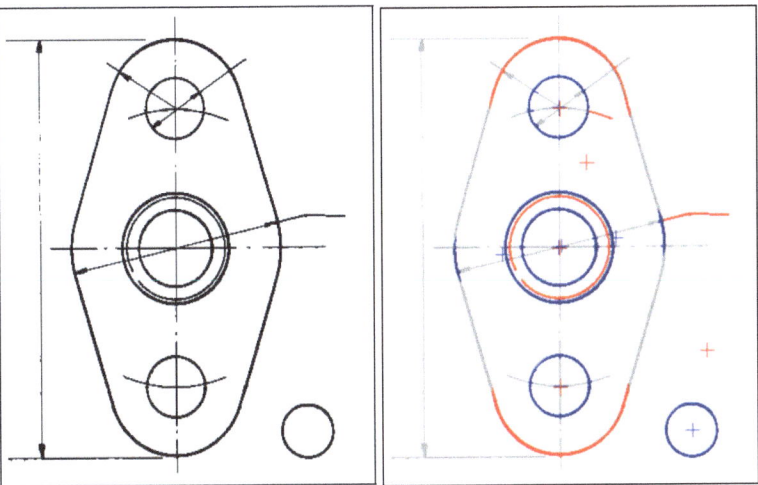

Fig. 2. Results for `2007-1.tif`: Left: input image. Right: final output.

P and S; (v) combining the arcs. Tests have been performed on several database images including those available in http://www.iupr.org/arcseg2007. Results of the algorithm on an image of thick digital curves is shown in Fig. 2.

4 Digital Discs

A digital disc, $\mathcal{D}^{\mathbb{Z}}(o,r)$, $r \in \mathbb{Z}^+$, is given by the union of the set of points defining the corresponding digital circle, $\mathcal{C}^{\mathbb{Z}}(o,r)$, and the set of grid points lying inside $\mathcal{C}^{\mathbb{Z}}(o,r)$. Each point (i,j) of $\mathcal{D}^{\mathbb{Z}}(\alpha,r)$ with center $\alpha(i_\alpha, j_\alpha) \in \mathbb{Z}^2$ has one-to-one correspondence with a unique point (i',j') of $\mathcal{D}^{\mathbb{Z}}(o,r)$, which is given by $i = i' + i_\alpha, j = j' + j_\alpha$. Hence, we can get a digital disc $\mathcal{D}^{\mathbb{Z}}(\alpha,r)$ from $\mathcal{D}^{\mathbb{Z}}(o,r)$, the latter being defined as follows:

$$\mathcal{D}^{\mathbb{Z}}(o,r) = \left\{ (i,j) \in \mathbb{Z}^2 : \ 0 \leqslant ii_c \leqslant i_c^2 \ \wedge \ |\max(|i_c|, |j_c|) - \delta| < \tfrac{1}{2} \right\}, \qquad (8)$$

where, $\delta = \sqrt{r^2 - (\min(|i_c|, |j_c|))^2}$. The condition $0 \leqslant ii_c \leqslant i_c^2$ relates a disc pixel (i,j) to a circle pixel (i_c, j_c), as $i_c \leqslant 0$ implies $0 \leqslant i \leqslant i_c$ and $i_c \geqslant 0$ implies $i_c \leqslant i \leqslant 0$.

An interesting property of a digital disc $\mathcal{D}^{\mathbb{Z}}(o,r)$ is that the union of the digital circles having radius in the range $[1,r]$ falls short of the set of pixels constituting $\mathcal{D}^{\mathbb{Z}}(o,r)$. That *absentee pixels* occur in a cover of digital disc with concentric digital circles, has been shown by Bera et al., in a recent study [5]. A pixel p is an absentee pixel (or simply an absentee) if and only if $p \in \mathcal{D}^{\mathbb{Z}}(o,r)$ and $p \notin \bigcup\limits_{s=1}^{r} \mathcal{C}^{\mathbb{Z}}(o,s)$. Figure 3 shows the absentee pixels for $r \leqslant 20$. It is evident that an absentee lies between two digital circles of consecutive radii, as stated in the following lemma.

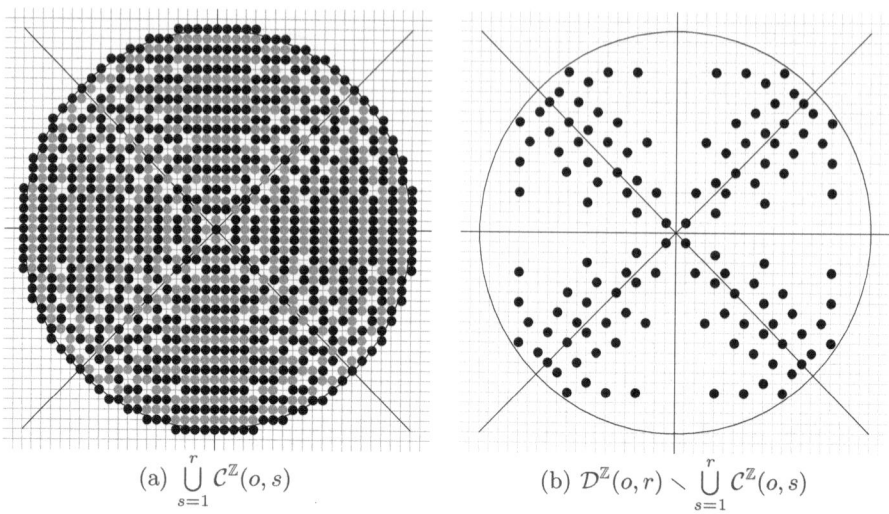

(a) $\bigcup\limits_{s=1}^{r} \mathcal{C}^{\mathbb{Z}}(o,s)$ (b) $\mathcal{D}^{\mathbb{Z}}(o,r) \smallsetminus \bigcup\limits_{s=1}^{r} \mathcal{C}^{\mathbb{Z}}(o,s)$

Fig. 3. Absentee pixels for $r \leqslant 20$

Lemma 3. *[5] If $p(i,j)$ is an absentee in Octant 1, then $(i,j-1) \in \mathcal{C}^{\mathbb{Z}}(o,r)$ and $(i,j+1) \in \mathcal{C}^{\mathbb{Z}}(o,r+1)$ for some $r \in \mathbb{Z}^{+}$.*

Bera et al. [5] have enhanced the notation from I_k corresponding to the kth run (Sec. 2) to $I_{r-j}^{(r)} = \left[u_{r-j}^{(r)}, v_{r-j}^{(r)} \right)$, where $u_{r-j}^{(r)} = r^2 - j^2 - j$ and $v_{r-j}^{(r)} = r^2 - j^2 + j$, which denotes the interval containing the squares of abscissae of the pixels in $\mathcal{C}^{\mathbb{Z}}(o,r)$ with ordinate j. So, the interval containing the squares of abscissae of the pixels in $\mathcal{C}^{\mathbb{Z}}(o,r+1)$ with ordinate j is given by $I_{r+1-j}^{(r+1)} = \left[(r+1)^2 - j^2 - j, (r+1)^2 - j^2 + j \right)$. Thus, the difference between the lower limit of $I_{r-j}^{(r)}$ and the upper limit of $I_{r+1-j}^{(r+1)}$ is given by

$$u_{r+1-j}^{(r+1)} - v_{r-j}^{(r)} = ((r+1)^2 - j^2 - j) - (r^2 - j^2 + j) = 2(r-j)+1. \qquad (9)$$

As $j \leqslant r$ in Octant 1, we have the following lemma.

Lemma 4. *[5] For $r > 0$, $I_{r-j}^{(r)}$ and $I_{r+1-j}^{(r+1)}$ are disjoint and $u_{r+1-j}^{(r+1)} > v_{r-j}^{(r)}$.*

Using Lemma 4, the existence of an absentee pixel can be decided, as stated in Lemma 5.

Lemma 5. *[5] A pixel $p(i,j)$ is an absentee if and only if i^2 lies in $J_{r-j}^{(r)} := \left[v_{r-j}^{(r)}, u_{r+1-j}^{(r+1)} \right)$ for some $r \in \mathbb{Z}^{+}$.*

An example of absentee is $(2,4)$, since for $j = 4$, we have $r = 4$ for which $v_{r-j}^{(r)} = r^2 - j^2 + j = 16 - 16 + 4 = 4$ and $u_{r+1-j}^{(r+1)} = (r+1)^2 - j^2 - j = 25 - 16 - 4 = 5$, thus giving $J_0^{(4)} = [4,5) = [4,4]$ in which lies the square number 4. On the contrary,

$(3, 4)$ is not an absentee, as for $j = 4$, there is no such r for which $J_{r-4}^{(r)}$ contains 3^2; in fact, for $j = 4$, we get the interval $I_{5-4}^{(5)} = [5^2 - 4^2 - 4, 5^2 - 4^4 + 4) = [5, 12]$ with $r = 5$, which contains 3^2, thereby making $(3, 4)$ a point on $\mathcal{C}^{\mathbb{Z}}(o, 5)$.

Although Lemma 5 provides a way to decide whether or not a given pixel is an absentee, it requires to find for which value(s) of r the existence of square numbers in $J_{r-j}^{(r)}$ has to be checked. The following theorem (proof given in [5]) tells exactly the value of r for which $J_{r-j}^{(r)}$ has to be tested for a given pixel $p(i, j)$.

Theorem 4. *The pixel (i, j) is an absentee if and only if $i^2 \in J_{r-j}^{(r)}$, where $r = \max \left\{ s \in \mathbb{Z} : s^2 < i^2 + j^2 \right\}$.*

To characterize the absentees as a whole, r is replaced by $j + k$, where k is the index of a run of pixels in $\mathcal{C}_1^{\mathbb{Z}}(o, r)$ with ordinate j. Hence, if $p(i, j)$ is a point in Octant 1 and lies on kth run of $\mathcal{C}_1^{\mathbb{Z}}(o, r)$, then

$$i^2 < (2k + 1)j + k^2; \tag{10}$$

and if p lies left of the $(k + 1)$th run of $\mathcal{C}_1^{\mathbb{Z}}(o, r + 1)$, then

$$i^2 < (2k + 1)j + (k + 1)^2. \tag{11}$$

On replacing $\langle i, j \rangle$ by $\langle x, y \rangle$ with k as a constant, Eqns. 10 and 11 correspond to two *open parabolic regions* in the real plane, given by

$$\begin{aligned} \underline{P}_k &: x^2 < (2k + 1)y + k^2, \\ \overline{P}_k &: x^2 < (2k + 1)y + (k + 1)^2. \end{aligned} \tag{12}$$

As a result, we have the following lemma.

Lemma 6. *All pixels in $F_k := P_k \cap \mathbb{Z}_1^2$ are absentees, where $P_k := \overline{P}_k \setminus \underline{P}_k = (2k + 1)y + k^2 \leqslant x^2 < (2k + 1)y + (k + 1)^2$.*

The family of all the half-open parabolic strips, $P_0, P_1, P_2, \ldots$, thus contains all the absentees in Octant 1, as stated in the following theorem.

Theorem 5. *Only and all the absentees of Octant 1 and Octant 8 lie in $\mathcal{F} := \left\{ P_k \cap \mathbb{Z}_1^2 : k = 0, 1, 2, \ldots \right\}$.*

5   Surface of Revolution

Digital circles can also be used to generate a digital surface of revolution, which can be subsequently processed to create interesting potteries, as shown in a recent work [52]. The existing graphic tools are mostly based on complex trigonometric procedures involving computations in the real space, which have to be tuned properly to suit the discrete nature of the 3D digital space [33,37]. For, the method of *circularly sweeping* a shape/polygon/polyline/generating curve about the axis of revolution is done in discrete steps using a sequence of transformation matrices, which requires a discrete approximation. The technique of *digital wheel-throwing*, as described by Kumar et al. [52], works purely in the digital domain and uses a few primitive integer computations only. Given a digital generatrix as an *irreducible digital curve segment*, the digital surface produced by the digital wheel-throwing is both connected and irreducible in digital-geometric sense. A successful rendition is finally ensured with quad decomposition, texture mapping, illumination, etc. Producing a monotone or a non-monotone digital surface of revolution is also feasible without destroying its digital connectivity and irreducibility by respective input of a monotone or a non-monotone digital generatrix.

The algorithm proposed by Kumar et al. [52] works with the following principle. A digital curve segment $\mathcal{G} := \{p_i : i = 1, 2, \ldots, n\}$ is a finite sequence of digital points (i.e., points with integer coordinates) [47], in which two points $p(x, y) \in \mathcal{G}$ and $p'(x', y') \in \mathcal{G}$ are *8-neighbors* of each other, if and only if $\max(|x - x'|, |y - y'|) = 1$. To ensure that $\mathcal{G}$ is *simple*, *irreducible*, and *open-ended*, each point $p_i(i = 2, 3, \ldots, n - 1)$ should have exactly two neighbors, and each of p_1 and p_n should have one, from $\mathcal{G}$. Thus, the chain code of a point $p_i \in \mathcal{G}$ w.r.t. its previous point $p_{i-1} \in \mathcal{G}$ is given by $c_i \in \{0, 1, 2, \ldots, 7\}$ [32]. In order to generate a digital surface of revolution, $\mathcal{S}$, a digital curve segment $\mathcal{G}$ is considered as the *digital generatrix*. The digital generatrix may be taken as input either as a sequence of chain codes or as a sequence of control points. For the latter, given $m(\geqslant 4)$ control points, the digital generatrix is constructed first as the digital irreducible curve segment that approximates the sequence of $m - 3$ uniform non-rational cubic B-spline segments interpolating the sequence of these control points [30,37].

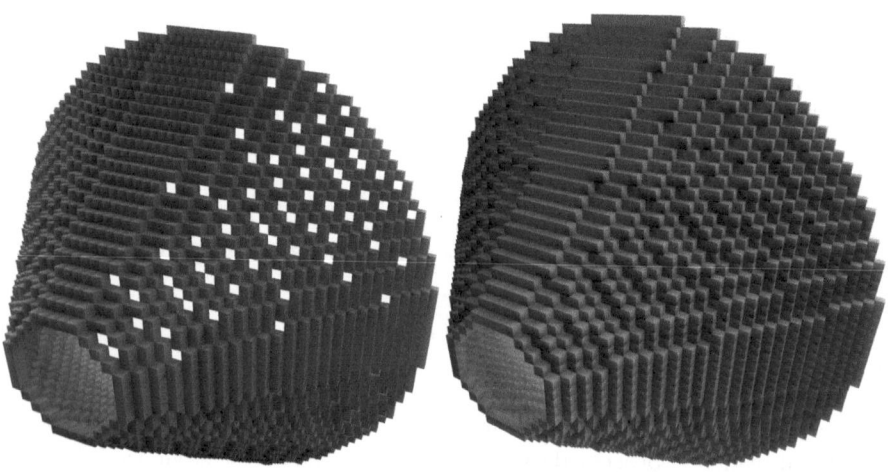

Fig. 4. Voxel mesh of a digital surface of revolution. The missing voxels (left) are detected and included to successfully create the digitally connected and irreducible surface (right).

Fig. 5. Some potteries created by wheel-throwing in digital space [52]

A wheel-thrown piece is generated by revolving the generatrix $\mathcal{G} := \{p_i : i = 1, 2, \ldots, n\}$ about an axis of revolution given by $\alpha : \langle z = -c, x = a \rangle$, where a and c are two positive integers. In order to achieve this, for each digital point $p_i \in \mathcal{G}$, the digital circle $\mathcal{C}^{\mathbb{Z}}_i$ is constructed by revolving p_i around the specified axis of revolution, α. However, as $\mathcal{G}$ is irreducible in nature, two consecutive points p_i and p_{i+1} have their distances, namely r_i and r_{i+1}, measured from α, differing by at most unity. If their distances from α is same, then the surface $\mathcal{C}^{\mathbb{Z}}_i \cup \mathcal{C}^{\mathbb{Z}}_{i+1}$ is digitally connected and irreducible. The problem arises when p_i and p_{i+1} have their respective distances from α differing by unity. Then there may arise some missing voxels trapped between $\mathcal{C}^{\mathbb{Z}}_i$ and $\mathcal{C}^{\mathbb{Z}}_{i+1}$, which results in digital disconnectedness in the surface $\mathcal{C}^{\mathbb{Z}}_i \cup \mathcal{C}^{\mathbb{Z}}_{i+1}$, as shown in Fig. 4. Detection of these missing voxels is performed to achieve a digitally connected surface, namely $S_\mathcal{G} := \mathcal{C}^{\mathbb{Z}}_1 \cup \mathcal{C}^{\mathbb{Z}}_2 \cup \ldots \cup \mathcal{C}^{\mathbb{Z}}_n$, using the respective projections, $\mathcal{C}^{\mathbb{Z}'}_i$ and

$\mathcal{C}^{\mathcal{Z}'}_{i+1}$, of $\mathcal{C}^{\mathcal{Z}}_i$ and $\mathcal{C}^{\mathcal{Z}}_{i+1}$ on the zx-plane. If $r_{i+1} > r_i$, w.l.o.g., then a "missing voxel" between $\mathcal{C}^{\mathcal{Z}}_i$ and $\mathcal{C}^{\mathcal{Z}}_{i+1}$ is formed only if there is a diagonal transition, and hence a change in run, from a point/pixel in $\mathcal{C}^{\mathcal{Z}'}_{i+1}$ giving rise to a "missing pixel" between $\mathcal{C}^{\mathcal{Z}'}_i$ and $\mathcal{C}^{\mathcal{Z}'}_{i+1}$. In other words, if a surface of revolution has its axis orthogonal to xy-plane and $r_i \neq r_{i+1}$, then the projections on xy-plane of the missing voxels lying between $\mathcal{C}^{\mathcal{Z}}_i$ and $\mathcal{C}^{\mathcal{Z}}_{i+1}$ have 1-to-1 correspondence with the missing pixels lying between $\mathcal{C}^{\mathcal{Z}'}_i$ and $\mathcal{C}^{\mathcal{Z}'}_{i+1}$.

6   Concluding Notes

The bizarre nature of digital circles, which marks its distinctiveness with real circles, has been made explicit in this paper. All the simple and axiomatic properties of real circles, such as chord property, sagitta property, properties related with intersection between two or more circles, cannot be directly used for digital circles — a fact that calls for visiting and revisiting the domain of digital circles with new theoretical paradigms in order to enrich its understanding and to invent novel algorithms for various practical applications. A relevant open problem in the context of this paper is to get a closed-form solution for the number of absentee pixels in a cover of digital disc of radius r with concentric digital circles. Another one is, given a sequence S of horizontal run-lengths corresponding to an irreducible digital curve, we have to decide whether there exists one or more real circles (with real- or integer-valued radius and center) whose digitization gives S. Regarding generation of digital surface of revolution, future possibility lies in generating irregular/distorted surfaces and subsurfaces of revolution, and in defining various digital-geometric operations on those digital surfaces so as to generate various 3D digital surfaces replicating their real-world counterparts.

References

1. Aken, J.R.V., Novak, M.: Curve-drawing algorithms for raster display. ACM Trans. Graphics 4(2), 147–169 (1985)
2. Andres, E.: Discrete circles, rings and spheres. Computers & Graphics 18(5), 695–706 (1994)
3. Andres, E., Jacob, M.: The discrete analytical hyperspheres. IEEE Trans. Visualization and Computer Graphics 3(1), 75–86 (1997)
4. Badler, N.I.: Disk generators for a raster display device. Computer Graphics and Image Processing 6, 589–593 (1977)
5. Bera, S., Bhowmick, P., Bhattacharya, B.B.: Absentee pixels in a cover of digital disc with concentric digital circles. TR-ISI/09/ACMU/1 (2009)
6. Bera, S., Bhowmick, P., Bhattacharya, B.B.: Detection of circular arcs in a digital image using chord and sagitta properties. In: Proc. Eighth Intl. Workshop on Graphics Recognition (GREC 2009), pp. 59–70 (2009)
7. Bhowmick, P., Bhattacharya, B.B.: Approximation of digital circles by regular polygons. In: Singh, S., Singh, M., Apte, C., Perner, P. (eds.) ICAPR 2005. LNCS, vol. 3686, pp. 257–267. Springer, Heidelberg (2005)

8. Bhowmick, P., Bhattacharya, B.B.: Number-theoretic interpretation and construction of a digital circle. Discrete Applied Mathematics 156(12), 2381–2399 (2008)
9. Biswas, S.N., Chaudhuri, B.B.: On the generation of discrete circular objects and their properties. Computer Vision, Graphics, and Image Processing 32(2), 158–170 (1985)
10. Blinn, J.F.: How many ways can you draw a circle? IEEE Computer Graphics and Applications 7(8), 39–44 (1987)
11. Bresenham, J.E.: A linear algorithm for incremental digital display of circular arcs. Communications of the ACM 20(2), 100–106 (1977)
12. Bresenham, J.E.: Run length slice algorithm for incremental lines. In: Earnshaw, R.A. (ed.) Fundamental Algorithms for Computer Graphics. NATO ASI Series, vol. F17, pp. 59–104. Springer, Heidelberg (1985)
13. Chan, Y.T., Thomas, S.M.: Cramer-Rao lower bounds for estimation of a circular arc center and its radius. Graphical Models and Image Processing 57(6), 527–532 (1995)
14. Chattopadhyay, S., Das, P.P., Ghosh-Dastidar, D.: Reconstruction of a digital circle. Pattern Recognition 27(12), 1663–1676 (1994)
15. Chen, T.C., Chung, K.L.: An efficient randomized algorithm for detecting circles. Computer Vision and Image Understanding 83(2), 172–191 (2001)
16. Chiu, S.H., Liaw, J.J.: An effective voting method for circle detection. Pattern Recognition Letters 26(2), 121–133 (2005)
17. Chung, W.L.: On circle generation algorithms. Computer Graphics and Image Processing 6, 196–198 (1977)
18. Coeurjolly, D., Gérard, Y., Reveillès, J.-P., Tougne, L.: An elementary algorithm for digital arc segmentation. Discrete Applied Mathematics 139, 31–50 (2004)
19. Damaschke, P.: The linear time recognition of digital arcs. Pattern Recognition Letters 16, 543–548 (1995)
20. Danielsson, P.E.: Comments on circle generator for display devices. Computer Graphics and Image Processing 7(2), 300–301 (1978)
21. Davies, E.: Machine Vision: Theory, Algorithms, Praticalities. Academic Press, London (1990)
22. Davies, E.R.: A modified Hough scheme for general circle location. Pattern Recogn. 7(1), 37–43 (1984)
23. Davies, E.R.: A high speed algorithm for circular object detection. Pattern Recognition Letters 6, 323–333 (1987)
24. Davies, E.R.: A hybrid sequential-parallel approach to accurate circle centre location. Pattern Recognition Letters 7, 279–290 (1988)
25. Dori, D., Liu, W.: Sparse pixel vectorization: An algorithm and its performance evaluation. IEEE Trans. Pattern Anal. Mach. Intell. 21(3) (1999)
26. Doros, M.: Algorithms for generation of discrete circles, rings, and disks. Computer Graphics and Image Processing 10, 366–371 (1979)
27. Doros, M.: On some properties of the generation of discrete circular arcs on a square grid. Computer Vision, Graphics, and Image Processing 28(3), 377–383 (1984)
28. Field, D.: Algorithms for drawing anti-aliased circles and ellipses. Computer Vision, Graphics, and Image Processing 33(1), 1–15 (1986)
29. Fisk, S.: Separating point sets by circles, and the recognition of digital disks. IEEE Trans. PAMI 8, 554–556 (1986)
30. Foley, J.D., Dam, A.V., Feiner, S.K., Hughes, J.F.: Computer Graphics — Principles and Practice. Addison-Wesley, Reading (1993)
31. Foresti, G.L., Regazzoni, C.S., Vernazza, G.: Circular arc extraction by direct clustering in a 3D Hough parameter space. Signal Processing 41, 203–224 (1995)

32. Freeman, H.: On the encoding of arbitrary geometric configurations. IRE Trans. Electronic Computers EC-10, 260–268 (1961)
33. Galyean, T.A., Hughes, J.F.: Sculpting: An interactive volumetric modeling technique. Computer Graphics (Proc. ACM Siggraph) 25(4), 267–274 (1991)
34. Gonzalez, R.C., Woods, R.E.: Digital Image Processing. Addison-Wesley, California (1993)
35. Haralick, R.M.: A measure for circularity of digital figures. IEEE Trans. Sys., Man & Cybern. 4, 394–396 (1974)
36. Hilaire, X., Tombre, K.: Robust and accurate vectorization of line drawings. IEEE Trans. Pattern Anal. Mach. Intell. 28(6), 890–904 (2006)
37. Hill Jr., F.S., Kelley, S.M.: Computer Graphics Using OpenGL. Prentice Hall, Englewood Cliffs (2007)
38. Ho, C.T., Chen, L.H.: A fast ellipse/circle detector using geometric symmetry. Pattern Recognition 28(1), 117–124 (1995)
39. Horn, B.K.P.: Circle generators for display devices. Computer Graphics and Image Processing 5(2), 280–288 (1976)
40. Hosur, P.I., Ma, K.-K.: A novel scheme for progressive polygon approximation of shape contours. In: Proc. IEEE 3rd Workshop on Multimedia Signal Processing, pp. 309–314 (1999)
41. Hsu, S.Y., Chow, L.R., Liu, C.H.: A new approach for the generation of circles. Computer Graphics Forum 12(2), 105–109 (1993)
42. Ioannoua, D., Hudab, W., Lainec, A.: Circle recognition through a 2D Hough Transform and radius histogramming. Image and Vision Computing 17, 15–26 (1999)
43. Jha, A., Bhowmick, P., Bhattacharya, B.B.: http://demonstrations.wolfram.com/ numbertheoreticconstructionofdigitalcircles/ (2009)
44. Kim, C.: Digital disks. IEEE Trans. PAMI 6, 372–374 (1984)
45. Kim, C.E., Anderson, T.A.: Digital disks and a digital compactness measure. In: Proc. 16th Annu. ACM Symp. Theory of Comput. (STOC), pp. 117–123 (1984)
46. Kim, H.S., Kim, J.H.: A two-step circle detection algorithm from the intersecting chords. Pattern Recognition Letters 22(6-7), 787–798 (2001)
47. Klette, R., Rosenfeld, A.: Digital Geometry: Geometric Methods for Digital Picture Analysis. Morgan Kaufmann Series in Computer Graphics and Geometric Modeling. Morgan Kaufmann, San Francisco (2004)
48. Klette, R., Rosenfeld, A.: Digital straightness: A review. Discrete Applied Mathematics 139(1-3), 197–230 (2004)
49. Kovalevsky, V.A.: New definition and fast recognition of digital straight segments and arcs. In: Proc. 10th Intl. Conf. Pattern Recognition (ICPR), pp. 31–34. IEEE CS Press, Los Alamitos (1990)
50. Kulpa, Z.: A note on "circle generator for display devices". Computer Graphics and Image Processing 9, 102–103 (1979)
51. Kulpa, Z., Kruse, B.: Algorithms for circular propagation in discrete images. Computer Vision, Graphics, and Image Processing 24(3), 305–328 (1983)
52. Kumar, G., Sharma, N.K., Bhowmick, P.: Creating wheel-thrown potteries in digital space. In: Proc. ArtsIT 2009 (to appear, 2009)
53. Leavers, V.: Survey: Which Hough transform? Computer Vision Graphics and Image Processing: Image Understanding 58, 250–264 (1993)
54. McIlroy, M.D.: Best approximate circles on integer grids. ACM Trans. Graphics 2(4), 237–263 (1983)

55. Megiddo, N.: Linear time algorithm for linear programming in $\mathbb{R}^3$ and related problems. SIAM J. Comput. 12, 759–776 (1983)

56. Mignosi, F.: On the number of factors of Sturmian words. Theoretical Computer Science 82(1), 71–84 (1991)

57. Nagy, B.: Characterization of digital circles in triangular grid. Pattern Recognition Letters 25(11), 1231–1242 (2004)

58. Nakamura, A., Aizawa, K.: Digital circles. Computer Vision, Graphics, and Image Processing 26(2), 242–255 (1984)

59. Pal, S., Bhowmick, P.: Determining digital circularity using integer intervals (communicated) (2009)

60. Pitteway, M.L.V.: Algorithm for drawing ellipses or hyperbolae with a digital plotter. The Computer Journal 10(3), 282–289 (1967)

61. Pitteway, M.L.V.: Integer circles, etc. — Some further thoughts. Computer Graphics and Image Processing 3, 262–265 (1974)

62. Pla, F.: Recognition of partial circular shapes from segmented contours. Computer Vision and Image Understanding 63(2), 334–343 (1996)

63. Rosin, P.L., West, G.A.W.: Detection of circular arcs in images. In: Proc. 4th. Alvey Vision Conf., Manchester, pp. 259–263 (1988)

64. Sauer, P.: On the recognition of digital circles in linear time. Computational Geometry 2, 287–302 (1993)

65. Shimizu, K.: Algorithm for generating a digital circle on a triangular grid. Computer Graphics and Image Processing 15(4), 401–402 (1981)

66. Song, J., Su, F., Tai, C.L., Cai, S.: An object-oriented progressive-simplification-based vectorization system for engineering drawings: Model, algorithm, and performance. IEEE Trans. Pattern Anal. Mach. Intell. 24(8), 1048–1060 (2002)

67. Sonka, M., Hlavac, V., Boyle, R.: Image Processing, Analysis, and Machine Vision. Chapman and Hall, Boca Raton (1993)

68. Suenaga, Y., Kamae, T., Kobayashi, T.: A high speed algorithm for the generation of straight lines and circular arcs. IEEE Trans. Comput. 28, 728–736 (1979)

69. Thomas, S.M., Chan, Y.T.: A simple approach for the estimation of circular arc center and its radius. Computer Vision, Graphics, and Image Processing 45(3), 362–370 (1989)

70. Worring, M., Smeulders, A.W.M.: Digitized circular arcs: Characterization and parameter estimation. IEEE Trans. PAMI 17(6), 587–598 (1995)

71. Wright, W.E.: Parallelization of Bresenham's line and circle algorithms. IEEE Computer Graphics and Applications 10(5), 60–67 (1990)

72. Wu, X., Rokne, J.G.: Double-step incremental generation of lines and circles. Computer Vision, Graphics, and Image Processing 37(3), 331–344 (1987)

73. Yao, C., Rokne, J.G.: Hybrid scan-conversion of circles. IEEE Trans. Visualization and Computer Graphics 1(4), 311–318 (1995)

74. Yuen, P.C., Feng, G.C.: A novel method for parameter estimation of digital arc. Pattern Recognition Letters 17(9), 929–938 (1996)

On the Convex Hull of the Integer Points in a Bi-circular Region

Valentin E. Brimkov

Mathematics Department, SUNY Buffalo State College,
Buffalo, NY 14222, USA
brimkove@buffalostate.edu

Abstract. Given a set $S \subseteq \mathbb{R}^2$, denote $S_{\mathbb{Z}} = S \cap \mathbb{Z}^2$. We obtain bounds for the number of vertices of the convex hull of $S_{\mathbb{Z}}$, where $S \subseteq \mathbb{R}^2$ is a convex region bounded by two circular arcs. Two of the bounds are tight bounds—in terms of arc length and in terms of the width of the region and the radii of the circles, respectively. Moreover, an upper bound is given in terms of a new notion of "set oblongness." The results complement the well-known $O(r^{2/3})$ bound [2] which applies to a disc of radius r.

Keywords: convex polygon, strictly convex set, facet complexity, convex hull, upper/lower/tight bound.

1 Introduction

In the present paper we consider the following problem. Let $S \subseteq \mathbb{R}^2$ be a convex region bounded by two circular arcs and denote $S_{\mathbb{Z}} = S \cap \mathbb{Z}^2$ (sometimes $S_{\mathbb{Z}}$ is called the *Gauss digitization* of S). Find bounds for the number of vertices of the convex hull $conv(S_{\mathbb{Z}})$ of $S_{\mathbb{Z}}$.

There are a number of motivations for our study. First, it is related to applications in integer programming, more precisely to the possibility of reducing a non-linear optimization problem of the form "Find $x \in \mathbb{Z}^n$ that maximizes/minimizes a linear (or convex) function cx on $S_{\mathbb{Z}}$" to an integer linear program of the form $\max\{cx|Ax \leq b\}$, where $A \in \mathbb{Z}^{m \times n}$, $c \in \mathbb{Z}^n$, $b \in \mathbb{Z}^m$, and $x \in \mathbb{Z}^n$. Upper and lower bounds on the number of vertices and facets of $conv(S_{\mathbb{Z}})$ may facilitate the complexity analysis of some algorithms for the above problem (see, e.g., [15,16] for related examples). One can also look for a convex polygon/polytope P that has a minimal possible number of facets and encloses the points of $S_{\mathbb{Z}}$ (and only those points) [8]. Such a polygon/polytope P is not necessarily the convex hull of $S_{\mathbb{Z}}$, but the number of vertices of the latter obviously provides an upper bound for the former. This is the so-called *optimal polyhedral reconstruction* problem, which is also important from a practical perspective. More specifically, such problems mainly originate from medical imaging where digital volumes of voxels result from scanning and MRI techniques. Since digital medical images involve a huge number of points, it may be problematic to apply traditional rendering or texturing algorithms in order to obtain satisfactory visualization. Moreover, one can face difficulties in storing or transmitting data of that size.

P. Wiederhold and R.P. Barneva (Eds.): IWCIA 2009, LNCS 5852, pp. 16–29, 2009.
© Springer-Verlag Berlin Heidelberg 2009

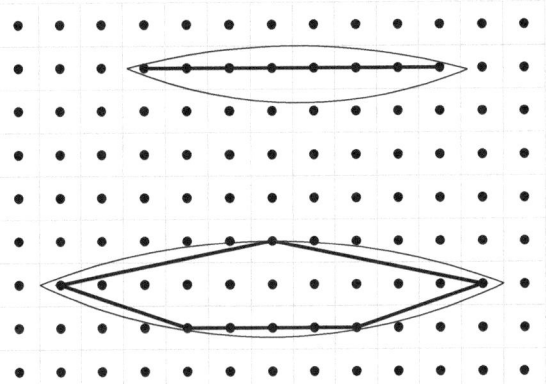

Fig. 1. "Thin" bi-circular regions, for which the convex hull of the contained integer points is small. *Top:* If the region width is less than two, the number of vertices of the convex hull of the integer points it contains may equal two, regardless of the diameter's length. *Bottom:* For a bit thicker region (of width less than three) the number of vertices is five, regardless of the diameter's length.

To overcome such kind of obstacles, one can try to transform a discrete data set to a polygon/polyhedron P, such that the number of its facets is as small as possible. Note that finding an enclosing polyhedron with a minimal number of facets can be computationally hard, in general, as shown in [7].

The problem considered is also related to some classical problems of discrete geometry and polyhedral combinatorics, such as problems about properties of the convex hull of a set of points [16,6,20,22], estimation of the number of integer points in a disc/ball [2,5,21], estimation of the maximal possible number of integer points on curves and surfaces [13,18], and properties of certain lattice polytopes [1,3,4,9,10,12,14,17,19,23], among others. More specifically, the results presented here are closely related to a well-known result of Balog and Bárány [2], who showed that if S is a disc of radius r, then the number of vertices of $conv(S_{\mathbb{Z}})$ is essentially $r^{2/3}$ as r approaches infinity. The upper bound follows from another classical result due to Jarnik [13], who has shown that any strictly convex plane curve of length s contains $O(s^{2/3})$ integer points.

The result of [2] was generalized to higher dimensions in [5] for S a hyperball (or a convex body in $\mathbb{R}^n$ satisfying certain conditions). The facet complexity (which is the number of facets in an optimal polyhedral reconstruction) of $S_{\mathbb{Z}}$ for a hyperball S was studied in [8]. To our knowledge, however, nontrivial results for other classes of convex sets that are far from discs/(hyper)balls are not available. In the present paper we obtain such a result.

Let S be a convex closed set whose boundary is a strictly convex closed curve $\gamma \subset \mathbb{R}^2$ of length s composed by two circular arcs. Denote for short $P = conv(S_{\mathbb{Z}})$, by $V(P)$ the set of vertices of P, and by $|V(P)|$ its cardinality. There is an evidence that if the diameter of S is much greater than its width (see Section 2 for these and other notions), then the order of $|V(P)|$ can be much lower than $O(s^{2/3})$ where s is the length of the boundary of S (see Fig. 1). How can the width/diameter ratio impact the upper or lower bounds on $|V(P)|$?

We consider the case where γ consists of two circular arcs that bound a convex plane region $S \subseteq \mathbb{R}^2$. We obtain bounds for the number of vertices of the convex hull of $S_{\mathbb{Z}}$. Two of the bounds are tight bounds—in terms of arc length and in terms of the width of the region and the radii of the circles, respectively. Moreover, an upper bound is given in terms of a new notion of "set oblongness."

In the next section we introduce some basic notions and notations and recall certain facts to be used in the sequel. In Section 3 we present the main results of the paper. We conclude with some remarks in Section 4.

2   Basic Notions, Notations, and Facts

For a closed set $A \subseteq \mathbb{R}^2$, $d(A)$ denotes its *diameter* defined as $d(A) = \max_{x,y \in A} ||x - y||$, where $||.||$ is the Euclidean norm. The *width* of A is denoted by $w(A)$ and defined as the minimal possible distance between two parallel lines that bound A (see Fig. 2).

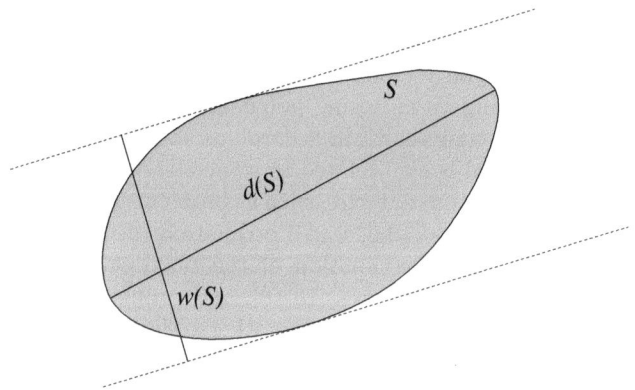

Fig. 2. Illustration to the notions of diameter and width. The two segments representing the diameter and the width of S are labeled by $d(S)$ and $w(S)$, respectively.

A straight line segment with end-points X and Y is denoted by $\overline{XY}$, and its length by $|XY|$.

A set $A \subseteq \mathbb{R}^2$ is called *strictly convex* if for every two points $p, q \in A$, $p \neq q$, the straight line segment $\overline{pq}$ is contained in A but does not belong to the boundary of A.

By $conv(A)$ we denote the convex hull of A and by $|A|$ its *cardinality*. Given a polygon $P \subset \mathbb{R}^2$, the set of its vertices is denoted by $V(P)$, so their number will be denoted by $|V(P)|$.

A closed disc D with center O and radius r is denoted by $D(O, r)$ and the corresponding circle by $C(O, r)$. The length of a circular arc γ is denoted by $|\gamma|$.

For two positive-valued functions $f(x)$ and $g(x)$ we write: $f(x) = O(g(x))$ iff there are constants $c > 0$ and x_0 such that $f(x) \leq c \cdot g(x)$ for all $x > x_0$; $f(x) = \Omega(g(x))$ iff there are constants $c > 0$ and x_0 such that $c \cdot g(x) \leq f(x)$ for all $x > x_0$; $f(x) = \Theta(g(x))$ iff there are constants $c_1 > 0$, $c_2 > 0$, and x_0 such that $c_1 \cdot g(x) \leq f(x) \leq c_2 \cdot g(x)$ for all $x > x_0$; $f(x) = o(g(x))$ iff for any constant $c > 0$ there is x_0, such that $0 \leq f(x) \leq c \cdot g(x)$ for all $x > x_0$. See [11] for further details.

Next, we recall in more detail the already-mentioned result of Balog and Bárány:

Fact 1. [2] *Let $D = D(O, r)$ be a disc in the plane. Let $K = conv(D_{\mathbb{Z}})$. For large enough r,*

$$c_1 r^{2/3} \leq |V(K)| \leq c_2 r^{2/3}, \tag{1}$$

where $c_1 \approx 0.33$ and $c_2 \approx 5.54$.

Remark 1. Note that the result remains true if $D \in C(d)$, where $C(d)$ is the family of convex bodies with C^2 boundary and radius of curvature at every point and every direction between $1/d$ and d, $d \geq 1$ [5].

The following is another useful fact.

Fact 2. [24] *Consider the class $\mathcal{F}$ of convex polygons which have the maximal number of vertices with respect to their diameter. Then the number of vertices of these polygons is asymptotically*

$$\frac{12s^{2/3}}{(2\pi(2 + \sqrt{2} \cdot \ln(1 + \sqrt{2})))^{2/3}} \approx 1.60739 \cdot s^{2/3},$$

where s is the perimeter of such a polygon.

3 Main Results

Let $S \subset \mathbb{R}^2$ be a plane region bounded by two circular arcs γ_1 and γ_2 of circles $C_1(O_1, r_1)$ and $C_2(O_2, r_2)$. We suppose that the central angles corresponding to γ_1 and γ_2 do not exceed π, i.e., γ_1 and γ_2 are at most semi-circles. Let these arcs intersect at points P and Q. We assume that $\overline{PQ}$ is a part of a horizontal grid-line.[1] P_1 and Q_1 will denote the leftmost and the rightmost integer points on the segment $\overline{PQ}$, respectively (see Fig. 5). Let M, N, and T be the mid-points of γ_1, γ_2, and $\overline{PQ}$, respectively. Denote by S^1 and S^2 the regions bounded by γ_1 and $\overline{PQ}$ and by γ_2 and $\overline{PQ}$ (so we have $S = S^1 \cup S^2$). Also, denote $|MT| = w_1$ and $|NT| = w_2$, which are clearly the widths of S^1 and S^2, respectively. See Fig. 3. The following are easy facts.

Fact 3. $w(S) = w_1 + w_2$

The proof easily follows by contradiction and elementary geometry arguments.

---

[1] As it will be discussed in Section 4, the results remain valid in the general case, as well, provided that certain technical modifications are made.

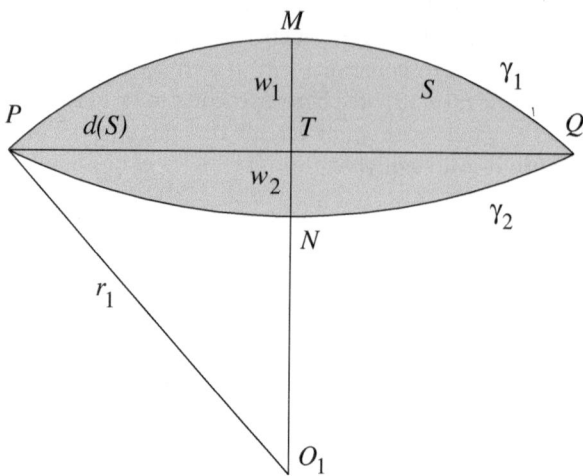

Fig. 3. Illustration to the considerations of Section 3

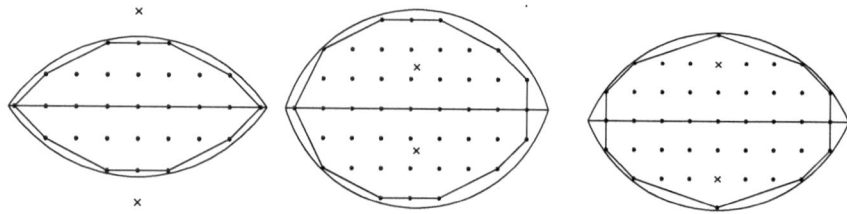

Fig. 4. Illustration to Fact 4. The figures on the left, middle, and right illustrate the cases where $k = 2, 3$, and 4, respectively. The crosses indicate the centers of the circles involved.

Fact 4. *A leftmost/rightmost integer point on $\overline{PQ}$ is either a vertex of $conv(S_{\mathbb{Z}})$ or lies on a (vertical) side of $conv(S_{\mathbb{Z}})$. Moreover, $V(conv(S_{\mathbb{Z}})) = V(conv(S_{\mathbb{Z}}^1)) \cup V(conv(S_{\mathbb{Z}}^2))$, and*

$$|V(conv(S_{\mathbb{Z}}))| = |V(conv(S_{\mathbb{Z}}^1))| + |V(conv(S_{\mathbb{Z}}^2))| - k, \qquad (2)$$

where $k = 2, 3$, or 4, depending on whether two, one, or none of the leftmost and rightmost integer points on $\overline{PQ}$ are vertices of $conv(S_{\mathbb{Z}})$. See Fig. 4.

Equation (2) suggests that we can restrict the considerations to region S^1. We will first obtain a tight bound in terms of the length of arc γ_1. We have

$$|\gamma_1| = \alpha_1 r_1, \text{ where } 0 < \alpha_1 = \angle PO_1M < \pi. \qquad (3)$$

As we are mainly interested in asymptotic results (in the spirit of Facts 1 and 2), we will focus on the case when $r_1 \to \infty$ (and $r_2 \to \infty$, as well). If we assume in this setting that α_1 is a fixed number in the interval $(0, \pi]$, then it follows that the bound of Fact 1 applies. This and other trivial cases are considered in Section 3.2. The nontrivial case is when the arc length $|\gamma_1|$ is increasing to infinity, while the ratio of $|\gamma_1|$ and the semi-circle length πr_1 is approaching zero. In view of equation (3), this would be possible only if $\alpha_1 \to 0$ and $\alpha_1 r_1 \to \infty$, as $\alpha_1 \to 0$ for $r_1 \to \infty$. This suggests to consider a model where the angle variable α_1 is a dependent variable that is a function of the radius r_1, i.e., $\alpha_1 = \alpha_1(r_1)$, with $\alpha_1(r_1) = o(1)$ as $r_1 \to \infty$. For technical convenience, we will assume that $\alpha_1(r_1)$ is monotone decreasing. We will also suppose that $\alpha_1(r_1)$ is differentiable. Thus, the arc length $|\gamma_1|$ can be considered as a function $\Gamma(r_1) = \bar{\gamma}_1(r_1, \alpha_1(r_1)) = \alpha_1(r_1) r_1$, with

$$\lim_{r_1 \to \infty} \alpha_1(r_1) r_1 = \infty, \qquad \lim_{r_1 \to \infty} \alpha_1(r_1) = 0. \qquad (4)$$

(For example, this is the case if we choose $\alpha_1(r_1) = 1/\ln r_1$ or $\alpha_1(r_1) = 1/r_1^t$ for $0 < t < 1$.) For the sake of brevity, further we will use the denotation $|\gamma_1|$ rather than $\Gamma(r_1)$ or $\gamma_1(r_1, \alpha_1(r_1))$.

3.1   Tight Bound in Terms of Arc Length

Let S^1 be the region defined above, and $d(conv(S_{\mathbb{Z}}^1))$ and $s(conv(S_{\mathbb{Z}}^1))$ the diameter and the perimeter of $conv(S_{\mathbb{Z}}^1)$, respectively. In obtaining the result of this section we will use the following lemmas.

Lemma 1. $s(conv(S_{\mathbb{Z}}^1)) = O(d(conv(S_{\mathbb{Z}}^1)))$.

Proof. Consider the polygon $conv(S_{\mathbb{Z}}^1)$ with horizontal side $\overline{P_1 Q_1}$ of length $|P_1 Q_1|$ and other sides of lengths $s_1, s_2, \ldots, s_k$ (see Fig. 5). We consider the case of Fig. 4, left. In addition, we assume that among the last k sides none is horizontal. The other possibilities can be handled analogously.

Obviously, for a sufficiently large r_1, $d(conv(S_{\mathbb{Z}}^1)) = |P_1 Q_1|$. Denote for short $s = s(conv(S_{\mathbb{Z}}^1))$. We have

$$s = \sum_{i=1}^{k} s_i + |P_1 Q_1| = \sum_{i=1}^{k} s_i + d(conv(S_{\mathbb{Z}}^1)).$$

Let us associate a right triangle with each side of $conv(S_{\mathbb{Z}}^1)$ (except for the horizontal one $\overline{P_1 Q_1}$), as illustrated in Fig. 5. Let s_i, h_i, and v_i be the hypotenuse, the horizontal, and the vertical leg of the ith triangle, respectively, for $1 \le i \le k$. Since $s_i < v_i + h_i$, we have

$$s < \sum_{i=1}^{k} v_i + \sum_{i=1}^{k} h_i + d(conv(S_{\mathbb{Z}}^1)).$$

Since $\sum_{i=1}^{k} h_i = |P_1 Q_1| = d(conv(S_{\mathbb{Z}}^1))$, we get

$$s < \sum_{i=1}^{k} v_i + 2d(conv(S_{\mathbb{Z}}^1)).$$

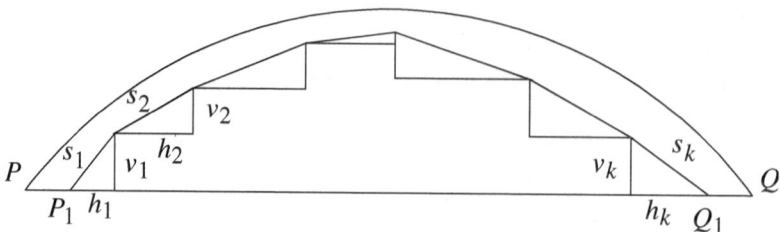

Fig. 5. Illustration to the proof of Lemma 1

Finally, since $\sum_{i=1}^{k} v_i = 2w(conv(S_{\mathbb{Z}}^1)) \leq 2d(conv(S_{\mathbb{Z}}^1))$ for sufficiently large r_1, we obtain

$$s < 4d(conv(S_{\mathbb{Z}}^1)),$$

which completes the proof. □

Lemma 2. $d(conv(S_{\mathbb{Z}}^1)) = \Omega(|\gamma_1|)$.

Proof. With reference to Fig. 3, consider the circle $C(T, |PQ|/2)$. Since the arc γ_1 does not exceed a semi-circle, we have

$$\pi \cdot \frac{|PQ|}{2} \geq |\gamma_1|.$$

Using the plain inequality

$$|PQ| < |P_1Q_1| + 2 = d(conv(S_{\mathbb{Z}}^1)) + 2,$$

we then consecutively obtain

$$\frac{\pi}{2}(d(conv(S_{\mathbb{Z}}^1)) + 2) > |\gamma_1|,$$

$$d(conv(S_{\mathbb{Z}}^1)) > \frac{2}{\pi}|\gamma_1| - 2,$$

and, eventually,

$$d(conv(S_{\mathbb{Z}}^1)) > c \cdot |\gamma_1|,$$

for an appropriate constant $c > 0$ and sufficiently large r_1. □

Now we can prove the following theorem.

Theorem 1. Let S^1 be the region defined above. Then $|V(conv(S_{\mathbb{Z}}^1))| = \Theta(|\gamma_1|^{2/3})$.

Proof. 1. We first show that $|V(conv(S_{\mathbb{Z}}^1))| = O(|\gamma_1|^{2/3})$. By Fact 2,

$$|V(conv(S_{\mathbb{Z}}^1))| = O\left((s(conv(S_{\mathbb{Z}}^1))^{2/3}\right). \tag{5}$$

By Lemma 1, we have

$$s(conv(S_{\mathbb{Z}}^1))^{2/3}) = O((d(conv(S_{\mathbb{Z}}^1)))^{2/3}).$$

Moreover, we obviously have that for a sufficiently large r_1,

$$d(conv(S_{\mathbb{Z}}^1)) = |P_1Q_1| < |\gamma_1|,$$

which implies

$$(d(conv(S_{\mathbb{Z}}^1)))^{2/3} = O(|\gamma_1|^{2/3}).$$

These last relations, coupled with (5), imply

$$|V(conv(S_{\mathbb{Z}}^1))| = O(|\gamma_1|^{2/3}).$$

2. Next we show that $|V(conv(S_{\mathbb{Z}}^1))| = \Omega(|\gamma_1|^{2/3})$. By Fact 2,

$$|V(conv(S_{\mathbb{Z}}^1))| = \Omega\left((s(conv(S_{\mathbb{Z}}^1))^{2/3}\right). \tag{6}$$

It is also clear that

$$s(conv(S_{\mathbb{Z}}^1)) = \Omega(d(conv(S_{\mathbb{Z}}^1))),$$

since the convex polygonal border of $conv(S_{\mathbb{Z}}^1)$ over $\overline{P_1Q_1}$ is obviously longer than $|P_1Q_1| = d(conv(S_{\mathbb{Z}}^1))$ (see Fig. 5). Moreover, by Lemma 2, we have

$$d(conv(S_{\mathbb{Z}}^1)) = \Omega(|\gamma_1|).$$

These relations, together with (6), imply

$$|V(conv(S_{\mathbb{Z}}^1))| = \Omega(|\gamma_1|^{2/3}),$$

which completes the proof.   □

3.2   Tight Bound in Terms of Width and Radius of Curvature

Once again, we are mostly interested in the case when $w_1 \to \infty$ and $w_1/r_1 \to 0$ as $r_1 \to \infty$. It turns out that postulates (4) for the angle α_1 assure these conditions, as we show next.

With reference to Fig. 3, $w_1 = r_1 - r_1 \cos \alpha_1$. With α_1 being considered as a function of r_1, we can write

$$w_1 = w_1(r_1) = r_1 - r_1 \cos(\alpha_1(r_1)) = r_1(1 - \cos(\alpha_1(r_1))).$$

As $\cos(\alpha_1(r_1)) \to 1$ (and, thus, $1 - \cos(\alpha_1(r_1)) \to 0$) if $\alpha_1(r_1) \to 0$ for $r_1 \to \infty$, we clearly have $w_1(r_1) = o(r_1)$ and $w_1(r_1)/r_1 \to 0$. It remains to show that $w_1(r_1) \to \infty$ as $r_1 \to \infty$.

Denoting $\phi(r_1) = 1 - \cos(\alpha_1(r_1))$ and $\psi(r_1) = 1/r_1$, we have $w_1(r_1) = \frac{\phi(r_1)}{\psi(r_1)}$. Obviously, both $\phi(r_1)$ and $\psi(r_1)$ converge to 0 as r_1 approaches infinity. Then, by l'Hôpital's theorem, we consecutively obtain

$$\lim_{r_1 \to \infty} w_1(r_1) = \lim_{r_1 \to \infty} \frac{\phi(r_1)}{\psi(r_1)} = \lim_{r_1 \to \infty} \frac{\phi'(r_1)}{\psi'(r_1)} =$$

$$\lim_{r_1 \to \infty} \frac{\alpha_1'(r_1) \sin(\alpha_1(r_1))}{-1/r_1^2} =$$

$$\lim_{r_1 \to \infty} \left(-r_1^2 \alpha_1'(r_1) \sin(\alpha_1(r_1)) \right) =$$

$$\lim_{r_1 \to \infty} \left(-r_1 \alpha_1'(r_1) \right) \left(r_1 \alpha_1(r_1) \right) \left(\sin(\alpha_1(r_1))/\alpha_1(r_1) \right) =$$

$$\lim_{r_1 \to \infty} -r_1 \alpha_1'(r_1) \lim_{r_1 \to \infty} r_1 \alpha_1(r_1) \lim_{r_1 \to \infty} \sin(\alpha_1(r_1))/\alpha_1(r_1).$$

In the last expression, $\lim_{r_1 \to \infty} \sin(\alpha_1(r_1))/\alpha_1(r_1) = 1$, while $\lim_{r_1 \to \infty} r_1 \alpha_1(r_1) = \infty$ (by definition). Thus, it follows that $w_1(r_1)$ may not go to infinity as $r_1 \to \infty$ only if $\lim_{r_1 \to \infty} -r_1 \alpha_1'(r_1) = 0$.[2] We will show that the latter cannot happen.

To see this, assume the opposite and consider the identity

$$(r_1 \alpha_1(r_1))' = \alpha_1(r_1) + r_1 \alpha_1'(r_1).$$

Both additives in the right-hand side converge to zero (the first one by definition, while the second one by assumption). So, the same applies to the left-hand side, i.e., $\lim_{r_1 \to \infty} (r_1 \alpha_1(r_1))' = 0$. Then $\lim_{r_1 \to \infty} r_1 \alpha_1(r_1)) = c$, where c is a constant, which contradicts (4).

After this preparation, denote $\alpha = \angle PO_1M$. From the right triangle $\triangle PO_1T$ (see Fig. 3) we have

$$\cos \alpha = \frac{r_1 - w_1}{r_1}.$$

Then, for the central angle 2α corresponding to γ_1 we have

$$2\alpha = 2 \arccos \frac{r_1 - w_1}{r_1}.$$

Further, for the length of arc γ_1 we have

$$|\gamma_1| = 2\alpha r_1 = 2r_1 \arccos \frac{r_1 - w_1}{r_1}.$$

Finally, using Theorem 1, we obtain that as r_1 increases,

$$|V(conv(S_{\mathbb{Z}}^1))| = \Theta \left(r_1 \arccos \frac{r_1 - w_1}{r_1} \right)^{2/3}. \tag{7}$$

Analogous considerations apply to region S^2. The above observations are summarized in the following theorem.

---

[2] Note that, since $\alpha_1(r_1)$ is monotone decreasing, $\alpha_1'(r_1) \leq 0$, so $-r_1 \alpha_1'(r_1) \geq 0$.

Theorem 2. $|V(conv(S_\mathbb{Z}))| = \Theta\left(\left(r_1 \arccos \frac{r_1 - w_1}{r_1}\right)^{2/3} + \left(r_2 \arccos \frac{r_2 - w_2}{r_2}\right)^{2/3}\right).$

Now denote

$$A(w_1, r_1) = \arccos \frac{r_1 - w_1}{r_1}.$$

Next we investigate more in detail the above bound (i.e., the behavior of $A(w_1, r_1)$), in a number of particular cases, depending on the relation between w_1 and r_1. First, we rule out two trivial extreme cases:

Case 1. If $w_1 = 0$, then $A(w_1, r_1) = \arccos 1 = 0$. This is the (impossible) degenerate case when $S^1 = \overline{P_1 Q_1}$, where P_1 and Q_1 are the leftmost and the rightmost integer points on the segment $\overline{PQ}$, respectively. (Note that $conv(S_\mathbb{Z}^1) = \overline{P_1 Q_1}$ also for $0 < w_1 < 1$.)

Case 2. If $w_1 = r_1$, then $A(w_1, r_1) = \arccos 0 = \frac{\pi}{2}$. Then we have

$$|V(conv(S_\mathbb{Z}^1))| = \Theta(r_1^{2/3}),$$

which is precisely the bound of Fact 1 applied to a semi-disc.

Next we consider other two trivial cases.

Case 3. Let w_1 and r_1 differ by a constant factor. We have $w_1 = \mu r_1$ for some constant $\mu < 1$. Then

$$A(w_1, r_1) = \arccos \frac{r_1 - \mu r_1}{r_1} = \arccos(1 - \mu),$$

which is a constant in the interval $(0, \frac{\pi}{2})$. This would change the bounds in (1) by a constant factor.

Case 4. Let $w_1 = \beta$, where β is a constant. In this case clearly $|V(conv(S_\mathbb{Z}^1))|$ is bounded by a constant, since the number of vertices cannot exceed the width β.

In the general case we have that w_1 is a non-constant function of r_1, with $w_1 = o(r_1)$, and that w_1 approaches infinity as $r_1 \to \infty$. We have

$$\lim_{r_1 \to \infty} A(w_1(r_1), r_1) = \lim_{r_1 \to \infty} \arccos \frac{r_1 - w_1}{r_1} = 0,$$

while

$$\lim_{r_1 \to \infty} r_1^{2/3} = \infty.$$

Hence, we have

$$(A(w_1(r_1), r_1))r_1)^{2/3} = o(r_1^{2/3}).$$

Note that both $\Theta(|\gamma_1|^{2/3})$ and $\Theta\left((A(w_1(r_1), r_1))r_1)^{2/3}\right)$ are tight bounds for $|V(conv(S_\mathbb{Z}^1))|$ (by Theorems 1 and 2, respectively). Moreover, $|\gamma_1|^{2/3} \to \infty$ as $r_1 \to \infty$. Thus, we have

$$\lim_{r_1 \to \infty} \Theta\left((A(w_1(r_1), r_1))r_1)^{2/3}\right) = \infty.$$

3.3   Upper Bound in Terms of Oblongness and Radius of Curvature

We start this section with a definition, which, we believe, may be of broader interest and importance.

Definition 1. *Let S be a compact subset of $\mathbb{R}^n$. Oblongness of S is the number*

$$\iota(S) = \frac{d(S) - w(S)}{d(S)}. \tag{8}$$

Clearly, the oblongness of a set S satisfies:

$$0 \leq \iota(S) \leq 1.$$

The larger $\iota(S)$, the "more oblong" S and viceversa. For example, the oblongness of a straight line segment I is 1, while the one of a disc D or a Reuleaux polygon is 0.

Below we provide an upper bound for $|V(conv(S_\mathbb{Z}))|$ in terms of $\iota(S)$.

Obviously, $d(S) = d(S_1) = d(S_2)$. With reference to Fig. 3, we have

$$d(S) = 2\sqrt{r_1^2 - (r_1 - w_1)^2}.$$

We look for a condition under which $r_1 \geq d(S)$, that is,

$$r_1 \geq 2\sqrt{r_1^2 - (r_1 - w_1)^2}, \quad \text{i.e.,}$$

$$r_1^2 - 8r_1w_1 + 4w_1^2 \geq 0.$$

Solving the above quadratic inequality with respect to r_1, we obtain that the inequality holds if and only if

$$r_1 \leq (4 - 2\sqrt{3})w_1 \approx 0.536 \cdot w_1 \quad \text{or} \quad r_1 \geq (4 + 2\sqrt{3})w_1 \approx 7.464 \cdot w_1.$$

The left inequality is out of interest since we always have $r_1 \geq w_1$. For $w_1 \leq r_1 < (4 + 2\sqrt{3})w_1$ we are in the framework of the trivial Case 3 considered in the previous section.

Let $r_1 \geq (4 + 2\sqrt{3})w_1$, which condition implies $r_1 \geq d(S)$. The last inequality is equivalent to

$$\frac{r_1 - w_1}{r_1} \geq \frac{d(S) - w_1}{d(S)}.$$

(Note that the above inequality is equivalent to $1 - w_1/r_1 \geq 1 - w_1/d(S)$ from where the claimed equivalence is obvious.)

Since $\arccos x$ is monotone decreasing, we have

$$\arccos \frac{r_1 - w_1}{r_1} \leq \arccos \frac{d(S) - w_1}{d(S)}.$$

Analogous considerations lead to the inequality

$$\arccos \frac{r_2 - w_2}{r_2} \le \arccos \frac{d(S) - w_2}{d(S)}.$$

Thus we obtain

$$|V(conv(S_{\mathbb{Z}}))| = O\left(\left(\arccos \frac{d(S) - w_1}{d(S)}\right)^{2/3} r_1^{2/3} + \left(\arccos \frac{d(S) - w_2}{d(S)}\right)^{2/3} r_2^{2/3}\right). \tag{9}$$

Since $w = w_1 + w_2$, it holds

$$\frac{d(S) - w}{d(S)} \le \frac{d(S) - w_1}{d(S)}$$

and

$$\frac{d(S) - w}{d(S)} \le \frac{d(S) - w_2}{d(S)}.$$

Since $\arccos x$ is non-increasing, we obtain

$$\arccos \frac{d(S) - w(S)}{d(S)} \ge \arccos \frac{d(S) - w_1}{d(S)}$$

and

$$\arccos \frac{d(S) - w(S)}{d(S)} \ge \arccos \frac{d(S) - w_2}{d(S)}.$$

Keeping (9) and Definition 1 in mind, we can formulate the following theorem.

Theorem 3. $|V(conv(S_{\mathbb{Z}}))| = O\left((\arccos \iota(S))^{2/3}(r_1^{2/3} + r_2^{2/3})\right).$

Theorem 3 shows that as the oblongness of a region is increasing, the upper bound on the number of vertices is decreasing.

4   Concluding Remarks

In this paper we obtained certain bounds for the number of vertices of the convex hull of $S_{\mathbb{Z}}$, where $S \subseteq \mathbb{R}^2$ is a convex region bounded by two circular arcs.

We recall that the presented results are obtained under the condition that the diameter $\overline{PQ}$ of the region S is a part of a grid-line. The main results remain valid in the general case, as well. However, regarding the proofs of some of the theorems, the above-mentioned condition appears as a restriction of the generality, as the proofs do not work without certain modifications. Such modifications, that are not hard to devise but are quite technical and space-consuming, will be included in the full-length journal version of the paper.

A challenging task is seen in looking for similar results for other classes of strictly convex sets with smooth boundary, such as ellipses.

Work in progress is aimed at extending the results to higher dimensions, where the considered region is an intersection of two hyperballs.

We also believe that further applications of the notion of oblongness can be found.

Acknowledgements

The author thanks the three anonymous referees for their useful remarks and suggestions, and Reneta Barneva for proof-reading.

References

1. Acketa, D.M., Žunić, J.: On the Maximal Number of Edges of Convex Digital Polygons Included into a $m \times m$-Grid. Journal of Combinatorial Theory (Ser. A) 69, 358–368 (1995)
2. Balog, A., Bárány, I.: On the Convex Hull of the Integer Points in a Disc. DIMACS Series, vol. 6, pp. 39–44 (1991)
3. Bárány, I.: Random points and lattice points in convex bodies. Bulletin of the American Mathematical Society 45(3), 339–365 (2008)
4. Bárány, I., Hove, R., Lovász, L.: On Integer Points in Polyhedra: A Lower Bound. Combinatorica 12(2), 135–142 (1992)
5. Bárány, I., Larman, D.G.: The Convex Hull of the Integer Points in a Large Ball. Math. Ann. 312, 167–181 (1998)
6. Boltianskii, V.G., Soltan, P.S.: Combinatorial Geometry of Various Classes of Convex Sets, Stiinţa, Chişinău (1978)
7. Brimkov, V.E.: Digitization Scheme that Assures Faithful Reconstruction of Plane Figures. Pattern Recognition 42, 1637–1649 (2009)
8. Brimkov, V.E., Barneva, R.P.: On the Polyhedral Complexity of the Integer Points in a Hyperball. Theoretical Computer Science 406, 24–30 (2008)
9. Brimkov, V.E., Dantchev, S.S.: Real Data – Integer Solution Problems within the Blum-Shub-Smale Computational Model. Journal of Complexity 13, 279–300 (1997)
10. Chirkov, A.Y.: On the Relation of Upper Bounds on the Number of Vertices of the Convex Hull of the Integer Points of a Polyhedron with its Metric Characteristics (in Russian). In: Alekseev, V.E., et al. (eds.) 2nd Int. Conf. Mathematical Algorithms, pp. 169–174. Nizhegorod Univ. Press (1997)
11. Garey, M.S., Johnson, D.S.: Computers and Intractability: a Guide to the Theory of NP-Completeness. Freeman & Co., San Francisco (1979)
12. Hayes, A.S., Larman, D.C.: The Vertices of the Knapsack Polytope. Discrete Applied Mathematics 6, 135–138 (1983)
13. Jarnik, V.: Über Gitterpunkte and konvexe Kurven. Math. Zeitschrift 24, 500–518 (1925)
14. McMullen, P.: The Maximum Numbers of Facets of a Convex Polytope. Mathematika 17, 179–184 (1970)
15. Papadimitriou, C., Steiglitz, K.: Combinatorial Optimization. Prentice-Hall, New Jersey (1982)
16. Preparata, F., Shamos, M.I.: Computational Geometry: An Introduction. Springer, New York (1985)
17. Rubin, D.S.: On the Unlimited Number of Faces in Integer Hulls of Linear Programs with a Single Constraint. Operations Research 18(5), 940–945 (1970)
18. Schmidt, W.M.: Integer Points on Curves and Surfaces. Monatshefte für Math. 99, 45–72 (1985)

19. Shevchenko, V.N.: On the Number of Extremal Points in Integer Programming. Kibernetika 4, 133–134 (1981)
20. Valentine, F.A.: Convex Sets. McGraw-Hill, New York (1964)
21. van der Corupt, J.G.: Verschärfung der Abschätzung beim Teilerproblem. Math. Annalen 87, 39–65 (1922)
22. van der Vel, M.: Theory of Convex Structures. Elsevier, Amsterdam (1993)
23. Veselov, S.I., Chirkov, A.Y.: Bounds on the Number of Vertices of Integer Polyhedra (in Russian). Discrete Analysis and Operations Research 14(2), 14–31 (2007)
24. Žunić, J.: Notes on Optimal Convex Lattice Polygons. Bull. London Math. Society 30, 377–385 (1998)

Multi-primitive Analysis of Digital Curves*

Alexandre Faure and Fabien Feschet

Univ. Clermont 1, LAIC, F-63172 Aubière, France
{afaure,feschet}@laic.u-clermont1.fr

Abstract. In this paper, we propose a new approach for the analysis and the decomposition of digital curves simultaneously into straight and circular parts. Both digital primitives are defined using a thickness parameter. Our method relies on the notion of Tangential Cover [8] which represents digital curves by the set of maximal primitives. The nature of the Tangential Cover allows for fast computation and makes our approach easily extendable, not only to other types of digital primitives, but also to thick digital curves [7]. The results are promising.

Keywords: Digital geometry, α-thickness, multi-primitive, tangential Cover, polygonalization.

1 Introduction

In the world of digital geometry, the extraction of geometric features from digital curves is a key problem. The most common approach is to extract straight parts of curves. Many different definitions of digital straightness have been given [12,15], and many different algorithms allow either representation or decomposition of digital curves into straight parts [12,5,14]. But straightness is far from the only meaningful feature of a digital curve. It is relevant to extend the study of digital curves to higher order primitives than straight ones. In particular, it is important to capture circular parts of a curve as well. Some publications tackle the issue of fitting both straight segments and circular arcs to digital curves at the same time. The famous approach of Rosin and West [16] relies on least square fitting. Another technique has been designed by Hilaire and Tombre [10], based on the notions of fuzzy digital segments and fuzzy digital arcs. Both of them provide vectorizations of curves. Rosin and West's approach is nonparametric whereas Hilaire and Tombre's one uses some parameters. Both are robust and accurate but suffer from a high time complexity. Moreover they are restricted to one pixel wide digital curves. Thus they require either perfect shapes such as technical drawings or some preprocessing such as thinning or skeletonization. With this paper, we wish to investigate on this problem and build a technique which would ultimately manage raw data without any of these usual preprocessings. The aim of this paper is rather identification of primitives than vectorization. Our proposed method

---

* This work was supported by the French National Agency of Research under contract GEODIB ANR-06-BLAN-0225.

P. Wiederhold and R.P. Barneva (Eds.): IWCIA 2009, LNCS 5852, pp. 30–42, 2009.

tackles the issue of one pixel wide curves as well, but can be easily extended to thick curves using [7]. We base our work on the notion of generalized Tangential Cover of a digital curve [7,8]. This tool allows us to extract all maximal subparts of a curve with regards to a geometric predicate of validity. This implies of course that we define notions of subparts and their associate predicates. In order to capture both straight and circular parts while tolerating irregularities, we use thickness parameters and thus define two types of digital primitives: the α-thick digital segments [3] and the α'-thick digital arcs. The latter is related to the notion of digital ring [2]. We compute the generalized Tangential Cover for both of these primitives and introduce a method which decomposes digital curves into circular and straight parts simultaneously. The paper is organized as follows: in the first section we recall the definition and usefulness of the generalized Tangential Cover. We also present the definitions of our two different types of primitives. The next section is concerned with our method for multi-primitive decomposition. Then some results are exhibited and discussed. The paper ends with a conclusion and some perspectives of future works.

2   Representation

2.1   Definitions

Let us first define the notion of 2-dimensional digital curve.

Definition 1. *[12] A 2D digital curve $\mathcal{C}$ of size n is an ordered list of 2D integer points P_i, $0 \leq i < n$, such that P_i has P_{i-1} and P_{i+1} as 8-neighbors. The curve is closed if and only if P_0 and P_{n-1} are 8-neighbors. The curve is simple if and only if P_i only has P_{i-1} and P_{i+1} as 8-neighbors (except possibly for P_0 and P_{n-1} when the curve is open).*

We wish to represent 2D digital curves using digital primitives. We expect such a representation to give us important geometric informations about digital curves. To do this, we introduce the Tangential Cover of a digital curve and we refer to [12] and [15] for definition and properties of digital segments.

Definition 2. *[8] The Tangential Cover (TC) of a digital curve $\mathcal{C}$ is the set of all maximal digital straight segments of connected subsets of $\mathcal{C}$ with respect to the inclusion order.*

The TC may be mapped onto a circular arcs graph. A 2D closed digital curve is homeomorphic to a circle. Since the points of the curve are indexed in order, it is possible to map each point of the curve to a point on the unit circle, as illustrated on Fig. 1. Each segment of the TC then corresponds to an arc on the circle. On Fig. 1 (left), let us consider that a segment exists between points 4 and 6. This segment is then mapped onto an arc of the circle.

This representation has proven itself a very useful and efficient tool for digital curves analysis. For instance, it has been used to solve the Min-DSS problem [9]. It has also been used to estimate global digital curvature [11] and tangents

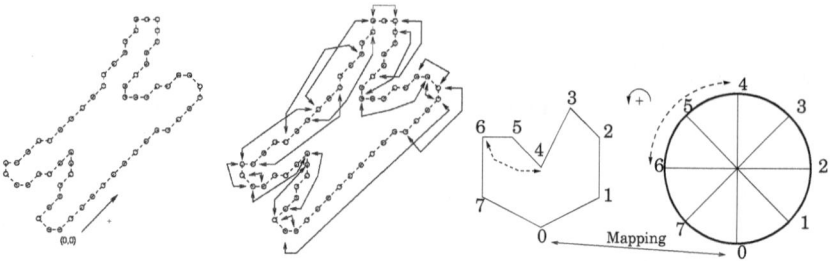

Fig. 1. (Left) the TC is the set of all maximal segments. (Right) the mapping onto the unit circle.

along digital contours [13]. In its primary form, the TC could only manage the extraction of maximal DSS (digital straight segments as in [12]). In previous works, we generalized the definition of the TC [7] in order to process different primitives. Let us recall the new definition.

Definition 3. *[7] Let C be a digital curve, and let P be a binary predicate of validity. The generalized Tangential Cover (gTC) of C is the set of all maximal valid subparts of C. Validity is intended with respect to P and maximality is intended with respect to the inclusion order.*

The genericity of the notion of *subpart* and its associated predicate of validity allows us to process various digital primitives. In this paper, our goal is to represent digital curves using circular and straight digital primitives simultaneously. In the following, we describe those two primitives.

2.2   α-Thick Digital Segments

The arithmetical definition of DSS, initiated by Reveills in [15], is a rather rigid one with regards to noise. If the processed shape is somewhat irregular, the recognized segments may be a lot shorter than expected. So we rely on an extended definition to overcome this drawback.

Definition 4. *[3] A set of points S is an α-thick digital segment if and only if its the isothetic thickness of its convex hull is less than a given real number α. The isothetic thickness of a convex hull is the minimum between its horizontal width and its vertical height.*

Figure 2 shows an example of an α-thick digital segment. This definition allows us to break the rigidity of arithmetical DSS. The value of α is a parameter and must be fixed in advance. An efficient algorithm which tests if a set of points is an α-thick digital segment is given in [3]. Its time complexity is $\mathcal{O}(\log n)$.

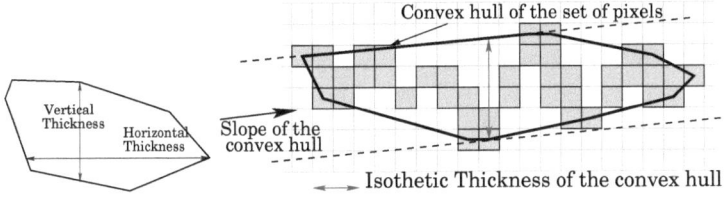

Fig. 2. (Left) the illustration of the isothetic thickness of a convex hull. (Right) an
α-thick digital segment, with $\alpha = 5$.

2.3 α'-Thick Digital Arcs

We wish to extend this notion of parametrized thickness to a circular digital primitive. In [10], Hilaire and Tombre define a fuzzy digital arc as a set of pixels "close enough" to a circular arc of the real plane, with some constraints. These constraints limit the recognition of digital arcs to rather thin ones, which does not suit our usage. Hence we use another approach. According to the work of Andrs [2], a digital ring is the set of all digital points comprised between two concentric euclidean circles (see Fig. 3). Our definition is related to the one of digital ring.

Definition 5. *A set of pixels S is an α'-thick digital arc if and only if they are completely enclosed between two Euclidean circles C_0 and C_1 with common center o and whose difference of radii $dr = r_1 - r_0$ is less than a given real number α', having $r_1 > r_0$.*

An example of an α'-thick digital arc is given on Fig. 3. Contrary to digital rings, some holes may appear in α'-thick digital arcs. In other words, it is not necessary for all digital points comprised between C_0 and C_1 to be in S for it to be an α'-thick digital arc. We wish to determine whether a given set of pixels S is part of an α'-thick digital arc. This problem is equivalent to finding the minimum-width enclosing annulus of S [4], and comparing its width to α'. There exists such an algorithm within the Computational Geometry Algorithm Library CGAL. Its time complexity is $\mathcal{O}(n)$.

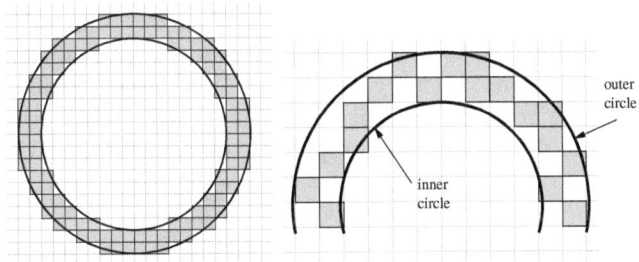

Fig. 3. (Left) a digital ring. (Right) an α'-thick digital arc, with $\alpha' = 2$.

2.4   Building the Tangential Cover

We detailed our way to compute the α-thick digital segments TC in [6]. The resulting time complexity is $\mathcal{O}(n \log n)$. To compute the TC for α'-thick digital arcs, we proceed as follows. We choose an arbitrary starting point on our input digital curve $\mathcal{C}$. Let us remark that this choice has no influence on the result. Since the input digital curve C is ordered, we add points in order to a current α'-thick digital arc $\mathcal{A}$. At each addition, we compute the minimum-width enclosing annulus thanks to CGAL, and store the obtained width w. When $w \geq \alpha'$, we stop the additions: the previous annulus was maximal with regards to α. We store the characteristics of the previous annulus: the center o, the radii r_0 and r_1, and the indices of the first and the last points of A (with regards to the ordering of $\mathcal{C}$). Then, we iteratively delete the first point of A and compute w until $w < \alpha'$. It is now possible to go on with the additions. This algorithm stops when the last point of the current maximal A is the same as the last point of the first computed maximal α'-thick digital arc, which we can now delete. The time complexity of this method is $\mathcal{O}(n^2)$. Let us remark that the Tangential Cover may as well be computed on open curves without any difference, except that the process stops as soon as the last point of the curve is reached. Since we are able to compute TCs for both circular and straight digital primitives, let us detail our method for multi-primitive analysis in the next section.

3   Methodology

3.1 α-Thick Digital Segments versus α'-Thick Digital Arcs

We have now extracted all maximal α-thick segments and α'-thick arcs of a digital curve $\mathcal{C}$. Our goal is to obtain a decomposition of $\mathcal{C}$ into both segments and arcs, using both Tangential Covers. This decomposition should capture geometric features of the curves. Obviously, straighter parts should be represented by segments and more circular ones should be represented by arcs. It resembles a polygonalization, but with several primitives. For given α and α' values, there exists a large amount of possible decompositions. The optimal decomposition(s) should be the one(s) which contain the minimum number of primitives. Our method should be able to compute the optimal decomposition, or at least reasonably approach it.

Let us first describe how to obtain a polygonalization with only one primitive using the Tangential Cover. A polygonalization is a decomposition of $\mathcal{C}$ into a series of consecutive geometric primitives. Such a decomposition may actually be seen as a path in the circular arcs graph corresponding to the TC (see section 2.1). This path may be computed easily. To do this, let us define the notion of *primitive pencil* $\mathcal{PP}(p)$ of a curve point with index p. $\mathcal{PP}(p)$ is the set of all primitives that contain p (see Fig. 4). The *front(P)* and *back(P)* elements are the beginning and end points of a primitive P. Remind that the graph is oriented. Given the orientation, we easily determine the primitive $P*$ of $\mathcal{PP}(p)$ that reaches the point with index $p_{\max} = front(P*)$ that is the farthest away

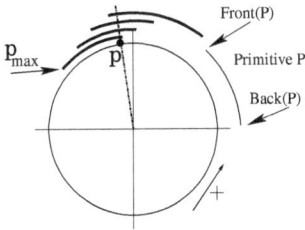

Fig. 4. The $front(P)$ and $back(P)$ elements are illustrated on a primitive P. The bold circular arcs represent the primitive pencil $\mathcal{PP}(p)$ of the point with index p.

from p on the curve. This point $p_{\max}$ defines a function $f()$, that is $f(p) = p_{\max}$. This $f()$ function is computed in linear time, since we only need to compute it for the beginning and ending points of each primitive.

The indices of the vertices of the polygonalization of $\mathcal{C}$ starting at point index p are obtained by iterates of the function $f()$. More formally, $vp_i = f^i(p)$ with vp_i being the index of the i-th vertex of the polygonalization, and with $f^i(p)$ being the i-th iterates of the function $f()$. We use this process for both Tangential Covers, building two functions $f_S()$ and $f_A()$ respectively for the segments TC and for the arcs TC. In this scope, a multi-primitive decomposition of a curve may be seen as a series of applications of the functions $f_X()$. To be more precise, such a decomposition is a word on the alphabet $\{f_A, f_S\}$.

A simple multi-primitive decomposition may be deduced straightaway. We map both Tangential Covers onto the same circular arcs graph and use the following greedy algorithm. We choose a starting point with index p_0. We have $f_S(p) = p_{Smax}$ and $f_A(p) = p_{Amax}$. If $p_{Amax} > p_{Smax}$ we choose the corresponding arc, else we choose the segment. The process stops when the point with index p_0 is reached again. This means that the decomposition is complete. The curve is then reconstructed using the chosen primitives and their associated

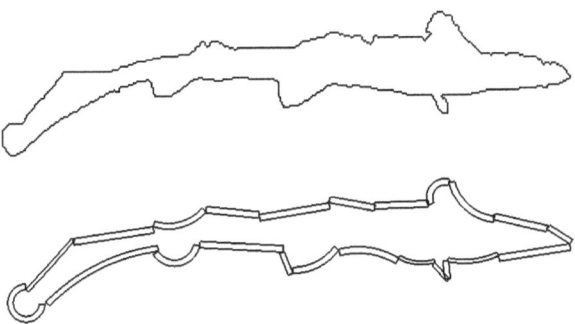

Fig. 5. The original curve (top) has been decomposed into α-thick digital segments and α'-thick digital arcs (bottom), with $\alpha = \alpha' = 4$ pixels

parameters. A result is presented on Fig. 5. In the case of open curves, the only difference is that the starting point is necessarily one of the endpoints, depending on the orientation.

However, this method does not allow us to control the quality of the solution (its proximity to the optimum). Moreover we do not have any information on the behavior of the decomposition with regards to the used primitives. Due to their construction, the two types of Tangential Covers contain all possible decompositions of the curve. To obtain a guarantee on the quality of the solution, a better method is to build the complete tree of all possible decompositions. We describe this process in the next subsection.

3.2   Building the Complete Tree

spaceskip=.27em pluse The idea is the following: from a given point with index p on the curve, we may either choose the longest segment or the longest arc covering p (longest meant as the result of the $f_X()$ function described in the previous section). Thus there exists two different possible decompositions starting from p. Then the same process is applied to both results, and so on. This process defines a binary tree. The root of the tree is the chosen starting point with index p_0 (see the next subsection). A node of the tree is a partial decomposition of the curve from p_0 to the current index p defined by any of the $f_X()$ functions. The process stops when $f_X(p) \geq p_0$, and the current node is then a leaf of the tree. Such a leaf is a complete decomposition of the curve, and a backtracking from the leaf to the root gives us the corresponding word on the alphabet $\{f_A, f_S\}$. We build the whole tree and analyze the results. We focus on the leaves with the smallest tree depth. Those leaves constitute a pool of minimal decompositions in terms of number of primitives. We show some statistics on Tab. 1. Three classical shapes are dealt with at different thickness values. We use a $(\sqrt{2})^i, i \in \{0, .., 4\}$ growth model for the thickness, having $\alpha = \alpha' = (\sqrt{2})^i$. For instance, for the chromosome shape and for $\alpha = \alpha' = 1$, there exists 4323 possible decompositions. Upon these decompositions, a pool of 20 of them contain only 9 primitives, which is the minimal value.

The building of the full tree allows us to better understand the behavior of the decompositions. However its main drawback is of course the combinatorial explosion that results. As exhibited on Tab. 1 for the leaf shape, which is a small curve (only 119 points) but rather irregular, the number of possible decompositions explodes for small α values. Thus its usage is limited to rather small curves for behavior studies only, and should be avoided otherwise. In the next section, we describe our way to avoid the full deployment of the tree while still obtaining a good, if not optimal, decomposition.

The Choice of the Starting Point. Choosing an arbitrary starting point index p_0 does not allow us to warrant the optimality of our decompositions. Anyway, we may improve the results by finding a "good" starting point. The best choice would be the endpoint of some primitive, in order to begin with a maximal primitive. But we are working with two types of primitives, so a good

Table 1. This table shows some statistics on the full decomposition tree. (*) To read the table: $\#D$ is the number of possible decompositions; $\#Min(D)$ is the number of minimal decompositions in terms of number of primitives; $|Min(D)|$ is the number of primitives of minimal decompositions. $\alpha = \alpha'$ for the whole table.

Curve	(*)	Chromosome (61 pts)	Leaf (119 pts)	Semicircle (103 pts)
$\alpha = 1$	$\#D$	4323	NA	3811
	$\#MinD$	20	NA	1
	$\|MinD\|$	9	NA	5
$\alpha = \sqrt{2}$	$\#D$	964	255744	810
	$\#MinD$	10	504	1
	$\|MinD\|$	8	15	5
$\alpha = 2$	$\#D$	150	11768	192
	$\#MinD$	18	88	1
	$\|MinD\|$	6	11	4
$\alpha = 2\sqrt{2}$	$\#D$	72	4480	127
	$\#MinD$	56	672	1
	$\|MinD\|$	6	11	4
$\alpha = 4$	$\#D$	25	175	32
	$\#MinD$	1	11	2
	$\|MinD\|$	3	6	3

starting point for one primitive is not necessarily good for the other one. We propose a quite simple method. $f_X(p) - p$ represents the length of the useful part of the primitive. For each point index p of the curve, we compute $Q_p = f_A(p) - p + f_S(p) - p$ and choose p_0 such that Q_{p_0} is maximal. This way, both primitives are taken into account for the choice of p_0.

3.3   A Partial Tree

We describe our method to avoid the full deployment of the tree while maintaining good results. Our approach is related to *branch and bound* techniques. We use a criterion to evaluate each node. All nodes which are not leaves and which sons have not been explored yet are considered *"open nodes"* suitable for evaluation. We only explore the node with the best criterion value upon the pool of open nodes. The process stops when the first leaf is built. We describe our criterion in the following.

Our Criterion. The criterion we use to evaluate the open nodes is based on the notion of "covering rate" $Cr(n)$ of a node n. Remind that each node is a partial decomposition of the curve (see section 3.2). Thus it covers a part of the curve. We compute the number of points $Q(n)$ covered until the current node, and divide this number by the tree depth $D(n)$ of the node. So the formula for the criterion is: $Cr(n) = \frac{Q(n)}{D(n)}$. It is the average value of the number of points contained in a primitive. The node with maximal $Cr()$ value upon the pool of

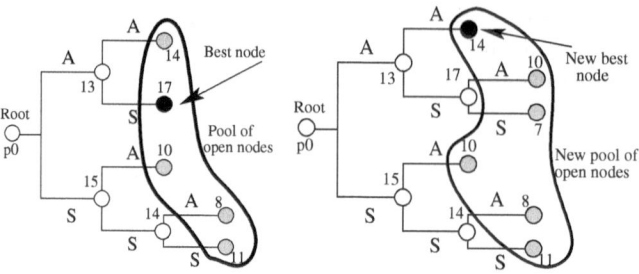

Fig. 6. Here is an example of the partial exploration of the tree based on our covering rate criterion

open nodes is the one to explore. Thus at each step, we deploy the branch that has the best chance to cover the entire curve with the best covering rate. For instance let us take a look at Fig. 6 (left). The decomposition starts with point index p_0. Each branch is labelled with the nature of its primitive (A for circular arc, S for segment), and each node is labelled with its covering rate $Cr()$. The pool of open nodes is represented, and the black node is the one that has the best covering rate. Hence it is the one that will be explored next. On the rightmost figure, the node has been explored and the pool of open nodes has been updated accordingly. The best node is now located in another branch. This process allows us to go back to previous unexplored nodes with lower depth if needed. $Cr(n)$ is computed in constant time at each node creation.

4   Experimental Results

We applied our method to a series of digital curves known as the *SQUID* database [1]. On Fig. 7 are exhibited several decompositions of a hippocampus shape. We used various thickness values, from $\alpha = \alpha' = 1$ to $\alpha = \alpha' = 4$ pixels. It is interesting to see how the local geometric features are captured at each different thickness value. For instance, the tail of the hippocampus is recognized as a circular part at thickness $\alpha = \alpha' = 2$, then it is again a series of straight parts for greater thickness values, and finally for $\alpha = \alpha' = 4$ the circular struc-ture reappears. Let us remark that the rate of circular arcs increases along with the thickness. Indeed, for small thickness values, the definition of digital circular arcs is too arithmetically constrained to result in long primitives when the input curves are somewhat irregular. A greater α value allows for a greater tolerance to irregularities, and more circular parts appear. This behavior is illustrated on the accompanying table of Fig. 7. We provide more similar experiments on the SQUID database on Fig. 8.

   To illustrate the quality of our greedy algorithm, we decomposed several sub-sampled digital shapes and compared the results using the full tree and the partial tree. These results are exhibited on Fig. 9. Those curves have been sub-sampled to approximatively 200 pixels each in order to deploy the full tree.

Thickness	$\alpha = 1$	$\alpha = \sqrt{2}$	$\alpha = 2$	$\alpha = 2\sqrt{2}$	$\alpha = 3$	$\alpha = 4$
Size of the decomposition	86	57	39	30	25	15
Ratio arcs/segments	2.3%	5.3%	12.8%	20%	28%	60%

Fig. 7. The digital curve is shown on top. Then we show the results of our multiprimitive decompositions for various thickness values increasing from left to right and from top to bottom (note that $\alpha = \alpha'$ for all decompositions). These values are exhibited on the bottom table, along with some statistics.

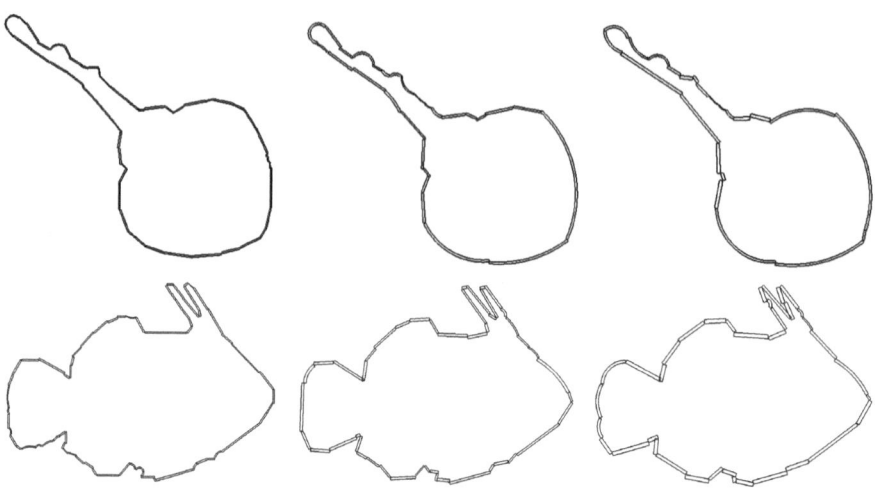

Fig. 8. Two different fish shapes have been decomposed using our method. From left to right, $\alpha = \sqrt{2}$, $\alpha = 2$, $\alpha = 2\sqrt{2}$.

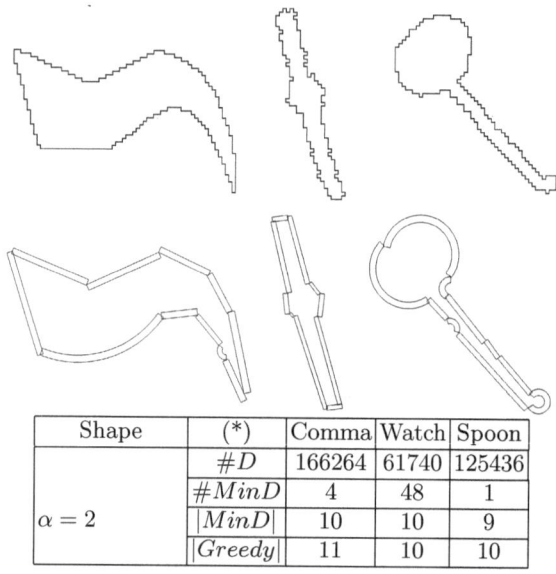

Shape	(*)	Comma	Watch	Spoon
	$\#D$	166264	61740	125436
	$\#MinD$	4	48	1
$\alpha = 2$	$\|MinD\|$	10	10	9
	$\|Greedy\|$	11	10	10

Fig. 9. Three subsampled digital shapes used to test the quality of our greedy algorithm. From left to right: "Comma", "Watch", "Spoon". $\alpha = \alpha' = 2$. (Top) the shapes. (Middle) decompositions using the greedy algorithm. (Bottom) statistics table: $\#D$, $\#Min(D)$ and $|Min(D)|$ have the same meaning as in Tab. 1; and $|Greedy|$ is the number of primitives of the greedy algorithm decompositions.

The accompanying statistics table of Fig. 9 shows that the greedy algorithm approaches or equals the minimal decomposition size. We conjecture that a greedy algorithm may guarantee an upper bound of one extra primitive with regards to the minimal size, but this has yet to be proven. On Fig. 10 are shown all four possible minimal decompositions of the "Comma" shape. The difference is in the bottom concave part. There, the decomposition seems to hesitate between circular or straight primitives. This might indicate that a greater thickness value is required to efficiently represent this part.

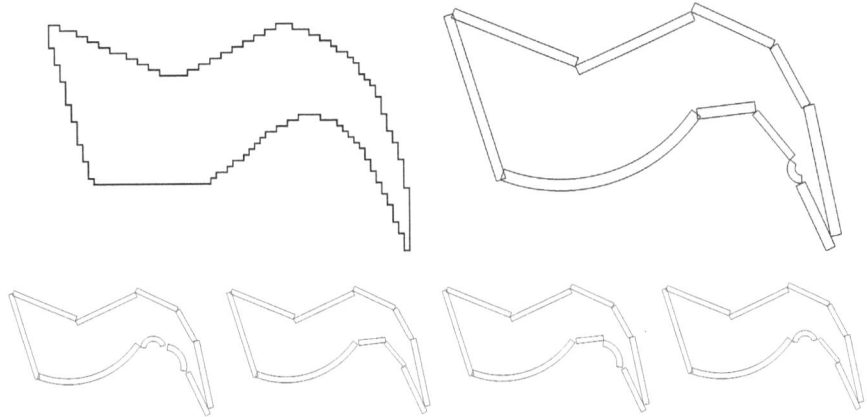

Fig. 10. (Top left) The subsampled "Comma" digital shape (Top right) The resulting multi-primitive decomposition using our greedy algorithm: 11 parts. (Bottom) The four minimal decompositions (10 parts) obtained using the full tree.

5   Conclusion

We have presented a method that analyzes digital curves using several different digital primitives. The main tool we have used is the generalized Tangential Cover, which performs the recognition of all maximal subparts of a digital curve with regards to a validity predicate. We have chosen to focus our analysis on two types of primitives: the α-thick digital segments and the α'-thick digital circular arcs, providing a definition for the latter. Then we have proposed an efficient partial exploration method to obtain decompositions of curves into both primitives simultaneously. In future works, we plan on solving the issue of the choice of the starting point, which is inherent to polygonalization problems. A good lead is the Min-DSS algorithm [9]. Then we will extend this method to the case of thick digital curves using [7]. We also plan to focus on the problem of sticking together connected primitives with close parameter values (i.e. slopes for segments or radii/centers for arcs). This could greatly improve the method by adding automatic local thickness adjustment.

References

1. Abbasi, S., Mokhtarian, F., Kittler, J.: The SQUID database, located at write time at (2002), http://www.ee.surrey.ac.uk/CVSSP/demos/css/demo.html
2. Andres, E.: Discrete Circles, Rings and Spheres. Computers and Graphics 18(5), 695–706 (1994)
3. Buzer, L.: Digital line recognition, convex hull, thickness, a unified and logarithmic technique. In: Reulke, R., Eckardt, U., Flach, B., Knauer, U., Polthier, K. (eds.) IWCIA 2006. LNCS, vol. 4040, pp. 189–198. Springer, Heidelberg (2006)
4. de Berg, M., Bose, P., Bremner, D., Ramaswami, S., Wilfong, G.T.: Computing Constrained Minimum-Width Annuli of Point Sets. In: Rau-Chaplin, A., Dehne, F., Sack, J.-R., Tamassia, R. (eds.) WADS 1997. LNCS, vol. 1272, pp. 392–401. Springer, Heidelberg (1997)
5. Debled-Rennesson, I., Reveillès, J.-P.: A Linear Algorithm for Segmentation of Digital Curves. International Journal of Pattern Recognition and Artificial Intelligence (IJPRAI) 9(4), 635–662 (1995)
6. Faure, A., Buzer, L., Feschet, F.: Tangential Cover for Thick Digital Curves. Pattern Recognition, doi:10.1016/j.patcog.2008.11.009
7. Faure, A., Feschet, F.: Robust Decomposition of Thick Digital Shapes. In: Brimkov, V.E., Barneva, R.P., Hauptman, H.A. (eds.) IWCIA 2008. LNCS, vol. 4958, pp. 148–159. Springer, Heidelberg (2008)
8. Feschet, F.: Canonical representations of discrete curves. Pattern Anal. Appl. 8(1-2), 84–94 (2005)
9. Feschet, F., Tougne, L.: On the min DSS problem of closed discrete curves. Discrete Applied Mathematics 151(1-3), 138–153 (2005)
10. Hilaire, X., Tombre, K.: Robust and Accurate Vectorization of Line Drawings. IEEE Transaction on Pattern Analysis and Machine Intelligence 28(6), 890–904 (2006)
11. Kerautret, B., Lachaud, J.-O.: Robust Estimation of Curvature along Digital Contours with Global Optimization. In: Coeurjolly, D., Sivignon, I., Tougne, L., Dupont, F. (eds.) DGCI 2008. LNCS, vol. 4992, pp. 334–345. Springer, Heidelberg (2008)
12. Klette, R., Rosenfeld, A.: Digital Geometry: Geometric Methods for Digital Picture Analysis. Computer Graphics and Geometric Modeling. Morgan Kaufman, San Francisco (2004)
13. Lachaud, J.O., Vialard, A., de Vieilleville, F.: Fast, accurate and convergent tangent estimation on digital contours. Image and Vision Computing 25(10) (2006)
14. Lindenbaum, M., Bruckstein, A.M.: On Recursive, O(N) Partitioning of a Digitized Curve into Digital Straight Segments. IEEE Trans. on Pattern Analysis and Machine Intelligence 15(9), 949–953 (1993)
15. Reveillès, J.P.: Géométrie discrète, calcul en nombres entiers et algorithmique, Thèse d'Etat (1991)
16. Rosin, P., West, G.: Nonparametric Segmentation of Curves into Various Representations. IEEE Trans. Pattern Anal. Mach. Intell. 17(12), 1140–1153 (1995)

What Does Digital Straightness Tell about Digital Convexity?

Tristan Roussillon[1,*], Laure Tougne[1], and Isabelle Sivignon[2]

[1] Université de Lyon,
Université Lyon 2, LIRIS, UMR5205, F-69676, FRANCE
{tristan.roussillon, laure.tougne}@liris.cnrs.fr
[2] Université de Lyon, CNRS
Université Lyon 1, LIRIS, UMR5205, F-69622, FRANCE
isabelle.sivignon@liris.cnrs.fr

Abstract. The paper studies local convexity properties of parts of digital boundaries. An online and linear-time algorithm is introduced for the decomposition of a digital boundary into convex and concave parts. In addition, other data are computed at the same time without any extra cost: the hull of each convex or concave part as well as the Bezout points of each edge of those hulls. The proposed algorithm involves well-understood algorithms: adding a point to the front or removing a point from the back of a digital straight segment and computing the set of maximal segments. The output of the algorithm is useful either for a polygonal representation of digital boundaries or for a segmentation into circular arcs.

Keywords: digital geometry, digital straightness, digital convexity, maximal segment, convex and concave parts.

1 Introduction

The paper studies local convexity properties of parts of digital boundaries. As shown in [9], the convexity of a digital boundary cannot be decided locally (where locally means in the 8-neighbourhood). Considering this fact, the following question has been raised in [10]: how far one can decide whether a part of a digital boundary is convex or not by a method that is *as local as possible*? Our answer is that a good neighbourhood for checking convexity is given by a segment that cannot be extended either at the front or at the back.

An online and linear-time algorithm is introduced for the decomposition of a digital boundary into convex and concave parts. The proposed algorithm uses well-understood algorithms: adding a point to the front [6] or removing a point from the back [14] of a digital straight segment. The core of the algorithm is similar to the one that computes the set of maximal segments [12,14]. Hence, one single scan of a digital boundary by a window corresponding to a digital

---

* Author supported by a grant from the DGA.

P. Wiederhold and R.P. Barneva (Eds.): IWCIA 2009, LNCS 5852, pp. 43–55, 2009.
© Springer-Verlag Berlin Heidelberg 2009

straight segment is sufficient to decompose the digital boundary into convex and concave parts. Our algorithm is on-line (contrary to [11,4]) and leads to a unique decomposition (contrary to [7]).

Moreover, other data are computed simultaneously. During the scan, the hull of each convex or concave part as well as the Bezout points of each edge of each hull are computed (we call *hull* a partial convex hull whose formal definition is given in Section 3.1). To do this, some operations are added to the maximal segments computation when the first and last leaning points of the current digital straight segment merge or split.

The link between the leaning points of the maximal segments and the hull of convex or concave parts has been investigated either for local estimators [8] or for faithful polygonalizations [10,5]. In this paper, the link between the hull of convex or concave parts and the leaning points of digital straight segments that are not necessarily maximal is also studied.

In Section 2, definitions of digital boundary, digital straight segment, leaning points and Bezout points are recalled in detail. The main contribution consists of Proposition 2 and Corollary 1 proved in Section 3.2. These propositions yield to Algorithm 2, which is an extended version of Algorithm 1. Applications of these algorithms are discussed in Section 4.

2   Preliminaries

2.1   Digital Object and Digital Boundary

A binary image I is viewed as a subset of points of $\mathbb{Z}^2$ that are located inside a rectangle of size $M \times N$. A digital object $O \in I$ is a 4-connected subset of $\mathbb{Z}^2$, without hole (Fig. 1.a). Its complementary set $\bar{O} = I\backslash O$ is the so-called background. The digital boundary B of O is defined as the 8-connected clockwise-oriented list of the digital points having at least one 4-neighbour in $\bar{O}$ (Fig. 1.b).

Each point of B is numbered according to its position in the list. The starting point, which is arbitrarily chosen, is denoted by B_0 and any arbitrary point of the list is denoted by B_k. A part $(B_i B_j)$ of B is the list of points that are ordered increasingly from index i to j (Fig. 1.c).

2.2   Digital Straight Line

Definition 1 (Digital straight line [15]). *The set of digital points (x, y) verifying $\mu \leq ax - by < \mu + max(|a|, |b|)$ belongs to the digital straight line (DSL) $\mathcal{D}(a, b, \mu)$ with slope $\frac{a}{b}$ and lower bound μ (with a, b, μ being integer such that $gcd(a, b) = 1$).*

The quantity $ax - by$, which is called the *remainder*, measures the distance between (x, y) and $\mathcal{D}$. Table 1 clusters the digital points of $\mathbb{Z}^2$ into seven groups according to their position with respect to $\mathcal{D}$. Note that merging the last two lines of Table 1 gives the two inequalities of definition 1.

Thanks to the vocabulary introduced in Table 1, two special kinds of points are easily defined:

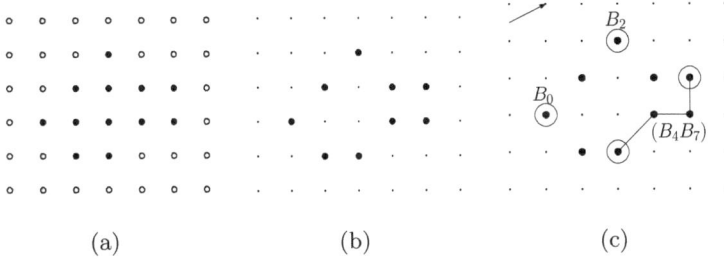

Fig. 1. A digital object depicted with black disks (a). Its digital boundary (b). Notation used (c).

Table 1. The digital points of $\mathbb{Z}^2$ are divided into seven groups according to their position with respect to the DSL $\mathcal{D}(a, b, \mu)$

position	on the left	on the right
strongly exterior	$ax - by < \mu - 1$	$ax - by > \mu + max(|a|, |b|)$
weakly exterior	$ax - by = \mu - 1$	$ax - by = \mu + max(|a|, |b|)$
weakly interior	$ax - by = \mu$	$ax - by = \mu + max(|a|, |b|) - 1$
strongly interior	$\mu < ax - by < \mu + max(|a|, |b|) - 1$	

Definition 2 (Leaning points and Bezout points (Fig. 2.a)). *The* leaning points *(resp.* Bezout points*) of a DSL $\mathcal{D}$ are defined as the points that are weakly interior (resp. exterior) to $\mathcal{D}$.*

The difference between two consecutive leaning points both located on the left or right of $\mathcal{D}$ is the vector $\boldsymbol{u} = (b, a)$ (Fig. 2.b).

The Bezout points are closely related to the leaning points. Indeed, the Bezout points that are on the left or right of $\mathcal{D}$ may be computed from the leaning points of the same side thanks to the well-known Bezout's identity: vector $\boldsymbol{v}$ (resp. $\boldsymbol{w}$) of Fig. 2.b is such that $det(\boldsymbol{u}, \boldsymbol{v})$ equals -1 (resp. $det(\boldsymbol{u}, \boldsymbol{w})$ equals 1) (Fig. 2.b). Moreover, the leaning points on the left or right of $\mathcal{D}$ maps to the Bezout points of the opposite side when shifted by a vector $\boldsymbol{s}$ that depends on the slope of $\mathcal{D}$. In the first octant, $\boldsymbol{s} = (0, 1)$ (Fig. 2.b).

2.3 Digital Straight Segment

Definition 3 (Digital straight segment). *A part $(B_i B_j)$ of B is a* digital straight segment *(DSS) if and only if there exists a DSL containing it.*

There are infinitely many DSL containing a DSS $(B_i B_j)$. However, there is always one DSL that is *strictly bounding* for $(B_i B_j)$.

Definition 4. *A DSL is* strictly bounding *for a DSS $(B_i B_j)$ if it has at least three leaning points belonging to $(B_i B_j)$. Moreover, at least one of them is a leaning point that is on the left of the strictly bounding DSL and at least one of them is a leaning point that is on the right of the strictly bounding DSL.*

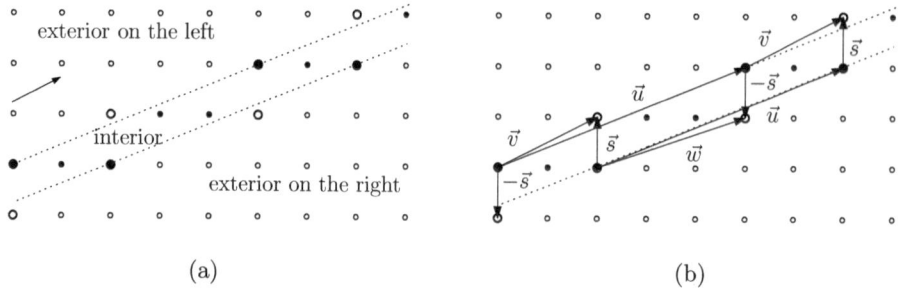

Fig. 2. The set of black disks lying between the two parallel dotted lines defines the DSL $\mathcal{D}(2,5,0)$. The large black and white disks depict the leaning (weakly interior) and Bezout (weakly exterior) points of $\mathcal{D}$. In (b), vectors v, w and s show that the Bezout points are closely related to the leaning points.

The recognition algorithm of Debled and Reveillès [6] returns the parameters a, b, μ of the strictly bounding DSL containing a DSS.

When speaking about a DSS (its slope for instance), we automatically refer to its strictly bounding DSL. Therefore, the exterior and interior points of a DSS (B_iB_j) are defined as the exterior and interior points of the strictly bounding DSL containing (B_iB_j). However, among all the leaning points of a DSS (B_iB_j), only those contained in the DSS, with an index ranging from i to j, are retained. Thus, the first (resp. last) leaning point is defined as the one with a minimal (resp. maximal) index.

Definition 5 (Maximal segment). *A DSS (B_iB_j) that cannot be extended at the front (resp. at the back), i.e. (B_iB_{j+1}) (resp. $(B_{i-1}B_j)$) is not a DSS, is said maximal at the front (resp. at the back). Moreover, a DSS that is both maximal at the front and maximal at the back is a maximal segment.*

There exist two algorithms to add [6] or remove [14] a point at one extremity of a DSS in constant time. Thanks to these two algorithms, the computation of the whole set of maximal segments of a digital boundary is done in linear time [12,14].

The set of maximal segments contains all DSS and all DSS segmentations (Fig. 8.a), one of which has the minimal number of segments [13]. Estimations of length [2], tangent [12,14] or curvature [12,3] may be derived from this set. Moreover, convex and concave parts are given by the slopes of the maximal segments [11,4].

3   Local Convexity and DSS

3.1   Definitions

Definition 6 (Convexity). *A digital object O is convex if and only if the digital points contained in belong to O only.*

According to definition 6, the object illustrated in Fig. 1.a is convex, whereas the object illustrated in Fig. 3.a is not convex.

Definition 7 hereafter is the analog of definition 6 for parts of digital boundaries and defines convex and concave parts with respect to the clockwise orientation of the digital boundary.

Definition 7 (Convex and concave parts (Fig. 3.b and 3.c)). *Let* $(B_i B_j)$ *be a part of a digital boundary* B. *The shortest polygonal line linking* B_i *and* B_j *located on the left (resp. right) of* $(B_i B_j)$ *is called the hull of* $(B_i B_j)$ *and is denoted by* $\overline{(B_i B_j)}$ *(resp.* $\underline{(B_i B_j)}$*).* $(B_i B_j)$ *is convex (resp. concave) if there is no digital point between the polygonal line linking the points of* $(B_i B_j)$ *and its hull* $\overline{(B_i B_j)}$ *(resp.* $\underline{(B_i B_j)}$*).*

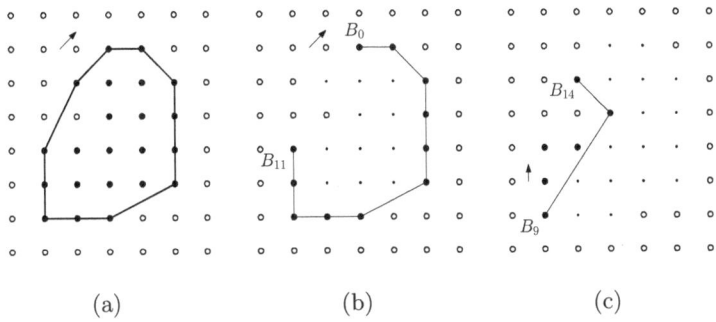

(a)   (b)   (c)

Fig. 3. In (a), the digital object depicted with black points is not convex because its convex hull contains one background point. In (b), the part $(B_0 B_{11})$ is convex because there is no digital point between $\overline{(B_0 B_{11})}$ and $(B_0 B_{11})$. Conversely, in (c), the part $(B_9 B_{14})$ is concave because there is no digital point between $\underline{(B_9 B_5)}$ and $(B_0 B_{14})$.

Similarly to the maximal segments, we define maximal convex and concave parts:

Definition 8 (Maximal convex or concave parts). *A convex part* $(B_i B_j)$ *that cannot be extended at the front (resp. at the back), i.e.* $(B_i B_{j+1})$ *(resp.* $(B_{i-1} B_j)$*) is not convex, is said maximal at the front (resp. at the back). Moreover, a maximal convex part is both maximal at the front and maximal at the back. Maximal concave parts are similarly defined.*

3.2 Main Results

A part $(B_i B_j)$ contains a part $(B_i B_l)$ (with $i < l < j$) that is supposed to be convex. The case where $(B_i B_l)$ is concave is symmetric.

Proposition 1 (proved in [7]) shows in which cases the convex part $(B_i B_l)$ is not maximal.

Proposition 1. *Let* $(B_k B_l)$ *(with* $i < k < l$*) be a DSS that is maximal at the back.* $(B_i B_{l+1})$ *is not convex if and only if* B_{l+1} *is strongly exterior to the left of* $(B_k B_l)$*.*

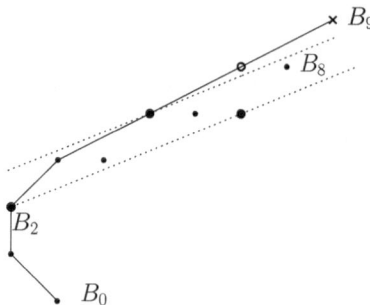

Fig. 4. $(B_2 B_8)$ is contained in a DSL of slope $\frac{2}{5}$. B_9 is strongly exterior to the left of this DSL, so $(B_0 B_9)$ is not convex.

Furthermore, as shown in Lemma 1, there is a link between leaning points of maximal segments and vertices of hulls. The part $(B_i B_j)$ is supposed to be convex. The case where $(B_i B_j)$ is concave is symmetric.

Lemma 1. *Let $(B_i B_j)$ contain a maximal segment $(B_k B_l)$. The leaning points on the left of $(B_k B_l)$ are vertices of $\overline{(B_i B_j)}$.*

Proof. Let $\frac{a}{b}$ be the slope of the strictly bounding DSL of $(B_k B_l)$. Since $(B_k B_l)$ is a maximal segment and $(B_i B_j)$ is convex, all the points of $(B_i B_j)$ that are not in $(B_k B_l)$ are on the right of the strictly bounding DSL of $(B_k B_l)$ (Fig. 5). Let $\mathcal{L}$ be a straight line of slope $\frac{a}{b}$ that is on the left of the strictly bounding DSL of $(B_k B_l)$. The first points hit by $\mathcal{L}$ while $\mathcal{L}$ is moving toward $(B_k B_l)$ are the leaning points on the left of $(B_k B_l)$. By definition, they are vertices of $\overline{(B_i B_j)}$ too.

The number of vertices of the convex hull of a convex boundary is greater that the number of its maximal segments [8]. Thus, we cannot retrieve all the vertices of $\overline{(B_i B_j)}$ from the leaning points of the maximal segments of $(B_i B_j)$. However, we can retrieve them in the course of the maximal segments computation, from the leaning points of segments that are not maximal, but either maximal at the front or at the back.

Proposition 2 and Corollary 1 define the two events that provide a way of finding, from a known vertex of $\overline{(B_i B_j)}$, the next vertex of $\overline{(B_i B_j)}$. Again, a part $(B_i B_j)$ contains a part $(B_i B_l)$ (with $i < l < j$) that is supposed to be convex. The case where $(B_i B_l)$ is concave is symmetric.

Proposition 2. *Let $(B_k B_l)$ (with $i < k < l$) be a DSS that is maximal at the back. If the point B_{l+1} is exterior to the right of the strictly bounding DSS of $(B_k B_l)$, then the last leaning point on the left of $(B_k B_l)$ is the next vertex of $\overline{(B_i B_j)}$.*

Proof. Let us denote by $\mathcal{L}$ the straight line passing through the first and last leaning points on the left of $(B_k B_l)$ (solid line of Fig. 6). By definition, $B_{l+1} + s$

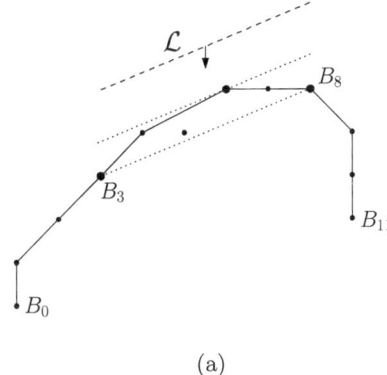

(a)

Fig. 5. The leaning point on the left of the maximal segment (B_3B_8) is a vertex of $\overline{(B_0B_{11})}$ because (B_0B_{11}) is convex

is located on $\mathcal{L}$ if B_{l+1} is *weakly* exterior to the right of $\mathcal{D}$ (Fig. 6) and strictly on the right of $\mathcal{L}$ if B_{l+1} is *strongly* exterior to the right of $\mathcal{D}$.

Let us denote by α the slope of $\mathcal{L}$. On the one hand, any straight line of slope greater than α that leaves the leaning points on the left of (B_kB_l) on its right side, leaves $B_{l+1} + \boldsymbol{s}$ on its right side too. Since (B_iB_j) is convex, such a line cannot contain an edge of $\overline{(B_iB_j)}$. On the other hand, any straight line that separates the leaning points on the left of (B_kB_l) from $B_{l+1} + \boldsymbol{s}$ has necessarily a slope lower than α (like the dashed line in Fig. 6).

As (B_kB_l) is maximal at the back, the first leaning point on the left of (B_kB_l) is a vertex of $\overline{(B_iB_j)}$ (Lemma 1). Thus, the last leaning point on the left of (B_kB_l) is the next vertex of $\overline{(B_iB_j)}$. $\square$

Corollary 1. *Let (B_kB_l) (with $i < k < l$) be a DSS that is maximal at the front. If (B_kB_l) has got only one leaning point and if $(B_{k+1}B_l)$ has got strictly more than one leaning point, then the last leaning point on the left of $(B_{k+1}B_l)$ is the next vertex of $\overline{(B_iB_j)}$.*

Fig. 7 illustrates Corollary 1.

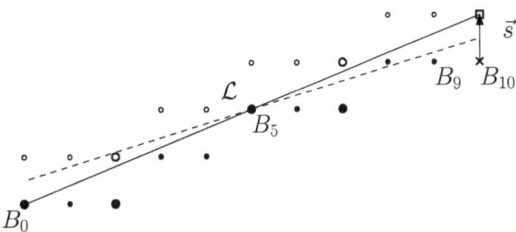

Fig. 6. B_{10} is weakly exterior to $\mathcal{D}(2,5,0)$. The first and last leaning points on the left of (B_0B_9) (B_0 and B_5) and $B_{10} + \boldsymbol{s}$ are collinear. Since B_0 is a vertex of $\overline{(B_0B_{10})}$ by hypothesis, B_5 is the next vertex of $\overline{(B_0B_{10})}$.

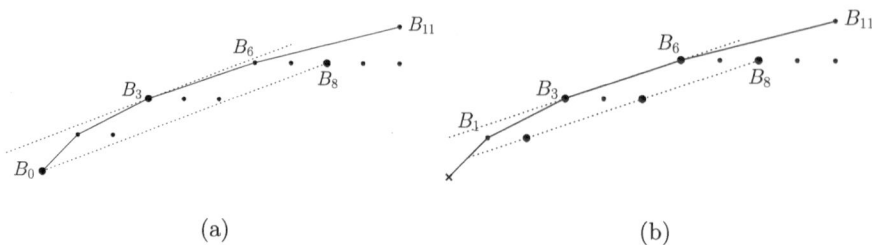

(a)   (b)

Fig. 7. B_0 is removed from the segment $(B_0 B_8)$ that is maximal at the front. Since B_3 is a vertex of $\overline{(B_0 B_{11})}$ by hypothesis, B_6 is the next vertex of $\overline{(B_0 B_{11})}$.

The proof of Corollary 1 is omitted because it is similar to the one of Proposition 2.

4   New and Revisited Algorithms

In this section, algorithms are derived from Proposition 1 and Proposition 2 in order to (i) decompose a digital boundary into convex and concave parts (Section 4.1), (ii) extract the hull of each convex or concave part (Section 4.2), (iii) compute polygonal representations respecting convex and concave parts (Section 4.3) and (iv) perform the preprocessing stage that optimizes the digital arc segmentation (Section 4.4).

4.1   Decomposition into Convex and Concave Parts

A simple online and linear-time algorithm to decompose a given part of boundary into convex and concave subparts is derived from Proposition 1 (Algorithm 1). For sake of clarity, we assume in Algorithm 1 that the first retrieved subpart is convex. The core of the algorithm is the scan of a part with a window corresponding to a DSS as maximal as possible, like in the maximal segments computation [12,14]. If a point that is strongly exterior to the left (resp. right) of the current DSS is found (line 6), then a convex (resp. concave) part is retrieved (line 7) and a new concave (resp. convex) part is searched (lines 8-9).

Contrary to [11] and [4], the algorithm is online, because the decomposition is not derived from the slopes of the maximal segments but is given in the course of the maximal segments computation.

In [1], the algorithm of [7] for testing the convexity is used in order to perform an online decomposition into convex and concave parts. But the decomposition is greedy and results in a set of pairwise disjoint parts of the boundary. Our decomposition is unique and results in a set of maximal convex and concave parts. Furthermore, in the algorithm of [7], each point is processed three times at most, according to the authors, whereas in Algorithm 1, because removing a point at the back of a DSS [14] is allowed, each point is processed twice at most.

Algorithm 1. Decomposition into convex and concave parts

Input: a part $(B_i B_j)$ of a boundary B
Output: The list $\mathcal{L}$ of convex and concave parts of $(B_i B_j)$

```
 1  L ← ∅ ;                          /* list of convex and concave parts */
 2  x ← true ;                       /* true if convex and false otherwise */
 3  k' ← i ;                         /* beginning of the current segment */
 4  k ← k' ;              /* beginning of the current convex or concave part */
 5  l ← i + 1 ;                      /* index of the current point */
 6  while l ≤ j do
 7  │   l ← l + 1 ;                  /* add a point to the front */
 8  │   if (B_k B_l) is not a DSS then
    │   │   /* side(x) returns 'left' if x is true, 'right' otherwise   */
 9  │   │   if B_l is strongly exterior and on the (side(x)) of (B_k B_{l-1}) then
10  │   │   │   L ← L + (B_{k'} B_{l-1}) ; /* add this part to the decomposition */
11  │   │   │   k' ← k;
12  │   │   │   x ← ¬x ;             /* from convex to concave and vice versa */
13  │   │   if B_l is strongly exterior and on the (side(¬x)) of (B_k B_{l-1}) then
14  │   │   │   while (B_k B_l) is not a DSS do
15  │   │   │   │   k ← k + 1 ;      /* remove a point from the back */

16  return L + (B_{k'} B_{l-1});
```

Fig. 8.b illustrates the decomposition. Notice that, as expected, the part $(B_{27} B_{38})$, which is contained both in the convex part $(B_0 B_{38})$ and the concave one $(B_{27} B_{66})$, is a maximal segment.

In the next subsection, we go further and propose an online and linear-time algorithm that provides the hull of each convex or concave parts.

4.2 Hull of Each Convex or Concave Parts

From Proposition 2 and Corollary 1, an online and linear-time algorithm is derived to extract the hull of each convex and concave part (Algorithm 2).

Moreover, the Bezout points of the edges of the hull, that is the Bezout points of the DSS whose extremities are the vertices of the edges of the hull, are naturally extracted at the same time. Indeed, as the edges are given by the leaning points on the left or right of a DSS, their Bezout points are computed from the opposite leaning points of the DSS, thanks to s (Section 2.2).

Algorithm 2 has an invariant : the first leaning point on the left or the right of the current segment is always a vertex of the hull of the current part. Since the current segment is either maximal at the front or at the back or both during the maximal segments computation, the assumptions of Proposition 1, Proposition 2 and Corollary 1 are fulfilled. Proposition 2 and Corollary 1 show that the invariant is valid and guarantee that Algorithm 2 is correct.

To process strictly convex or concave parts is straightforward. However, a change of convexity brings trickier issues because two hulls enclose a part that is both convex and concave. As described in Section 4.3, the last points of the

---

Algorithm 2. Hull of each convex or concave part and its Bezout points

---

Input: a part $(B_i B_j)$ of a boundary B
Output: The list $\mathcal{L}$ of leaning and Bezout points

```
1  x ← true ;                    /* true if convex and false otherwise */
2  k ← i ;                       /* beginning of the current segment */
3  l ← i+1 ;                     /* index of the current point */
4  L ← Bk ;                      /* list of leaning and Bezout points */
5  while l ≤ j do
6  │  l ← l+1 ;                              /* add a point to the front */
   │  /* side(x) means 'left' if x is true, 'right' otherwise        */
7  │  if (Bk Bl) is a DSS then
8  │  │  if Bl is weakly exterior and on the (side(¬x)) of (Bk Bl−1) then
9  │  │  │  L ← L + the list of Bezout points on the (side(¬x)) of (Bk Bl−1)
   │  │  │  that are between the first and last opposite leaning points;
10 │  │  │  L ← L + the last leaning point on the (side(x)) of (Bk Bl−1);
11 │  else
12 │  │  if Bl is strongly exterior and on the (side(x)) of (Bk Bl−1) then
13 │  │  │  x ← ¬x ;            /* from convex to concave and vice versa */
14 │  │  if Bl is strongly exterior and on the (side(¬x)) of (Bk Bl−1) then
15 │  │  │  while (Bk Bl) is not a DSS do
16 │  │  │  │  k ← k+1 ;            /* remove a point from the back */
17 │  │  │  │  if (Bk−1 Bl−1) has a unique leaning point on the (side(x)) and
   │  │  │  │  (Bk Bl−1) has more than one leaning point on the same side
   │  │  │  │  then
18 │  │  │  │  │  L ← L + the list of Bezout points on the (side(¬x)) of
   │  │  │  │  │  (Bk Bl−1) between the first and last opposite leaning points;
19 │  │  │  │  │  L ← L + the last leaning point on the (side(x)) of (Bk Bl−1);

20 return L;
```

---

first hull and the first points of the second hull are not stored in the list in order to correctly link the two hulls.

4.3 Polygonal Representations

The output of Algorithm 2 may be modified to get different meaningful lists of points. In Fig. 8.c, the set of black points depicts the leaning points retrieved by Algorithm 2. The black polygonal line that goes through the black points is a polygonal line respecting convex and concave parts. Indeed, in the strictly convex and concave parts, the polygonal line is equal to their hull. In the parts that are both convex and concave, the first leaning point of the maximal segment of inflection is linked with the last one. For instance, in the maximal segment of inflection $(B_{27}B_{38})$, B_{29} is linked with B_{36}.

In Fig. 8.d, the set of black points depicts the leaning points retrieved by Algorithm 2 in the convex parts. The set of white points depicts the leaning points

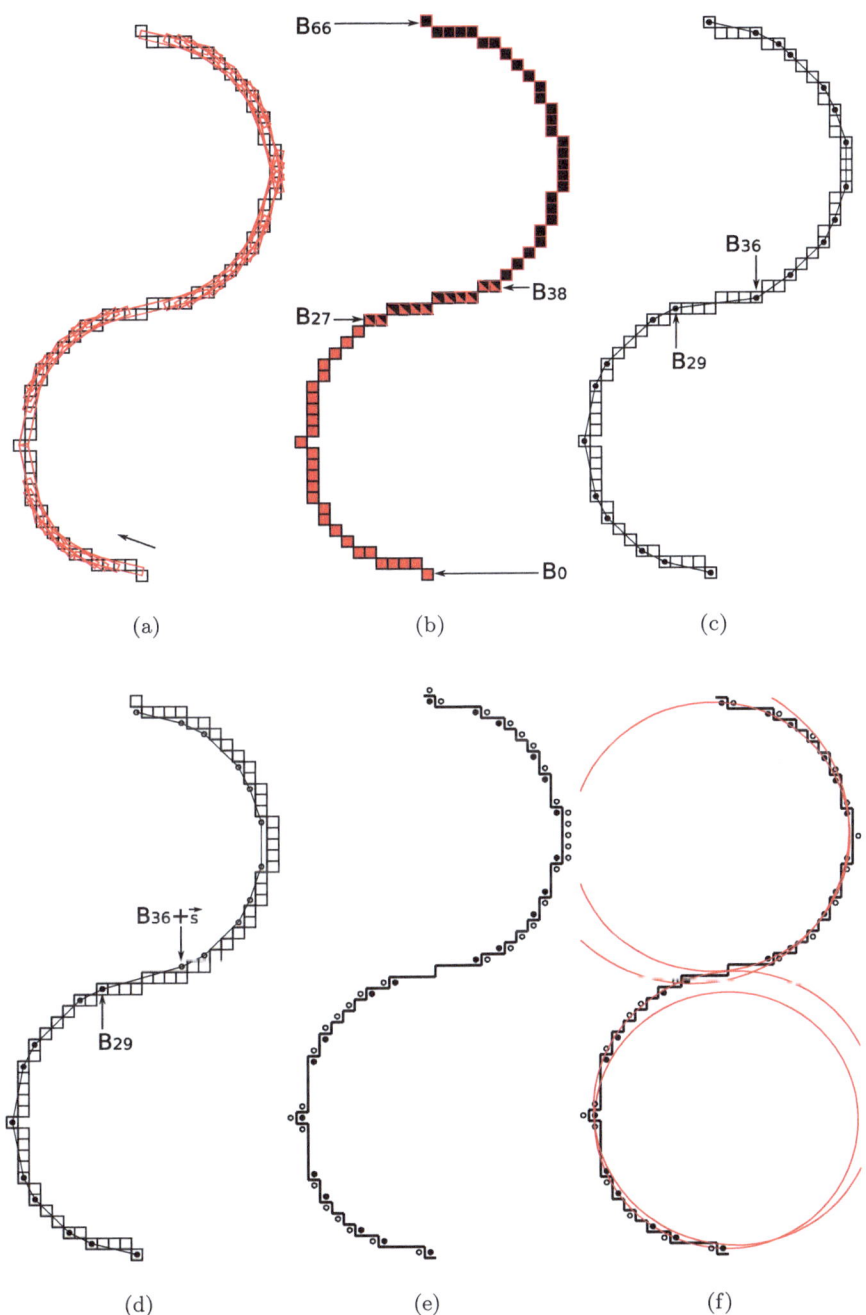

Fig. 8. (a) Maximal segments, (b) Convex and concave parts, (c) Polygonal representation respecting convex and concave parts, (d) Minimum-Length Polygon, (e) Minimum-Length Polygon with its associated Bezout points, (e) List of sufficient points for the digital arc segmentation

retrieved by Algorithm 2 in the concave parts, but shifted by s. For instance, as the DSS from which B_{36} has been extracted is in the first octant, $B_{36} + s = B_{36} + (0, 1)$. The black polygonal line that goes through the whole set of points faithfully represents the convex and concave parts. It is not hard to show that this polygonal line is actually the minimum-length polygon (MLP) between O and $\bar{O}$ [16]. The MLP is reversible if two digitization schemes are considered at the same time: Object Boundary Quantification (OBQ) for convex parts and Background Boundary Quantification (BBQ) for concave parts. In other words, the MLP of O equals the MLP of $\bar{O}$. Thanks to these properties, the MLP is a good perimeter estimator as experimentally shown in [2].

4.4   Digital Arc Segmentation

The list of points extracted thanks to Algorithm 2 is useful for circular arc segmentation too. A common approach to the circular arc recognition is to search Euclidean circles separating O from $\bar{O}$. Elementary algorithms that retrieve the set of separating circles are computationally expensive. That is why an optimisation has been proposed in [1]. For each convex or concave part, the set of points that have to be enclosed by the separating circles can be reduced to the hull of the part. Moreover, the set of points that have not to be enclosed by the separating circles can be reduced to some of the Bezout points of the edges of the hull: those that are located near the bisector of each edge.

In Fig 8.e and 8.f, the black polygonal line separates O from $\bar{O}$. The set of black points depicts the vertices of the MLP. In Fig 8.e, the set of white points depicts the Bezout points of the edges of the MLP. In Fig 8.f, only those that are located near the bisector of each edge are depicted. An approach to the circular arc segmentation is to iteratively search Euclidean circles separating the points of the MLP from some of its Bezout points (Fig 8.f).

In [1], the extraction of these points consists of three steps. First, the procedure of [7] is used to decompose the digital curve into convex and concave parts. Then, each part is decomposed into DSS [6]. Finally, the hull and Bezout points of each segment are computed with the extended Euclid's algorithm. Algorithm 2 provides in one scan the hull and Bezout points of each convex or concave part. Taking only the Bezout points located near the bisector of the two leaning points instead of all (lines 7 and 16 of Algorithm 2) is sufficient to simply perform in one scan what is done in three steps in [1].

5   Conclusion

Algorithm 1 and Algorithm 2 are both similar to the algorithm that computes the set of maximal segments [14]. As a consequence we can merge these algorithms to get in one scan: (i) the set of maximal segments (Fig. 8.a) (ii) the convex and concave parts (Fig. 8.b) (iii) the hull of each convex or concave part and their Bezout points (Fig. 8.e). These data are useful for polygonal representation (Fig. 8.c and 8.d) as well as for decomposition into circular arcs (Fig. 8.f).

The algorithms presented in this paper are considerably neater than previous ones [7,4,1] because only one scan by a window corresponding to a segment maximal either at the front or at the back is performed. Each point is processed in a constant time twice at most and thus, the whole algorithm is of order $\mathcal{O}(n)$ for a part of a digital boundary having n points.

References

1. Coeurjolly, D., Gérard, Y., Reveillès, J.-P., Tougne, L.: An Elementary Algorithm for Digital Arc Segmentation. Discrete Applied Mathematics 139(1-3), 31–50 (2004)
2. Coeurjolly, D., Klette, R.: A Comparative Evaluation of Length Estimators of Digital Curves. IEEE Trans. on Pattern Analysis and Machine Intelligence 26, 252–257 (2004)
3. Coeurjolly, D., Miguet, S., Tougne, L.: Discrete Curvature Based on Osculating Circle Estimation. In: Arcelli, C., Cordella, L.P., Sanniti di Baja, G. (eds.) IWVF 2001. LNCS, vol. 2059, p. 303. Springer, Heidelberg (2001)
4. Dorksen-Reiter, H., Debled-Rennesson, I.: Convex and Concave Parts of Digital Curves. In: Klette, R., Kozera, R., Noakes, L., Weickert, J. (eds.) Geometric Properties from Incomplete Data, vol. 31, pp. 145–159. Springer, Heidelberg (2005)
5. Dorksen-Reiter, H., Debled-Rennesson, I.: A linear Algorithm for Polygonal Representations of Digital Sets. In: Reulke, R., Eckardt, U., Flach, B., Knauer, U., Polthier, K. (eds.) IWCIA 2006. LNCS, vol. 4040, pp. 307–319. Springer, Heidelberg (2006)
6. Debled-Rennesson, I., Reveillès, J.-P.: A linear algorithm for segmentation of digital curves. Intern. Journal of Pattern Recognition and Artificial Intelligence 9, 635–662 (1995)
7. Debled-Rennesson, I., Rémy, J.-L., Rouyer-Degli, J.: Detection of the Discrete Convexity of Polyominoes. Discrete Applied Mathematics 125, 115–133 (2003)
8. de Vieilleville, F., Lachaud, J.-O., Feschet, F.: Maximal Digital Straight Segments and Convergence of Discrete Geometric Estimators. Journal of Mathematical Image and Vision 27(2), 471–502 (2007)
9. Eckhardt, U.: Digital Lines and Digital Convexity. In: Bertrand, G., Imiya, A., Klette, R. (eds.) Digital and Image Geometry. LNCS, vol. 2243, pp. 209–227. Springer, Heidelberg (2002)
10. Eckhardt, U., Dorksen-Reiter, H.: Polygonal Representations of Digital Sets. Algorithmica 38(1), 5–23 (2004)
11. Feschet, F.: Canonical Representations of Discrete Curves. Pattern Analysis and Applications 8, 84–94 (2005)
12. Feschet, F., Tougne, L.: Optimal Time Computation of the Tangent of a Discrete Curve: Application to the Curvature. In: Bertrand, G., Couprie, M., Perroton, L. (eds.) DGCI 1999. LNCS, vol. 1568, pp. 31–40. Springer, Heidelberg (1999)
13. Feschet, F., Tougne, L.: On the Min DSS Problem of Closed Discrete Curves. Discrete Applied Mathematics 151, 138–153 (2005)
14. Lachaud, J.-O., Vialard, A., de Vieilleville, F.: Fast, Accurate and Convergent Tangent Estimation on Digital Contours. Image and Vision Computing 25, 1572–1587 (2007)
15. Reveillès, J.-P.: Géométrie Discrète, calculs en nombres entiers et algorithmique. thèse d'etat, Université Louis Pasteur (1991)
16. Sklansky, J., Chazin, R.L., Hansen, B.J.: Minimum-perimeter Polygons of Digitized Silhouettes. IEEE Transactions on Computers 21(3), 260–268 (1972)

Hierarchical Discrete Medial Axis
for Sphere-Tree Construction

Alain Broutta, David Coeurjolly, and Isabelle Sivignon

Université de Lyon CNRS
Laboratoire LIRIS - UMR 5205 F-69622 Villeurbanne Cedex, France
{alain.broutta,david.coeurjolly,isabelle.sivignon}@liris.cnrs.fr

Abstract. In discrete geometry, the Distance Transformation and the Medial Axis Extraction are classical tools for shape analysis. In this paper, we present a Hierarchical Discrete Medial Axis, based on a pyramidal representation of the object, in order to efficiently create a sphere-tree, which has many applications in collision detection or image synthesis.

Keywords: discrete medial axis, bounding volumes hierarchies, sphere-trees, regular pyramids.

1 Introduction

In interactive environments, hierarchical representations are classical tools in many applications, *e.g.* in image synthesis, multi-resolution representations and interference detection, as allowing a fast interaction location. The Bounding Volume Hierarchies (BVH) consist in coverage with an increasing number of simple volumes (spheres [10,16,19], axis-aligned bounding boxes [20], oriented bounding boxes [12], ...) on different levels, ensuring that each part of the object covered by a children node must be covered by the parent node. Hence the collision detection between two hierarchical models is computed by a recursive process in which overlaps between pairs of primitive volumes are tested. The choice of the primitive volume is important for the BVH properties and sphere-trees are very interesting for interference detection [12]. Nevertheless, as they are bad estimators of the object geometry, the hierarchy needs an effective construction of the tree, reducing the error between the object and the associated set of spheres.

In computational geometry, first algorithms for sphere-tree construction were based on Octree data structures which consist in recursive spatial subdivisions of the object [10,13,16]. Later, efficient algorithms were proposed, based on the object Medial Axis (MA) [4], which corresponds to a skeletal representation of the object. MA extraction is based in a Voronoi Diagram [6,7,14], and the sphere-tree is produced by complex optimization heuristics to reduce the number of spheres. The error is controled by an approximated Hausdorff distance.

In discrete geometry, the study of multiresolution representation of digital objects has been carried out using a homotopic thinning [18]. However this approach should be considered as a medial axis filtering instead of a hierarchical

P. Wiederhold and R.P. Barneva (Eds.): IWCIA 2009, LNCS 5852, pp. 56–67, 2009.
© Springer-Verlag Berlin Heidelberg 2009

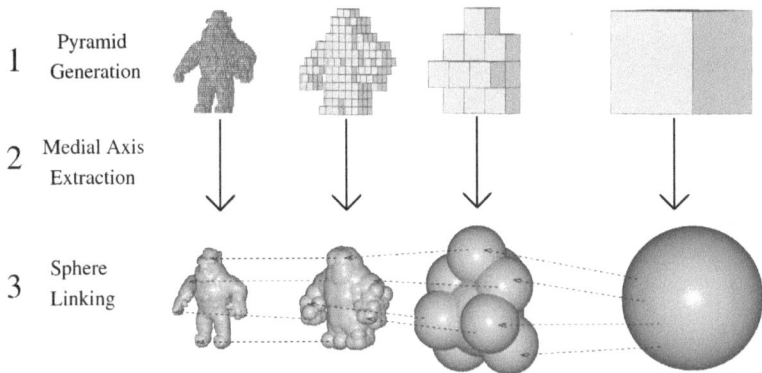

1 Pyramid Generation

2 Medial Axis Extraction

3 Sphere Linking

Fig. 1. The main stages of the sphere-tree construction

reduction of the set of balls. Our goal is to develop a hierarchical structure which is flexible with respect to the reversibility of the construction. Indeed the Discrete Medial Axis (DMA) is a convenient tool to represent objects in digital space, thanks to its reversibility: from the DMA balls, we can exactly reconstruct the original shape. The exact Euclidian DMA can be efficiently computed from Distance Transformation (DT) [9]. Furthermore, a discrete approach benefits from the exact computation of the error between the object and the hierarchy (at each level) with a Hamming distance (the difference volume between the objects).

In this paper, we present an original method for sphere-tree construction (sketched in Figure 1): the set of spheres at each level is obtained by a DMA extraction at different object resolution levels in a regular pyramid. Preliminaries are presented in Section 2. The sphere-tree construction is achieved by linking the spheres on consecutive levels. This method is first defined for a reversible model for volume synthesis (Sections 3.1), however it can be easily adapted for interactive environments (Section 3.2). Experiments in Section 4 also show that this last modification reduces the error. Finally we obtain a d-dimensional generic sphere-tree computation in linear time for any discrete distance and including pyramidal model.

2 Preliminaries

As sketched in the introduction, the methods is based on a regular pyramid, where set of spheres are produced by a DMA extraction on each level. This section presents these preliminary stages.

2.1 Pyramidal Model

In image analysis, the pyramidal structure is a convenient tool [8,15]. In dimension 2, a pyramid $\mathcal{P}$ of depth $N+1$ can be defined by a set of 2D images $\{\mathcal{F}_0, ..., \mathcal{F}_N\}$. The regular pyramid construction is a bottom-up process: $\mathcal{F}_N$ is

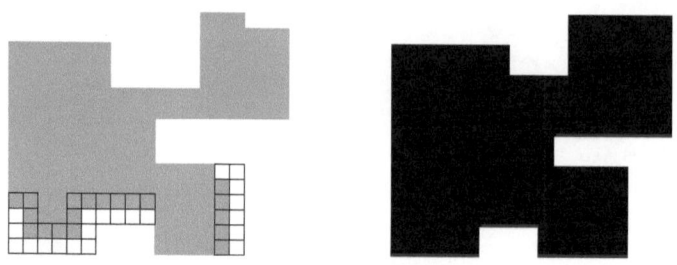

Fig. 2. Consecutive levels on a regular pyramid with the OR-Model

the original image, the upper levels are its representations at lower resolutions in a quad-tree approach; the pixel size at each level is 4 times bigger than at next level. The pixel color is based on the 4 pixels and computed by a transfer function (cf Figure 2).

We can generalize this process to any dimension d with an integer factor f for the voxel size expansion. In other words, a voxel v' at level $L-1$ contains f^d voxels $\{v_1, ..., v_{f^d}\}$ at level L. So, the transfer function, denoted $\mathcal{M}$ (standing for Model), is a mapping between these voxels:

Definition 1. $\mathcal{M}$ *is a model between* $\mathcal{F}_L$ *and* $\mathcal{F}_{L-1}(\mathcal{F}_{L-1} = \mathcal{M}(\mathcal{F}_L))$ *if*
$$\forall v', \ \exists\{v_1, ..., v_{f^d}\} \setminus \mathcal{F}_{L-1}(v') = \mathcal{M}(\mathcal{F}_L(v_1), ...\mathcal{F}_L(v_{f^d}))$$

In the proposed bounding volume hierarchy approach, the original object should be completely covered by each level of the hierarchy. So, we used the OR-Model: each voxel at level $L-1$ belongs to $\mathcal{F}_{L-1}$ when at least one of the f^d voxels that it contains (at level L) belongs to $\mathcal{F}_L$. Figure 2 shows an application of this model on 2D images.

2.2   Discrete Medial Axis and Discrete Power Diagram

Discrete Medial Axis. The Medial Axis is a shape descriptor first presented by Blum in 1967 [4] in order to simulate the wavefront propagation from its boundary (*prairie fire model*). The medial axis is defined by (1) the locus of points equidistant from the two closest boundary points, or (2) the locus of centers of maximal balls included in the object (a ball is *maximal* if it is not included in any other ball included in the object) [17].

In binary images, the Discrete Medial Axis (DMA) can be efficiently extracted from a Distance Transformation (DT). It consists in labeling each voxel of an object with distance of the closest voxel of the complement (background). In other words, the DT value at a voxel v corresponds to the radius of the largest discrete ball centered in s included in the object. From the Euclidean DT, linear time algorithms exist to extract the set of maximal balls [9].

Discrete Power Diagram. In computational geometry, the power diagram (also known as the *Laguerre diagram*) is a generalization of the Voronoi Diagram [3]. This tool is widely used for ball interaction computation and surface reconstruction [1,2,5]. We consider a set of sites $\mathcal{S} = \{s_i\}$, such that each point s_i is associated with a radius r_i. The power $\sigma_i(p)$ of a point p according to the site s_i is given by:

$$\sigma_i(p) = d(p, s_i) - r_i \tag{1}$$

If $\sigma_i(p) < 0$, p belongs to the ball of center s_i and radius r_i. The power diagram is based on the metric induced by σ and is a decomposition of the object into cells $\mathcal{C} = \{c_i\}$ associated with each site s_i such that:

$$c_i = \{p \in \mathbb{R}^d : \sigma_i(p) < \sigma_j(p), i \neq j\} \tag{2}$$

In discrete geometry, the power labeling is defined as the power diagram labeling of grid points. More precisely, we assign to each grid point the index of the cell it belongs to. In [9], authors have illustrated the links between MA balls and power diagram cells.

3   Sphere-Tree Construction

As the set of maximal spheres can be extracted at each level of the pyramid, it remains to link the spheres of consecutive levels to complete the sphere-tree construction. The following section first presents a graph construction by a simple linking process, and its reduction in different sphere-trees.

3.1   Power Diagram and Sphere-DAG

In order to link a sphere s at level N and a sphere t at level $N - 1$, a simple intersection test can be used to express the covering of a part of s by t. We note $t \rightarrow s$ an edge between t and s:

$$t \rightarrow s \Leftrightarrow t \in AM_{N-1} \wedge s \in AM_N \wedge t \cap s \neq \emptyset \tag{3}$$

As these edges are always oriented from a sphere t to a sphere s at next level, the graph defined by the equation is a Direct Acyclic Graph (DAG), where nodes are DMA spheres at different levels. The DAG construction needs to test the intersection between two spheres of consecutive levels, so computation time is in $O(n^2)$ for n spheres at the lower level. However, in a correct hierarchy, each part of the object does not need to be covered by a large set of spheres. Hence, this Sphere-DAG is too exhaustive for an efficient sphere-tree simplification. In order to reduce the number of spheres of the complete Sphere-DAG, we use the discrete power diagram, defined in Section 2.2. Indeed it is a voxel labeling of discrete points by the sphere which best covers it. Let v be a voxel of $\mathcal{F}_N$ at level N. As the upper representation $\mathcal{F}_{N-1}$ at level $N - 1$ is built with a model $\mathcal{M}$, v is included in a voxel v'. We denote $v' = \mathcal{R}(v)$. Including models, like the "OR" model defined in Section 2.1, ensure that $\mathcal{F}_{N-1}(v') = 1$. So we can compare the power diagrams $\mathcal{C}_N$ (from $\mathcal{F}_N$) and $\mathcal{C}_{N-1}$ (from $\mathcal{F}_{N-1}$). If v belongs to the cell $\mathcal{C}_N(s_i)$ and v' to $\mathcal{C}_{N-1}(t_j)$, the sphere s_i (at level N) covers a part of $\mathcal{F}_N$ which includes the voxel v, and the representation of this part in $\mathcal{F}_{N-1}$ is covered by the sphere t_j. In other words, t_j covers a part of s_i, and we represent it by linking t_j and s_i.

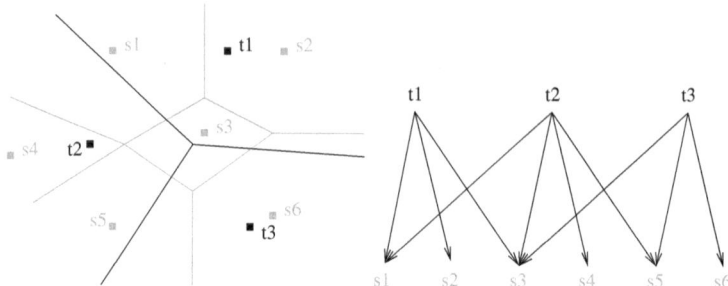

Fig. 3. Overlapping of power diagrams $\mathcal{C}_N$ and $\mathcal{C}_{N-1}$ and the associated sphere DAG between $\mathcal{F}_{N-1}$ and $\mathcal{F}_N$. For example, the cell s_1 in $\mathcal{C}_N$ is covered by both t_1 and t_2 in $\mathcal{C}_{N-1}$, so we have $t_1 \rightarrow s_1, t_2 \rightarrow s_1$.

Definition 2. *t is a parent sphere for s ($t \rightarrow s$) in the Sphere-DAG if*
$$\exists v \in \mathcal{F}_N \setminus (v \in \mathcal{C}_N(s) \wedge \mathcal{R}(v) \in \mathcal{C}_{N-1}(t))$$

By an overlapping of $\mathcal{C}_N$ and $\mathcal{C}_{N-1}$ we can detect all relations by only one scan on each voxels at level N (cf Figure 3). The algorithm 1 is generic for objects in dimension d and for any including pyramidal model.

Algorithm 1. Generic Algorithm for Sphere-DAG Computation

1: **Input:** $\mathcal{F}_N$ the original object,
2:      $\mathcal{AM}_N$ its Discrete Medial Axis and $\mathcal{C}_N$ its Discrete Power Diagram
3: **Output:** $\mathcal{T} = \{\mathcal{AM}_0, \mathcal{AM}_1, \ldots, \mathcal{AM}_N\}$ the Sphere-tree
4: **while** $|\mathcal{AM}_N| > 1$ **do**
5:     $\mathcal{F}_{N-1} \leftarrow \mathcal{M}(\mathcal{F}_N)$ {with $\mathcal{M}$ the pyramidal Model function}
6:     Extraction of $\mathcal{AM}_{N-1}$ and $\mathcal{C}_{N-1}$
7:     **for all** v such as v is a voxel $\in \mathcal{F}_N$ **do**
8:         $v' \leftarrow \mathcal{S}(v)$
9:         **if** $(v \in \mathcal{C}_N(s))$ **and** $(v' \in \mathcal{C}_{N-1}(t))$ **then**
10:             Linking the sphere $t \in AM_{N-1}$ with $s \in AM_N$
11:         **end if**
12:     **end for**
13:     $N \leftarrow N - 1$
14: **end while**

Theorem 1. *The generic Sphere-DAG construction process is linear in the number of voxels in $\mathcal{F}_N$.*

Proof. We consider that the original object $\mathcal{F}_N$ is composed of m voxels (or c^d voxels if the object is bounded by a cube of side c in dimension d). The Medial Axis extraction and power diagram computation are in $O(c^d)$ [9]. Furthermore, the construction of $\mathcal{F}_{N-1}$ and the linking between the two sets of spheres can be also computed in one scan of voxels in $\mathcal{F}_N$. So the computation of one iteration of the while loop ($\mathcal{F}_{N-1}$ construction and sphere linking) is in $O(c^d)$. Next, we work with the object $\mathcal{F}_{N-1}$, whose size is f^d times smaller, and the computation of $\mathcal{F}_{N-2}$ is in $O(\frac{c^d}{f^d})$, and so on. Thus the overall computation time for the whole hierarchy is given by the geometric series $c^d + \frac{c^d}{f^d} + \frac{c^d}{(f^d)^2} + ... = c^d \sum_{i=0}^{N} \frac{1}{(f^d)^i}$ bounded by $\frac{c^d}{1-\frac{1}{f^d}} = c^d \frac{f^d}{f^d-1}$. Hence the generic Sphere-DAG computation process is linear in the number of voxels in $\mathcal{F}_N$. $\square$

3.2   Reversible and Extended Sphere-Trees

Usually, in a bounding volume hierarchy each node has to cover the union of parts of the object covered by its children nodes, and not the whole volume of each child. Here, our spheres result from DMA extraction, ensuring the fact that they cover a part of the object without error. So, each sphere has to be completely covered by (at least) one parent sphere at upper level. However, this condition implies an expansion of parent spheres and a modification of the representation. Two methods can be distinguished, respecting either the pyramidal reversibility or the BVH properties.

Reversible Sphere-Tree. To compute a sphere-tree from the sphere-DAG, we extract a spanning tree, keeping the parent sphere for each node which best covers it. We can easily determine if a sphere $s_1(c_{s_1}, r_{s_1})$ is covered by another sphere $s_2(c_{s_2}, r_{s_2})$ comparing the radius r_{s_2} with the distance between centers $d(c_{s_1}, c_{s_2})$ added to the radius r_{s_1}:

$$s_1 \subseteq s_2 \Leftrightarrow d(c_{s_1}, c_{s_2}) + r_{s_1} - r_{s_2} \leq 0 \tag{4}$$

However it is not sufficient if we want to know if the sphere is covered by more than one sphere. We reformulate the previous covering relation considering the σ function defined in the power diagram description (in Section 2.2):

Definition 3. *let $s(c_s, r_s)$ a sphere at level N and $t(c_t, r_t)$ at level $N - 1$, the covering power $\sigma'_t(s)$ is given by*

$$\sigma'_t(s) = d(\mathcal{S}(c_s), c_t) - r_t + r_s$$

If $\sigma'_t(s) \leq 0$ then s is entirely covered by t. More precisely, the intersection between s and t increases when $\sigma'_t(s)$ is small. So, for a children sphere s with p parents $t_1, ..., t_p$, we choose the parent t_j where $\sigma'_{t_j}(s)$ is minimal, in order to get the best covering of s. In fact, we can immediately determine the best parent: for a sphere s_i with radius r_{s_i} we can only search the minimum of the quantity $d(\mathcal{S}(c_{s_i}), c_{t_j}) - r_{t_j}$ for each parent t_j. With the computation of the power diagram $\mathcal{C}_{N-1}$, it corresponds to the function $\sigma_{t_j}(\mathcal{S}(c_{s_i}))$ of the point $\mathcal{S}(c_{s_i})$ for the site t_j. The minimum of σ is reached at t_j if the point $\mathcal{S}(c_{s_i})$ is included in the cell associated to t_j. Hence, for each sphere s_i at level N, we just need to detect the position of $\mathcal{S}(c_{s_i})$, which represents the center c_{s_i} at level $N - 1$. If $\mathcal{S}(c_{s_i})$ belongs to the cell associated to the sphere t_j then t_j is the parent sphere of s_i. Figure 4 shows this computation in the overlapped power diagrams of Figure 3.

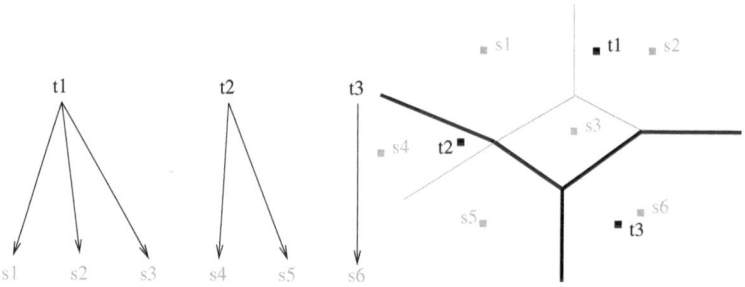

Fig. 4. Sphere-tree reduction of the previous DAG. We can also create the recovering diagram, extending each cell in $\mathcal{C}_{N-1}$ (associated with a sphere t_j) as the union of children cells.

Extended Sphere-Tree. In order to respect the inclusion property, we could replace each parent sphere by the minimal bounding sphere of its children. However, the minimal covering sphere computation is not very efficient since the problem is related to the minimal enclosing ball of a set of points in dimension d [11]. In order to maintain the reversibility of multiresolution representations of the original object, we propose an original approach, which consists in extending the radius of parent spheres.

Theorem 2. *Let a sphere t and its set of children spheres $\{s_i\}$. The sphere t is a minimal bounding sphere centered at c_t for its children spheres if its radius r_t is extended by $r'_t = \sigma'_t(max)$, where max is the child sphere with maximal σ'_t.*

Proof. For two spheres s, t, from Definition 3 we know that t entirely covers s if $\sigma'_t(s) \leq 0$. Moreover, if $\sigma'_t(s) = 0$ we have $d(\mathcal{S}(c_s), c_t) + r_s = r_t$, so the sphere t is the minimal bounding sphere for s centered at c_t. Let r'_t the quantity to add to r_t in order to have a covering of s by t. So, we have:

$$d(\mathcal{S}(c_s), c_t) - (r_t + r'_t) + r_s = 0 \Leftrightarrow r'_t = d(\mathcal{S}(c_s)), c_t - r_t + r_s \Leftrightarrow r'_t = \sigma'_t(s)$$

So $\sigma'_t(s)$ defines the extension quantity. Now, for each parent sphere t_j, we search among its children s_i the sphere s_{max}, where $\sigma'_t(s_{max})$ is maximal. Extending the radius with this value, we substract $\sigma'_t(s_{max})$ at all $\sigma'_t(i)$, so we have $\sigma'_t(i) \leq 0$ for all children spheres s_i and $\sigma'_t(max) = 0$. Hence t becomes the minimal bounding sphere of its children centered at c_t. □

The Algorithm 2 adds this extending process to the reversible sphere-tree computation.

As radii have been extended, we have to modify $\mathcal{F}_{N-1}$ by a reverse reconstruction of the object with the new set of spheres. For n spheres in $\mathcal{AM}_N$, the linking computation and the extension of the spheres are in $O(n)$, and the reconstruction process in $O(c^d)$. As the number of spheres n is lower than the size of the object (c^d), an iteration is still performed in $O(c^d)$. So this process is also in $O(c^d)$ like the Sphere-DAG Computation (cf Theorem 1).

Algorithm 2. Generic Algorithm for Exact Sphere-tree Computation

---

1: **Input:** $\mathcal{F}_N$ the original object,
2:      $\mathcal{AM}_N$ its Discrete Medial Axis and $\mathcal{C}_N$ its Discrete Power Diagram
3: **Output:** $\mathcal{T} = \{\mathcal{AM}_0, \mathcal{AM}_1, \ldots, \mathcal{AM}_N\}$ the Sphere-tree
4: **while** $|\mathcal{AM}_N| > 1$ **do**
5:    $\mathcal{F}_{N-1} \leftarrow \mathcal{M}(\mathcal{F}_N)$ {with $\mathcal{M}$ a bounding model}
6:    Extraction of $\mathcal{AM}_{N-1}$ and $\mathcal{C}_{N-1}$
7:    **for each** sphere $s : (c(s), r(s)) \in \mathcal{AM}_N$ **do**
8:       **if** $\mathcal{S}(c(s)) \in \mathcal{C}_{N-1}(t)$ **then**
9:          t is the parent sphere for s
10:      **end if**
11:    **end for**
12:    **for each** sphere $t : (c(t), r(t)) \in \mathcal{AM}_{N-1}$ **do**
13:       **if** t has no child **then**
14:          $\mathcal{AM}_{N-1} \leftarrow \mathcal{AM}_{N-1} - \{t\}$
15:       **else**
16:          $r' \leftarrow max(\sigma'_t(s))$ {for all s child of t}
17:          $r(t) \leftarrow r(t) + r'$
18:       **end if**
19:    **end for**
20:    $\mathcal{F}_{N-1} \leftarrow \cup_t \{\, t \in \mathcal{AM}_{N-1} \,\}$
21:    $N \leftarrow N - 1$
22: **end while**

From the Algorithm 2, we have a sphere-tree which respects the covering condition. Nevertheless, a modification of spheres radii is performed, although the spheres were first defined to ensure the distribution of error along the object. So the extension of radii disturbs the reversibility property on each sphere-tree level. On the other hand, when the value $\sigma'_t(s_{max})$ is negative, the parent spheres radii decreases, generating an improvement of representation tightness. Moreover, as we simplify the sphere-tree by deleting the spheres without child, this algorithm may reduce the depth of the tree. The following section illustrates these observations on real images. In order to reduce the error with the original object, the radii extension is evaluated for each node with its set of leaves (*i.e.* spheres of the original object).

4   Experiments

This section presents a comparison between the Reversible and the Extended Algorithms, with experiments on several 3D discrete objects in .vol or .longvol formats, defined in the **simplevol** library. Distance Transformation, Reduce Discrete Medial Axis Extraction and Discrete Power Diagram are computed thanks to the **MAEVA** Toolkit[1].

---

[1] Simplevol and MAEVA Toolkit are available on http://gforge.liris.cnrs.fr/

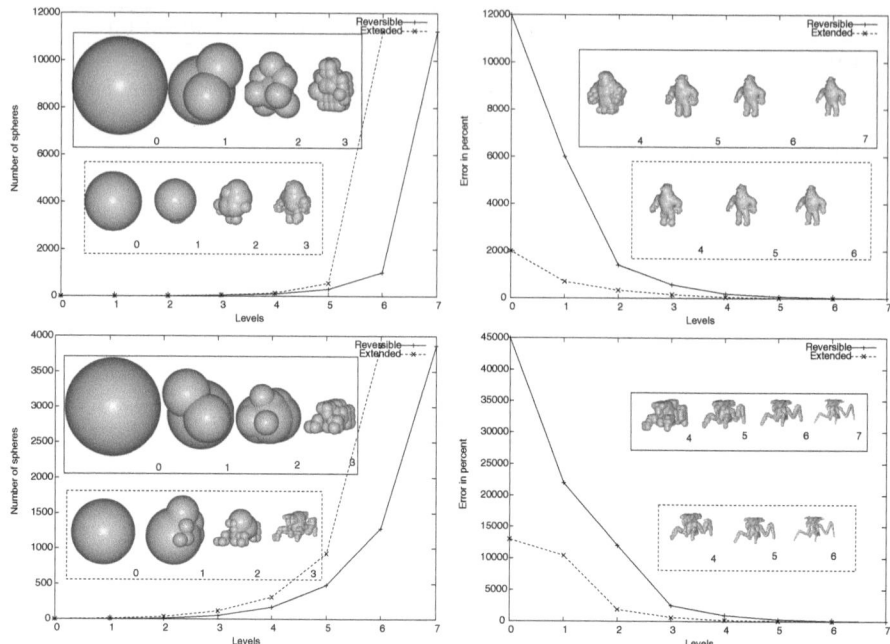

Fig. 5. Experiments for `Al.100.vol` and `ArachnidWarrior.100.vol`.
The left graph shows that the decreasing of the sphere number is faster on the Extended
case, and the depht of the hierarchy is smaller. The dynamic extension process also
produces an error reduction as we can see on the right graph.

Figure 5 shows differences between both algorithms on objects `Al.100.vol`
and `ArachnidWarrior.100.vol`[2]. The left graph presents the number of spheres
at each level. The right one details the percentage of error for each representation
over the original object. This error is computed with the Hamming Distance
which can be efficiently computed in our discrete model. As the experimental
objects have different number of voxels, we prefer to represent the Hamming
distance by a percentage representing the error volume added to the original
object.

4.1   Reversible Algorithm

As said in Section 2.1, the representation $\mathcal{F}_{L-1}$ is f^d times smaller than $\mathcal{F}_L$ in
number of voxels, as we can see in Figure 6. Thus, the associated Medial Axis
contains fewer spheres. However, the distance is computed between two voxel
centers in the discrete approach (here we use the square-euclidean distance), but
this distance also depends on the voxel sizes at each resolution level. That's the
reason why we scale the spheres to the finer resolution level by a reconstruction

---

[2] These objects are available on `http://www.tc18.org/`

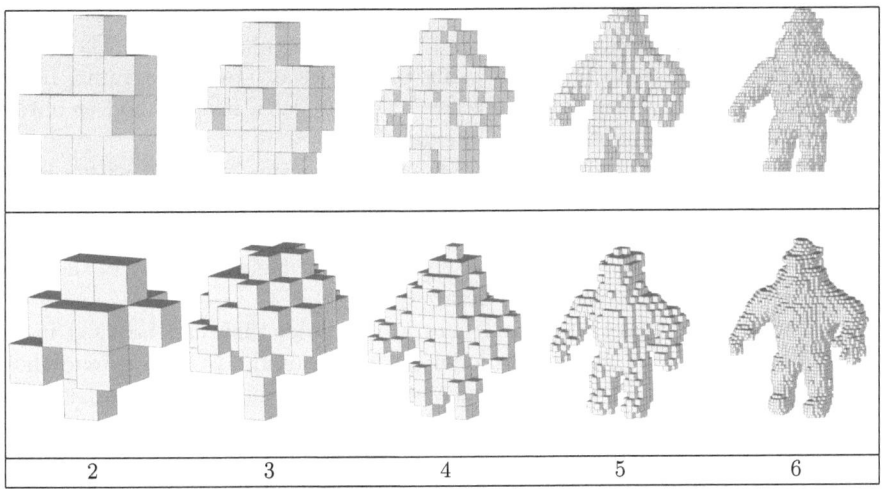

Fig. 6. Comparison between the reversible pyramid and the dynamic reconstruction of the extended version

process. Moreover, at upper levels, the variability of radii is lower as the interval of radii becomes smaller. This range remains low during the reconstruction because all sphere radii are increased by the same value (at one level). So the sphere-tree looks like an octree, as many spheres have the same radius. Nevertheless, the reconstruction ratio at an upper level is very high, and the spheres overestimate the object geometry on this level.

4.2   Extended Algorithm

The experiments for the extended sphere-tree construction show that the dynamic reconstruction before the extension of radius simplifies the upper levels (see Figure 6). As we delete spheres without child, the reduction of the sphere number is faster than in the reversible algorithm. Moreover, in Algorithm 2 we proposed a reconstruction of the upper level just after the extension of the radius of its spheres. The experiments show that this process controls the error increasing which is higher in the reversible case. In order to reduce the incidence of the final reconstruction process, we propose to fit one more time the spheres by $\sigma'_t(s_{max})$ after the increasing. As the reconstruction has produced spheres which overestimate the object, this final extension reduces the error.

5   Conclusion and Future Works

In this paper, we have presented an original method for a sphere-tree construction in discrete geometry. Its construction is based on the Discrete Medial Axis of the object, as best algorithms in computational geometry [6,7,14], but we benefit from the fact that we can efficiently extract a reversible skeleton. Hence we extract reversible sets of spheres from levels of a regular pyramid. The sphere linking at different levels is solved using properties of power diagrams. In order to ensure the covering conditions, we also propose a fast method to obtain bounding spheres of nodes, with radii extensions.

Moreover in discrete geometry, we can exactly measure the error with a Hamming distance instead of a Hausdorff estimation. Experiments show that in the reversible algorithm the error is greater when we return at the original resolution. However, the extended algorithm solves this problem and reduces the error increasing, as we build a dynamic hierarchy.

The methods we have presented here are generic, they can be used in dimension d, and for any including model of the pyramid. We may extend these methods for generic models. We can also imagine other heuristics in order to optimize the sphere-tree. For example, we could replace the extending treatment by others minimal bounding sphere computations. However, the future works could be oriented to the topology maintenance of the object, using adaptative pyramids or morphological models, in order to reduce the error of the reversible method.

References

1. Amenta, N., Choi, S., Kolluri, R.K.: The power crust, unions of balls, and the medial axis transform. Comput. Geom. 19(2-3), 127–153 (2001)
2. Attali, D., Edelsbrunner, H.: Inclusion-exclusion formulas from independent complexes. Discrete & Computational Geometry 37(1), 59–77 (2007)
3. Aurenhammer, F.: Power diagrams: Properties, algorithms, and applications. SIAM Journal on Computing 16(1), 78–96 (1987)
4. Blum, H.: A transformation for extracting new descriptors of shape. In: Whaten-Dunn, W. (ed.) Models for the Perception of Speech and Visual Form, pp. 362–380. MIT Press, Cambridge (1967)
5. Boissonnat, J.-d., Cerezo, A., Devillers, O., Duquesne, J., Yvinec, M.: An algorithm for constructing the convex hull of a set of spheres in dimension d. CGTA: Computational Geometry: Theory and Applications 6 (1996)
6. Bradshaw, G., O'Sullivan, C.: Sphere-tree construction using dynamic medial-axis approximation. In: Spencer, S.N. (ed.) Proceedings of the 2002 ACM SIGGRAPH Symposium on Computer Animation (SCA 2002), July 21–22, pp. 33–40. ACM Press, New York (2002)
7. Bradshaw, G., O'Sullivan, C.: Adaptive medial-axis approximation for sphere-tree construction. ACM Transactions on Graphics 23(1), 1–26 (2004)
8. Cignoni, P., Puppo, E., Scopigno, R.: Multiresolution representation and visualization of volume data. IEEE Trans. on Visualization and Computer Graphics 3(4), 352–369 (1997)
9. Coeurjolly, D., Montanvert, A.: Optimal separable algorithms to compute the reverse euclidean distance transformation and discrete medial axis in arbitrary dimension. IEEE Trans. on Pattern Analysis and Machine Intelligence 29(3) (March-August 22, 2007)
10. Dingliana, C.: Real-time collision detection and response using sphere-trees. Technical report, March 02 (1999)

11. Goodman, J.E., O'Rourke, J. (eds.): Handbook of Discrete and Computational Geometry. CRC Press, Boca Raton (1997)
12. Gottschalk, S., Lin, M.C., Manocha, D.: OBBTree: A hierarchical structure for rapid interference detection. In: Computer Graphics (Annual Conference Series), vol. 30, pp. 171–180 (1996)
13. He, T., Kaufman, A.: Collision detection for volumetric objects. In: IEEE Visualization 1997 (October 1997)
14. Hubbard, P.: Approximating polyhedra with spheres for time-critical collision detection. ACM Transactions on Graphics 15(3), 179–210 (1996)
15. Jolion, J.M., Montanvert, A.: The adaptive pyramid: a framework for 2D image analysis. Computer Vision, Graphics, and Image Processing. Image Understanding 55(3), 339–349 (1992)
16. Palmer, I.J., Grimsdale, R.L.: Collision detection for animation using sphere-trees. Computer Graphics Forum 14(2), 105–116 (1995)
17. Pfaltz, J., Rosenfeld, A.: Computer representation of planar regions by their skeletons. Communications of the ACM 10(2), 119–122 (1967)
18. Prevost, S., Lucas, L., Bittar, E.: Multiresolution and shape optimization of implicit skeletal model. In: Skala, V. (ed.) WSCG 2001 Conference Proceedings (2001)
19. Quinlan, S.: Efficient distance computation between non-convex objects. In: Straub, E., Spencer Sipple, R. (eds.) Proceedings of the International Conference on Robotics and Automation, May 1994, vol. 4, pp. 3324–3330. IEEE Computer Society Press, Los Alamitos (1994)
20. van den Bergen, G.: Efficient collision detection of complex deformable models using AABB trees. Journal of Graphics Tools: JGT 2(4), 1–14 (1997)

Quasi-Affine Transformation in 3-D: Theory and Algorithms

David Coeurjolly[1], Valentin Blot[2], and Marie-Andrée Jacob-Da Col[3]

[1] Université de Lyon, CNRS, LIRIS, UMR5205, F-69622, France
david.coeurjolly@liris.cnrs.fr
[2] Ecole Normale Supérieure de Lyon
valentin.blot@ens-lyon.fr
[3] LSIIT-UMR 7005, Pôle API Bd Sébastien Brant, Illkirch, F-67412, France
dacolm@iutlpa.u-strasbg.fr

Abstract. In many applications and in many fields, algorithms can considerably be speed up if the underlying arithmetical computations are considered carefully. In this article, we present a theoretical analysis of affine transformations in dimension 3. More precisely, we investigate the arithmetical paving induced by the transformation to design fast algorithms.

Keywords: quasi-affine transform, periodic tiling, arithmetic, image transformation.

1 Introduction

In many computer vision and image processing applications, we are facing new constraints due to the image sizes both in dimension with 3-D and 3-D+t medical acquisition devices, and in resolution with VHR (Very High Resolution) satellite images. This article deals with high performance image transformations using quasi-affine transforms (QATs for short), which can be viewed as a discrete version of general affine transformations. QAT can approximate rotations and scalings, and in some specific cases, QAT may also be one-to-one and onto mappings from $\mathbb{Z}^n$ to $\mathbb{Z}^n$, leading to exact computations. In dimension 2, the QAT appeared in several articles [9,5,8,4,6]. In higher dimension, theoretical results have been demonstrated [7,2,1]. More precisely, authors have demonstrated the arithmetical and periodic structures embedded in $n-$dimensional QAT, leading to generic transformation algorithms. To implement these generic algorithms, several elements have to be fixed when considering a specific dimension. In this paper, we detail the computation of the minimal periods in dimension 3 leading to efficient transformation algorithms. Due to the space limitation, the proofs are available in the technical report [2]). In Section 2, we first present some definitions and results on $n-$D QAT. Section 3 focuses on the parameter computation in 3-D. Finally, Section 4 evaluate all the algorithms compared to a classical backward-mapping technique [3].

P. Wiederhold and R.P. Barneva (Eds.): IWCIA 2009, LNCS 5852, pp. 68–81, 2009.

2   Preliminaries

In this section, we present definitions and results obtained for QAT in higher dimension. These results have been independently proved by [7] and [2,1]. In the following, we consider notations introduced in [2,1]. Let n denote the dimension of the considered space, V_i the ith coordinate of vector V, and $M_{i,j}$ the $(i,j)^{th}$ coefficient of matrix M. We use the notation $\gcd(a, b, \ldots)$ for the greatest common divisor of an arbitrary number of arguments, and $\text{lcm}(a, b, \ldots)$ for their least common multiple. Let $\left[\frac{a}{b}\right]$ denote the quotient of the euclidean division of a by b, that is the integer $q \in \mathbb{Z}$ such that $a = bq + r$ satisfying $0 \le r < |b|$ regardless of the sign of b[1]. We also consider the straightforward generalization of these operators to $n-$dimensional vectors (e.g. $\left[\frac{V}{b}\right]$ is a vector where each component is the quotient of the division by b).

Definition 1. *A quasi-affine transformation is a triple* $(\omega, M, \boldsymbol{V}) \in \mathbb{Z} \times M_n(\mathbb{Z}) \times \mathbb{Z}^n$ *(we assume that* $\det(M) \neq 0$*). The associated application is :*

$$\mathbb{Z}^n \longrightarrow \mathbb{Z}^n$$

$$X \longmapsto \left[\frac{MX + V}{\omega}\right]$$

Definition 2. *The inverse of a QAT* (ω, M, V) *is the QAT:*

$$(\det(M), \omega \, \text{com}(M)^t, -\text{com}(M)^t V), \tag{1}$$

where M^t *denotes the transposed matrix and* $\text{com}(M)$ *the co-factor matrix of* M *(Remind that* $M \, \text{com}(M)^t = \text{com}(M)^t M = \det(M) I_n$*.).*

The associated affine application of the inverse of a QAT is therefore the inverse of the affine application associated to the QAT. However, due to the nested floor function, the composition $f \cdot f^{-1}$ is not the identity function in the general case. Let us recall the well-known Bezout Identity:

$$\forall (a, b) \in \mathbb{Z}^2, \exists (u, v) \in \mathbb{Z}^2 / au + bv = \gcd(a, b).$$

In Section 3, we have to consider a generalized form of the Bezout identity in dimension 3:

Proposition 1 ([2]). $\forall (a, b, c) \in \mathbb{Z}^3, \exists (u, v, w) \in \mathbb{Z}^3 / au + bv + cw = \gcd(a, b, c).$

We present now several results and definitions that have been presented for $n-$dimensional QAT. All these results are given in [2,1] but we present here the main theorems, which will be used in the rest of the paper. First, the key feature of $n-$D QAT is that it contains a periodic paving structure.

Definition 3 (Tile). *Let* f *be a QAT. For* $Y \in \mathbb{Z}^n$*, we denote:*

$$P_Y = \{X \in \mathbb{Z}^n / f(X) = Y\}, \tag{2}$$

P_Y *is called order 1 tile of index* Y *of* f*.*

---

[1] $\left\{\frac{a}{b}\right\}$ denotes the corresponding remainder $\left\{\frac{a}{b}\right\} = a - b \left[\frac{a}{b}\right]$.

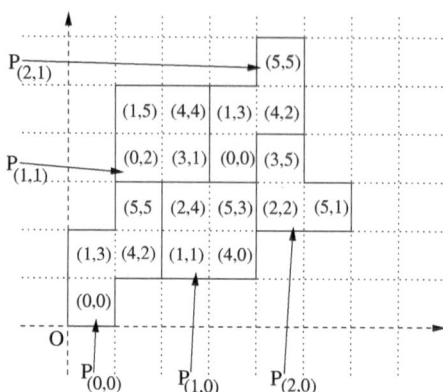

Fig. 1. Example of tiles and remainders

Definition 4. P_Y *is said arithmetically equivalent to* P_Z *(denoted* $P_Y \equiv P_Z$ *) if:*

$$\forall X \in P_Y, \exists X' \in P_Z / \left\{ \frac{MX + V}{\omega} \right\} = \left\{ \frac{MX' + V}{\omega} \right\}. \tag{3}$$

Definition 5. P_Y *and* P_Z *are said geometrically equivalent if:*

$$\exists v \in \mathbb{Z}^n / P_Y = T_v P_Z, \tag{4}$$

where T_v *denotes the translation of vector* v.

The following theorem exhibits a relation between the notions of arithmetically and geometrically equivalent tiles.

Theorem 1 ([2,1]). *If* $P_Y \equiv P_Z$, *then* P_Y *and* P_Z *are geometrically equivalent. Since* $P_Y \equiv P_Z$, *there exists* $X \in P_Y$ *and* $X' \in P_Z$ *such that:*

$$\left\{ \frac{MX + V}{\omega} \right\} = \left\{ \frac{MX' + V}{\omega} \right\}.$$

Then $v = X - X'$ *is the translation vector:*

$$P_Y = T_v P_Z.$$

In Figure 1 we show some tiles of the QAT defined by $(6, \begin{pmatrix} 3 & 1 \\ -1 & 3 \end{pmatrix}, \begin{pmatrix} 0 \\ 0 \end{pmatrix})$ (a point of $\mathbb{Z}^2$ is represented by a unit square whose bottom-left corner corresponds to the represented point). For each point X in a tile we provide its corresponding remainder $\left\{ \frac{MX+V}{\omega} \right\}$. Tiles $P_{(2,1)}$ and $P_{(0,0)}$ are arithmetically equivalent, therefore they are also geometrically equivalent. It should also be noted that tiles $P_{(1,0)}$ and $P_{(1,1)}$ are geometrically equivalent but they are not arithmetically equivalent .

Definition 6 (Periodicity notations, [2,1]). *For all* $0 \leq i < n$, *We define the set* $\mathcal{A}_i$ *as follows:*

$$\mathcal{A}_i = \{\alpha \in \mathbb{N}^* / \exists (\beta_j)_{0 \leq j < i} \in \mathbb{Z}^i, \forall (y_0, \dots, y_{n-1}) \in \mathbb{Z}^n,$$

$$P_{y_0, \dots, y_i + \alpha, \dots, y_{n-1}} \equiv P_{y_0 + \beta_0, \dots, y_{i-1} + \beta_{i-1}, y_i, \dots, y_{n-1}}\}$$

Furthermore, let us consider $\alpha_i = \min(\mathcal{A}_i)$. *We define* $\{\beta_j^i\}_{0 \leq j < i} \in \mathbb{Z}^i$ *and* $U_i \in \mathbb{Z}^n$ *such that*

$$\forall (y_0, \dots, y_{n-1}) \in \mathbb{Z}^n, P_{y_0, \dots, y_i + \alpha_i, \dots, y_{n-1}} = T_{U_i} P_{y_0 + \beta_0^i, \dots, y_{i-1} + \beta_{i-1}^i, y_i, \dots, y_{n-1}}.$$

The key theorem for the QAT periodic structure can now be presented:

Theorem 2 (Periodicity, [2,1]). *The set of QAT tiles is* $n-$*periodic, in other words*

$$\forall 0 \leq i < n, \mathcal{A}_i \neq \emptyset$$

Let us suppose that quantities α_i, β_j^i and U_i are given. The following theorem allows us to obtain P_Y as the translation of a canonical tile P_{Y^0}.

Theorem 3 ([2,1]). $\forall (y_0, \dots, y_{n-1}) \in \mathbb{Z}^n$, *we have* $P_{y_0, \dots, y_{n-1}} = T_W P_{y_0^0, \dots, y_{n-1}^0}$ *with*

$$W = \sum_{i=0}^{n-1} w_i U_i \quad and \quad \forall n > i \geq 0, \begin{cases} w_i = \left[\dfrac{y_i + \sum_{j=i+1}^{n-1} w_j \beta_i^j}{\alpha_i} \right] \\ y_i^0 = \left\{ \dfrac{y_i + \sum_{j=i+1}^{n-1} w_j \beta_i^j}{\alpha_i} \right\} \end{cases}.$$

In [2,1], we have proved that canonical tiles P_{Y^0} are associated to grid points of a special tile called *super-tile*.

Definition 7 (Super-tile, [2,1]). *A super-tile of a QAT is the set* $\mathcal{P}$ *such that*

$$\mathcal{P} = \bigcup_{0 \leq Y^0 < (\alpha_0, \dots, \alpha_{n-1})} P_{Y^0}$$

Theorem 4 ([2,1]). $\mathcal{P}$ *is the tile* $P_{(0, \dots, 0)}$ *of the QAT defined by:*

$$\left(\omega \operatorname{lcm}_{0 \leq i < n}(\alpha_i), \begin{pmatrix} \theta_0 & \cdots & 0 \\ \vdots & \ddots & \vdots \\ 0 & \cdots & \theta_{n-1} \end{pmatrix} M, \begin{pmatrix} \theta_0 & \cdots & 0 \\ \vdots & \ddots & \vdots \\ 0 & \cdots & \theta_{n-1} \end{pmatrix} V \right),$$

with $\forall 0 \leq i < n - 1, \theta_i = \dfrac{\operatorname{lcm}_{0 \leq j < n-1}(\alpha_j)}{\alpha_i}$.

Figure 2 illustrates tiles of the QAT $(84, \begin{pmatrix} 12 & -11 \\ 18 & 36 \end{pmatrix}, \begin{pmatrix} 0 \\ 0 \end{pmatrix})$ in $\mathbb{Z}^2$ with 15 arithmetically distinct tiles (the tiles with same color are arithmetically equivalent). In this example, for all $i, j \in \mathbb{N}$, $P_{(i+5,j)} \equiv P_{(i,j)}$ and $P_{(i+2,j-3)} \equiv P_{(i,j)}$. The set

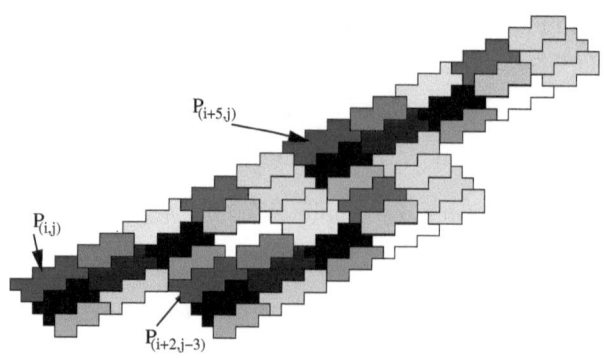

Fig. 2. Periodicity of thes tiles of a 2D QAT

$\{P_{(i,j)}|i = 0, 1, 2, 3, 4, j = 0, 1, 2\}$ contains exactly once all distinct tiles so it is a super-tile of the QAT.

In [2,1], we have demonstrated that if we consider the Hermite Normal Form of the QAT matrix M, then efficient tile construction can be designed. In the following, let $MH = T$ denote the Hermite Normal Form (with $\det(H) = \pm 1$ and T upper triangular). Note that this form always exists for nonsingular integer square matrix.

Theorem 5 (Tile Construction, [2,1]). $\forall Y \in \mathbb{Z}^n$, let $MH = T$ be the Hermite Normal Form of the QAT matrix M,

$$P_Y = \{HX \ / \ \forall n > i \geq 0, A_i(X_{i+1}, \ldots, X_{n-1}) \leq X_i < B_i(X_{i+1}, \ldots, X_{n-1})\}$$

$$\text{With } A_i(X_{i+1}, \ldots, X_{n-1}) = -\left\lceil \frac{-\omega Y_i + \sum_{j=i+1}^{n-1} T_{i,j} X_j + V_i}{T_{i,i}} \right\rceil,$$

$$B_i(X_{i+1}, \ldots, X_{n-1}) = -\left\lceil \frac{-\omega(Y_i + 1) + \sum_{j=i+1}^{n-1} T_{i,j} X_j + V_i}{T_{i,i}} \right\rceil.$$

In Algorithm 1, we give the generic algorithm applying a contracting QAT f to an image $\mathcal{A}$ (see Fig. 3). The principle is that we give to each pixel Y of image $\mathcal{B}$ the average color of the tile P_Y in image $\mathcal{A}$. If f is a dilating QAT, we obtain a similar algorithm in which we replace f with f^{-1}, and then we give the color of each pixel Y of image $\mathcal{A}$ to each pixel of P_Y in image $\mathcal{B}$ (see Fig. 3 for an illustration in 2-D). In both algorithms, some elements cannot be computed in arbitrary dimension n. Indeed, even if there exist algorithms to compute the Hermite Normal Form of an arbitrary square integer matrix [10], there is no generic algorithm to obtain the minimal periodicities $\{\alpha_i\}$ (see discussion in Sect. 5). In the next section, we focus on the minimal periodicity computation in dimension 3.

Algorithm 1. Generic QAT algorithm for a contracting QAT

Input: a contracting QAT $f := (\omega, M, V)$, an image $\mathcal{A} : \mathbb{Z}^n \to \mathbb{Z}$
Output: a transformed image $\mathcal{B} : \mathbb{Z}^n \to \mathbb{Z}$
Compute the Hermite Normal Form of the matrix M;
Determine the minimal periodicities $\{\alpha_i\}$ and vectors $\{U_i\}$;
Use Theorems 4 and 5 to compute the canonical tiles in the super-tile $\mathcal{P}$;
foreach $Y \in \mathcal{B}$ **do**
    Find Y^0 and W such that $P_Y = T_W P_{Y^0}$;
    foreach $Z \in P_{Y^0}$ **do**
        $c \leftarrow \mathcal{A}(T_W Z)$;   // we read the color in the initial image
        $sum \leftarrow sum + c$;
    $\mathcal{B}(Y) \leftarrow sum/|P_{Y^0}|$;   // we set the color

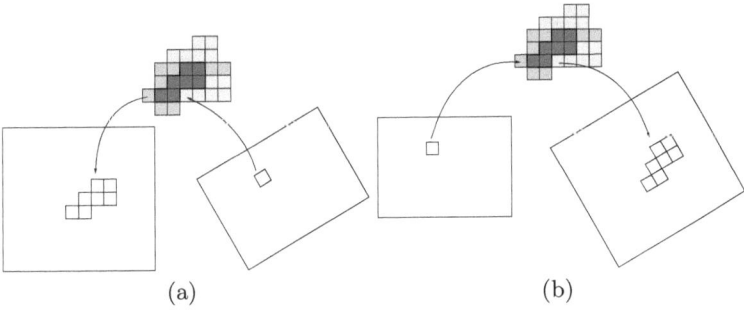

           (a)   (b)

Fig. 3. Illustration in dimension 2 of the QAT algorithm when f is contracting (a) and dilating (b). In both cases, we use the canonical tiles contained in the super-tile to speed-up the transformation.

3   QAT in Dimension 3

In dimension 3, we consider the following framework: we first define the Hermite Normal Form, the minimal periods and then we detail the transformation algorithm.

3.1   Hermite Normal Form and Tile Construction

Let us consider a QAT (ω, M, V) with $M = \begin{pmatrix} a_0 & b_0 & c_0 \\ d_0 & e_0 & f_0 \\ g_0 & h_0 & i_0 \end{pmatrix}$ and $V = \begin{pmatrix} j_0 \\ k_0 \\ l_0 \end{pmatrix}$.

In [2], we present explicit formulas to compute the Hermite Normal Form in 3-D. In the following, we define $H = H_1 H_2 H_3 H_4$ and $MH = T = \begin{pmatrix} a & b & c \\ 0 & d & e \\ 0 & 0 & f \end{pmatrix}$.

Thanks to Hermite decomposition, we have $a > 0, d > 0$ and $f > 0$. To construct the tile of index (i, j, k) and thanks to Theorem 5, we have:

$$A_2 = - \left[\frac{-\omega k + l_0}{f} \right], B_2 = - \left[\frac{-\omega(k+1) + l_0}{f} \right]$$

$$A_1(z) = -\left[\frac{-\omega j + k_0 + ez}{d}\right], B_1(z) = -\left[\frac{-\omega(j+1) + k_0 + ez}{d}\right]$$

$$A_0(y, z) = -\left[\frac{-\omega i + j_0 + by + cz}{a}\right], B_0(y, z) = -\left[\frac{-\omega(i+1) + j_0 + by + cz}{a}\right]$$

Algorithm 2. Tile construction in 3-D

$A_2 \leftarrow -\left[\frac{-\omega k + l_0}{f}\right]$;

$B_2 \leftarrow -\left[\frac{-\omega(k+1) + l_0}{f}\right]$;

for $z \leftarrow A_2$ **to** B_2 - 1 **do**

  $A_1 \leftarrow -\left[\frac{-\omega j + k_0 + ez}{d}\right]$;

  $B_1 \leftarrow -\left[\frac{-\omega(j+1) + k_0 + ez}{d}\right]$;

  for $y \leftarrow A_1$ **to** B_1 - 1 **do**

    $A_0 \leftarrow -\left[\frac{-\omega i + j_0 + by + cz}{a}\right]$;

    $B_0 \leftarrow -\left[\frac{-\omega(i+1) + j_0 + by + cz}{a}\right]$;

    for $x \leftarrow A_0$ **to** B_0 - 1 **do**

      $H\begin{pmatrix} x \\ y \\ z \end{pmatrix} \in P_{i,j,k}$;

3.2 Minimal Periodicity and Super-Tile Construction

In dimension 3, we need to compute the periodicity along each dimension. Let us first denote $a'_h = \frac{a}{\gcd(a,\omega)}, w'_h = \frac{\omega}{\gcd(a,\omega)}, Y = \begin{pmatrix} w'_h \\ 0 \\ 0 \end{pmatrix}$

Theorem 6 (Horizontal Periodicity). *Let* $\alpha_h = a'_h$ *and* $U = HY$. *Then* $\alpha_h > 0$, $P_{i+\alpha_h,j,k} \equiv P_{i,j,k}$ *and* $\forall (i, j, k) \in \mathbb{Z}^3$, $P_{i+\alpha_h,j,k} = T_U P_{i,j,k}$.

Proof. The proof is detailed in [2].

Theorem 7. *The period* α_h *is a minimal horizontal period, i.e.* $\alpha_h = \alpha_0$.

Proof. The proof is given in [2].

Concerning the vertical period, let:

$$d'_v = \frac{d}{\gcd(d, \omega)}, w'_v = \frac{\omega}{\gcd(d, \omega)}, a'_v = \frac{a}{\gcd(a, bw'_v, \omega)}, \phi = \frac{bw'_v}{\gcd(a, bw'_v, \omega)},$$

$$w''_v = \frac{\omega}{\gcd(a, bw'_v, \omega)}, \alpha'_v = \gcd(a'_v, w''_v),\ u_1 \text{ and } v_1 \text{ are such that}: a'_v u_1 + w''_v v_1 =$$

$$\gcd(a'_v, w''_v)(= \alpha'_v), \beta_0 = -\phi v_1, Y = \begin{pmatrix} -\phi u_1 \\ w'_v \alpha'_v \\ 0 \end{pmatrix}$$

Theorem 8 (Vertical Periodicity). *Let* $\alpha_v = d'_v \alpha'_v, U = HY$. *Then* $\alpha_v > 0$, $P_{i,j+\alpha_v,k} \equiv P_{i+\beta_0,j,k}$ *and* $\forall (i,j,k) \in \mathbb{Z}^3$, $P_{i,j+\alpha_v,k} = T_U P_{i+\beta_0,j,k}$.

Proof. The proof is given in [2].

Theorem 9. *The period* α_v *is a minimal vertical period, i.e.* $\alpha_v = \alpha_1$.

Proof. The proof is detailled in [2].

For the last period, let us consider:

$$f'_d = \frac{f}{\gcd(\omega, f)}, \omega'_d = \frac{\omega}{\gcd(\omega, f)}, d'_d = \frac{d}{\gcd(d, e\omega'_d, \omega)}, \phi = \frac{e\omega'_d}{\gcd(d, e\omega'_d, \omega)},$$

$$\omega''_d = \frac{\omega}{\gcd(d, e\omega'_d, \omega)}, u_1 \text{ and } v_1 \text{ are such that}: d'_d u_1 + \omega''_d v_1 = \gcd(d'_d, \omega''_d),$$

$$\psi = c\omega'_d \gcd(d'_d, \omega''_d) - b\phi u_1, a'_d = \frac{a}{\gcd(a, \psi, \omega, \frac{\omega''_d b}{\gcd(d'_d, \omega''_d)})}, \psi' = \frac{\psi}{\gcd(a, \psi, \omega, \frac{\omega''_d b}{\gcd(d'_d, \omega''_d)})},$$

$$\omega'''_d = \frac{\omega}{\gcd(a, \psi, \omega, \frac{\omega''_d b}{\gcd(d'_d, \omega''_d)})}, \chi = \frac{\frac{\omega''_d b}{\gcd(d'_d, \omega''_d)}}{\gcd(a, \psi, \omega, \frac{\omega''_d b}{\gcd(d'_d, \omega''_d)})},$$

$$\alpha''_d = \gcd(a'_d, \chi, \omega'''_d), \alpha'_d = \alpha''_d \gcd(d'_d, \omega''_d),$$

$$u_2, v_2 \text{ and } w_2 \text{ are such that}: a'_d u_2 + \chi v_2 + \omega'''_d w_2 = \gcd(a'_d, \chi, \omega'''_d)(= \alpha''_d),$$

$$k = -\psi' v_2, \beta_0 = -\psi' w_2, \beta_1 = -\phi v_1 \alpha''_d - k\frac{d'_d}{\gcd(d'_d, \omega''_d)}, Y = \begin{pmatrix} -\psi' u_2 \\ -\phi u_1 \alpha''_d + k\frac{\omega''_d}{\gcd(d'_d, \omega''_d)} \\ \alpha'_d \omega'_d \end{pmatrix}$$

Theorem 10 (Depth Periodicity). *Let* $\alpha_d = \alpha'_d f'_d, U = HY$. *Then*

$$\alpha_d > 0 \quad P_{i,j,k+\alpha_d} \equiv P_{i+\beta_0,j+\beta_1,k} \quad \text{and} \quad \forall (i,j,k) \in \mathbb{Z}^3, P_{i,j,k+\alpha_d} = T_U P_{i+\beta_0,j+\beta_1,k}$$

Proof. The proof is detailed in [2].

Theorem 11. *The period* α_d *is a minimal depth period, i.e.* $\alpha_d = \alpha_2$.

Proof. The proof is detailed in [2].

Based on these periods, we can construct the super-tile and all the initial period tiles. To design a transformation algorithm, for each point $X \in \mathcal{P}$, we need to determine the tile index Y to which X belongs to. Since $X \in \mathcal{P}_Y \Leftrightarrow \left[\frac{MX+V}{\omega}\right] = Y$, Algorithm 3 details the initial period tile construction with scanning points in $\mathcal{P}$. The computational cost of Alg. 3 exactly corresponds to the number of tiles in the initial period.

Proposition 2. *The number of tiles of the initial period is* $\omega'_d \omega''_d \omega'''_d$.

In the Proposition statement, we do not give the closed formula. However, $\omega'_d \omega''_d \omega'''_d$ is equal to ω^3 divided by a product of three gcd().

Algorithm 3. Super-tile and initial period tiles construction in 3-D

$A_2' \leftarrow -\left[\frac{l_0}{f}\right]$;
for $z \leftarrow A_2'$ **to** $A_2' + \frac{w\alpha_2}{f}$ - *1* **do**
  $A_1' \leftarrow -\left[\frac{k_0 + ez}{d}\right]$;
  for $y \leftarrow A_1'$ **to** $A_1' + \frac{w\alpha_1}{d}$ - *1* **do**
    $A_0' \leftarrow -\left[\frac{j_0 + by + cz}{a}\right]$;
    for $x \leftarrow A_0'$ **to** $A_0' + \frac{w\alpha_0}{a}$ - *1* **do**
$$Y \leftarrow \left[\frac{T\begin{pmatrix}x\\y\\z\end{pmatrix}+V}{\omega}\right] ;$$
$$H\begin{pmatrix}x\\y\\z\end{pmatrix} \in P_Y ;$$

Proof. The proof is detailed in [2].

Using Theorems 4 and 5, we have

$$\theta_0 = \frac{\mathrm{lcm}(\alpha_0, \alpha_1, \alpha_2)}{\alpha_0}, \theta_1 = \frac{\mathrm{lcm}(\alpha_0, \alpha_1, \alpha_2)}{\alpha_1}, \theta_1 = \frac{\mathrm{lcm}(\alpha_0, \alpha_1, \alpha_2)}{\alpha_2},$$

and

$$P = \left\{ H\begin{pmatrix}x\\y\\z\end{pmatrix} / A_2' \leq z < B_2', A_1'(z) \leq y < B_1'(z) \text{ and } A_0'(y, z) \leq x < B_0'(y, z) \right\},$$

with $A_2' = -\left[\frac{l_0}{f}\right], A_1'(z) = -\left[\frac{k_0 + ez}{d}\right], A_0'(y, z) = -\left[\frac{j_0 + by + cz}{a}\right], B_2' = A_2' + \frac{w\alpha_2}{f},$
$B_1'(z) = A_1'(z) + \frac{w\alpha_1}{d}$, and $B_0'(y, z) = A_0'(y, z) + \frac{w\alpha_0}{a}$ (see [2] for details).

Figure 4 shows the 16 distinct tiles of the QAT $\left(11, \begin{pmatrix} 6 & -2 & 3 \\ 2 & -2 & -3 \\ 4 & 4 & -4 \end{pmatrix}, \begin{pmatrix} 0 \\ 0 \\ 0 \end{pmatrix}\right)$ in $\mathbb{Z}^3$

and its corresponding super-tile.

3.3   QAT Algorithm in 3-D

To obtain the overall QAT algorithm, we need to find both the initial period tile index and the translation vector associated to a given tile $P_{i,j,k}$. Hence, thanks to Theorem 3, we have

$$\forall (i, j, k) \in \mathbb{Z}^3, P_{i,j,k} = T_W P_{i^0, j^0, k^0} \text{ with } W = w_0 U_0 + w_1 U_1 + w_2 U_2$$

$$\text{and } 0 \leq k^0 = \left\{\frac{k}{\alpha_2}\right\} < \alpha_2, w_2 = \left[\frac{k}{\alpha_2}\right], \quad 0 \leq j^0 = \left\{\frac{j + w_2\beta_1^2}{\alpha_1}\right\} < \alpha_1, w_1 = \left[\frac{j + w_2\beta_1^2}{\alpha_1}\right],$$

$$0 \leq i^0 = \left\{\frac{i + w_1\beta_0^1 + w_2\beta_0^2}{\alpha_0}\right\} < \alpha_0, w_0 = \left[\frac{i + w_1\beta_0^1 + w_2\beta_0^2}{\alpha_0}\right].$$

Fig. 4. All distinct tiles of a QAT in $\mathbb{Z}^3$ and its super-tile

4   Experiments

The algorithms were implemented in both 2D and 3D, with different refinements in order to be able to compare the implementations. For details on the 2-D algorithms, refer to [2]. The `backward mapping` (B. M. for short) implementation let

---

Algorithm 4. QAT Algorithm in 3-D.

Input: A QAT (ω, M, V) and an image $g : [0, t_0] \times [0, t_1] \times [0, t_2] \to \mathbb{Z}$
Output: an image $h : [min_0, max_0] \times [min_1, max_1] \times [min_2, max_2] \to \mathbb{Z}$
Compute min_i and max_i quantities from t_i;
if f *dilating* **then**

> $f \leftarrow f^{-1}$;
> Compute the Hermite Normal Form of the matrix M;
> Compute the minimal periodicities $\{\alpha_0, \alpha_1, \alpha_2\}$ and vectors $\{\boldsymbol{U}_0, \boldsymbol{U}_1, \boldsymbol{U}_2\}$;
> Use Algorithm 3 to compute the canonical tiles in the super-tile $\mathcal{P}$;
> **for** $i \leftarrow 0$ **to** $t_0 - 1$ **do**
>> **for** $j \leftarrow 0$ **to** $t_1 - 1$ **do**
>>> **for** $k \leftarrow 0$ **to** $t_2 - 1$ **do**
>>>> Compute W, i^0, j^0, k^0 ;
>>>> $h(T_W P_{i^0, j^0, k^0}) \leftarrow g(i, j, k)$;

else

> Compute the Hermite Normal Form of the matrix M;
> Compute the minimal periodicities $\{\alpha_0, \alpha_1, \alpha_2\}$ and vectors $\{\boldsymbol{U}_0, \boldsymbol{U}_1, \boldsymbol{U}_2\}$;
> Use Algorithm 3 to compute the canonical tiles in the super-tile $\mathcal{P}$;
> **for** $i \leftarrow min_0$ **to** $max_0 - 1$ **do**
>> **for** $j \leftarrow min_1$ **to** $max_1 - 1$ **do**
>>> **for** $k \leftarrow min_2$ **to** $max_2 - 1$ **do**
>>>> Compute W, i^0, j^0, k^0 ;
>>>> $h(i, j, k) \leftarrow g(T_W P_{i^0, j^0, k^0})$;

Table 1. Comparative evaluation in 2-D and 3-D. The last table presents the PSNR evaluation of the composition $f \cdot f^{-1}$.

	2-D - instructions (time in sec.)			
	B.M.	simple	Periodicity	noMultiply
contracting	1 607 774 (0.036)	64 536 315 (0.06)	29 578 702 (0.036)	27 679 044 (0.036)
isometry	63 058 160 (0.112)	57 619 374 (0.064)	39 682 795 (0.056)	35 875 892 (0.044)
dilating	391 622 017 (0.404)	185 956 768 (0.12)	87 490 567 (0.084)	83 472 387 (0.078)

	3-D - instructions (time in sec.)		
	B.M.	simple	Periodicity
contracting	15 864 982 (0.02)	47 303 861 (0.052)	12 865 125 (0.012)
isometry	750 102 224 (0.416)	51 121 827 (0.068)	15 234 007 (0.016)
dilating	170 072 035 547 (79.637)	2 479 676 409 (1.384)	7 760 893 011 (0.632)

	2-D (PSNR in dB)		3-D (PSNR in dB)	
	B.M.	Periodicity	B.M.	Periodicity
contracting	24.764	23.4823	17.8026	17.0304
isometry	27.7619	25.8052	19.4115	15.9481
dilating	31.2331	30.8375	20.4435	16.7862

us compare the tile periodicity method with the widely used **backward mapping** method [3]. The **simple** implementation does not use tiles periodicity and uses algorithm 2 for each tile. The **periodicity** implementation uses the periodicity and the algorithm 4. The **noMultiply** implementation additionally uses a method presented in [4] which uses a handling of remains instead of computing a matrix product in 3. The experiments are performed on an Intel© Centrino© Duo T2080 (2 x 1.73 GHz) in monothread and we give on one hand the time of computation and on the other hand the number of elementary instructions. The QATs used are the following : In 2-D:

$$\left(\omega, \begin{pmatrix} 3 & -4 \\ 4 & 3 \end{pmatrix}, \begin{pmatrix} 0 \\ 0 \end{pmatrix} \right) \text{ where } \omega = \begin{cases} 10 \text{ for the contracting application} \\ 5 \text{ for the isometry} \\ 2 \text{ for the dilating application} \end{cases}$$

In 3-D:

$$\left(\omega, \begin{pmatrix} 9 & -20 & -12 \\ 12 & 15 & -16 \\ 20 & 0 & 15 \end{pmatrix}, \begin{pmatrix} 0 \\ 0 \\ 0 \end{pmatrix} \right) \text{ where } \omega = \begin{cases} 100 \text{ for the contracting application} \\ 25 \text{ for the isometry} \\ 4 \text{ for the dilating application} \end{cases}$$

The pictures are of size : 200 x 171 in 2-D and 10 x 10 x 10 in 3-D (simple cube).

Figure 5 illustrates the results in dimension 2. As expected, when comparing B.M. and Periodicity, results are similar for both contracting and isometry QATs. Differences appear when dilating QAT is considered. Indeed, since a unique color is associated to a tile in the Periodicity algorithm, the

Fig. 5. Results in dimension 2: $(a-d)$ Contracting QAT (B.M. $(a-b)$ and Periodicity $(c-d)$); $(e-h)$ Isometry (B.M. and Periodicity); $(i-l)$ Dilating (B.M. and Periodicity). (m) illustrates the tile structure of the dilating QAT on the same square as in (l).

transformed image contains sharp edges (Fig 5-(l)) On the other hand, the interpolation process in the B.M. algorithm makes the image blurred. To compare the time efficiency (Table 4), we have considered two quantities: the total number of elementary operations of the main loop[2] and the overall computational time in seconds. Table 4 and Figure 6 present the results in dimension 3. For the sake of clarity, we have only considered an input binary image but the transformation algorithms can be applied to 3-D color images. We have also performed a peak signal-to-noise ratio (PSNR for short, given in decibel dB) computation between the input image in 2-D and 3-D, and the result of the composition $f.f^{-1}$. This test has been designed to evaluate the propagation of the error through the transformations with a signal processing tool. As presented in Table 4, the distortion induced by the proposed method is always smaller than the one induced by the backward mapping technique. Note that to have relevant measurements, we have used a density 3-D volume for the 3-D test (see Fig. 6-(h)).

---

[2] obtained with the `valgrind` profiling tool.

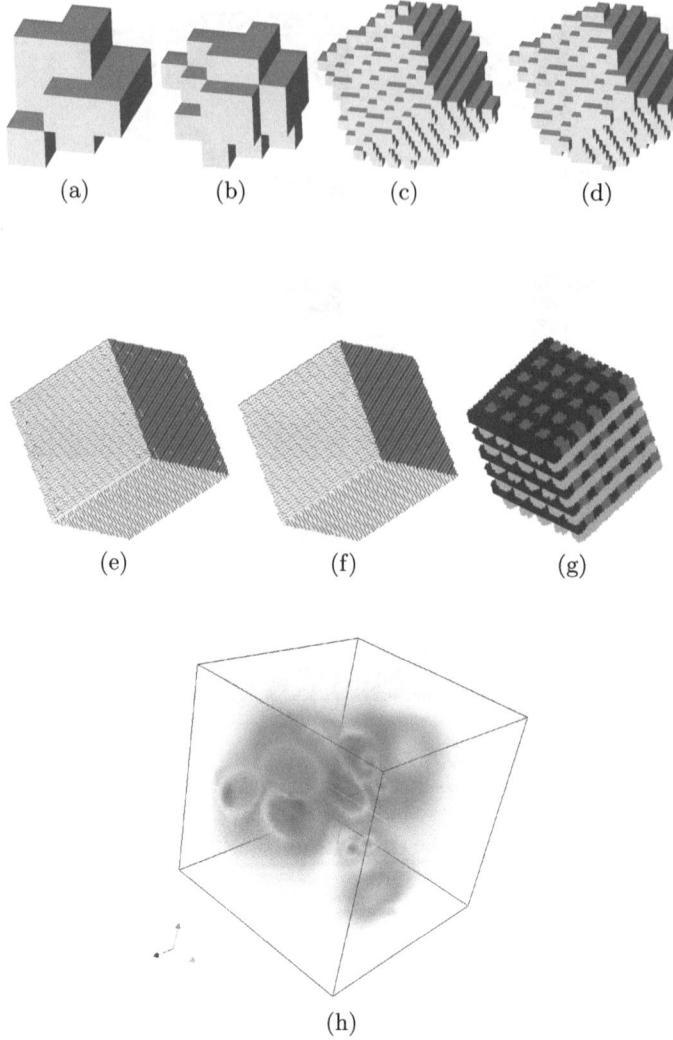

Fig. 6. Results in dimension 3: $(a - b)$ Contracting (B.M. and Periodicity), $(c - d)$ Isometry and $(e - f)$ Dilating (B.M. and Periodicity). (g) illustrates the tile structure of the dilating QAT. (h) is the $32 \times 32 \times 32$ density volume used in the $f.f^{-1}$ composition test (from the TC18 data-set http://www.tc18.org).

5   Conclusion and Future Works

In this paper, we have first re-demonstrated an existing result in dimension 2 with our formalism and provide a generalization in dimension 3 of fast image transformations with QAT. Compared to a classical image transformation technique, we have also illustrated the strength of such arithmetical methods to speed up transformations in higher dimensions. As detailed above and based on the generic $n - D$ transformation algorithm proposed in [7,2,1], we had to perform specific computation to obtain the minimal periods in dimension 2 and 3. A very challenging future work is to define a framework to compute these minimal periods when a specific dimension is considered. In order to achieve this goal and instead of having explicit formulas, we plan to investigate an algorithmic solution based on the $n - D$ canonical tile counting algorithm proposed in [7]. Furthermore, in dimension 2 and 3, comparisons to other transformation algorithms, such as Fourier based technique, are also of interest.

References

1. Blot, V., Coeurjolly, D.: Quasi-affine transform in higher dimension. In: 15th International Conference on Discrete Geometry for Computer Imagery. LNCS. Springer, Heidelberg (submitted)
2. Blot, V., Coeurjolly, D.: Quasi-affine transform in higher dimension. Technical Report RR–LIRIS-2009-010, Laboratoire LIRIS (April 2009), http://liris.cnrs.fr/publis?id=3853
3. Foley, J.D., van Dam, A., Feiner, S.K., Hughes, J.F.: Computer Graphics: Principles and Practice, 2nd edn. Addison-Wesley, Reading (1990)
4. Jacob, M.: Applications quasi-affines. PhD thesis, Université Louis Pasteur, Strasbourg, France (1993)
5. Jacob, M.: Transformation of digital images by discrete affine applications. Computers & Graphics 19(3), 373–389 (1995)
6. Jacob-Da Col, M.: Applications quasi-affines et pavages du plan discret. Theoretical Computer Science 259(1-2), 245–269 (2001), http://dpt-info.u-strasbg.fr/~jacob/articles/paving.pdf
7. Jacob-Da Col, M.A., Tellier, P.: Quasi-linear transformations and discrete tilings. Theoretical Computer Science 410(21-23), 2126–2134 (2009)
8. Nehlig, P.: Applications quasi affines: pavages par images réciproques. Theoretical Computer Science 156(1-2), 1–38 (1996)
9. Nehlig, P., Ghazanfarpour, D.: Affine Texture Mapping and Antialiasing Using Integer Arithmetic. Computer Graphics Forum 11(3), 227–236 (1992)
10. Storjohann, A., Labahn, G.: Asymptotically fast computation of hermite normal forms of integer matrices. In: ISSAC 1996: Proceedings of the 1996 international symposium on Symbolic and algebraic computation, pp. 259–266. ACM, New York (1996)

Farey Sequences and the Planar Euclidean Medial Axis Test Mask

Jérôme Hulin and Édouard Thiel*

Laboratoire d'Informatique Fondamentale de Marseille (LIF, UMR 6166),
Aix-Marseille Université,
163, Avenue de Luminy, Case 901, 13288 Marseille cedex 9, France
{Jerome.Hulin,Edouard.Thiel}@lif.univ-mrs.fr
http://www.lif.univ-mrs.fr/~thiel/

Abstract. The Euclidean test mask $T(r)$ is the minimum neighbourhood sufficient to detect the Euclidean Medial Axis of any discrete shape whose inner radius does not exceed r. We establish a link between $T(r)$ and the well-known Farey sequences, which allows us to propose two new algorithms. The first one computes $T(r)$ in time $\mathcal{O}(r^4)$ and space $\mathcal{O}(r^2)$. The second one computes for any vector $\overrightarrow{v}$ the smallest r for which $\overrightarrow{v} \in T(r)$, in time $\mathcal{O}(r^3)$ and constant space.

Keywords: medial axis, Farey sequence, Euclidean distance.

1 Introduction

The Medial Axis is a geometrical tool, which is widely used in numerous fields of image analysis [12]. The basic idea is simple: consider a subset S of a space E; the medial axis MA of S is the set of centres (and radii) of all maximal disks in S. A disk D is maximal in S if D is included in S and if D is not included in any other disk included in S. Thus MA(S) is a covering of S and so, the list of centres and radii of MA(S) is a reversible coding, a key point in many applications.

Another important point of interest about MA is that detecting the centres of maximal disks can be very efficiently achieved using a distance transform of S. The distance transform DT(S) is a copy of S where each point is labeled with its distance to the closest point of the background $E \setminus S$. The geometric properties of MA, as well as the computation of DT, both depend on the distance function. The main distances used in discrete geometry are the Euclidean distance d_E [4], generally squared to get integer values in $E = \mathbb{Z}^n$, and the chamfer (or weighted) norms [14].

The general method to extract MA from DT, applicable to d_E and chamfer norms in $\mathbb{Z}^n$, is known as the LUT method [1, 2]. A precomputation [10, 11] gives a neighbourhood T and a look-up table LUT; then, to know if a point $p \in S$ is a medial axis point, it is sufficient to check the neighbours $p + \overrightarrow{v}$ for all $\overrightarrow{v} \in T$ and compare their DT values to the values stored in LUT. Lately, a fast algorithm using H-polytopes was proposed for 2D and 3D chamfer norms in [7, 8].

---

* Work supported in part by ANR grant BLAN06-1-138894 (projet OPTICOMB).

P. Wiederhold and R.P. Barneva (Eds.): IWCIA 2009, LNCS 5852, pp. 82–95, 2009.

Instead of computing T, we may use $T = \mathbb{Z}^n$, but the MA extraction would be excessively long. On the other hand, if T is too small, the MA extraction may choose non maximal disks. Therefore, we aim at computing a minimal T with respect to a class of shapes: Given a distance d in $\mathbb{Z}^n$, n fixed, and a radius r, we denote by $T(r)$ the minimum test neighbourhood sufficient to extract the MA of all n-dimensional shapes whose inner radius are not greater than r.

Concerning the Euclidean distance, we have proved in [6] that in any dimension, $T(r)$ is unique and tends to the set of visible vectors when r tends to infinity. We have recently tackled the search of arithmetical and geometrical properties concerning $T(r)$ and the appearence radii of vectors in the 2-dimensional case. We have presented recent results concerning 5×5 chamfer norms in [5]; in this paper we present new properties in the Euclidean case.

When r grows, new vectors (that is to say, neighbours) are inserted from time to time in $T(r)$. Given a vector $\overrightarrow{v}$, we denote by $r_{app}(\overrightarrow{v})$ the appearance radius of $\overrightarrow{v}$, that is, the smallest r for which $\overrightarrow{v} \in T(r)$.

In order to incrementally find the new vectors of $T(r)$, the method proposed in [10, 11] consists in checking, for every point p in a disk of radius r, disks inclusion relations in directions $p + \overrightarrow{v}$, for all $\overrightarrow{v} \in T(r)$. In this paper, we show that it is sufficient to test the inclusion relations with two neighbours, instead of all the neighbours of $T(r)$. These two points are indeed the neighbours of p in some Farey sequence [3].

This property has two consequences: first, the computation of $T(r)$ can be sped up significantly. Second, these two points to test are independent from $T(r)$, so we are able to compute the appearance radius $r_{app}(\overrightarrow{v})$ of any visible vector $\overrightarrow{v}$ without computing T up to $r_{app}(\overrightarrow{v})$.

In Section 2 we introduce the definitions which are used in Section 3, where we establish the link between Farey sequences and the appearance radii of visible vectors. Then we present the two algorithms in Sections 4 and 5. We conclude by observations and conjectures in Section 6.

2 Preliminaries

2.1 The Discrete Space $\mathbb{Z}^n$

Throughout the paper, we work in the discrete space $\mathbb{Z}^n$ (we will at some point fix $n = 2$). We consider $\mathbb{Z}^n$ both as an n-dimensional $\mathbb{Z}$-module (a discrete vector space) and as its associated affine space; and we write $\mathbb{Z}_*^n = \mathbb{Z}^n \setminus \{0\}$. We denote by $(v_1, \ldots, v_n)$ the Cartesian coordinates of a given vector $\overrightarrow{v}$. A vector $\overrightarrow{Op} \in \mathbb{Z}^n$ (or a point $p \in \mathbb{Z}^n$) is said to be *visible* if the line segment connecting O to p contains no other point of $\mathbb{Z}^n$, i.e., if the coordinates of p are coprime.

We call Σ^n the group of axial and diagonal symmetries in $\mathbb{Z}^n$ about centre O. The cardinal of the group is $\#\Sigma^n = 2^n \, n!$ (which is 8, 48 and 384 for $n = 2, 3$ and 4). An n-dimensional *shape* S is by definition a subset of $\mathbb{Z}^n$. A shape S is said to be *G-symmetrical* if for every $\sigma \in \Sigma^n$ we have $\sigma(S) = S$. The *generator* of a set $S \subseteq \mathbb{Z}^n$ is $G(S) = \{ (p_1, \ldots, p_n) \in S : 0 \leqslant p_n \leqslant p_{n-1} \leqslant \ldots \leqslant p_1 \}$.

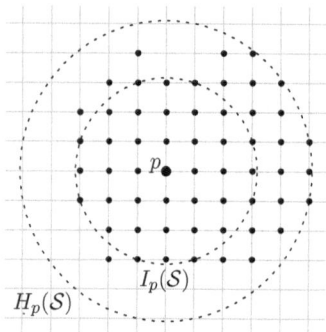

8	5	4	5	8
5	2	1	2	5
4	1	O	1	4
5	2	1	2	5
8	5	4	5	8

Fig. 1. $H_p(\mathcal{S})$ and $I_p(\mathcal{S})$ for a shape $\mathcal{S}$ ($\mathcal{S}$ is represented by bullets)

Fig. 2. Balls of squared Euclidean radii 1 (shaded) and 2 (delimited by the thick line). Values indicate the distance to O.

2.2 Balls and Medial Axis

Let d be a distance on $\mathbb{Z}^n$. The *ball* of centre $p \in \mathbb{Z}^n$ and radius $r \in \mathbb{R}$ is $\mathcal{B}(p, r) = \{q \in \mathbb{Z}^n : d(p, q) \leqslant r\}$. To shorten notation, we let B_r stand for $\mathcal{B}(O, r)$. Since we consider discrete closed balls, any ball B has an infinite number of real radii in a left-closed interval $[r_1, r_2[$, with $r_1, r_2 \in \mathrm{Im}(d)$. We define the *representable radius* of a given ball B to be the radius of B which belongs to $\mathrm{Im}(d)$.

Let $\mathcal{S}$ be a shape and $p \in \mathbb{Z}^n$. We define $I_p(\mathcal{S})$ to be the largest ball of centre p, included in $\mathcal{S}$ (if $p \notin \mathcal{S}$, we have $I_p(\mathcal{S}) = \emptyset$). Analogously, we define $H_p(\mathcal{S})$ to be the smallest ball of centre p, which contains $\mathcal{S}$. We also define $\mathcal{R}_p(B)$ to be the representable radius of $H_p(B)$. See an example in Fig. 1. The *inner radius* of a given shape $\mathcal{S}$, denoted by $\mathrm{rad}(\mathcal{S})$, is the representable radius of a largest ball included in $\mathcal{S}$. We denote by $\mathrm{CS}^n(r)$ the class of all n-dimensional shapes whose inner radius are less than or equal to r. A ball included in a shape $\mathcal{S}$ is said to be *maximal* in $\mathcal{S}$ if it is not included in any other ball included in $\mathcal{S}$. The *Medial Axis* (MA) of a shape $\mathcal{S}$ is the set of centres (and radii) of all maximal balls of $\mathcal{S}$:

$$p \in \mathrm{MA}(\mathcal{S}) \Leftrightarrow p \in \mathcal{S} \text{ and } \forall q \in \mathcal{S} \setminus \{p\}, \ I_p(\mathcal{S}) \nsubseteq I_q(\mathcal{S}). \tag{1}$$

Finally, if a point $q \in \mathcal{S}$ satisfies $I_p(\mathcal{S}) \subseteq I_q(\mathcal{S})$, we say that q *forbids* p from $\mathrm{MA}(\mathcal{S})$, see Fig. 3.

2.3 The Medial Axis Test Mask $\mathcal{T}$

For the coherence of what follows, we assume d is a translation invariant distance. We define the *test mask* $\mathcal{T}(r)$ to be the minimum neighbourhood sufficient to detect locally, for all $\mathcal{S}$ in $\mathrm{CS}^n(r)$ and all $p \in \mathcal{S}$, if p is a point of $\mathrm{MA}(\mathcal{S})$:

$$\begin{cases} \forall \mathcal{S} \in \mathrm{CS}^n(r), \forall p \in \mathcal{S}, \left(p \notin \mathrm{MA}(\mathcal{S}) \Rightarrow \exists \vec{v} \in \mathcal{T}(r), \ I_p(\mathcal{S}) \subseteq I_{p-\vec{v}}(\mathcal{S})\right); \\ \mathcal{T}(r) \text{ has minimum cardinality.} \end{cases} \tag{2}$$

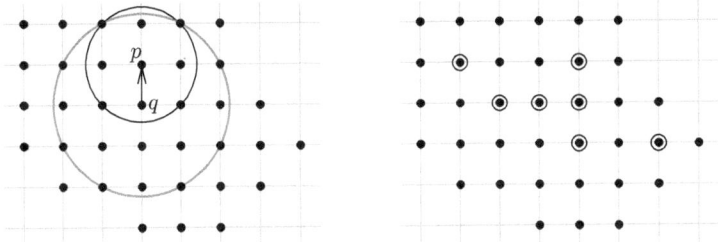

Fig. 3. Left: A shape $\mathcal{S}$ (bullets). Since $I_p(\mathcal{S}) \subseteq I_q(\mathcal{S})$, the point q forbids p from $\mathrm{MA}(\mathcal{S})$. Right: the medial axis of $\mathcal{S}$ (circled points).

We have shown the unicity of $\mathcal{T}(r)$ for all $r \geqslant 0$ in [6]. As a corollary, if the considered distance is G-symmetrical, then so is $\mathcal{T}(r)$. Therefore, by abuse of notation we write $\mathcal{T}(r)$ instead of $\mathrm{G}(\mathcal{T}(r))$. Finally, the *appearance radius* $r_{app}(\overrightarrow{v})$ of a given vector $\overrightarrow{v}$ is the smallest radius r for which $\overrightarrow{v} \in \mathcal{T}(r)$.

Once we have a pre-computed test mask $\mathcal{T}(r)$, it is quite straightforward to compute the MA of a given $\mathcal{S} \in \mathrm{CS}^n(r)$: first, compute for each $p \in \mathcal{S}$, the ball $I_p(\mathcal{S})$. $\mathrm{DT}(p)$ is the distance from p to $\mathbb{Z}^n \setminus \mathcal{S}$, so $\mathrm{Ip(S)}$ is equal to the open ball of centre p and radius $\mathrm{DT(p)}$. Second, test for each point $p \in \mathcal{S}$ whether $I_p(\mathcal{S})$ is included in some $I_{p - \overrightarrow{v}}(\mathcal{S})$, with $\overrightarrow{v} \in \mathcal{T}(r)$.

Actually, the hard part consists in computing $\mathcal{T}(r)$, as we will see. According to (2), finding the appearance radius of a given vector $\overrightarrow{v}$ consists in solving the following **keyhole problem**: Given a vector $\overrightarrow{v}$, find the smallest positive r s.t. there is no ball B satisfying $I_{O+\overrightarrow{v}}(B_r) \subsetneq B \subsetneq B_r$. The term *key* refers to the ball B that we are trying to insert between B_r and $I_{O+\overrightarrow{v}}(B_r)$.

It is easy to check that the above problem is equivalent to finding the smallest positive $\mathcal{R}_{O+\overrightarrow{v}}(B_r)$ s.t. there is no ball B satisfying $B_r \subsetneq B \subsetneq H_{O+\overrightarrow{v}}(B_r)$. We would like to test as few balls (keys) B as possible when solving the keyhole problem.

Definition 1 (Set of keys). *Given a vector $\overrightarrow{v} \in \mathbb{Z}^n$, a set A is called a set of keys of $\overrightarrow{v}$ iff A is a neighbourhood sufficient to solve the keyhole problem with parameter $\overrightarrow{v}$. Precisely, A is a set of keys of $\overrightarrow{v}$ iff*

$$\overrightarrow{0}, \overrightarrow{v} \notin A \text{ and } \forall r < r_{app}(\overrightarrow{v}), \exists \overrightarrow{u} \in A \text{ s.t. } H_{O - \overrightarrow{u}}(B_r) \subseteq H_{O - \overrightarrow{v}}(B_r).$$

We aim at finding a set of keys of $\overrightarrow{v}$ having minimal cardinality. To do so, we define a domination relation as follows:

Definition 2 ($\overrightarrow{v}$-Domination). *We say that a vector $\overrightarrow{u}$ is $\overrightarrow{v}$-dominated by a vector $\overrightarrow{u}'$ ($\overrightarrow{u} \prec_{\overrightarrow{v}} \overrightarrow{u}'$ for short) iff*

$$\forall r \geqslant 0, \left(H_{O - \overrightarrow{u}'}(B_r) \not\subseteq H_{O - \overrightarrow{v}}(B_r) \Rightarrow H_{O - \overrightarrow{u}}(B_r) \not\subseteq H_{O - \overrightarrow{v}}(B_r) \right).$$

Thus, if $\vec{u} \preccurlyeq_{\vec{v}} \vec{u}'$, then $\vec{u}$ can be replaced by $\vec{u}'$ in any set of keys of $\vec{v}$. Notice that $\preccurlyeq_{\vec{v}}$ is not a total order, however it is clearly reflexive and transitive.

2.4 Farey Sequences

Let k be a positive integer. The *Farey sequence* of order k is the sequence of all irreducible fractions between 0 and 1, whose denominators do not exceed k, arranged in increasing order. For example, the Farey sequence of order 5 is

$$F_5 = \left\{ \frac{0}{1}, \frac{1}{5}, \frac{1}{4}, \frac{1}{3}, \frac{2}{5}, \frac{1}{2}, \frac{3}{5}, \frac{2}{3}, \frac{3}{4}, \frac{4}{5}, \frac{1}{1} \right\}.$$

Let $0 < \frac{a}{b} < 1$ be an irreducible fraction. We call *predecessor* of $\frac{a}{b}$ the term which precedes $\frac{a}{b}$ in F_b; we denote it by $\operatorname{pred}(\frac{a}{b})$. Similarly, the *successor* of $\frac{a}{b}$, denoted by $\operatorname{succ}(\frac{a}{b})$, is the term which follows $\frac{a}{b}$ in F_b. For example, the predecessor and the successor of $\frac{3}{4}$ are respectively $\frac{2}{3}$ and $\frac{1}{1}$. Cauchy proved that if $\frac{a}{b}$ and $\frac{c}{d}$ are two consecutive terms in some Farey sequence, then $bc - ad = 1$ (see [3] for details). Conversely, if $0 \leqslant \frac{a}{b} < \frac{c}{d} \leqslant 1$ are two irreducible fractions satisfying $bc - ad = 1$, then there is a Farey sequence in which $\frac{a}{b}$ and $\frac{c}{d}$ are neighbours. As a corollary, if $\frac{a}{b}, \frac{e}{f}, \frac{c}{d}$ are three consecutive terms in a Farey sequence, then $\frac{e}{f}$ is the *mediant* of $\frac{a}{b}$ and $\frac{c}{d}$, that is to say $\frac{e}{f} = \frac{a+c}{b+d}$. For instance $\frac{1}{4}, \frac{1}{3}, \frac{2}{5}$ are three consecutive terms in F_5, so $\frac{1}{3} = \frac{1+2}{4+5}$.

In this paper, we will use a geometric interpretation of Farey sequences: each irreducible fraction $0 \leqslant \frac{y}{x} \leqslant 1$ can be associated with a visible point p (or a vector $\overrightarrow{Op}$) in $G(\mathbb{Z}^2)$, with coordinates (x, y). By abuse of notation, we may write $p \in F_k$ or $\overrightarrow{Op} \in F_k$ if $\frac{y}{x}$ belongs to the Farey sequence F_k. Accordingly, the Farey sequence of order k is the sequence of all visible vectors of $G(\mathbb{Z}^2)$ whose abscissas do not exceed k, arranged counterclockwise from the x-axis. As an example, the points of F_6 are depicted in Fig. 4.

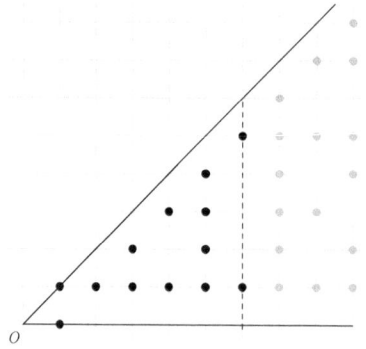

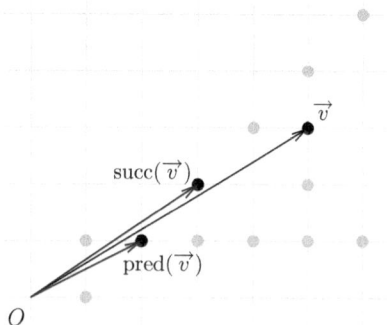

Fig. 4. Points of the Farey sequence of order 6 (in black) among the visible points in $G(\mathbb{Z}^2)$ (in grey)

Fig. 5. The predecessor and the successor of $(5, 3)$ are respectively $(2, 1)$ and $(3, 2)$

Let $\vec{u}$ and $\vec{v}$ be two consecutive vectors in some Farey sequence. Since $|u_1v_2 - u_2v_1| = 1$, the set $\{\vec{u}, \vec{v}\}$ is a basis of $\mathbb{Z}^2$. Moreover, the triangle $(O, O + \vec{u}, O + \vec{v})$ has area $1/2$, and it contains only three lattice points — its vertices. Also, let $\vec{v}$ be a visible vector whose predecessor and successor are respectively denoted by $\vec{u}$ and $\vec{w}$. Since $\vec{v}$ is the mediant of $\vec{u}$ and $\vec{w}$, we have $\vec{v} = \vec{u} + \vec{w}$ (see Fig. 5 for an example).

3   Farey Sequences and $\vec{v}$-Domination

From now on, we use the squared Euclidean distance d_E^2 in the space $\mathbb{Z}^2$. The variable $R \in \mathbb{N}$ will always denote a squared Euclidean radius. Also, we write $R_{app}(\vec{v})$ for the squared Euclidean appearance radius of $\vec{v}$: $R_{app}(\vec{v}) = r_{app}^2(\vec{v})$. For abbreviation, we write pq instead of $d_E^2(p, q)$. Nevertheless, since d_E^2 is not a norm, the standard notation $\|\vec{v}\|$ will denote the Euclidean norm of $\vec{v}$. Notice that d_E^2 is not even a metric, however it is important to point out that the medial axis with d_E and d_E^2 are identical, since these two distances yields the same set of balls.

Throughout this section, $\vec{v}$ denotes a visible vector in $G(\mathbb{Z}^2)$ different from $(1, 0)$ and $(1, 1)$. We set $p = O + \vec{v}$, and we denote by q and q' (respectively) the predecessor and the successor of p. Also, the points $p'(p_1, 0)$ and $p''(p_1, p_1)$ are the points with same abscissa as p and minimal (resp. maximal) ordinate in $G(\mathbb{Z}^2)$, see Fig. 6.

Lemma 1. *If t is a lattice point inside the triangle (Opp') (different from O and p) then q belongs to the triangle (Opt). Besides, if t is a lattice point in the triangle (Opp'') (different from O and p) then q' belongs to the triangle (Opt).*

Proof. We prove the Lemma in the case where $t \in (Opp')$. Since q is the predecessor of p, there is no vector in the cone $\overrightarrow{Oq}\mathbb{N} + \overrightarrow{Op}\mathbb{N}$ and whose abscissa is not greater than that of p. Suppose that q is not inside the triangle (Opt); the vector $\overrightarrow{pt}$ must belong to the cone $\overrightarrow{pq}\mathbb{N} + \overrightarrow{pO}\mathbb{N}$, see Fig. 6. The point p is the mediant of q and q' so $\overrightarrow{Oq} = \overrightarrow{qp}$ (and $\overrightarrow{Oq} = \overrightarrow{q'p}$). By symmetry, we deduce that $\overrightarrow{tp}$ belongs to the cone $\overrightarrow{Op}\mathbb{N} + \overrightarrow{Oq'}\mathbb{N}$; furthermore its abscissa is not greater than that of $\overrightarrow{Op}$. It follows that there is a point between p and q' in the Farey sequence of order p_1, which contradicts our assumption $q' = \text{succ}(p)$.
By symmetry, the same reasoning applies to the case where $t \in (Opp'')$.   □

The preceding Lemma states that all lattice points in the triangle $(Op'p'')$ belong to the shaded area drawn in Fig. 7 (with the exception of O and p). This geometrical property allows to establish $\vec{v}$-domination relations between points of $(Op'p'')$:

Lemma 2. *Every lattice point inside the triangle (Opp') is $\vec{v}$-dominated by q. Besides, every lattice point inside the triangle (Opp'') is $\vec{v}$-dominated by q'.*

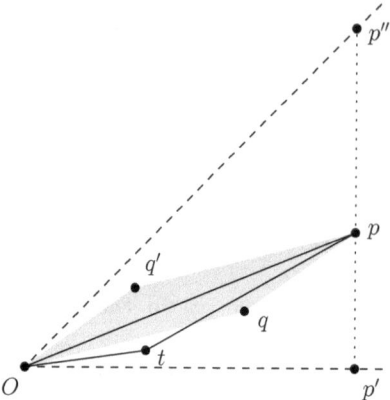
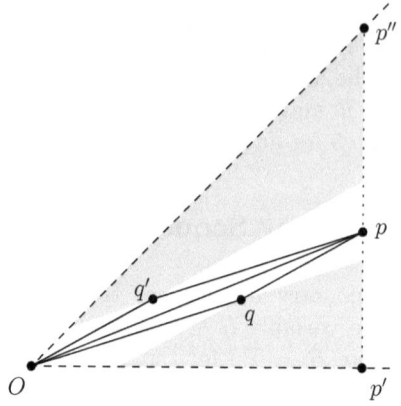

Fig. 6. Impossible configuration for the proof of Lemma 1. A visible point p, its predecessor q and successor q' (the shaded parallelogram has area 1 and contains only 4 lattice points — its vertices). For the sake of clarity, the angle $\widehat{Oqp}$ is exaggerated.

Fig. 7. The shaded areas (together with their boundaries) contain all the lattice points within the triangle $(Op'p'')$, with the exception of O and p

Proof. We examine the case of a point $t \in (Opp'')$. Lemma 1 claims that $q' \in (Opt)$, i.e., t belongs to the upper shaded area in Fig. 7. Now, let B be a ball of centre O. According to Def. 2, it remains to prove that $H_{q'}(B) \not\subseteq H_p(B)$ implies $H_t(B) \not\subseteq H_p(B)$. Actually, it turns out that any point z in $H_{q'}(B) \setminus H_p(B)$ also belongs to $H_t(B)$, as we will see. Let x be a point of the ball B which maximizes the distance to q', as illustrated in Fig. 8, left. We now need to establish three inequalities:

- By definition of x, the representable radius of $H_{q'}(B)$ is $q'x$. Moreover, $z \in H_{q'}(B)$, thus $q'z \leqslant q'x$.
- The point x belongs to B, unlike z (since $z \notin H_p(B)$ and $B \subseteq H_p(B)$). Accordingly, $Ox \leqslant Oz$.
- x belongs to $H_p(B)$ (because $x \in B$), but z does not, so $px \leqslant pz$.

These three inequalities give information about the position of O, p and q' with respect to the perpendicular bisector of the line segment $[xz]$: the points O and p lie on one side, while q' lies on the other, see Fig. 8, right. Moreover, q' is inside (Opt), so t must lie on the same side of the bisector as q'. It follows that t is closer to z than x. However x belongs to $H_t(B)$, therefore z also belongs to $H_t(B)$, which is the desired conclusion.

Again, similar arguments apply in the case $t \in (Opp')$, to show $t \preccurlyeq_{\vec{v}} q$. □

We have shown in [6] that the set $\Diamond\vec{v} = \{\vec{u} \in G(\mathbb{Z}^n) : \vec{v} - \vec{u} \in G(\mathbb{Z}^n_*)\}$ (roughly, a parallelogram) is a set of keys of any visible vector $\vec{v}$ in $G(\mathbb{Z}^n)$. The

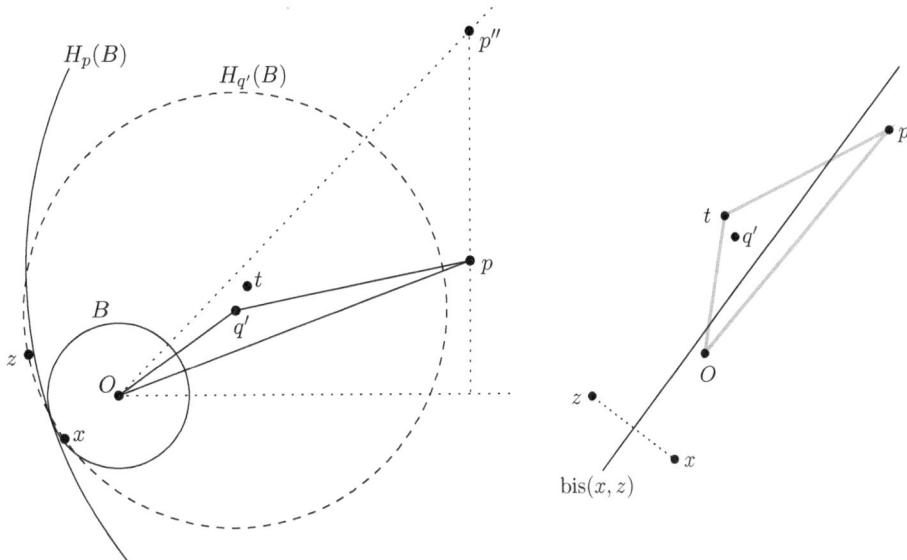

Fig. 8. Illustration for the proof of Lemma 2. Left: a point p and its successor q'. If for a ball B of centre O there is a point $z \in H_{q'}(B) \setminus H_p(B)$, then $z \in H_t(B)$. Right: position of the points O, p, q' and t with respect to the perpendicular bissector of $[xz]$.

set $\Diamond \overrightarrow{v}$ is included in the triangle $(Op'p'')$, so Lemma 2 shows that any vector in $\Diamond \overrightarrow{v}$ is $\overrightarrow{v}$-dominated either by $\text{pred}(\overrightarrow{v})$, or by $\text{succ}(\overrightarrow{v})$. We have thus proved:

Theorem 1. *Let $\overrightarrow{v}$ be a visible vector in $G(\mathbb{Z}^2)$, different from $(1,0)$ and $(1,1)$. The set $\{\text{pred}(\overrightarrow{v}), \text{succ}(\overrightarrow{v})\}$ is a set of keys of $\overrightarrow{v}$.*

4   Computing the Appearance Radius of a Given Vector

Theorem 1 will allow us to compute efficiently the appearance radius of any visible vector $\overrightarrow{v}$. The point is that we do not need to compute the test mask up to $R_{app}(\overrightarrow{v})$; we only need to test the inclusion of two balls — in direction $\text{pred}(\overrightarrow{v})$ and $\text{succ}(\overrightarrow{v})$.

We first examine the case of the vectors $(1,0)$ and $(1,1)$. It is easy to check that $\mathcal{T}(1) = \{(1,0)\}$ and $\mathcal{T}(2) = \{(1,0),(1,1)\}$ (balls of radii 1 and 2 are drawn in Fig. 2):

- The smallest representable non-zero value for d_E^2 is 1. Since $p(1,0)$ is in B_1 and does not belong to $MA(B_1)$ (the medial axis of a ball is its center), it follows that the appearance radius of $(1,0)$ is 1.
- The next representable integer is 2. The point $q(1,1)$ is not in $MA(B_2)$, however the point $(1,0)$ does not forbid q from $MA(B_2)$ because $I_p(B_2)$ has radius 0. So $(1,1)$ belongs to $\mathcal{T}(2)$.

We now proceed to the general case. For any visible vector $\vec{v}$ in $G(\mathbb{Z}^2)$ different from $(1,0)$ and $(1,1)$, set $p = O + \vec{v}$, and let q and q' denote the predecessor and the successor of p. We know from Theorem 1 that the geometrical configuration for the appearance of $\vec{v}$ is obtained for the smallest R for which $H_q(B_R) \not\subseteq H_p(B_R)$ and $H_{q'}(B_R) \not\subseteq H_p(B_R)$.

Alg. 1 works as follows: for all balls B_R in order of increasing radius R, it first computes $R_p = \mathcal{R}_p(B_R)$ and $R_q = \mathcal{R}_q(B_R)$ on line 4. The function $\mathrm{Rcov}(R, \vec{z})$ computes, for any $R \geq 0$ and $\vec{z} \in \mathbb{Z}^2$, the covering radius of B_R in direction $\vec{z}$, that is to say, $\mathcal{R}_{O+\vec{z}}(B_R)$. Since $\overrightarrow{qp} = \overrightarrow{Oq'}$, we have $H_q(B_R) \not\subseteq H_p(B_R) \Leftrightarrow \mathrm{Rcov}(R_q, \overrightarrow{Oq'}) > R_p$. So, if $\mathrm{Rcov}(R_q, \overrightarrow{Oq'}) > R_p$, it remains to test whether $H_{q'}(B_R) \not\subseteq H_p(B_R)$; this is done in the same manner on line 6. In this case, i.e., if $\mathrm{Rcov}(R_{q'}, \overrightarrow{Oq}) > R_p$ then neither q nor q' forbids O from $\mathrm{MA}(B_p)$, hence $H_p(B_R)$ is a shape of smallest inner radius for which the inclusion test in direction $\overrightarrow{Op}$ is mandatory. The existence of such a configuration is guaranteed by the fact that $\vec{v}$ is visible [6].

Algorithm 1. Comp_Rapp

Input: a visible vector $\vec{v} = \overrightarrow{Op}$
Output: the appearance radius of $\vec{v}$

1 **if** $\vec{v} = (1,0)$ **then return** 1 ; **if** $\vec{v} = (1,1)$ **then return** 2 ;
2 $q \leftarrow \mathrm{pred}(p)$; $q' \leftarrow \mathrm{succ}(p)$; $R \leftarrow 0$;
3 **loop**
4      $R_p \leftarrow \mathrm{Rcov}(R, \vec{v})$; $R_q \leftarrow \mathrm{Rcov}(R, \overrightarrow{Oq})$;
5   **if** $\mathrm{Rcov}(R_q, \overrightarrow{Oq'}) > R_p$ **then**
6      $\lfloor$ $R_{q'} \leftarrow \mathrm{Rcov}(R, \overrightarrow{Oq'})$; **if** $\mathrm{Rcov}(R_{q'}, \overrightarrow{Oq}) > R_p$ **then** **return** R_p ;
7      $R \leftarrow R + 1$;

Let us examine the complexity of Alg. 1. First, the computation of $\vec{u} = \mathrm{pred}(\vec{v})$ and $\vec{w} = \mathrm{succ}(\vec{v})$ is easy: since $u_1 v_2 - u_2 v_1 = 1$, we are reduced to finding positive integers x and y satisfying $v_2 x - v_1 y = 1$. This can be achieved using the well-known extended Euclidean algorithm, in time $\mathcal{O}(\log(u_2))$.

The function $\mathrm{Rcov}(R, \vec{v})$ consists in computing the maximum value of $\|\overrightarrow{Ox} + \vec{v}\|^2$ for all points x on the boundary of B_R. As R is the squared Euclidean radius of B_R, there are about $\sqrt{R}$ points on the boundary of B_R. We use a simple Bresenham-Pitteway algorithm [9] to scan the boundary of B_R; hence $\mathrm{Rcov}(R, \vec{v})$ can be computed in time $\mathcal{O}(\sqrt{R})$.

Besides, the main loop is composed of at most 5 calls to Rcov with parameter R, thus the global complexity of the algorithm is roughly $\sum_{R=0}^{R_{app}(\vec{v})} \sqrt{R}$. We can find a tight upper bound of this sum by using an integral: $\sum_{R=0}^{S} \sqrt{R} \leq \int_1^{S+1} \sqrt{R}$, which approaches $\frac{2}{3} S^{3/2}$ when $S \to \infty$. In consequence, Alg. 1 runs in time $\mathcal{O}(R_{app}^{3/2}(\vec{v})) = \mathcal{O}(r_{app}^3(\vec{v}))$. Furthermore, it is clear that this algorithm uses $\mathcal{O}(0)$ space.

5   An Algorithm for $\mathcal{T}(R)$

In this section we propose an algorithm to compute $\mathcal{T}(R_{max})$ for any $R_{max} > 0$. This algorithm does not rely on the computation of DTs or LUTs, and need not scan the boundary of a ball when computing a covering radius. The basic idea is the following: we link each vector $\overrightarrow{v} \in G(\mathbb{Z}^2)$ to a vector $next(\overrightarrow{v}) \in G(\mathbb{Z}^2)$ whose norm is greater than or equal to that of $\overrightarrow{v}$ (see Fig. 9). Thus, it is easy to localize the new points of a ball B_R, when R increases.

When invoked with parameter R_{max}, the algorithm is looking for the appearance configuration of all vectors $\overrightarrow{v}$ whose norm are not greater than $\sqrt{R}$, for all representable integers R s.t. $R \leqslant R_{max}$. To do so, each visible vector also has three pointers $vecR(\overrightarrow{v})$, $testP(\overrightarrow{v})$ and $testS(\overrightarrow{v})$:

- $vecR(\overrightarrow{v})$ is a vector whose squared norm is equal to the radius of the largest open ball of center $O + \overrightarrow{v}$ included in B_R, i.e., the radius of $I_{O+\overrightarrow{v}}(B_R)$.
- $testP(\overrightarrow{v})$ (resp. $testS(\overrightarrow{v})$) is the vector $\overrightarrow{z} \in G(\mathbb{Z}^2)$ having minimum norm which satisfies $O + \overrightarrow{v} + \overrightarrow{z} \notin I_{O+pred(\overrightarrow{v})}(B_R)$ (resp. $\notin I_{O+succ(\overrightarrow{v})}(B_R)$).

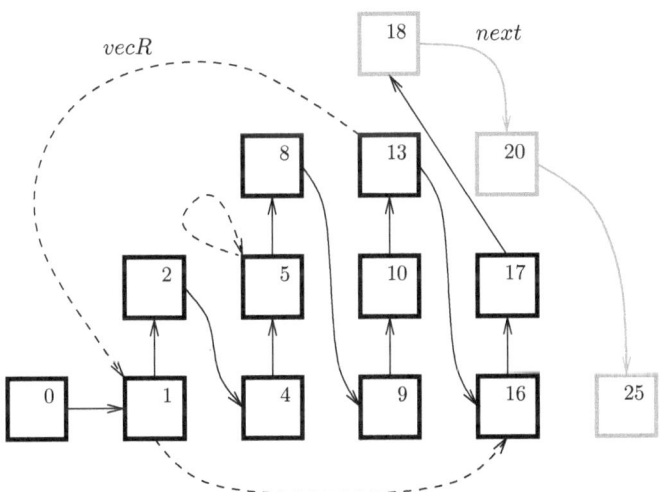

Fig. 9. Linked list of points inside the open ball B of radius 18 (black squares). Each point p has a pointer $next(p)$ to a point whose squared norm is the next representable integer (link represented by a solid arrow). The grey squares are the points of $G(\mathbb{Z}^2)$ on the outside of B, at the top of all columns around the boundary of B. The first grey point in the list gives the radius of the smallest ball strictly larger than B. Also, each visible point p has a pointer $vecR(p)$ (depicted by a dashed arrow) to the first point in the linked list whose squared norm is the radius of $I_p(B)$. For the sake of clarity, only 3 pointers $vecR$ are shown.

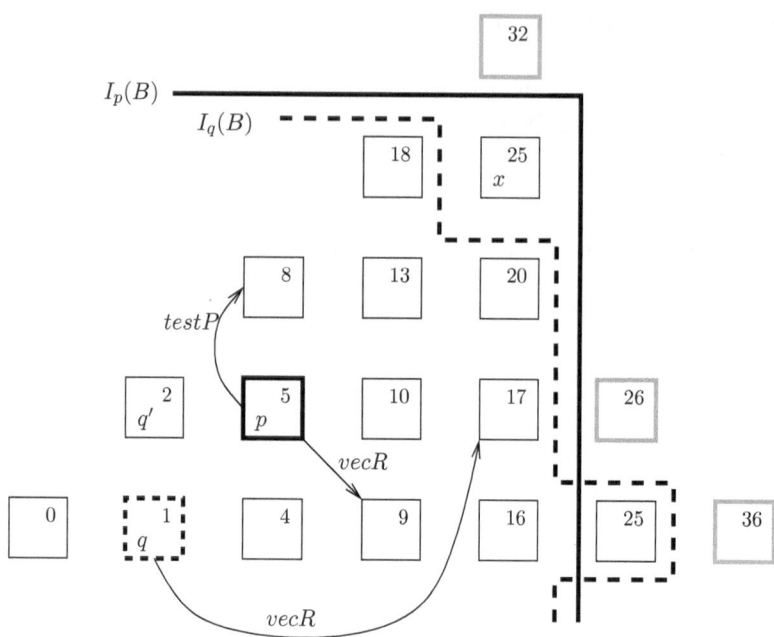

Fig. 10. Configuration for the open ball B of radius 26 (black squares). q is the predecessor of $p\,(2,1)$; $I_p(B)$ and $I_q(B)$ are open balls of respective radii 9 and 17. testP$(p) = \overrightarrow{px}$ is the smallest vector in $G(\mathbb{Z}^2)$ for which $x \notin I_q(B)$. Here $px < 9$, so $I_p(B) \not\subseteq I_q(B)$.

The property $I_{O+\vec{v}}(B_R) \not\subseteq I_{O+pred(\vec{v})}(B_R)$ is therefore equivalent to $\|testP(\vec{v})\| \leqslant \|vecR(\vec{v})\|$. Similarly, $I_{O+\vec{v}}(B_R) \not\subseteq I_{O+succ(\vec{v})}(B_R) \Leftrightarrow \|testS(\vec{v})\| \leqslant \|vecR(\vec{v})\|$. See Fig. 9 and 10 for examples of links between vectors.

Let us give some details about Alg. 2. The vector $outV$ is the first vector in the linked list of vectors, whose norm is greater than $\sqrt{R}$. From line 5 to line 7, we update the list of vectors outside B_R (the grey points in Fig. 9 and 10): it is sufficient to insert the vector located above $outV$, and the vector on the right of $outV$ if its abscissa is 0. The new vectors must be inserted according to their norm. This is done by the procedure insert_next. When a new vector is inserted in the linked list, its pointers $vecR$, $testP$ and $testS$ are set to $(1,0)$.

For each vector $\vec{v}$ inside B_R, we update the pointers $vecR$ at lines $9-10$, and we update the pointers $testP$ and $testS$ at lines $14-17$. Then, the test of appearance of $\vec{v}$ is done at line 18. Notice that for a given radius R, we need to scan the vectors $\vec{v}$ in order of increasing norm (or, alternatively, in order of increasing abscissa) since the predecessor and successor of $\vec{v}$ must both have their $vecR$ pointers updated.

Concerning the complexity of this algorithm: let us denote by $up(\vec{v}, R)$ the number of instructions required to update $vecR(\vec{v})$, $testP(\vec{v})$ and $testS(\vec{v})$

Algorithm 2. Comp_T

Input: a positive integer R_{max}
Output: the test mask $T(R_{max})$, with the appearance radii of each vector

1 $T \leftarrow \{((1,0),1),((1,1),2)\}$;
2 $outV \leftarrow Vector(2,0)$;
3 **while** $\|outV\|^2 \leqslant R_{max}$ **do**
4 $\quad R \leftarrow \|outV\|^2$;
5 $\quad$ **while** $\|outV\|^2 = R$ **do**
6 $\quad\quad$ insert_next($outV$) ;
7 $\quad\quad outV \leftarrow$ next($outV$) ;

8 $\quad$ **foreach** *visible vector v in B_R, in order of increasing norm* **do**
9 $\quad\quad$ **while** $\|v + \text{vecR}(v)\|^2 \leqslant R$ **do**   /* update of $vecR(v)$ */
10 $\quad\quad\quad$ vecR(v) $\leftarrow$ next(vecR(v)) ;
11 $\quad\quad$ **if** $v \notin T$ **then**   /* update of $testP(v)$ and $testS(v)$ */
12 $\quad\quad\quad u \leftarrow$ Farey predecessor of v ;
13 $\quad\quad\quad w \leftarrow$ Farey successor of v ;
14 $\quad\quad\quad$ **while** $\|w + \text{testP}(v)\| < \|\text{vecR}(u)\|$ **do**
15 $\quad\quad\quad\quad$ testP(v) $\leftarrow$ next(testP(v)) ;
16 $\quad\quad\quad$ **while** $\|u + \text{testS}(v)\| < \|\text{vecR}(w)\|$ **do**
17 $\quad\quad\quad\quad$ testS(v) $\leftarrow$ next(testS(v)) ;
18 $\quad\quad\quad$ **if** $\|\text{testP}(v)\| \leqslant \|\text{vecR}(v)\|$ **and** $\|\text{testS}(v)\| \leqslant \|\text{vecR}(v)\|$ **then**
19 $\quad\quad\quad\quad T \leftarrow T \cup \{(v,R)\}$;   /* tests if $v \in T(R)$ */

20 **return** T ;

---

from R to the smallest representable integer greater than R. For each radius R, the time required to scan the vectors in B_R is roughly $\sum_{\|\vec{v}\|^2=1}^{R} up(\vec{v}, R)$; the procedure insert_next is negligible since it only has to scan the boundary of B_R to insert new points. It follows that the global time complexity of the algorithm is roughly

$$\sum_{R=1}^{R_{max}} \sum_{\|\vec{v}\|^2=1}^{R} up(\vec{v}, R) \leqslant \sum_{R=1}^{R_{max}} \sum_{\|\vec{v}\|^2=1}^{R_{max}} up(\vec{v}, R) = \sum_{\|\vec{v}\|^2=1}^{R_{max}} \sum_{R=1}^{R_{max}} up(\vec{v}, R).$$

We now point out the fact that for any $\vec{v}$, the sum of the $up(\vec{v}, R)$ from $R = 1$ to R_{max} is $\mathcal{O}(R_{max})$, because the updates are done using a linked list. Hence, Alg. 2 runs in time

$$\mathcal{O}\left(\sum_{\|\vec{v}\|^2=1}^{R_{max}} R_{max} \right) = \mathcal{O}(R_{max}^2).$$

Besides, it is clear that this algorithm has space complexity linear in the number of points inside $B_{R_{max}}$, hence it runs in space $\mathcal{O}(R_{max})$.

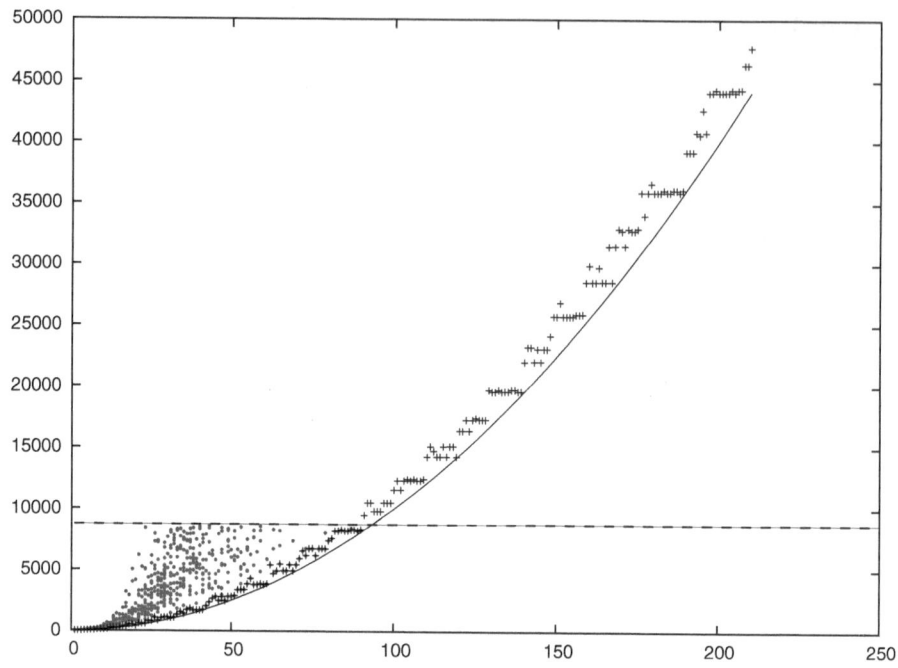

Fig. 11. Appearance radii of some visible vectors $\overrightarrow{v}$. Along the x-axis: abscissa v_1 of $\overrightarrow{v}$. Along the y-axis: $r_{app}(\overrightarrow{v})$, the Euclidean appearance radius of $\overrightarrow{v}$. The dots represent the 540 vectors of $\mathcal{T}(r = 8600)$; the crosses represent the vectors $(x, 1)$ for all $x \leqslant 210$. All the points we have found are above the function $y = x^2$.

6   Conclusion and Experiments

We have established a link between the Farey sequences and covering relations between Euclidean disks. Accordingly, we have proposed two algorithms: the first one computes the appearance radius of any visible vector $\overrightarrow{v}$ in time $\mathcal{O}\big(r^3_{app}(\overrightarrow{v})\big)$, the second one computes the test mask $\mathcal{T}(r)$ in time $\mathcal{O}(r^4)$.

The source code of our two algorithms is freely available in C language [13]. We illustrate outputs of these algorithms in Fig. 11. The dots in the bottom left corner have been computed by Alg. Comp_T; they represent the 540 vectors of $\mathcal{T}(r = 8600)$, which are necessary and sufficient to compute the medial axis of all the shapes that can be drawn within an image of size 17200×17200 pixels. In other words, we know all the vectors that are located below the dashed line (there is no vector in the bottom right corner). Prior to this, $\mathcal{T}(r)$ was known up to $r = 4800$ [10].

We have noticed that the vectors which are close to the x-axis, i.e., whose coordinates are $(x, 1)$ with $x \in \mathbb{N}_*$, have a small appearance radii compared to other vectors. So we used Alg. Comp_Rapp to compute the appearance radii of these vectors up to $x = 210$, see the crosses in Fig. 11. The results do not disprove the conjecture we expressed in [6], which suggests that the Euclidean appearance radius $r_{app}(\vec{v})$ of any visible vector $\vec{v} \in \mathbb{Z}^2$ is greater than the squared abscissa of $\vec{v}$. Moreover, these new data suggest that the bound is tight.

Also, we observed an interesting phenomenon: each time a vector $\vec{v}$ appears in some $\mathcal{T}(r)$, we noticed that either the predecessor or the successor of $\vec{v}$ had already appeared. Furthermore, the smallest of these two vectors had appeared. We therefore conjecture the following: if $\vec{v}$ belongs to some $\mathcal{T}(r)$, then the smallest vector among $\{\mathrm{pred}(\vec{v}), \mathrm{succ}(\vec{v})\}$ belongs to $\mathcal{T}(r)$.

Obviously, proving any of the above conjectures constitute a natural follow-up of this work, and would allow to speed up significantly the computation of the test mask.

References

1. Borgefors, G.: Centres of maximal disks in the 5-7-11 distance transform. In: 8th SCIA, Tromsø, pp. 105–111 (1993)
2. Borgefors, G., Ragnemalm, I., Sanniti di Baja, G.: The Euclidean DT, finding the local maxima and reconst. the shape. In: 7th SCIA, Aalborg, pp. 974–981 (1991)
3. Hardy, G.H., Wright, E.M.: An introduction to the theory of numbers, 5th edn. Oxford Science Pub. (1979)
4. Hirata, T.: A unified linear-time algorithm for computing distance maps. Information Processing letters 58(3), 12–133 (1996)
5. Hulin, J., Thiel, E.: Appearance Radii in Medial Axis Test Masks for Small Planar Chamfer Norms. In: Brlek, S., Reutenauer, C., Provençal, X. (eds.) DGCI 2009. LNCS, vol. 5810, pp. 434–445. Springer, Heidelberg (2009)
6. Hulin, J., Thiel, E.: Visible vectors and discrete Euclidean medial axis. Discrete and Computational Geometry (in press) (avail. online December 10, 2008)
7. Normand, N., Évenou, P.: Medial axis lookup table and test neighborhood computation for 3D chamfer norms. Pattern Recogn. 42(10), 2288–2296 (2009)
8. Normand, N., Évenou, P.: Medial axis LUT comp. for chamfer norms using H-polytopes. In: Coeurjolly, D., Sivignon, I., Tougne, L., Dupont, F. (eds.) DGCI 2008. LNCS, vol. 4992, pp. 189–200. Springer, Heidelberg (2008)
9. Pitteway, M.L.V.: Algorithm for Drawing Ellipses or Hyperbolae with a Digital Plotter. Computer J. 10(3), 282–289 (1967)
10. Rémy, E., Thiel, E.: Exact Medial Axis with Euclidean Distance. Image and Vision Computing 23(2), 167–175 (2005)
11. Rémy, E., Thiel, E.: Medial axis for chamfer distances: computing look-up tables and neighbourhoods in 2D or 3D. Pattern Recog. Letters 23(6), 649–661 (2002)
12. Siddiqi, K., Pizer, S.M. (eds.): Medial Representations. Comp. Imaging and Vision, vol. 37. Springer, Heidelberg (2008)
13. Source code of the algorithms:
    http://pageperso.lif.univ-mrs.fr/~hulin/iwcia09/
14. Thiel, E.: Géométrie des distances de chanfrein. Habilitation à Diriger les Rech. Univ. de la Méditerranée (December 2001),
    http://www.lif.univ-mrs.fr/~thiel/hdr/

Neighborhood Sequences on nD Hexagonal/Face-Centered-Cubic Grids

Benedek Nagy[1] and Robin Strand[2]

[1] Department of Computer Science, Faculty of Informatics,
University of Debrecen, Debrecen, Hungary
nbenedek@inf.unideb.hu
[2] Centre for Image Analysis, Uppsala University, Sweden
robin@cb.uu.se

Abstract. The two-dimensional hexagonal grid and the three-dimensional face-centered cubic grid can be described by intersecting $\mathbb{Z}^3$ and $\mathbb{Z}^4$ with a (hyper)plane. Corresponding grids in higher dimensions (nD) are examined. In this paper, we define distance functions based on neighborhood sequences on these, higher dimensional generalizations of the hexagonal grid. An algorithm to produce a shortest path based on neighborhood sequences between any two gridpoints is presented. A formula to compute distance and condition of metricity are presented for neighborhood sequences using two types of neighbors. Distance transform as an application of these distances is also shown.

Keywords: Digital geometry, nD grids, Neighborhood sequences.

1 Introduction

It is well-known that the hexagonal grid has several advantages over the two-dimensional Cartesian grid, the square grid. Most importantly, since the hexagonal grid is a densest packing, fewer samples are needed to represent a two-dimensional signal (an image) when it is sampled on the hexagonal grid. Also, since there is a larger number of closest neighbors and these neighbors are such that the same adjacency relation can be used for object grid points and background grid points, and because its low rotational dependency the hexagonal grid is a good choice for two-dimensional image processing [7,9,10,26,29].

One argument for using Cartesian grids in image processing is that many functions defined on these grids are separable, which allows fast and efficient image processing. Consider, e.g., the important discrete Fourier transform. By using redundancy of the matrices used for computing the Fourier transform and the separability on the square grid, a fast Fourier transform is derived, see, e.g., [4], where the well-known Cooley-Tukey algorithm is presented. The so-obtained algorithm has computation time $O\left(N \log_2 N\right)$, where $N = 2^k$ for some k. By utilizing the 6-fold symmetry of the hexagonal grid and applying a vector-radix Cooley-Tukey algorithm, the discrete Fourier transform can be computed in $O\left(N \log_7 N\right)$, where $N = 7^k$ for some k, on the hexagonal grid [10]. In fact, the

P. Wiederhold and R.P. Barneva (Eds.): IWCIA 2009, LNCS 5852, pp. 96–108, 2009.

number of operations needed to compute the Fourier transform on the hexagonal grid is roughly 60%(!) fewer on the hexagonal grid compared to the square grid, [10].

The densest packing in three dimensions is the face-centered cubic (fcc) grid. The fcc grid has many of the advantages over the three-dimensional Cartesian grid, the cubic grid, see [3,6,21,25].

In this paper, we will consider n-dimensional generalizations of the hexagonal and fcc grids [3,8,12,14,17,19,24]. In [3,6,7], the hexagonal and fcc grids are obtained by intersecting Cartesian grids by a plane/hyperplane. For example, the hexagonal grid is obtained by intersecting $\mathbb{Z}^3$ with the plane $x + y + z = 0$. In this paper, we consider $(n-1)$-dimensional grids obtained by intersecting $\mathbb{Z}^n$ with a hyperplane. This grid is called A_{n-1} in [3]. We will present some digital distance functions on these grids. The distance between two points is defined as the shortest path using local steps corresponding to neighboring grid points.

In digital images, distance functions defined as the minimal cost-path are often considered [20]. There are basically two approaches – either weights are used between neighboring grid points [1] or a neighborhood sequence is used to define which adjacency relation is allowed in each step of the path [13,15,18,27,28]. It is also possible to combine the two approaches to get even lower rotational dependency [23,24,28].

Distance functions have also been defined for high dimensional and non-Cartesian grids, see e.g. [5,11,16,22].

To handle non-Cartesian grids efficiently in a computer, one needs a good coordinate system. The points of the hexagonal grid can be addressed by two integers [9]. There is a more elegant solution using three coordinate values with zero sum reflecting the symmetry of the grid [7]. Opposed to [24], where a general theory for distance functions is presented, we will use the zero-sum coordinates in this paper. Since integer coordinates are used with this approach, clear and elegant expressions are obtained.

In this paper we deal with three-dimensional grids and a family of grids in higher dimensions. Since the fcc grid is the densest packing in three dimensions, it appears frequently in nature. For example, the fcc grid is the structure of, e.g., gold (Au), silver (Ag), calcium (Ca) and aluminium (Al).

In the figures, each grid point is sometimes illustrated by the corresponding picture element (pixel in 2D and voxel in 3D). The picture element of a given grid point is the Voronoi region of the grid point. In Figure 1, the Voronoi regions for the fcc grid is shown.

2   Grids Described by Intersecting Cartesian Grids

In [7], the hexagonal grid is described as the intersection of $\mathbb{Z}^3$ and the plane $x + y + z = 0$. The so-obtained coordinates has some advantages. Most importantly, the points in the hexagonal grid are addressed by integer coordinates, which fits the digital geometry framework. In Section 3, we will define distance based on neighborhood sequences and for this, a convenient way to describe

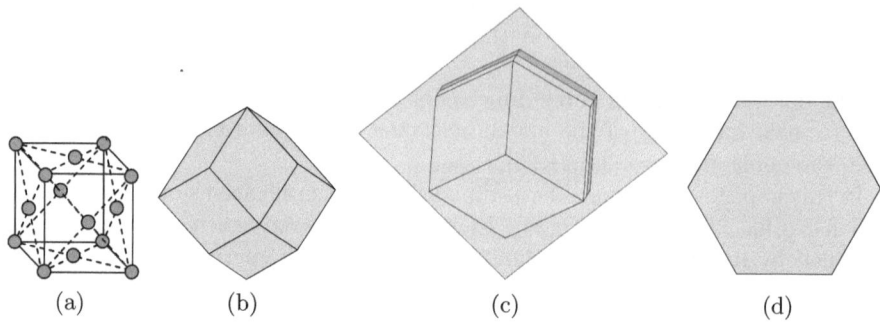

Fig. 1. A unit cell of the fcc grid is shown in (a). Voronoi region (voxels) in an fcc grid (b) and a hexagonal grid (d). In (c), the fcc voxels is intersected with the plane $x + y + z = 0$ (a planar patch is shown). In this intersection, the hexagonal grid is embedded.

neighboring points is needed. We will see that when describing the hexagonal grid as an intersecting plane, there is a natural way to describe the neighboring grid points. One can use a symmetric description with three values by adding a third coordinate value to have zero-sum vectors. The three values are dependent, but the description (including the neighborhood structure) is more simple and more elegant.

In [6], the four-dimensional Cartesian grid $\mathbb{Z}^4$ was intersected with the hyperplane $x + y + z + w = 0$ to obtain the fcc grid. In [3], some properties of the grid obtained by intersecting $\mathbb{Z}^n$ with a hyperplane with zero sum for any n are presented. Here, we will consider this generalization and define distance functions on these grids.

Other high-dimensional grids can be obtained by considering not only intersections of $\mathbb{Z}^n$ with zero-sum hyperplanes, but also other hyperplanes. In [12,14], it is noted that the two-dimensional hexagonal and triangular grids can be obtained by the union of integer points of some hyperplanes in three dimensions. In [17,19], we showed that some well-known three-dimensional grids can be obtained in a similar manner by unions of hyperplanes intersecting $\mathbb{Z}^4$.

In geometry, the term (point) lattice is frequently used to describe regularly spaced arrays of points. A lattice can be defined by a finite set of linearly independent vectors over $\mathbb{Z}$ ($\{v_1, ..., v_n\}$, basis) as $\left\{ \sum_{i=1}^{n} a_i v_i \mid a_i \in \mathbb{Z} \right\}$. A lattice may be viewed as a regular tiling of a space by a primitive cell. The Bravais lattices are those kinds of lattices used in crystallography and solid state physics.

Further, in this section, we recall some results from [14,12] and introduce our definitions and notations.

In $\mathbb{Z}^n$, we have the following natural neighborhood relations:

Definition 1. *Let* $\mathbf{p} = (p(1), p(2), \ldots, p(n))$ *and* $\mathbf{q} = (q(1), q(2), \ldots, q(n))$ *be two points in* $\mathbb{Z}^n$. *For any integer* k *such that* $0 \leq k \leq n$, $\mathbf{p}$ *and* $\mathbf{q}$ *are* $\mathbb{Z}^n$ k-*neighbors if* $|p(i) - q(i)| \leq 1$ *for* $1 \leq i \leq n$ *and* $\sum_{i=1}^{n} |p(i) - q(i)| \leq k$.

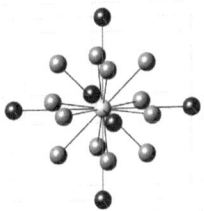

Fig. 2. Neighbors up to order two are shown. The grid points represent an fcc grid. The figure show eight unit-cell size of the fcc grid.

When constructing the grids, we will consider intersections between $\mathbb{Z}^k$ and hyperplanes with normal direction $(1, 1, \ldots, 1)$. Let $\mathcal{Q}^k = \{(x_1, x_2, \ldots, x_k) \in \mathbb{Z}^k : x_1 + x_2 + \cdots + x_k = 0\}$.

Let us start with the cubic grid $\mathbb{Z}^3$. Three natural neighborhood relations on this grid are given by Definition 1. It is well-known that the grid points of $\mathbb{Z}^3$ on a plane with $x + y + z = 0$ form a hexagonal grid [7]. The neighbor relation of the hexagonal grid has only one type of natural neighborhood, the one that corresponds to the grid points in the 2-neighborhood of $\mathbb{Z}^3$ satisfying $x + y + z = 0$.

Since $\mathcal{Q}^3$ is the hexagonal grid and $\mathcal{Q}^4$ is the face-centered-cubic grid, the grids defined by $\mathcal{Q}^n$ are the generalizations of them. Therefore we may use the terms nD *hexagonal grids* and/or nD *fcc grids* for $\mathcal{Q}^{n+1}$.

In Figure 2 $4 \times 4 \times 4$ size part of the fcc grid is shown, where every gridpoint has integer coordinate values.

3   Distances Based on Neighborhood Sequences on the Grids with a Projected Hyperplane in Higher Dimensions

In the first part of this section we described the n-dimensional generalizations of the previously recalled hexagonal and face-centered-cubic grids. Now we will define and analyze digital distances based on some neighborhood sequences.

We can define some types of neighborhood structures on these grids generalizing the concept from the previous lower dimensional grids. Let $\mathbf{p} = (p(1), p(2), \ldots, p(n)), \mathbf{q} = (q(1), q(2), \ldots, q(n)) \in \mathcal{Q}^n$ be two points. Then their difference vector: $\mathbf{w} = \mathbf{p} - \mathbf{q}$ is defined as $(p(1) - q(1), p(2) - q(2), \ldots, p(n) - q(n))$. Two points having difference vector with values only from the set $\{-1, 0, 1\}$ can be considered as neighbors. For instance the difference vector $(1, -1, 0, \ldots, 0)$ corresponds to a closest neighbor. The neighbor points with four ± 1 elements in their difference vector are second closest neighbors point-pair. We will define distances based on neighborhood sequences in higher dimensions using these two neighborhood relations. In this way we generalize the distances of the fcc grid to higher dimension. We may also define a more extended neighborhood structure depending on the dimension n of the space. In this way a further generalization of our distances can be obtained. The neighbor points with difference vector with $2k$ elements of ± 1 are defined as strict k-neighbors. Therefore, the next formal definition of the neighborhood structure from [19] is natural.

Definition 2. *Let* $\mathbf{p} = (p(1), ..., p(n))$ *and* $\mathbf{q} = (q(1), ..., q(n))$ *two points of* $\mathcal{Q}^n$. *Then* $\mathbf{p}$ *and* $\mathbf{q}$ *are* $\mathcal{Q}^n$ *k-neighbors if* $\sum |p(i) - q(i)| \leq 2k$ *and* $|p(i) - q(i)| \leq 1$ *for every* $0 < i \leq n$.

We remark that there are $\lfloor \frac{n}{2} \rfloor$ types of neighbors in $\mathcal{Q}^n$, where the floor function is used. If there is an equality in the first condition, then the points are *strict k-neighbors*. In special cases $n = 3, 4$ the hexagonal and the fcc grids are obtained by their natural neighborhood structure (only one type of neighborhood in the hexagonal grid and two types of neighbor relations on the fcc grid).

In the next subsection we define formally the distances based on neighborhood sequences on these higher dimensional grids with the general neighborhood structure. After this we analyze the distances in detail, specially the ones using only the two closest neighborhood in neighborhood sequences.

3.1   Neighborhood Sequences on $\mathcal{Q}^n$

Let $\mathbf{p}$ and $\mathbf{q}$ be two points of $\mathcal{Q}^n$. The sequence $B = (b(i))_{i=1}^{\infty}$, where the values of $b(i)$ are possible neighborhood relations of the nD hexagonal grid ($0 < b(i) \leq \lfloor \frac{n}{2} \rfloor$) for all $i \in \mathbb{N}$, is called a neighborhood sequence (on the nD hexagonal grid).

A movement is called a $b(i)$-step when we move from a point $\mathbf{p}$ to a point $\mathbf{q}$ and they are $b(i)$-neighbors. Let $\mathbf{p}, \mathbf{q}$ be two points and B be a neighborhood sequence. The point-sequence $\mathbf{p} = \mathbf{p}_0, \mathbf{p}_1, \ldots, \mathbf{p}_m = \mathbf{q}$, in which we move from $\mathbf{p}_{i-1}$ to $\mathbf{p}_i$ by a $b(i)$-step ($1 \leq i \leq m$), is called a B-path from $\mathbf{p}$ to $\mathbf{q}$. The length of this path is the number of its steps, i.e., m. The B-distance $d(\mathbf{p}, \mathbf{q}; B)$ from $\mathbf{p}$ to $\mathbf{q}$ is defined as the length of the shortest B-path(s) between them. As usual in digital geometry, the shortest path may not be unique.

In the next subsection we give an algorithm that produces a shortest path between any two points $\mathbf{p}$ and $\mathbf{q}$ of $\mathcal{Q}^n$ using a given neighborhood sequence B. Then we also give a formula to compute these path-based distances when only 1's and 2's are allowed in the neighborhood sequence, furthermore some properties of these distances will also be analyzed.

3.2   Algorithm for Producing a Shortest Path

The next algorithm provides a shortest path from $\mathbf{p}$ to $\mathbf{q}$ using an arbitrary neighborhood sequence B.

Algorithm 1 is a greedy algorithm working in a linear time of the sum of the coordinate differences of the points.

3.3   Formula for B-Distance

In this section we give a formula to compute the B-distance of any two points of the nD fcc grid, when B has elements only from the set $\{1, 2\}$.

Algorithm 1. Producing a shortest B-path Π in $\mathcal{Q}^n$ and computing the B-distance

---

Input: $\mathbf{p} = (p(1), ..., p(n))$, $\mathbf{q} = (q(1), ..., q(n))$ and the neighborhood sequence $B = (b(i))_{i=1}^{\infty}$.

Output: A shortest B-path from $\mathbf{p}$ to $\mathbf{q}$ and the B-distance.

Initialization: Set $i = 0$, $\mathbf{p_0} = \mathbf{p}$. Let Π contain the start point $\mathbf{p}$.

while $\mathbf{p}_i \neq \mathbf{q}$ **do**

    Let $i = i + 1$ and $\mathbf{p}_i = \mathbf{p}_{i-1}$;

    Let j_p and j_n be the number of positive and negative values among $q(h) - p_i(h)$, respectively;

    Let $j = \min\{j_p, j_n, b(i)\}$;

    Let the values $q(h) - p_i(h)$ are permuted in a monotonous order and let us modify the first and last j values:

        (a) Let $p_i(h) = p_i(h) + 1$ if h is a coordinate among the j largest positive values of the set of $q(h) - p_i(h)$;

        (b) Let $p_i(h) = p_i(h) - 1$ if h is a coordinate among the j smallest (negative) values of the set of $q(h) - p_i(h)$;

    Concatenate the point $\mathbf{p}_i$ to the path Π.

end

The result is Π as a shortest B-path and i is the B-distance of $\mathbf{p}$ and $\mathbf{q}$.

Theorem 1. *The B-distance of the points $\mathbf{p}$ and $\mathbf{q}$ in $\mathcal{Q}^n$ can be computed as*

$$d(\mathbf{p}, \mathbf{q}; B) = \max_{i=1}^{n} \left\{ |p(i) - q(i)|, k \left| \frac{\sum_{i=1}^{n} |p(i) - q(i)|}{2} > \sum_{j=1}^{k-1} b(j) \right. \right\}$$

with a neighborhood sequence B containing only elements of the set $\{1, 2\}$.

Proof. Since a coordinate can be changed by at most one in a step in a B-path, $d(\mathbf{p}, \mathbf{q}; B) \geq \max_{i=1}^{n} \{|p(i) - q(i)|\}$. On the other side, in a step at most $2b(i)$ values can be changed, therefore $d(\mathbf{p}, \mathbf{q}; B) \geq \max \left\{ k \left| \sum_{i=1}^{n} |p(i) - q(i)| > \sum_{j=1}^{k-1} 2b(j) \right. \right\}$.

By technical calculation it can be proven that the maximum of these two values gives the actual B-distance. $\square$

Remark 1. Observe that the formula holds for the hexagonal grid ($n = 3$ and, since there is only one type of neighborhood relation, $b(i) = 1, i \in \mathbb{N}$). In this case $d(\mathbf{p}, \mathbf{q}) = \max_{i=1,2,3} \{|p(i) - q(i)|\} = \frac{\sum_{h=1}^{3} |p(h) - q(h)|}{2}$ [9,11].

The formula holds for the fcc grid, i.e., the formula presented in [22,16] is equivalent to our:

Proposition 1. *The formula* $d(\mathbf{p}, \mathbf{q}; B) =$

$$= \min\left\{ k \in \mathbb{N} \,\middle|\, k \geq \max\left\{ \frac{\sum_{i=1}^{3} |p(i) - q(i)|}{2}, \max_{i=1,2,3}\left\{|p(i) - q(i)| - 2^k\right\} \right\} \right\}$$

for the fcc grid with $2^k = |\{i : b(i) = 2, 1 \leq i \leq k\}|$
is equivalent to our formula with $n = 4$.

Proof. Let the point (x, y, z) be a point in the fcc grid, i.e., such that $x + y + z$ is even. The linear transformation

$$a = (x + y - z)/2$$
$$b = (x - y + z)/2$$
$$c = (-x + y + z)/2$$
$$d = -(x + y + z)/2$$

is orthogonal, see [19]. Also, $a + b + c + d = 0$, so $(a, b, c, d) \in \mathcal{Q}^4$, so it is nothing but a mapping between the different representations of the fcc grid.
(i) First we note that

$$\max\{|a|, |b|, |c|, |d|\} = \max\left\{ \frac{|x+y-z|}{2}, \frac{|x-y+z|}{2}, \frac{|-x+y+z|}{2}, \frac{|x+y+z|}{2} \right\}$$
$$= \frac{|x| + |y| + |z|}{2}. \tag{1}$$

(ii) Assume that $x \geq y \geq z \geq 0$ (the other cases follow by symmetry).
(a) $x \geq y + z$ In this case, $|a| + |b| + |c| + |d| = 2x = 2\max\{x, y, z\}$. We have

$$\max\left\{ k \,\middle|\, \frac{|a| + |b| + |c| + |d|}{2} > \sum_{j=1}^{k-1} 2b(j) \right\}$$

$$= \min\left\{ k \,\middle|\, \frac{|a| + |b| + |c| + |d|}{2} \leq \sum_{j=1}^{k} 2b(j) \right\}$$

$$= \min\left\{ k \,\middle|\, x \leq k + 2^k \right\}$$

$$= \min\left\{ k \,\middle|\, k \geq \max\{x, y, z\} - 2^k \right\}.$$

(b) $x < y + z$ This case gives $|a| + |b| + |c| + |d| = x - y - z$, which implies that $\frac{|x|+|y|+|z|}{2} > x$.

Now we sum up our formulas and get

$$d(\mathbf{0}, (a, b, c, d); B) = \max \left\{ |a|, |b|, |c|, |d|, k \left| \frac{|a| + |b| + |c| + |d|}{2} > \sum_{j=1}^{k-1} b(j) \right. \right\}$$

$$= \max \left\{ \frac{|x| + |y| + |z|}{2}, \min \left\{ k \, \left| \, k \geq \max\{x, y, z\} - 2^k \right. \right\} \right\}$$

$$= \min \left\{ k \, \left| \, k \geq \max \left\{ \frac{|x| + |y| + |z|}{2}, \max\{x, y, z\} - 2^k \right. \right\} \right\},$$

which, with $\mathbf{p} - \mathbf{q} = (x, y, z)$ is the formula in the proposition. $\square$

When the extended neighborhood is used in $\mathcal{Q}^n$ ($n \geq 6$) allowing not only 1's and 2's in B, we have the following conjecture.

Conjecture 1. The B-distance of $\mathbf{p}$ and $\mathbf{q}$ in $\mathcal{Q}^n$ can be determined in the following way. Let $W_+ = \{i | q(i) - p(i) > 0\}$ and $W_- = \{i | q(i) - p(i) < 0\}$ be two sets of coordinates. Then W_+ and W_- give two subspaces of $\mathbb{Z}^n$. Let $\mathbf{p}_+$ and $\mathbf{q}_+$ the images of $\mathbf{p}$ and $\mathbf{q}$ in the subspace defined by W_+, and similarly $\mathbf{p}_-$ and $\mathbf{q}_-$ in the subspace defined by W_-. Let $d_+ = d(\mathbf{p}_+, \mathbf{q}_+; B)$ in the subspace of $\mathbb{Z}^n$ defined by W_+, and similarly, $d_- = d(\mathbf{p}_-, \mathbf{q}_-; B)$ using the corresponding formula for $\mathbb{Z}^n$ from [15,18]. Then $d(\mathbf{p}, \mathbf{q}; B) = \max\{d_+, d_-\}$ give the B-distance in the nD fcc grid.

When only the two closest neighbor relations can be used in B the Conjecture 1 coincides with Theorem 1.

3.4   Metrical Properties

A distance function is called a metric, if it satisfies the following three properties:
(1) positive definiteness: $d(\mathbf{p}, \mathbf{q}) \geq 0$ for any point-pair $(\mathbf{p}, \mathbf{q})$ and $d(\mathbf{p}, \mathbf{q}) = 0$, if and only if $\mathbf{p} = \mathbf{q}$;
(2) symmetry: $d(\mathbf{p}, \mathbf{q}) = d(\mathbf{q}, \mathbf{p})$ for any point-pair $(\mathbf{p}, \mathbf{q})$;
(3) triangular inequality: $d(\mathbf{p}, \mathbf{q}) + d(\mathbf{q}, \mathbf{r}) \geq d(\mathbf{p}, \mathbf{r})$ for any point-triplet $(\mathbf{p}, \mathbf{q}, \mathbf{r})$.
In this section we deal mostly with neighborhood sequences using only the closest two neighborhoods.

Definition 3. *For any neighborhood sequence* $B = (b(i))_{i=1}^{\infty}$, *the sequence* $B(j) = (b(i))_{i=j}^{\infty}$ *is the j-shifted sequence of B.*

We say, that a neighborhood sequence B_1 is faster than B_2 if the B_1-distance is not larger than the B_2-distance for any point-pairs. Based on Theorem 1 one can prove the following statement.

Proposition 2. *For the nD-fcc grids, a neighborhood sequence B_1 is faster than B_2 if and only if*

$$\sum_{i=1}^{j} b_1(i) \geq \sum_{i=1}^{j} b_2(i) \quad \text{for all } j \in \mathbb{N}.$$

This can be written in the following equivalent condition: $2_{B_1}^j \geq 2_{B_2}^j$ *for all* $j \in \mathbb{N}$ *(with the notation* $2^k = |\{i : b(i) = 2, 1 \leq i \leq k\}|).$

Theorem 2. *The B-distance (using only the closest two neighborhoods) on $\mathcal{Q}^n$ is a metric if and only if B has the following property: $B(i)$ is faster than B for all $i \in \mathbb{N}$.*

Proof. Since the first two properties of the metricity (positive definiteness and symmetry) are automatically fulfilled, the triangular inequality is crucial and can technically be proven using Theorem 1. □

Proposition 2 gives a computationally efficient way of deciding if a B-distance is a metric or not. For instance, the periodic neighborhood sequence repeating $(1, 1, 2, 1, 2)$ generates a metric.

Allowing extended neighborhood, i.e., more neighbors than the closest two in B we have the following conjecture.

Conjecture 2. The B-distance is metrical on $\mathcal{Q}^n$ if it is metrical on $\mathbb{Z}^{2n}$.

If Conjecture 1 holds, then the proof of Conjecture 2 goes in the same manner as it works in $\mathbb{Z}^n$, see [13,15].

4   Distance Transform

In this section we apply the digital distances defined above in an image processing algorithm, namely to produce distance transform.

Algorithm 2. Computing the distance transform DT for distances based on neighborhood sequences by wave-front propagation on $\mathcal{Q}^n$

Input: The neighborhood sequence B and an object $X \subset \mathcal{Q}^n$.
Output: The distance transforms DT.
Initialization: Set $DT(\mathbf{p}) \leftarrow 0$ for grid points $\mathbf{p} \in \overline{X}$ and $DT(\mathbf{p}) \leftarrow \infty$ for grid points $\mathbf{p} \in X$. For all grid points $\mathbf{p} \in \overline{X}$ adjacent to X: push $(\mathbf{p}, DT(\mathbf{p}))$ to the list L of ordered pairs sorted by increasing $DT(\mathbf{p})$.
while L *is not empty* **do**
    foreach $\mathbf{p}$ *in L with smallest $DT(\mathbf{p})$* **do**
        Pop $(\mathbf{p}, DT(\mathbf{p}))$ from L;
        foreach $\mathbf{q}$: $\mathbf{q}, \mathbf{p}$ *are $b(DT(\mathbf{p})+1)$-neighbors* **do**
            if $DT(\mathbf{q}) > DT(\mathbf{p}) + 1$ **then**
                $DT(\mathbf{q}) \leftarrow DT(\mathbf{p}) + 1$;
                Push $(\mathbf{q}, DT(\mathbf{q}))$ to L;
            end
        end
    end
end

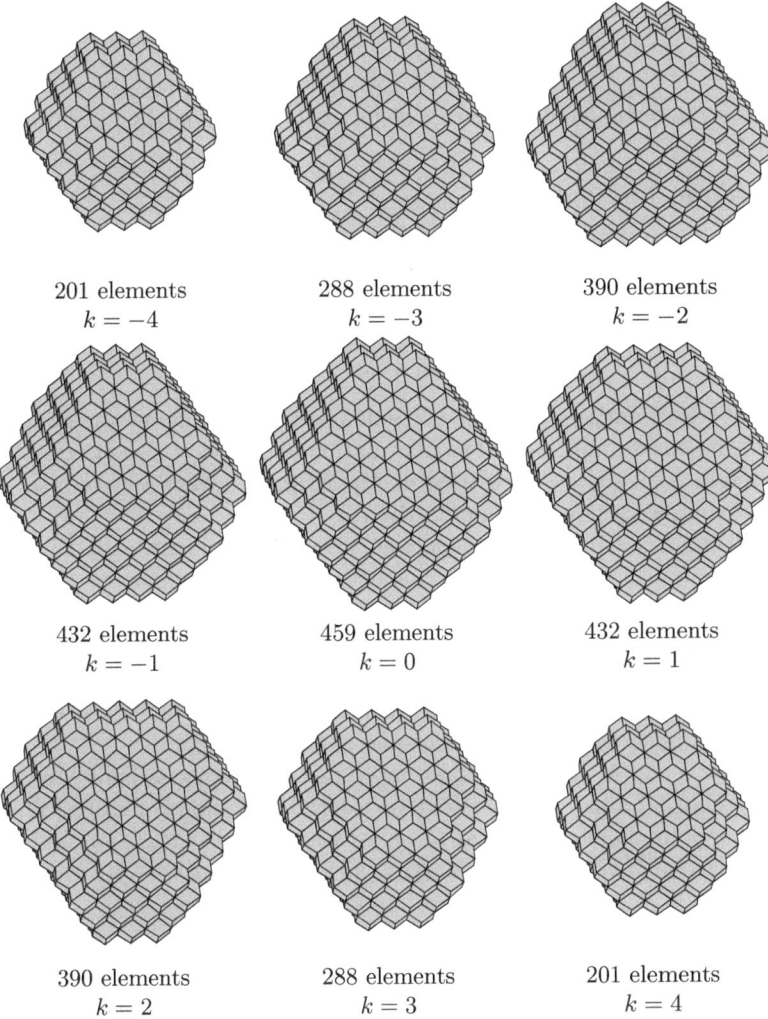

201 elements	288 elements	390 elements
$k = -4$	$k = -3$	$k = -2$
432 elements	459 elements	432 elements
$k = -1$	$k = 0$	$k = 1$
390 elements	288 elements	201 elements
$k = 2$	$k = 3$	$k = 4$

Fig. 3. A ball of radius 4 in the four-dimensional grid $\mathcal{Q}^5$ generated by $B = (1, 2)$ is illustrated by showing the subsets $\{\mathbf{p} : p(1) + p(2) + p(3) + p(4) + p(5) = 0, p(5) = k\}$ for some values of k. Each obtained grid point is shown as a rhombic dodecahedron, since for each k, an fcc grid is obtained.

Definition 4 (Image). *The image domain is a finite subset of $\mathcal{Q}^n$ denoted $\mathcal{I}$. We call the function $F : \mathcal{I} \longrightarrow \mathbb{R}_0^+$ an image.*

Note that real numbers are allowed in the range of F.

Definition 5 (Object and background). *We denote the object X and the background $\overline{X}$. These sets have the following properties:*

1. $X \subset \mathcal{I}$ and $\overline{X} \subset \mathcal{I}$
2. $X \cap \overline{X} = \emptyset$
3. $X \cup \overline{X} = \mathcal{I}$.

We denote the distance transform for distances based on neighborhood sequences with DT.

Definition 6. *The* distance transform DT *of an object* $X \subset \mathcal{I}$ *is the mapping*

$$DT : \mathcal{I} \to \mathbb{R} \text{ defined by}$$
$$\mathbf{p} \mapsto d\left(\mathbf{p}, \overline{X}; B\right), \text{ where}$$
$$d\left(\mathbf{p}, \overline{X}; B\right) = \min_{\mathbf{q} \in \overline{X}} \{d\left(\mathbf{p}, \mathbf{q}; B\right)\}.$$

Note that the distance transform DT holds information about the minimal distance to the background as well as information about what size of neighborhood is allowed in each step. This is used in Algorithm 2, where a wave-front propagation technique is used to compute the DT.

Algorithm 2 can be used to generate balls in $\mathcal{Q}^n$ for any n by thresholding the distance transform, where only a single grid point is in the background.

4.1   Digital Balls

One of the easiest and most usual ways to analyze a distance function d is to analyze the digital balls, i.e., (hyper)spheres based on d.

In Figure 3 and 4, balls of radius 4 are shown. In Figure 4, balls generated by some neighborhood sequences in $\mathcal{Q}^4$ (the fcc grid) are shown. Figure 3 illustrates balls in the four-dimensional grid $\mathcal{Q}^5$ generated by $B = (1, 2)$. To illustrate the four-dimensional ball, the subsets $\{\mathbf{p} : p(1) + p(2) + p(3) + p(4) + p(5) = 0, p(5) = k\}$ for $k = -4 \ldots 4$ are shown.

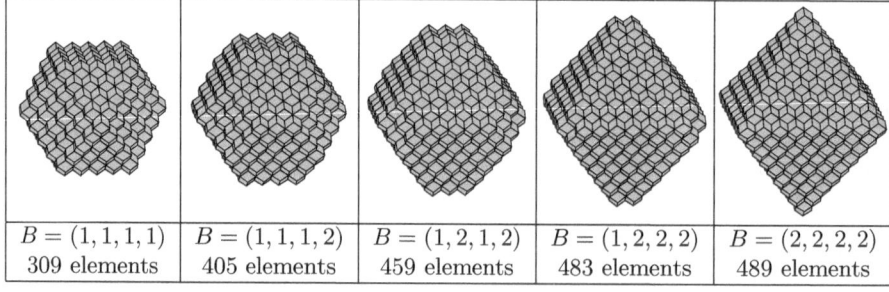

$B = (1,1,1,1)$	$B = (1,1,1,2)$	$B = (1,2,1,2)$	$B = (1,2,2,2)$	$B = (2,2,2,2)$
309 elements	405 elements	459 elements	483 elements	489 elements

Fig. 4. Balls of radius 4 in $\mathcal{Q}^3$ (the fcc grid) obtained by some neighborhood sequences

5   Conclusions and Future Work

Non-traditional grids are used in image processing and computer graphics. Some results on the hexagonal grids can be found in [2,26,29]. Non-standard grids are also used in higher dimensions [8,21]. The hexagonal grid and the fcc grid are the most dense packing with the highest kissing number in 2D and 3D, respectively. The dual of the fcc grid gives a thinnest covering in 3D [3]. Their extensions, that were analysed in this paper, are also important grids. The dual of the nD fcc grids is a thinnest covering in 4D, 5D, 6D, 7D, 8D, 12D and 16D [3], therefore our work is not only for theoretic interest, but it is worth to consider some of these grids in applications.

Their advantage and the easy-to-handle property of the classical integer lattice $\mathbb{Z}^n$ can be used in a common frame by our method. Since there are various neighbors in the grid, one may use distances varying them along a path. In this way a large class of flexible digital distances are defined and used, namely distances based on neighborhood sequences. Some results as an algorithm to generate a shortest path between two points, and hence, compute their distance were presented for the general case. In a restricted case, using only the two closest neighborhood relations, further properties were presented, such as formula for the distance and conditions of metrical distances. Some problems left open using more extended neighborhood structure, we addressed some of them as conjectures, their proofs are topics of future work.

Our result can also be applied in digital geometry and in several applications when (digital) distances can/should be used. As an example the distance transform is presented.

References

1. Borgefors, G.: Distance transformations in digital images. Computer Vision, Graphics, and Image Processing 34(3), 344–371 (1986)
2. Brimkov, V.E., Barneva, R.R.: "Honeycomb" vs square and cubic models. Electronic Notes in Theoretical Computer Science 46 (2001)
3. Conway, J.H., Sloane, N.J.A., Bannai, E.: Sphere-packings, lattices, and groups. Springer, New York (1988)
4. Cooley, J.W., Lewis, P., Welch, P.: The fast Fourier transform and its applications. IEEE Trans. on Education 12(1), 28–34 (1969)
5. Fouard, C., Strand, R., Borgefors, G.: Weighted distance transforms generalized to modules and their computation on point lattices. Pattern Recognition 40(9), 2453–2474 (2007)
6. Her, I.: Description of the F.C.C. lattice geometry through a four-dimensional hypercube. Acta Cryst. A 51, 659–662 (1995)
7. Her, I.: Geometric transformations on the hexagonal grid. IEEE Trans. Image Processing 4(9), 1213–1222 (1995)
8. Ibanez, L., Hamitouche, C., Roux, C.: A vectorial algorithm for tracing discrete straight lines in N-dimensional generalized grids. IEEE Trans. Visualization and Computer Graphics 7, 97–108 (2001)

9. Luczak, E., Rosenfeld, A.: Distance on a hexagonal grid. Trans. Computers C-25(5) , 532–533 (1976)
10. Middleton, L., Sivaswamy, J.: Hexagonal image processing: A practical approach. Springer, New York (2005)
11. Nagy, B.: Shortest paths in triangular grids with neighbourhood sequences. Journal of Computing and Information Technology 11(2), 111–122 (2003)
12. Nagy, B.: A family of triangular grids in digital geometry. In: Guo, M. (ed.) ISPA 2003. LNCS, vol. 2745, pp. 101–106. Springer, Heidelberg (2003)
13. Nagy, B.: Distance functions based on neighbourhood sequences. Publicationes Mathematicae Debrecen 63(3), 483–493 (2003)
14. Nagy, B.: Generalized triangular grids in digital geometry. Acta Math. Acad. Paed. Nyíregyháziensis 20, 63–78 (2004)
15. Nagy, B.: Metric and non-metric distances on $\mathbb{Z}^n$ by generalized neighbourhood sequences. In: Proc. of the 4th ISPA, pp. 215–220 (2005)
16. Nagy, B., Strand, R.: Approximating Euclidean distance using distances based on neighbourhood sequences in non-standard three-dimensional grids. In: Reulke, R., Eckardt, U., Flach, B., Knauer, U., Polthier, K. (eds.) IWCIA 2006. LNCS, vol. 4040, pp. 89–100. Springer, Heidelberg (2006)
17. Nagy, B., Strand, R.: A connection between $\mathbb{Z}^n$ and generalized triangular grids. In: Bebis, G., Boyle, R., Parvin, B., Koracin, D., Remagnino, P., Porikli, F., Peters, J., Klosowski, J., Arns, L., Chun, Y.K., Rhyne, T.-M., Monroe, L. (eds.) ISVC 2008, Part II. LNCS, vol. 5359, pp. 1157–1166. Springer, Heidelberg (2008)
18. Nagy, B.: Distance with generalized neighbourhood sequences in nD and ∞D. Discrete Applied Mathematics 156, 2344–2351 (2008)
19. Nagy, B., Strand, R.: Non-traditional grids embedded in $\mathbb{Z}^N$. International Journal of Shape Modeling (in press, 2009)
20. Rosenfeld, A., Pfaltz, J.L.: Distance functions on digital pictures. Pattern Recognition 1, 33–61 (1968)
21. Strand, R., Stelldinger, P.: Topology preserving marching cubes-like algorithms on the face-centered cubic grid. In: Proc. ICIAP 2007, pp. 781–788 (2007)
22. Strand, R., Nagy, B.: Distances based on neighbourhood sequences in non-standard three-dimensional grids. Discrete Applied Mathematics 155(4), 548–557 (2007)
23. Strand, R.: Weighted distances based on neighbourhood sequences. Pattern Recognition Letters 28(15), 2029–2036 (2007)
24. Strand, R.: Distance functions and image processing on point-lattices: with focus on the 3D face- and body-centered cubic grids. PhD thesis, Uppsala University, Sweden (2008)
25. Takahashi, T., Yonekura, T.: Isosurface construction from a data set sampled on a face-centered-cubic lattice. In: Proc. of ICCVG 2002, vol. 2, pp. 754–763 (2002)
26. Wüthrich, C.A., Stucki, P.: An algorithm comparison between square- and hexagonal-based grids. Graph. Model Im. Proc. 53(4), 324–339 (1991)
27. Yamashita, M., Honda, N.: Distance functions defined by variable neighborhood sequences. Pattern Recognition 5(17), 509–513 (1984)
28. Yamashita, M., Ibaraki, T.: Distances defined by neighborhood sequences. Pattern Recognition 19(3), 237–246 (1986)
29. Yong-Kui, L.: The generation of straight lines on hexagonal grids. Comput. Graph. Forum 12(1), 21–25 (1993)

Neighborhood Sequences in the Diamond Grid – Algorithms with Four Neighbors

Benedek Nagy[1] and Robin Strand[2]

[1] Department of Computer Science, Faculty of Informatics, University of Debrecen,
PO Box 12, 4010 Debrecen, Hungary
nbenedek@inf.unideb.hu
[2] Centre for Image Analysis, Uppsala University,
Box 337, SE-75105 Uppsala, Sweden
robin@cb.uu.se

Abstract. In digital image processing digital distances are useful; distances based on neighborhood sequences are widely used. In this paper the diamond grid is considered, that is the three-dimensional grid of Carbon atoms in the diamond crystal. This grid can be described by four coordinate values using axes of the directions of atomic bonds. In this way the sum of the coordinate values can be either zero or one. An algorithm to compute a shortest path defined by a neighborhood sequence between any two points in the diamond grid is presented. The metric and non-metric properties of some distances based on neighborhood sequences are also discussed. The constrained distance transformation and digital balls obtained by some distance functions are presented.

Keywords: digital geometry, non-standard grids, neighborhood sequences, diamond grid, distance transform.

1 Introduction

In many applications (see [10]) it is worth to consider other grids than the square one. With processing power of computers and capabilities of graphics devices increasing rapidly, the time is ripe to consider using non-square sampling for image processing and computer vision and graphics. The hexagonal based models have some advantages compared to the square based models ([1,2,11,27]). According to [26], the theoretical study of non-square based structures is of interest as well, because it provides unified treatments of large classes of models for digital images. The 2D triangular grids can also be used [6,17]. Some well-known nontraditional grids in 3D are the face-centered cubic, body-centered cubic ([29,4,28]) and other Bravais lattices ([3]). The diamond grid is also an option for a compact description of the space. The digital geometry on the diamond grid was started in [19] using the two types of closest neighbors on the grid. These neighbors correspond to the covalent atomic bonds, and for the same-type of closest atoms in gallium-arsenide (it has also same grid structure as the diamond, silicone, zinc-blende, etc.), respectively. They are also the two

P. Wiederhold and R.P. Barneva (Eds.): IWCIA 2009, LNCS 5852, pp. 109–121, 2009.

types of face-neighbors of the Voronoi cells. In [22] the theory developed further, involving the third closest neighbors, that are the Voronoi corner-neighbors. (See Figure 1.) In this paper we allow to use also the fourth order neighbors. This is done by allowing steps to neighboring points at a distance equal to or less than the size of the unit-cell of the crystal.

In digital geometry points with integer coordinate values are used to describe the digital space (grid). In some cases it is more elegant and simple to use more coordinate values than the dimension of the grid. In this way the coordinates are not independent, but the description may reflect symmetry and therefore it is more easy to use. Her described the hexagonal (2D) grid with three coordinate values with zero sum ([7,8]). Similarly the 3D face-centered cubic grid is described by four values with zero sum in [9]. The hexagonal and the triangular grids are dual of each-other (in graph-theoretic sense), therefore the triangular grid can also be described by three coordinates [14,15,16]. Moreover the 3D diamond grid (i.e., the atomic structure of the diamond crystal) is described by four coordinate values that sum up to either zero or one [20,21]. In this paper we will use this description. Moreover by the four coordinate values it is natural to consider four types of neighbor relations on the grid.

A distance transform (DT) is obtained by assigning the distance from the closest background grid point to each object grid point. DTs are frequently used in image processing. In [24], the DT is defined using the classical city block and chessboard distances, defined as the shortest path between two grid points using only four and eight neighbors, respectively. One disadvantage with these distance functions is that the rotational dependency is high. Computing DTs using the city block and chessboard distances is efficient, see [24] for a sequential algorithm. In [25], it is noted that the rotational dependency can be reduced by mixing the city block and chessboard distances using a neighborhood sequence (ns). The literature on distances based on neighborhood sequences is rich; a theory for periodic neighborhood sequences not connected to any specific neighborhood relations in $\mathbb{Z}^n$ is presented in [30,31] and further developed for the natural neighborhood structure, by the so-called octagonal neighborhood sequences in [5]. Results for general (not necessarily periodic) neighborhood sequences are presented in [13].

Distances based on neighborhood sequences in the triangular grid are examined and analyzed in [12,17,18]. The generalizations of the hexagonal grid in three dimensions are the face-centered cubic (FCC) and body-centered cubic (BCC) grids, while the generalization of the triangular grid in three dimensions is the diamond grid, as mentioned in [20,21]. Distances based on neighborhood sequences in the FCC and BCC grids are analyzed and computed in [28].

In [19,22] distances based on neighborhood sequences in the diamond grid using the closest two and three neighbors are examined. In this paper we make a step forward: we describe the grid by four natural neighborhoods. The structure of the paper is the following. In the next section we fix our notation, recall and give some basic definitions and properties of the neighborhood structure of the diamond grid. In Section 3 we present an algorithm for shortest path between any two points of the grid. In Section 4 some interesting examples, such as some non-metric distances are shown. It is easy to find neighborhood sequences such that the triangular inequality and the symmetry are violated by these distances. In Section 5 the algorithm of the (constrained) distance transform is presented. Digital distances can be illustrated by generated digital balls, some of them are shown in Section 6. Finally some conclusions close the paper.

2   Notation and Definitions

The following definitions of the diamond grid (i.e., tetrahedral packing, as it is also used in crystallography and solid state physics) are used (based on [3] by a rescaling to integer coordinates):

$$\{(x, y, z) : x, y, z \in \mathbb{Z}, x \equiv y \equiv z \pmod 2 \text{ and } x + y + z \in \{0, 1\} \pmod 4\}.$$

Observe that each grid point has only integer coordinates in this way. The unit cell has size $4 \times 4 \times 4$ in this frame. This description is used in [19,22].

The diamond grid is a union of two FCC grids, and an FCC grid can be given by four zero-sum coordinate values ([9]). Based on this fact, now an alternative description of the diamond grid follows from [20,21]. In this paper we will use this new description.

Let us consider the points $p = (p(1), p(2), p(3), p(4))$ with $\sum_{i=1}^{4} p(i) \in \{0, 1\}$. These points represent a diamond grid. We call the points with zero-sum coordinates *even* and the points with 1-sum coordinate values *odd*. They are on two parallel hyperplanes of the 4D digital space $\mathbb{Z}^4$. One can consider the following neighborhood structure:

Definition 1. *Two points p and q are adjacent (neighbor) if*

$$|p(i) - q(i)| \leq 1 \text{ for every } i \in \{1, 2, 3, 4\}.$$

The neighbor points p and q are m-neighbors ($m \in \{1, 2, 3, 4\}$), if

$$\sum_{i=1}^{4} |p(i) - q(i)| \leq m.$$

In case of equality p and q are strict m-neighbors.

For any value of m the result is exactly the $\mathbb{Z}^4$ m-neighbors (in the four dimensional hypercubic grid) that are present in the hyperplanes with coordinate sum 0 and 1. Note that the coordinates of the grid points using this definition can be transformed into the coordinates of the grid points seen as a subset of $\mathbb{Z}^3$ and vice versa by using the transformation matrix in [20,21].

We consider path-generated distances, therefore, the terms 1-step, 2-step, 3-step and 4-step will also be used instead of step to a 1-neighbor, step to a 2-neighbor, step to a 3-neighbor and step to a 4-neighbor, respectively.

The concept of 1-neighbors are related to the covalent bonds of a crystal (diamond, silicon or GaAs). They are also quite natural when the grid points are represented by their Voronoi neighborhood: the 1-neighbors are face-neighbors. The 2-neighbors can also be introduced by the natural concept of Voronoi regions; they are bounded by two kinds of polygons: hexagons and triangles, see Figure 1(a). The 1-neighbors are connected through hexagons, while the strict 2-neighbors are through triangles, see Figure 1(b) and (c). The strict 2-neighbors are also the same type closest atoms in GaAs. The strict 3-neighbors are the Voronoi corner neighbors (in the same way as at the cubic grid), the two voxels share a vertex (Figure 1(d)). The strict 4-neighbors are at neighbor corner positions of the unit-cell (connected by a side of the cell). Strict 1 and 3-neighbors are on different hyperplanes; strict 2 and 4-neighbors are on the same hyperplane.

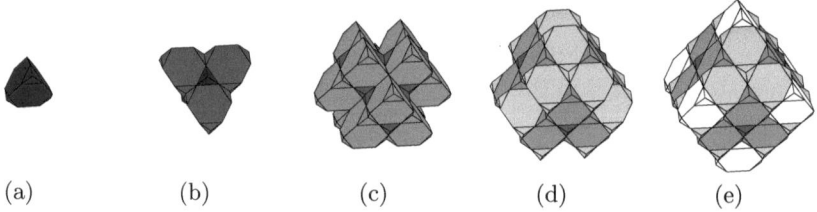

(a)   (b)   (c)   (d)   (e)

Fig. 1. A Voronoi voxel (a), its 1-neighbors (b), its 2-neighbors (c), its 3-neighbors (d), and its 4-neighbors (e)

The neighborhood relations are visualized in Figure 1 and in Figure 2. In Figure 1 the Voronoi region (the voxel) corresponding to a grid point and the regions corresponding to its neighbors can be seen. The strict neighbors are shown with new (lighter) color in each subfigure. In Figure 2 a part of the grid (connecting 1-neighbor points) and a point (an atom) with all its 1-neighbors, strict 2-neighbors, strict 3-neighbors and strict 4-neighbors are shown.

The diamond grid is not a *point-lattice*: the sum of the coordinate values of two odd points gives a point that is outside of the grid, therefore it can not be

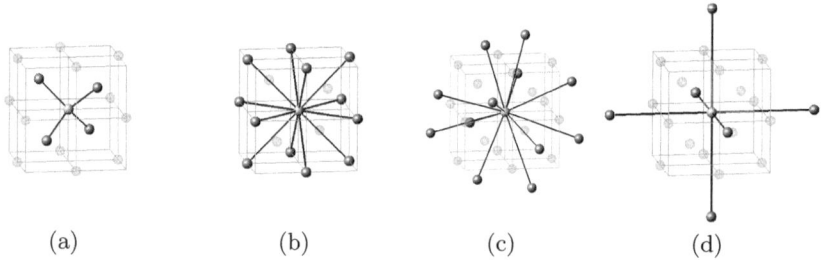

(a)   (b)   (c)   (d)

Fig. 2. A unit-cell of the diamond grid showing the 1-neighbors (a), strict 2-neighbors (b), strict 3-neighbors (c), and strict 4-neighbors (d) of the center points (atom)

represented by a set of basis vectors ([3]). It has two types of points (such as the two-dimensional triangular grid that is also not a point-lattice, but the theory of neighborhood sequences is developed for this grid as well, [12,17,18]). There are two local neighborhoods depending on the coordinate-sum (i.e., parity) of the points. They are reflection-symmetric.

Note that as we have already mentioned for crystal GaAs, both even and odd points form FCC-grids themselves. Since the diamond grid is not a point-lattice the positions of the neighbor points depend on the type of the point.

Remark 1. The strict 2-neighbors are exactly those points that can be reached by two consecutive 1-steps (if they are not exactly opposite to each other). We can say that a 2-step contains two 1-steps.

Remark 2. The strict 3 and 4-neighbors are exactly those points that can be reached by three/four consecutive 1-steps without steps in the same and in the opposite direction. Therefore the 3 and 4-neighbors form a special subclass of the points that can be reached by three/four consecutive 1-steps.

For instance, the points $o = (0,0,0,0)$ and $p = (1,-1,1,0)$ are 3-neighbors, but the point $r = (2,-1,0,0)$ is not a 3-neighbor of o even if, it can be reached from o by three consecutive 1-steps. This happens when two of the 1-steps go exactly to the same direction. (See Figure 3 (a) also.) Similarly, $o = (0,0,0,0)$ and $s = (1,-1,1,-1)$ are 4-neighbors. But the sequence of 1-neighbor points

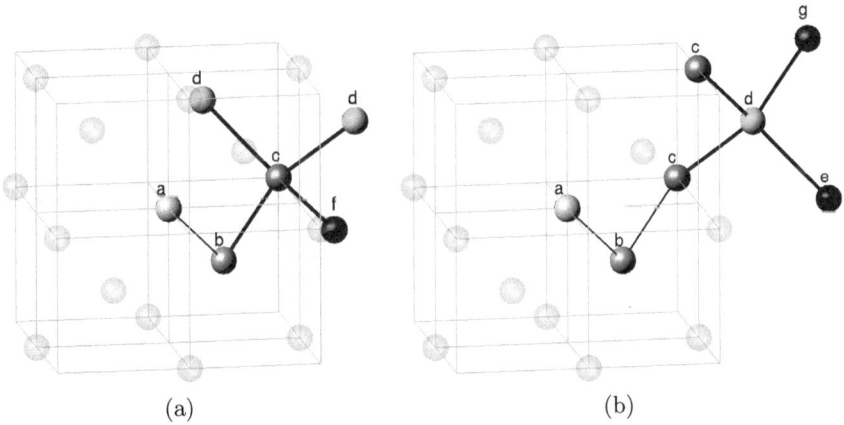

(a)   (b)

Fig. 3. Points that can be reached by three and four 1-steps can be seen in (a) and (b), respectively. In (a) ball c is a strict 2-neighbor of ball a, while balls d are strict 3-neighbors of ball a, ball f can be reached by 3 consecutive 1-steps, but it is not a neighbor of ball a, since there are two steps (the first and the third) to the same direction. In (b) ball d is a strict 3-neighbor of the ball a, balls c are strict 2-neighbors of ball a, they can be obtained by 4 consecutive 1-steps, but with two steps exactly to the opposite direction, ball e is a strict 4-neighbor of ball a, while ball g is not neighbor of ball a.

$(0,0,0,0)$, $(1,0,0,0)$, $(1,-1,0,0)$, $(1,-1,1,0)$, $(0,-1,1,0)$ leads to not a strict 4-neighbor point pair, since there are steps to the opposite/inverse direction of each other. However, the points $(0,0,0,0)$ and $(0,-1,1,0)$ are strict 2-neighbors, and therefore they are non strict(!) 4-neighbors, as well.

All the presented neighborhood relations are symmetric relations among the points of the grid, i.e., if p is (strict) m-neighbor of q, then q is (strict) m-neighbor of p.

A neighborhood sequence B is a sequence of integers $b(i) \in \{1,2,3,4\}$, $B = (b(i))_{i=1}^{\infty}$. If B is periodic, i.e., if for some fixed $l \in \mathbb{N}$ $(l > 0)$, $b(i) = b(i+l)$ is valid for all $i \in \mathbb{N}$, then B is written $B = (b(1), b(2), \dots, b(l))$.

A *path* in a grid is a sequence $(p = p_0), p_1, \dots, (p_m = q)$ of adjacent grid points. A path is a B-*path* of length m if, for all $i \in \{1,2,\dots,m\}$, p_{i-1} and p_i are $b(i)$-neighbors. The B-distance $d(p,q;B)$ is defined as the length of the shortest B-path(s) from p to q. Usually in digital geometry the shortest path is not unique. The distance function generated by B is denoted $d(B)$.

3   Shortest Paths

Let $w_{p,q} = (p(1)-q(1), p(2)-q(2), p(3)-q(3), p(4)-q(4))$, the difference vector of the points p and q.

In this section we give an algorithm that produces a shortest path between any two points in an effective way. In some image processing algorithms it is sufficient/necessary to know a shortest path from a point to another one. In the Algorithm 1, we used the agreement that $+0 = -0 = 0$. The length i of the path gives the B-distance of p and q.

Lemma 1. *Algorithm 1 terminates for any two points p,q and neighborhood sequence B.*

Proof. It is easy to see that the value $\sum_{i=1}^{4} |w(i)|$ is strictly monotonously decreasing in every loop of the algorithm having only integer values, till it obtains the value zero. In this way the algorithm terminates for any input.   □

The (1)-distance, i.e., the B-distance with $B = (1)$ allowing only 1-steps in a path, of any two points p and q is given by

$$d(p, q; (1)) = \sum_{i=1}^{4} |p(i) - q(i)|.$$

Note that at strict m-steps the (1)-distance decreased by m, but in some cases 3-steps and 4-steps are not strict even the (1)-distance is large. Observe that in some cases having a 3-step we use a strict 2-neighbor, even if the endpoint is not reached. Moreover instead of a strict 4-step sometimes a strict 3-step, sometimes a strict 2-step can be in the path. This fact relates to the fact that not all 3 or 4 consecutive 1-steps can be viewed/replaced by a 3 or 4-step (Remark 2).

Algorithm 1. Computing a shortest path based on neighborhood sequences in the diamond grid
Input: *the start and end points p, q and the neighborhood sequence B.*
Output: *one of the shortest B-paths from p to q and the distance $d(p, q; B)$.*

1 let $p = p_0$ *and* $i = 0$.
2 if $p_i = q$ then *we have a shortest path* $(p_0, ..., p_i)$ *with length i.* end.
3 let $i = i + 1$, $p_i = p_{i-1}$ *and* $w = w_{q, p_i}$.
4 in case of $b(i) = 1$:
5   let k *be the index of the (one of the) largest absolute values of the positive/negative values of w if p_i is odd/even;*
6   let $|p(k)| = |p(k)| - 1$ *with the same sign as $p(k)$;* go to 2.
7 in case of $b(i) = 2$:
8   if *there is only one non zero element in w,* then *go to 5.*
9   let k *and l be the index of (one of) the largest positive and negative value(s) in w, respectively;*
11   let $|p(k)| = |p(k)| - 1$ *with the same sign as $p(k)$, and*
       let $|p(l)| = |p(l)| - 1$ *with the same sign as $p(l)$;* go to 2.
12 in case of $b(i) = 3$:
13   if *the number of positive/negative values of w is less than two while p_i is odd/even,* then *go to 8.*
14   if *p_i is even/odd,* then
       let k, l *be the indices of the two largest positive/negative values, and j be the index of the largest negative/positive value of w;*
15   let $|p(k)| = |p(k)| - 1$ *with the same sign as $p(k)$;*
       let $|p(l)| = |p(l)| - 1$ *with the same sign as $p(l)$, and*
       let $|p(j)| = |p(j)| - 1$ *with the same sign as $p(j)$,* go to 2.
16 in case of $b(i) = 4$:
17   if *w has two positive and two negative values* then
18   let *the $|p(k)| = |p(k)| - 1$ with the same sign as $p(k)$ for every $k \in \{1, 2, 3, 4\}$;* go to 2.
19   else go to 13.

It is a simple observation that Algorithm 1 generates a B-path in a greedy way.

In the next section some interesting properties of these distances will be analyzed.

4   Examples for B-Distances

First we recall the definition of metric. A function f is a metric if it satisfies the following conditions:

1. $\forall p, q : f(p, q) \geq 0$ and $f(p, q) = 0$ if and only if $p = q$
2. $\forall p, q : f(p, q) = f(q, p)$ (symmetry)
3. $\forall p, q, r : f(p, q) + f(q, r) \geq f(p, r)$ (triangular inequality)

As one follow the work of Algorithm 1 it is clear that all 1-steps are strict 1-steps, moreover but only the case when the endpoint is reached by a 1-step, all 2-steps are strict 2-steps.

In this way, the following proposition can be proven.

Proposition 1. *Let $B = (b(i))_{i=1}^{\infty}$ be a neighborhood sequence such that $b(i) \in \{1,2\}$ for every value of i. Then the B-distance of the points p and q is given by*

$$d(p, q; B) = \min \left\{ k \left| \sum_{i=1}^{4} |p(i) - q(i)| \leq \sum_{i=1}^{k} b(i) \right. \right\}.$$

Note here that this result is the 4D form of the result presented in [19]. These B-distances are symmetric and the triangular inequality hold if and only if the number of 2's is not more in any prefix of the neighborhood sequence B than in any same-length consecutive subsequence of B. (This is presented and proved in a formal way in [19] and [22].)

Now some further B-distances are shown allowing 3 and 4-neighborhood. First we deal with distances using the same neighborhood criterion in each step.

Theorem 1. *The (m)-distance $(m \in \{1, 2, 3, 4\})$ is a metric on $\mathbb{D}$.*

Proof. Let p and q be two points of the diamond grid. It is trivial, that their distance is non-negative integer. Moreover it is 0 if and only if $p = q$ and there is no step needed to connect them. Let $\Pi_1 = (p = p_0), p_1, p_2, ..., p_{n-1}, (p_n = q)$ be a minimal (m)-path from p to q. Then its inverse $\Pi_1' = p_n, p_{n-1}, ..., p_2, p_1, p_0$ is also an (m)-path from q to p with the same length as Π_1. Therefore $d(p, q; (m)) \geq d(q, p; (m))$. Now let $\Pi_2 = (q = q_0), q_1, ..., q_{m-1}, (q_m = p)$ be a minimal (m)-path from q to p. Its inverse path $\Pi_2' = q_m, p_{m-1}, ..., q_1, q_0$ is an (m)-path from p to q with the same length as Π_2, so by definition $d(q, p; (m)) \geq d(p, q; (m))$. Therefore the (m)-distance is symmetric. Let us consider the triangular inequality. Let $p, q, r \in \mathbb{D}$ three arbitrary points. Let Π_{pq}, Π_{qr} and Π_{pr} are minimal (m)-paths from p to q, from q to r and from p to r, respectively. The concatenation $(p = p_0), p_1, ..., (p_n = q = q_0), q_1, ..., (q_z = r)$ of Π_{pq} and Π_{qr} is an (m)-path from p to r, therefore $d(p, q; (m)) + d(q, r; (m)) = n + z \geq d(p, r; (m))$. The theorem is proved. □

Actually the distances of the previous theorem are path generated graph metrics. If one allows to use various neighbors in a path, i.e., there are distinct neighborhoods in a neighborhood sequence, the previous metrical property may be destroyed. Since the possible steps in a B-path governed by the neighborhood sequence B, the concatenation of two B-paths is not necessarily B-path with the same neighborhood sequence.

Now we show by an example that two of the metrical properties are not trivially satisfied by distances based on neighborhood sequences in the diamond grid.

Example 1. Let $B = (4, 3, 2, 1)$. Let $p = (0, 0, 0, 0)$ and $q = (4, -3, -2, 1)$. Then there is a shortest B-path $(0, 0, 0, 0), (1, -1, -1, 1), (2, -2, -1, 1), (3, -3, -1, 1),$

$(4, -3, -1, 1)$, $(4, -3, -2, -1)$ from p to q with length 5. There is a shortest B-path $(4, -3, -2, 1)$, $(3, -2, -1, 0)$, $(2, -1, 0, 0)$, $(1, 0, 0, 0)$, $(0, 0, 0, 0)$ from q to p with length 4. Let $r = (1, 1, -1, -1)$ and $s = (2, 2, -2, -2)$. Then $d(p, r; B) = 1$, $d(r, s; B) = 1$ and $d(p, s; B) = 3$ (a shortest B-path: $(0, 0, 0, 0)$, $(1, 1, -1, -1)$, $(2, 2, -2, -1)$, $(2, 2, -2, -2)$). Thus this B-distance does not satisfy neither the symmetry nor the triangular inequality. Let $t = (4, -2, -1, -1)$, then a shortest B-path from p to t is: $(0, 0, 0, 0)$, $(1, -1, 0, 0)$, $(2, -2, 0, 0)$, $(3, -2, -1, 0)$, $(4, -2, -1, 0)$, $(4, -2, -1, -1)$. In this path neither the 4, nor the 3-step were strict, actually both were 2-steps.

The non-symmetric property comes from the fact, that in some cases only a 2-step (or a 3-step) can be used even if there is a value 3 or 4 in the neighborhood sequence, even in cases when none of the differences of the coordinate values are 0. This fact is strictly connected to the non-lattice property of the grid. In this way, some of the shortest paths have no inverse in the sense that the same-type of strict neighbors are used in a B-path backward direction.

5   The (Constrained) Distance Transform

In this section we implement the distance transform and the constrained distance transform using distances based on neighborhood sequences with possible use of the four types of neighborhood relation.

The Euclidean distance function is used in many image processing-applications since it has minimal rotational dependency. However, in some applications, distance functions defined by the minimal cost-path between pixels, i.e., path-based distances, are still preferred.

The distance transform is obtained by finding the minimal distance to each object grid point from the closest source grid point. In a *constrained* distance transform (CDT), there are grid points, *obstacles*, that are not allowed in any minimal cost-path. The algorithm for computing the CDT is faster and less complex for path-based distances compared to the Euclidean distance. The CDT is an image in which each object point is labeled with the constrained distance to their respective closest source pixel. When there are no obstacle grid points, obviously, the standard distance transform is obtained.

The CDT based on a path-based distance function can be computed using standard minimal cost-path techniques for weighted graphs resulting in a linear (in the number of object pixels) time algorithm. In [23], the Dijkstra's graph search algorithm is used which works for arbitrary weighted graphs. This algorithm has complexity $O(n \log n)$, where n is the number object pixels and the $\log n$ factor is due to the sorting. When only integer weights are used, the sorting can be done in constant time by bucket sort and the complexity is $O(n)$. For distances based on neighborhood sequences, unit weights are used, i.e., all possible steps increase the length of the path by 1, so the algorithm has linear time complexity. Algorithm 2 computes the constrained distance transform based on distances by neighborhood sequences on the diamond grid using an algorithm similar to the Dijkstra algorithm.

Algorithm 2. Computing CDT for distances based on neighborhood sequences in the diamond grid
Input: *A neighborhood sequence B, an object S and source pixels 0.*
Output: *The distance transform, DT.*
Auxiliary data structures: *A list $\mathcal{L}$ of active pixels.*

1. let $DT(0) = 0$ *and* $DT(p) = \infty$ *for all other points* $p \in S$.
2. Insert the source pixels 0 in $\mathcal{L}$.
3. while $\mathcal{L}$ *is not empty*
4.   Find p in $\mathcal{L}$ with lowest $DT(p)$
5.   Remove p from $\mathcal{L}$
6.   for each $q \in S$ *that are* $b(DT(p)+1)$*-neighbors with p:*
7.   if $DT(p) + 1 < DT(q)$ then
8.   let $DT(q) = DT(p) + 1$
9.   Insert q in $\mathcal{L}$
10.   endif
11.   endfor
12. endwhile

Some results are shown in Figure 4. A source point is added to the center of the bottom surface of the cube. The cube has a tunnel in which the distance values can propagate. Figure 4 shows how some different neighborhood sequences influence the constrained DT.

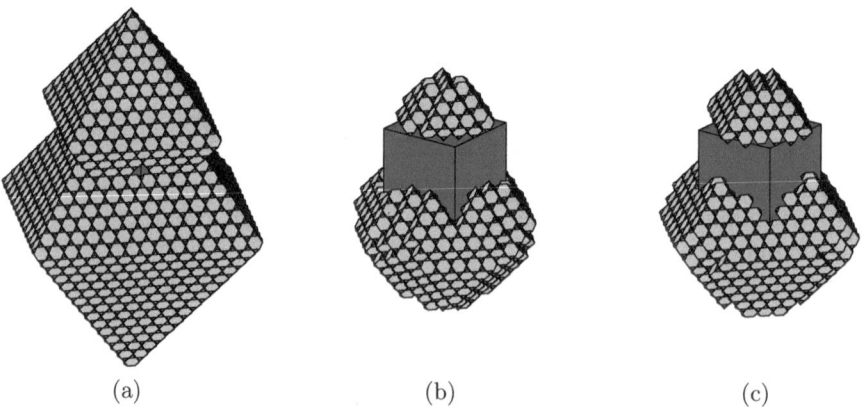

(a)   (b)   (c)

Fig. 4. Constrained DT. Voxels corresponding to grid points at distance less than 9 from a single source point in the lower part of the cube. Distance functions given by $B = (4)$ (a), $B = (1,4)$ (b), $B = (1,2,3,4)$ (c) are used.

6   Digital Balls

The shapes of the digital ball illustrate the digital distance functions. For example, the rotational dependency can be illustrated by showing the digital ball – "round" (in the Euclidean sense) balls means low rotational dependent distance function. The balls in Fig. 5 are obtained by computing the DT from a single point.

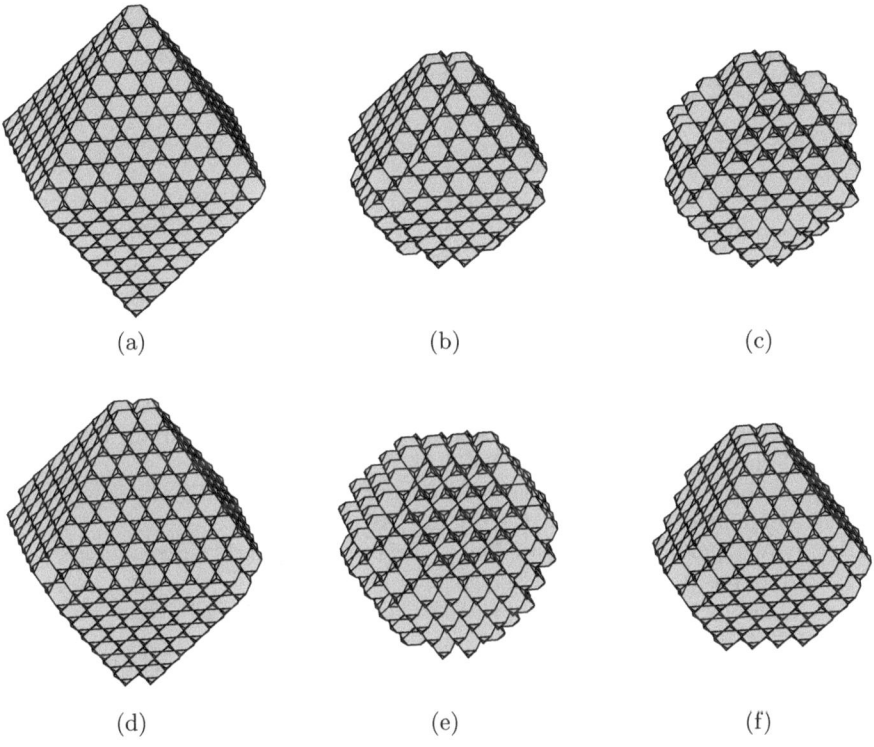

(a)   (b)   (c)

(d)   (e)   (f)

Fig. 5. Balls of radius 4 generated by $B = (4)$ (a), $B = (1,4)$ (b), $B = (4,1)$ (c), $B = (3,4)$ (d), $B = (4,3,2,1)$ (e), $B = (1,2,3,4)$ (f)

The B-distance, and so the digital ball depends on the order of the used steps, as one may observe on Figure 5. A B-distance is symmetric if and only if the sequence of generated balls are the same using the radius long prefixes of the sequence in the opposite order.

7   Conclusions

In the diamond grid, each point has four types of neighbors that can be defined in a straightforward natural manner, their distances are not more than the unit cell. In this paper, we have considered path-based distance functions defined by neighborhood sequences in which all four neighborhood types are allowed. We have presented algorithms for computing shortest B-paths and constrained distance transforms. The distance transform is used to generate digital balls, which have been used to illustrate distance functions obtained by different neighborhood sequences. Also, examples high-lighting interesting properties, such as non-metricity, have been presented.

There are some unsolved problems related to the results presented in this paper: formula for distance, condition of metrical distances, and approximation of the Euclidean ball. We note that the non-lattice property and the non-symmetric distance functions presented here may have some interesting applications.

References

1. Borgefors, G.: Distance Transformations on Hexagonal Grids. Pattern Recognition Letters 9, 97–105 (1989)
2. Brimkov, V.E., Barneva, R.P.: Analytical Honeycomb Geometry for Raster and Volume Graphics. The Computer Journal 48(2), 180–199 (2005)
3. Conway, J.H., Sloane, N.J.A., Bannai, E.: Sphere-packings, lattices, and groups. Springer, New York (1988)
4. Csébfalvi, B.: Prefiltered Gaussian reconstruction for high-quality rendering of volumetric data sampled on a body-centered cubic grid. IEEE Visualization, 40 (2005)
5. Das, P.P., Chakrabarti, P.P., Chatterji, B.N.: Distance functions in digital geometry. Information Sciences 42, 113–136 (1987)
6. Deutsch, E.S.: Thinning algorithms on rectangular, hexagonal and triangular arrays. Communications of the ACM 15, 827–837 (1972)
7. Her, I.: A symmetrical coordinate frame on the hexagonal grid for computer graphics and vision. ASME J. Mech. Design 115, 447–449 (1993)
8. Her, I.: Geometric transformations on the hexagonal grid. IEEE Trans. on Image Processing 4(9), 1213–1222 (1995)
9. Her, I.: Description of the F.C.C. lattice geometry through a four-dimensional hypercube. Acta Cryst. A 51, 659–662 (1995)
10. Klette, R., Rosenfeld, A.: Digital geometry - Geometric methods for digital picture analysis. Morgan Kaufmann Publ., San Francisco (2004)
11. Middleton, L., Sivaswamy, J.: Framework for practical hexagonal-image processing. Journal of Electronic Imaging 11, 104–114 (2002)
12. Nagy, B.: Shortest Paths in Triangular Grids with Neighbourhood Sequences. Journal of Computing and Information Technology 11(2), 111–122 (2003)
13. Nagy, B.: Distance functions based on neighbourhood sequences. Publicationes Mathematicae Debrecen 63(3), 483–493 (2003)
14. Nagy, B.: A Family of Triangular Grids in Digital Geometry. In: Proc. of the 3rd ISPA, International Symposium on Image and Signal Processing and Analysis, Rome, Italy, pp. 101–106 (2003)
15. Nagy, B.: A symmetric coordinate frame for hexagonal networks. In: Proc. of ISTCS 2004, Theoretical Computer Science - Information Society, Ljubljana, Slovenia, pp. 193–196 (2004)
16. Nagy, B.: Generalized triangular grids in digital geometry. Acta Mathematica Academiae Paedagogicae Nyíregyháziensis 20, 63–78 (2004)

17. Nagy, B.: Characterization of digital circles in triangular grid. Pattern Recognition Letters 25(11), 1231–1242 (2004)
18. Nagy, B.: Distances with Neighbourhood Sequences in Cubic and Triangular Grids. Pattern Recognition Letters 28, 99–109 (2007)
19. Nagy, B., Strand, R.: Neighborhood sequences in the diamond grid. In: Barneva, R.P., Brimkov, V.E. (eds.) Image Analysis - From Theory to Applications, Research Publishing, Singapore, Chennai, pp. 187–195 (2008)
20. Nagy, B., Strand, R.: A connection between $\mathbb{Z}^n$ and generalized triangular grids. In: Bebis, G., Boyle, R., Parvin, B., Koracin, D., Remagnino, P., Porikli, F., Peters, J., Klosowski, J., Arns, L., Chun, Y.K., Rhyne, T.-M., Monroe, L., et al. (eds.) ISVC 2008, Part II. LNCS, vol. 5359, pp. 1157–1166. Springer, Heidelberg (2008)
21. Nagy, B., Strand, R.: Non-traditional grids embedded in $\mathbb{Z}^n$. Int. Journal of Shape Modeling (2009) (accepted for publication)
22. Nagy, B., Strand, R.: Neighborhood sequences in the diamond grid: algorithms with 2 and 3 neighbors. Int. Journal of Imaging Systems and Technology 19(2), 146–157 (2009)
23. Piper, J., Granum, E.: Computing distance transformations in convex and non-convex domains. Pattern Recognition 20(6), 599–615
24. Rosenfeld, A., Pfaltz, J.L.: Sequential Operations in Digital Picture Processing. Journal of the ACM 13(4), 471–494 (1966)
25. Rosenfeld, A., Pfaltz, J.L.: Distance Functions on Digital Pictures. Pattern Recognition 1, 33–61 (1968)
26. Rosenfeld, A.: Digital Geometry: Introduction and Bibliography. Techn. Rep. CS-TR-140/CITR-TR-1 (Digital Geometry Day 1997), Computer Science Dept., The University of Auckland, CITR at Tamaki Campus (1997)
27. Snyder, W.E., Qi, H., Sander, W.A.: A coordinate system for hexagonal pixels. Proc. of SPIE, Medical Imaging Pt.1-2, 716–727 (1999)
28. Strand, R., Nagy, B.: Distances Based on Neighbourhood Sequences in Non-Standard Three-Dimensional Grids. Discrete Applied Mathematics 155(4), 548–557 (2007)
29. Takahashi, T., Yonekura, T.: Isosurface construction from a data set sampled on a face-centered-cubic lattice. In: Proc. of ICCVG 2002, vol. 2, pp. 754–763 (2002)
30. Yamashita, M., Honda, N.: Distance Functions Defined by Variable Neighborhood Sequences. Pattern Recognition 5(17), 509–513 (1984)
31. Yamashita, M., Ibaraki, T.: Distances Defined by Neighborhood Sequences. Pattern Recognition 19(3), 237–246 (1986)

Hinge Angles for 3D Discrete Rotations[*]

Yohan Thibault[1], Akihiro Sugimoto[2], and Yukiko Kenmochi[1]

[1] Université Paris-Est, LIGM, UMR UPEMLV/ESIEE/CNRS 8049, France
{thibauly,y.kenmochi}@esiee.fr
[2] National Institute of Informatics, Japan
sugimoto@nii.ac.jp

Abstract. In this paper, we focus on 3D rotations on grid points computed by using only integers. For that purpose, we study the intersection between the 3D half-grid and the rotation plane. From this intersection, we define 3D hinge angles which determine a transit of a grid point from a voxel to its adjacent voxel during the rotation. Then, we give a method to sort all 3D hinge angles with integer computations. The study of 3D hinge angles allows us to design a 3D discrete rotation.

Keywords: discrete geometry, discrete rotation, half-grid, multi-grid, hinge angle, integer computation.

1 Introduction

Rotations in the 3D space are required in many applications for computer imagery, such as image processing [13], computer vision [7,10] and computer graphics [6]. A rotation in the 3D Euclidean space can be in general represented in two different typical ways. One is to represent a rotation as a combination of three particular rotations around the three axes of the coordinate system in concern [14]. The other is to represent a rotation by a rotation axis together with an angle around the axis [7,10]. Even if the representations of a rotation are different, computed rotation results are the same as far as the space is continuous. However, this is not the case for the discrete space. Namely, depending on the rotation representation, the computed rotation result can change in the discrete space [2]. As is the case of 2D rotations, computing a 3D rotation once in the discrete space brings displacement from that in the continuous space; computing 3D rotations many times causes difficulty in analyzing inherited displacements during the computation. Accordingly, representing a 3D rotation by a rotation axis together with an angle around the axis is more preferable in the 3D discrete space. Besides, it is known that such axis-angle representation is useful for 3D rotation estimation from given image sets, which is one of important problems in computer vision [7,10].

In [8], the author proved that a 2D discrete image have a finite number of different rotation. This number is directly related to the number of hinge angles.

---

[*] This work was supported by grant from Région Ile-de-France and the French Foreign Ministry.

P. Wiederhold and R.P. Barneva (Eds.): IWCIA 2009, LNCS 5852, pp. 122–134, 2009.

The extension of hinge angles for the 3D space allows to obtain the same result. This result can be useful for some problems such as the pose problem. In order to obtain the optimal solution, the pose problem need to try all different rotation, which is impossible in continuous space since there exists an infinity of rotation [7]. Using hinge angles for this problem can help to reach the optimal solution.

This paper presents a study of the rotation in the 3D discrete space. Since we admit only integer computations, we assume that our rotation center is a grid point such as the origin, and that a rotation axis has integer coordinates. In the 2D case, hinge angles are known to correspond to the discontinuity caused by discretisation of the rotation in the continuous plane [1,8,12]. Intuitively, hinge angles determine a transit of a grid point from a pixel to its adjacent pixel during the rotation. In other words, two rotations with nearby angles transform the same grid point to two adjacent pixels because discrete rotations around a given center are locally continuous with regard to the angle. Hinge angles are their discontinuity angles. Computing hinge angles using integers alone allows us to compute 2D discrete rotations without any approximation errors, which designs the 2D discrete rotation. Extending these to the 3D case, we design a 3D discrete rotation. In the 3D case, however, depending on the rotational axis, we have a variety of transitions of a grid point across voxels. How to capture these transitions systematically is a big issue.

We define hinge angles for 3D rotations so that they determine a transit of a grid point from a voxel to its adjacent voxel. To compute the hinge angles using integers alone, we pay our attentions to the intersection of voxels and a given rotation plane to define convexels. The variety of transitions of a grid point across voxels in 3D results in the variation of convexels in 2D. We then compute the hinge angles by using the intersection between the convexel and a locus of the rotation in the rotation plane. Finally, we give a method to sort all the possible hinge angles in concern to design a 3D discrete rotation. Differently from 2D discrete rotations [2,8,3], few attempts on 3D discrete rotations have been reported [13,6]. In particular, to our best knowledge, this is the first work on 3D discrete rotation using integer computation without digitization errors.

2 Hinge Angles

Hinge angles for 2D rotations are defined to represent the discontinuities of rotations in the discrete plane [8]. Hinge angles determine a transit of a grid point from a pixel to its adjacent pixel during the rotation. To characterize those hinge angles, the 2D half-grid plays an important role.

Definition 1. *The half-grid in the plane is the set of lines, each of which is represented by one of $x = i + \frac{1}{2}$ and $y = i + \frac{1}{2}$ where $i \in \mathbb{Z}$.*

In other words, the 2D half-grid represents the border between two adjacent pixels. From the definition of the 2D half-grid, we define the hinge angles in the plane.

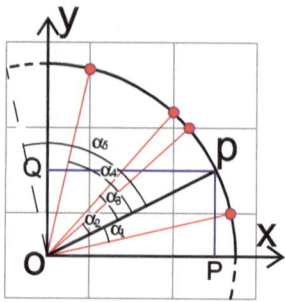

Fig. 1. All hinge angles in the first quadrant for the grid point $p = (2,1)^\top$, such that $\alpha_1 \approx -13.68°$, $\alpha_2 \approx 15.59°$, $\alpha_3 \approx 21.31°$, $\alpha_4 \approx 50.52°$, $\alpha_5 = \frac{\pi}{2} - \alpha_1 \approx 76.32°$

Definition 2. *An angle α is called a hinge angle if at least one point in $\mathbb{Z}^2$ exists such that its image by the Euclidean rotation with α around the origin is on the 3D half-grid.*

Figure 1 illustrates some examples of hinge angles for the grid point $(2,1)^\top$. It presents all the hinge angles in the first quadrant. Note that hinge angles in other quadrants are obtained by symmetry with respect to the x-axis and/or y-axis from those in Figure 1.

To extend the definition of hinge angles into the 3D case, we first define the half-grid in the 3D space. Similarly to the 2D half-grid, the 3D half-grid defines the limit between two adjacent voxels in the 3D discrete space.

Definition 3. *The half-grid in the 3D space is the set of planes, each of which is represented by one of $x = i + \frac{1}{2}, y = i + \frac{1}{2}$ and $z = i + \frac{1}{2}$ where $i \in \mathbb{Z}$.*

Introducing the definition of the half-grid in the 3D space allows us to define hinge angles in 3D as a natural extension of hinge angles in 2D. As mentioned in the introduction, we only consider here 3D rotations whose rotation axes have integer coordinates and go through the origin. Hereafter, we call such an axis an integer-axis.

Definition 4. *An angle α is called a hinge angle if at least one point in $\mathbb{Z}^3$ exists such that its image by the Euclidean rotation with α around an integer-axis is on the half-grid.*

Similarly to the case of 2D, for a grid point p in 3D, an angle α is a hinge angle if and only if the discretised point of the rotation result of p with angle $\alpha + \epsilon$ becomes different from that with angle $\alpha - \epsilon$ for any $\epsilon > 0$.

Differently from the case of 2D rotations, we need not only a rotation angle but also a rotation axis in order to specify a 3D rotation. This requires investigation of the intersection between voxels and a plane determined by a given rotation axis because a variety of transitions of a grid point across voxels exist depending on the plane. To capture this variety, we introduce the multi-grid in the next section, which allows us to compute hinge angles in the 3D rotation plane.

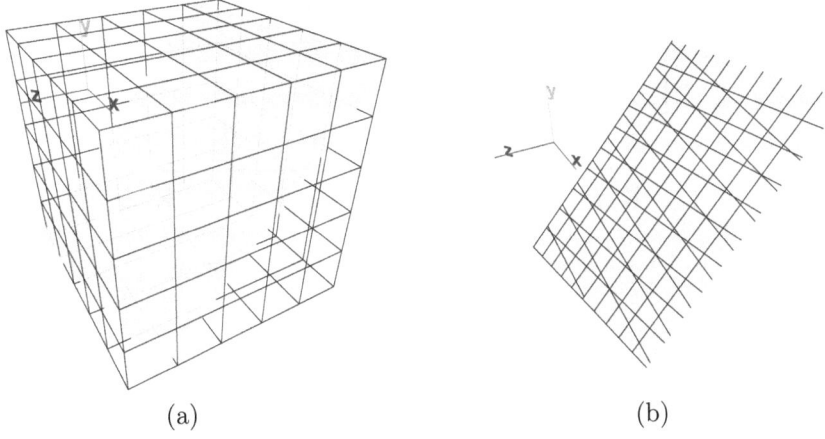

(a)   (b)

Fig. 2. The 3D half-grid cut by a plane (a), and its multi-grid (b)

3   Hinge Angles in a Rotation Plane

A 3D rotation is represented by a rotation axis and a rotation angle. Accordingly, the 3D rotation of a given grid point is considered as the 2D rotation of the point in the rotation plane that is defined as the plane going through the grid point and having the normal vector parallel with the rotation axis. In this section, we capture hinge angles in a given rotation plane.

3.1   Multi-grid

When a rotation plane in 3D is given, we consider the intersection between the plane and the half-grid in the 3D space as illustrated in Figure 2(a). As we see, the intersection consists of three different sets of parallel lines as illustrated in Figure 2(b), except for cases where the normal of the rotation plane is parallel to one of the axes defining the coordinate system of the 3D space. As such exceptional cases provide only two different sets of parallel lines, which are identical with those of the 2D half-grid, we here do not take into account those cases. In other words, in such cases, 3D rotations become identical with 2D rotations. We call the three different sets of parallel lines a multi-grid, which is used for characterizing hinge angles for 3D rotations instead of the 2D half-grid for 2D rotations.

In a multi-grid, the interval between parallel lines having the same directional vector is regular. Normalizing the interval allows us to represent a set of parallel lines having the same directional vector as

$$\{(x, y)^\top \in \mathbb{R}^2 | ax + by + c + k = 0, a^2 + b^2 \neq 0, k \in \mathbb{Z}, a, b, c \in \mathbb{R}\}. \quad (1)$$

The integer parameter k denotes the index number of each parallel line. Figure 3 gives a geometrical explanation of the parameters of a set of parallel lines. For

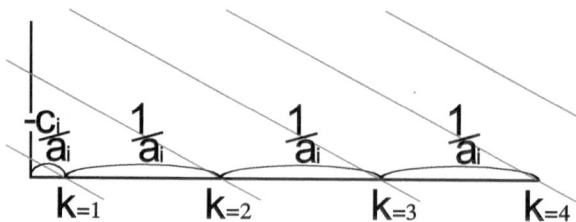

Fig. 3. Parallel lines of a set $\mathcal{M}_i^{\mathcal{A},p}$ and geometric interpretation of their parameters

example, if a point $(x, y)^\top$ is on one of the parallel lines, $(x - \frac{k}{a}, y)^\top$ is on the k-th next line, providing that $a \neq 0$. Now we can give a formal definition of a multi-grid.

Definition 5. *Let* $\mathcal{M}_i = \{(x, y)^\top \in \mathbb{R}^2 | a_i x + b_i y + c_i + k = 0, a_i^2 + b_i^2 \neq 0, k \in \mathbb{Z}, a_i, b_i, c_i \in \mathbb{R}\}$ *for* $i = 1, 2, 3$ *be each set of parallel lines induced from a given rotation plane and the 3D half-grid. Then the multi-grid* $\mathcal{M}$ *is the union of* $\mathcal{M}_i$: $\mathcal{M} = \cup_i \mathcal{M}_i$.

The multi-grid in the rotation plane forms various convex polygons surrounded by lines, which we call convexels. Depending on the rotation plane, we have a variety of shapes of convexels. The shapes of convexels are investigated in [4] under the context of the intersection of a voxel and a plane, and the number of vertices of a convexel can be 3, 4, 5 or 6 as illustrated in Figure 4. The convexel is the counterpart of the squared pixel defined in the 2D plane. We remark that when the normal of the rotation plane is parallel with one of the axes defining the coordinate system of the 3D space, the notion of convexels coincides with the notion of pixels.

The following proposition allows us to characterize hinge angles in the framework of multi-grids, instead of the half-grid.

Proposition 1. *Let* **p** *be a grid point and* **p**′ *be the result of the rotation of* **p** *by an angle* α *around an integer-axis. If* α *is a hinge angle, then* **p**′ *is on the multi-grid.*

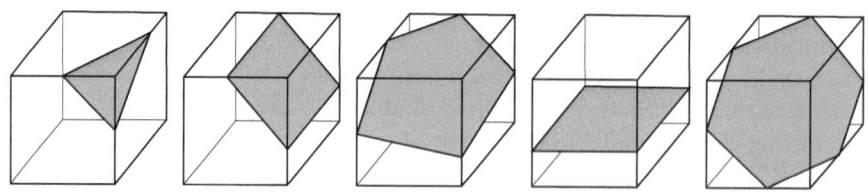

Fig. 4. The five different shapes of convexels, which are constructed as the intersections between a rotation plane and voxels [4]

3.2   Hinge Angles and Multi-grids

In 2D, there exists a property on hinge angles ensuring that the locus of rotation of a grid point cannot contain the intersection of two lines belonging to the half-grid [8]. In this section, we show that the same property for the hinge angles in the rotation plane holds. Namely, we show that, for a given multi-grid, the locus of the rotation of a grid point does not go through any vertex of convexels in the multi-grid.

Hereafter, we denote by $\mathcal{M}^{\mathcal{A},\boldsymbol{p}}$ the multi-grid defined by a rotation plane with a normal vector $\mathcal{A} = (a_x, a_y, a_z)^\top$ going through point $\boldsymbol{p} = (p_x, p_y, p_z)^\top$ and by $\boldsymbol{A}$ the axis of rotation of directional vector $\mathcal{A}$ going through the origin.

Proposition 2. Let $\mathcal{M}^{\mathcal{A},\boldsymbol{p}}$ be a multi-grid where $\mathcal{A} \in \mathbb{Z}^3$ and $\boldsymbol{p} \in \mathbb{Z}^3$. Then, the locus of the rotation of $\boldsymbol{p}$ around $\boldsymbol{A}$ does not go through any vertex of the convexels on $\mathcal{M}^{\mathcal{A},\boldsymbol{p}}$.

Proof. The equation of the rotation plane $\mathcal{P}$ of $\mathcal{M}^{\mathcal{A},\boldsymbol{p}}$ is $a_x x + a_y y + a_z z - \mathcal{A}.\boldsymbol{p} = 0$. Let $\boldsymbol{p}' = (p'_x, p'_y, p'_z)^\top$ be a point which belongs to the locus of the rotation of $\boldsymbol{p}$ in $\mathcal{P}$ around $\boldsymbol{A}$. Let us assume that $\boldsymbol{p}'$ is also a vertex of a convexel of $\mathcal{M}^{\mathcal{A},\boldsymbol{p}}$, so that it belongs to two planes of the 3D half-grid. Thus we can set, without loss of generality, that $p'_x = k_x + \frac{1}{2}$ and $p'_y = k_y + \frac{1}{2}$ where $k_x, k_y \in \mathbb{Z}$, and then $\boldsymbol{p}' = (k_x + \frac{1}{2}, k_y + \frac{1}{2}, p'_z)^\top$.

The locus of the rotation of $\boldsymbol{p}$ is the intersection between $\mathcal{P}$ and the sphere $\mathcal{S}$: $x^2 + y^2 + z^2 - (p_x^2 + p_y^2 + p_z^2) = 0$. Thus, by the assumption that $\boldsymbol{p}'$ belongs to $\mathcal{P}$ and $\mathcal{S}$, we have

$$a_x \left(k_x + \frac{1}{2} \right) + a_y \left(k_y + \frac{1}{2} \right) + a_z p'_z - \mathcal{A}.\boldsymbol{p} = 0, \tag{2}$$

$$\left(k_x + \frac{1}{2} \right)^2 + \left(k_y + \frac{1}{2} \right)^2 + p_z'^2 - (p_x^2 + p_y^2 + p_z^2) = 0. \tag{3}$$

From (2) we see that p'_z must be a rational number, so that there exists a pair of integers λ_1, λ_2 such that $p'_z = \frac{\lambda_1}{\lambda_2}$ and $\gcd(\lambda_1, \lambda_2) - 1$. From (3) we see that $p_z'^2 + \frac{1}{2}$ must be an integer. Thus we have

$$p_z'^2 = \frac{\lambda_1^2}{\lambda_2^2} = k + \frac{1}{2}, \tag{4}$$

where $k, \lambda_1, \lambda_2 \in \mathbb{Z}$. Since $\gcd(\lambda_1, \lambda_2) = 1$ and $2\lambda_1^2 = (2k+1)\lambda_2^2$, λ_2 must be even. Setting $\lambda_2 = 2\lambda'_2$ where $\lambda'_2 \in \mathbb{Z}$, we then obtain $2\lambda_1^2 = 4(2k+1)\lambda_2'^2$ and deduce that λ_1 must be also even, which contradicts the assumption that $\gcd(\lambda_1, \lambda_2) = 1$.

Therefore we can conclude that there is not such a point $\boldsymbol{p}'$. □

This proposition shows that two adjacent convexels, where the transition of $\boldsymbol{p}$ during its rotation is done between them, always share a convexel edge. In other words, the circle of $\boldsymbol{p}$ around $\boldsymbol{A}$ passes through a sequence of voxels such that any successive voxels are connected by their common face.

4   Integer Representation of Hinge Angles

The goal of this section is to show how to obtain the unique representation of an integer quintuplet for a 3D hinge angle, namely an injective map from hinge angles to quintuplets. We first explain how to obtain the equations representing lines in a multi-grid provided that the rotation plane is given. Then, we represent 3D hinge angles using five integers and then explain how to decode them to obtain the hinge angle. Note that the rotation plane is defined by a normal vector A and a point p on the plane, and we assume in the following that $\mathcal{A}$ and p are given. Thus a multi-grid is denoted by $\mathcal{M}^{\mathcal{A},p} = \cup_{i=1,2,3}\mathcal{M}_i^{\mathcal{A},p}$.

4.1   Multi-grid Line Equations

In Section 3, we explained that we do not consider the cases where the vector $\mathcal{A}$ is collinear with an axis defining the coordinate system in concern, so that lines in $\mathcal{M}_i^{\mathcal{A},p}$ are not orthogonal to those in $\mathcal{M}_j^{\mathcal{A},p}$ where $i \neq j$. To simplify the derivation of line equations in $\mathcal{M}_1^{\mathcal{A},p}$, we introduce a base where lines in $\mathcal{M}_1^{\mathcal{A},p}$ are parallel with the y-axis. Note that the same discussion can be applied to $\mathcal{M}_2^{\mathcal{A},p}$ and $\mathcal{M}_3^{\mathcal{A},p}$.

Let $\mathcal{A} = (a_x, a_y, a_z)^\top$ and $p = (p_x, p_y, p_z)^\top$ in the standard orthonormal base B and the plane $\mathcal{P}$ with normal vector $\mathcal{A}$ that going through p. Assuming lines in $\mathcal{M}_1^{\mathcal{A},p}$ come from the intersection between $\mathcal{P}$ and the planes of the 3D half-grid that are parallel to yz-plane, we obtain the directional vector v_1 of lines in $\mathcal{M}_1^{\mathcal{A},p}$ as $v_1 = \mathcal{A} \wedge e_1$ where $e_1 = (1,0,0)^\top$. We set $v_2 = \frac{v_1 \wedge \mathcal{A}}{\|\mathcal{A}\|}$, which is linearly independent of v_1. Note that each of v_1 and v_2 is orthogonal with $\mathcal{A}$.

We introduce a new base B_1 in such a way that v_1 and v_2 respectively become $u_1 = (1,0)^\top$ and $u_2 = (0,1)^\top$ in $\mathcal{P}$. The transformation from B to B_1 is realized by

$$M_{BB_1} = \begin{pmatrix} 0 & \frac{a_z}{a_y^2+a_z^2} & \frac{-a_y}{a_y^2+a_z^2} \\ -\psi & \frac{\psi a_x a_y}{a_y^2+a_z^2} & \frac{\psi a_x a_z}{a_y^2+a_z^2} \end{pmatrix}, \tag{5}$$

where $\psi = \frac{1}{\sqrt{a_x^2+a_y^2+a_z^2}}$. We remark that $\mathcal{A}$ is transformed to $(0,0)^\top$ by M_{BB_1}; the rotation center in $\mathcal{P}$ thus becomes the origin of B_1.

Applying M_{BB_1} to the plane $x = k(k \in \mathbb{Z})$ induces a line in $\mathcal{M}_1^{\mathcal{A},p}$ whose equation is

$$\psi(a_y^2 + a_z^2)y + k - \mathcal{A}.p\psi^2 a_x = 0, \tag{6}$$

where $k \in \mathbb{Z}$.

Changing the roles between $\mathcal{M}_1^{\mathcal{A},p}$ and $\mathcal{M}_2^{\mathcal{A},p}$ ($\mathcal{M}_3^{\mathcal{A},p}$, resp.) and between e_1 and $e_2 = (0,1,0)^\top$ ($e_3 = (0,0,1)^\top$ resp.), we obtain the line equations for $\mathcal{M}_2^{\mathcal{A},p}$ and $\mathcal{M}_3^{\mathcal{A},p}$ with the bases B_i and the transformation matrices M_{BB_i} for $i = 2,3$:

$$\psi(a_x^2 + a_z^2)y + k - \mathcal{A}.p\psi^2 a_y = 0, \tag{7}$$
$$\psi(a_x^2 + a_y^2)y + k - \mathcal{A}.p\psi^2 a_z = 0, \tag{8}$$

where $k \in \mathbb{Z}$.

We remark that if $\mathcal{A}$ is collinear with one of the axes defining the coordinate system of the 3D space, M_{BB_i} degenerates: the rank of M_{BBi} becomes 1. In such cases, 3D rotations become identical with 2D rotations, which are not our concern here.

4.2   Quintuplet Integer Representation of Hinge Angles

A hinge angle in a given rotation plane is represented by a quintuplet of integers (p_x, p_y, p_z, i, k) given by $\alpha(p_x, p_y, p_z, i, k)$. Namely, we keep the coordinates of $\boldsymbol{p}$ and the information required for obtaining the coordinates of the arriving point $\boldsymbol{i}$ after the rotation of $\boldsymbol{p}$ by α. The first three integers p_x, p_y, p_z represent, in the old basis B, the coordinates of $\boldsymbol{p}$. The fourth integer i indicates the index number for the set $\mathcal{M}_i^{\mathcal{A},\boldsymbol{p}}, i = 1, 2, 3$, where the hinge angle α is defined. The last integer k represents the index number of the line in $\mathcal{M}_i^{\mathcal{A},\boldsymbol{p}}$.

From these five integers, we can obtain, in the basis B_i, the coordinates $(P_x, P_y)^\top$ and $(I_x, I_y)^\top$ of points $\boldsymbol{p}$ and $\boldsymbol{i}$ which define α as represented in Figure 5. The coordinates of $\boldsymbol{p}$ in B_i are obtained by applying M_{BB_i} to the coordinates of $\boldsymbol{p}$ in B. The coordinate I_y is obtained from one of (6), (7) or (8) depending on i. Since $\boldsymbol{i}$ belongs to the locus of the rotation of $\boldsymbol{p}$, we have $I_x^2 + I_y^2 = P_x^2 + P_y^2$ and deduce the value of I_x.

In general, I_x can take two values which define two points. To discriminate two different hinge angles corresponding to these two points by their integer quintuplets, we add the positive sign to the fourth integer for the greater I_x and the negative sign for another value of I_x. If $I_x = 0$, then the two points merge into one. This particular case where the locus of the rotation of $\boldsymbol{p}$ is tangent with the k-th line of $\mathcal{M}_i^{\mathcal{A},\boldsymbol{p}}$ does not define a hinge angle since there is no transition of convexel. Hereafter, the representation of hinge angles by integer quintuplets will be denoted by $\alpha(p_x, p_y, p_z, \pm i, k)$.

Note that according to Proposition 2 we know that any hinge angle cannot rotate a grid point to the intersection between two lines of a multi-grid. Therefore, two different integer quintuplets cannot represent the same hinge angle.

4.3   Rational Multi-grids

In Section 4.1, we obtained (6)-(8) for the lines in the multi-grid $\mathcal{M}^{\mathcal{A},\boldsymbol{p}}$. In general, parameters of these equations belong to $\mathbb{R}$. However in order to use only integers during computation, we need these parameters to belong to $\mathbb{Q}$.

If all elements of M_{BB_i} belong to $\mathbb{Q}$, then all the parameters of (6)-(8) become rational, so that $P_x, P_y, I_y \in \mathbb{Q}$. Even if $I_x \notin \mathbb{Q}$, $I_x^2 \in \mathbb{Q}$ because $I_x^2 = P_x^2 + P_y^2 - I_y^2$. Accordingly, computation can be realized as far as we use I_x^2. In order to obtain rational parameters in M_{BB_i} we set $\mathcal{A}$ to be a prime Pythagorean vector; a vector $\boldsymbol{v} = (i_1, i_2 \ldots, i_n)^\top, i_1, i_2, \ldots, i_n \in \mathbb{Z}$, is a Pythagorean vector if $\|\boldsymbol{v}\| = \lambda$ where $\lambda \in \mathbb{Z}$, and a Pythagorean vector $\boldsymbol{v}$ is prime if $\gcd(i_1, i_2, \ldots, i_n, \lambda) = 1$. Note that a rotation axis whose directional vector is such a prime Pythagorean vector is called a Pythagorean axis. This assumption ensures that $\mathcal{A}.\boldsymbol{p}$ becomes

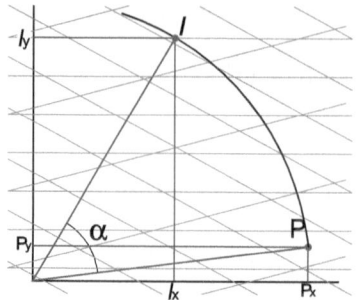

Fig. 5. A hinge angle α for a point p in a rational multi-grid

a rational value. We call a multi-grid defined from a Pythagorean axis a rational multi-grid.

In the following, we assume that all hinge angles are obtained in a rational multi-grid. In the next section we explain how rational multi-grids allow us to compare a pair of hinge angles using only integer computations in order to present a 3D discrete rotation algorithm.

5   3D Discrete Rotations around a Pythagorean Axis

In this section, we develop a 3D discrete rotation. This rotation is the extension to the 3D space of the 2D discrete rotation given in [12]. In order to estimate the complexity of the algorithm described in this section, we need to estimate the number of hinge angles existing for the image to rotate. In the case of 2D hinge angles, the upper bound of the number of different hinge angles for an image is given in [1,8]; we will use a similar method. Our algorithm, to be efficient, needs to compare a pair of hinge angles in constant time. Besides, in order to keep integer computation, we need that the comparison is done using only integers. Note that the hinge angle comparison in constant time is not indeed trivial due to our integer computation constraint. After giving the upper bound of the number of hinge angles for an image, we will show how to compare a pair of hinge angles in constant time and with integer computation, and then present a 3D discrete rotation algorithm for a given image.

In Section 5 we assume that $\mathcal{A}$ is given to be a prime Pythagorean vector.

5.1   Comparing Hinge Angles with Integer Computations

From the integer representation of hinge angles we can obtain the sine and cosine of a hinge angle $\alpha(p_x, p_y, p_z, \pm i, k)$. The following equations are derived from Figure 5:

$$\cos \alpha = \frac{P_x I_x + P_y I_y}{P_x^2 + P_y^2},\tag{9}$$

$$\sin \alpha = \frac{P_x I_y - P_y I_x}{P_x^2 + P_y^2}, \tag{10}$$

where P_x, P_y, I_x, I_y are values obtained as described in Section 4.2.

Proposition 3. *Let α_1 and α_2 be two hinge angles defined for $\mathcal{A}$. Then it is possible to decide if $\alpha_1 > \alpha_2$ using only integer computations.*

Proof.   *Let $\alpha_1 = \alpha(p_{1x}, p_{1y}, p_{1z}, \pm i_1, k_1)$ and $\alpha_2 = \alpha(p_{2x}, p_{2y}, p_{2z}, \pm i_2, k_2)$. Comparing α_1 and α_2 is equivalent to comparing their sines and cosines which are given in (9) and (10). First we compare the signs of both sine and cosine between α_1 and α_2. If the sines and cosines of both angles have different signs, then we can conclude whether $\alpha_1 > \alpha_2$ without other computation. Otherwise, without loss of generality, we can assume that both α_1 and α_2 belong to $[0, \frac{\pi}{2}]$, so that $\cos \alpha_i \geq 0$ and $\sin \alpha_i \geq 0$ for both $i = 1, 2$. As the method for comparing two sines is similar to the one for comparing two cosines, we will only show the later one.*

If α_1 is greater than α_2, $\cos \alpha_2 - \cos \alpha_1 > 0$. Thus we have, from (9),

$$(P_{1x}^2 + P_{1y}^2)(P_{2x} I_{2x} + P_{2y} I_{2y}) > (P_{2x}^2 + P_{2y}^2)(P_{1x} I_{1x} + P_{1y} I_{1y}). \tag{11}$$

For simplicity, let $A_1 = (P_{1x}^2 + P_{1y}^2) P_{2x} I_{2x}$, $B_1 = (P_{1x}^2 + P_{1y}^2) P_{2y} I_{2y}$, $A_2 = (P_{2x}^2 + P_{2y}^2) P_{1x} I_{1x}$ and $B_2 = (P_{2x}^2 + P_{2y}^2) P_{1y} I_{1y}$. Note that $A_1^2, B_1, A_2^2, B_2 \in \mathbb{Q}$. Now (11) is rewritten as

$$A_1 + B_1 > A_2 + B_2. \tag{12}$$

Squaring both sides of (12) since they are not negative, and moving rational values to the left-hand side and irrational values to the right-hand side, we obtain

$$A_1^2 + B_1^2 - A_2^2 - B_2^2 > 2A_2 B_2 - 2A_1 B_1. \tag{13}$$

If the left-hand side and the right-hand side of (13) do not have the same sign, then we can conclude whether $\alpha_1 > \alpha_2$ or $\alpha_2 > \alpha_1$. We can check the sign of both sides of (13) with integer computations since the left-hand side contains only rational numbers and the sign of the right side is the same as $A_2^2 B_2^2 - A_1^2 B_1^2$ which also contains only rational numbers. If signs of both sides are the same, assuming that they are positives, we square both sides of (13) to obtain

$$(A_1^2 + B_1^2 - A_2^2 - B_2^2)^2 - 4A_1^2 B_1^2 - 4A_2^2 B_2^2 > -8A_1 B_1 A_2 B_2. \tag{14}$$

If the sign of the left-hand side of (14) is positive, we can deduce that $\alpha_2 > \alpha_1$. Otherwise, taking the square of each side gives us

$$\left[(A_1^2 + B_1^2 - A_2^2 - B_2^2)^2 - 4A_1^2 B_1^2 - 4A_2^2 B_2^2 \right]^2 < 64 A_1^2 B_1^2 A_2^2 B_2^2. \tag{15}$$

We note that we can easily verify whether (15) is satisfied with integer computation alone. If (15) is true, $\alpha_1 < \alpha_2$; otherwise $\alpha_2 < \alpha_1$. The same logic can be applied to the case where the signs of the both sides of (13) are negative. $\square$

Thanks to Proposition 2, a hinge angle cannot have two quintuplet integer representations. Thus we can conclude that the comparison of a pair of hinge angles α_1, α_2 is always possible with integer computation if they have different quintuplets.

Note that the comparison of a pair of hinge angles is done in constant time because in the worst case, we only have to check (13),(14) and (15). It seems obvious that the comparison is done in constant time. However, keeping integer computation may increase the number of necessary comparisons. From Proposition 3, we now have a guarantee of a constant number for each comparison.

In the 2D case, is it known that the comparison between a hinge angle and a Pythagorean angle can be also done using only integers during its computation and in constant time. An angle is a Pythagorean angle if its sine and cosine belong to $\mathbb{Q}$ [8]. In the 3D case, the similar proposition is still valid. This proposition is required for our algorithm of 3D discrete rotation.

Proposition 4. *Let α be a hinge angle and θ be a Pythagorean angle defined for $\mathcal{A}$. Then it is possible to decide if $\alpha > \theta$ in constant time with integer computations.*

The proof of Proposition 4 is similar to the proof on comparison for a hinge angle and a Pythagorean angle in the plane [12]. Due to the page limitation, we skip the proof.

5.2   Upper Bound of the Number of 3D Hinge Angles

In the 2D case, the upper bound of the number of hinge angles for a given image of size $m \times m$ is known to be $O(m^3)$ [8]. This is obtained by computing the bound for the furthest point from the origin in the image and multiplying this bound by the number of points in the image. To compute the number of hinge angles in the 3D case, we will use a similar method.

In the 3D case, we assume that an image of size $m \times m \times m$ is given. The number of hinge angles for a given point $\boldsymbol{p}$ depends on the distance between $\boldsymbol{p}$ and the axis of rotation $\boldsymbol{A}$. Therefore, we define the distance function $d(\boldsymbol{p})$ that is the Euclidean distance between $\boldsymbol{A}$ and $\boldsymbol{p}$. Then, the rotation of $\boldsymbol{p}$ around $\boldsymbol{A}$ intersects at most $3\lfloor d(\boldsymbol{p}) \rfloor$ planes of the half-grid and defines at most $6\lfloor d(\boldsymbol{p}) \rfloor$ different hinge angles. Because $d(\boldsymbol{p}) \leq \sqrt{3}m$, the upper bound of the number of hinge angles for any point in the image is $6\sqrt{3}m$. Accordingly, we can conclude that the upper bound of the number of hinge angles for a given image of size $m \times m \times m$ is $6\sqrt{3}m^4$; thus $O(m^4)$.

5.3   3D Discrete Rotations Induced by Hinge Angles

In this section, we explain how to design a discrete rotation of a 3D digital image using hinge angles for a given Pythagorean axis of rotation. This method is the 3D extension of the 2D discrete rotation described in [12]. As input of such discrete rotation, we have a digital image $\mathcal{I}$ of size $m \times m \times m$, a vector $\mathcal{A}$, supposed to

Algorithm 1. Rotation of a 3D image around an integer-axis whose direction is $\mathcal{A}$ by a Pythagorean angle θ

Input: An image $\mathcal{I}$, a vector $\mathcal{A}$, a Pythagorean angle θ
Output: A rotated image $\mathcal{I}'$
1: $L <-$ An empty list;
2: **for all** Points p in $\mathcal{I}$ **do**
3:     Compute the three generic equations (6)-(8) of $\mathcal{M}_1^{\mathcal{A},p}, \mathcal{M}_2^{\mathcal{A},p}, \mathcal{M}_3^{\mathcal{A},p}$;
4:     **for all** k-th lines in $\mathcal{M}_i^{\mathcal{A},p}, i = 1, 2, 3$ **do**
5:         Compute all hinge angles corresponding to p and the k-th line
6:         of $\mathcal{M}_i^{\mathcal{A},p}$ and add $\alpha(p_x, p_y, p_z, \pm i, k)$ to the list L;
7:     **end for**
8:     Sort hinge angles corresponding to p;
9:     Search in L the greatest hinge angle α which is smaller than θ;
10:    Copy the image value from p to the rotated point with α in $\mathcal{I}'$
11: **end for**
12: **return** $\mathcal{I}'$;

---

be a prime Pythagorean vector, and an angle θ supposed to be Pythagorean [8]. The assumption that the rotation axis is a Pythagorean axis and the angle is a Pythagorean angle does not restrict the field of possible rotations because in [11] it is proved that the Pythagorean vectors are dense in the 3D space and in [5] it is proved that any angle can be approximated with a small difference $\epsilon > 0$ by a Pythagorean angle. The output is a rotated digital image $\mathcal{I}'$.

The rotation algorithm is described in Algorithm 1. For each point $p = (p_x, p_y, p_z)^\top$ in the image, the algorithm computes the corresponding multi-grid $\mathcal{M}^{\mathcal{A},p}$ and search for each line k-th in $\mathcal{M}_i^{\mathcal{A},p}, i = 1, 2, 3$, a pair of hinge angles $\alpha(p_x, p_y, p_z, \pm i, k)$. Then we stock and sort all hinge angles corresponding to p using Proposition 3. The algorithm searches in the sorted list the greatest hinge angle α which is smaller than θ using Proposition 4. This operation can be done using only integer computations thanks to our assumption that θ is a Pythagorean angle. Finally the new point after the rotation of p by α is generated in $\mathcal{I}'$.

The time complexity of this algorithm is $O(m^4 \log m)$. The computation and the sorting of all hinge angles for each point is done in $O(m \log m)$ operations because the comparison between two hinge angles is done in constant time according to Proposition 3. Searching the largest hinge angle α smaller than θ is done in $O(m)$ operations because the comparison between a hinge angle and a Pythagorean angle can be performed in constant time according to Proposition 4. Therefore, the time complexity of this algorithm is $O(m^4 \log m)$ because we repeat m^3 times the previous operations.

Note that Algorithm 1 describes a classic rotation. In [8], the author describes a 2D incremental rotation that allows obtaining all possible configurations of 2D discrete rotations for a given image. Hinge angles on rotation planes also allow us to design a 3D incremental discrete rotation.

6   Conclusion

In this paper, we extended the notion of hinge angles, introduced for 2D discrete rotation in [8,12] to the 3D. Extension of hinge angles from the 2D to 3D space involves many problems because most of properties of 2D hinge angles are not valid for 3D hinge angles. In order to regard hinge angles in the 3D space similarly to the 2D ones, we introduced the notion of a multi-grid that is the intersection of the 3D half-grid and a rotation plane. By redefining the hinge angles on the rotation plane, which are the extension of hinge angles for the rotation in 2D, we showed a subgroup of the multi-grids where all parameters are rational, called rational multi-grids. This subgroup allows us to compare two hinge angles on rotation planes in constant time by using integer computations. It also allows us to design a 3D discrete rotation.

In [9], the authors studied particular configurations induced by 2D discrete rotations, which provided their periodicity and neighborhood properties. Our future work will be, by using multi-grids and 3D hinge angles, to extend their work to 3D discrete rotations.

References

1. Amir, A., Butman, A., Crochemore, M., Landau, G.M., Schaps, M.: Two-dimensional pattern matching with rotations. Theor. Comput. Sci. 314(1-2), 173–187 (2004)
2. Andres, E.: Cercles discrets et rotations discrètes. PhD thesis, Université Louis Pasteur, Strasbourg, France (1992)
3. Andres, E.: The quasi-shear rotation. In: Miguet, S., Ubéda, S., Montanvert, A. (eds.) DGCI 1996. LNCS, vol. 1176, pp. 307–314. Springer, Heidelberg (1996)
4. Andres, E., Sibata, C., Acharya, R., Shin, K.: New methods in oblique slice generation. In: SPIE Medical Imaging 1996. SPIE, vol. 2707, pp. 580–589 (1996)
5. Anglin, W.S.: Using pythagorean triangles to approximate angles. American Mathematical Monthly 95(6), 540–541 (1988)
6. Chen, B., Kaufman, A.: 3d volume rotation using shear transformations. Graphical Models 62, 308–322 (2000)
7. Hartley, R., Kahl, F.: Global optimization through rotation space search. International Journal of Computer Vision 82(1), 64–79 (2009)
8. Nouvel, B.: Rotations discrètes et automates cellulaires. PhD thesis, Ecole Normale Supèrieure de Lyon, France (2006)
9. Nouvel, B., Rémila, E.: Configurations induced by discrete rotations: periodicity and quasi-periodicity properties. Discrete Applied Mathematics 147(2-3), 325–343 (2005)
10. Schmidt, J., Niemann, H.: Using quaternions for parametrizing 3-d rotations in unconstrained nonlinear optimization. In: VMV 2001: Proceedings of the Vision Modeling and Visualization Conference, Aka GmbH, pp. 399–406 (2001)
11. Thibault, Y., et al.: Density of pythagorean n-tuples (in preparation)
12. Thibault, Y., Kenmochi, Y., Sugimoto, A.: Computing upper and lower bounds of rotation angles from digital images. Pattern Recognition 42(8), 1708–1717 (2009)
13. Toffoli, T., Quick, J.: Three-dimensional rotations by three shears. CVGIP: Graphical Model and Image Processing 59(2), 89–95 (1997)
14. Voss, K.: Discrete Images, Objects and Functions in Z^n. Springer, Heidelberg (1993)

Surface Thinning in 3D Cubical Complexes*

John Chaussard and Michel Couprie

Université Paris-Est, Laboratoire d'Informatique Gaspard-Monge,
Équipe A3SI, ESIEE Paris, France
`john.chaussard@gmail.com`, `coupriem@esiee.fr`

Abstract. We introduce a parallel thinning algorithm with directional substeps based on the collapse operation, which is guaranteed to preserve topology and to provide a thin result. Then, we propose two variants of a surface-preserving thinning scheme, based on this parallel directional thinning algorithm. Finally, we propose a methodology to produce filtered surface skeletons, based on the above thinning methods and the recently introduced discrete λ-medial axis.

Keywords: Topology, cubical complex, thinning, skeleton, collapse.

1   Introduction

The notion of skeleton plays a major role in shape analysis and recognition. Informally, the skeleton of an object X is a subset of X which is 1) thin, 2) centered in X and 3) topologically equivalent to X. (To be more precise, we say that a transformation Ψ "preserves topology" if X is homotopy-equivalent to $\Psi(X)$ for any X.) To extract skeletons from geometric objects, different methods have been proposed, relying on different frameworks: discrete geometry [8,13,19,22,14], digital topology [12,28,27,21], mathematical morphology [24,26], computational geometry [2,3,20], and partial differential equations [25].

In this paper, we focus on skeletons in the discrete 3-dimensional cubical space. In such spaces, topology preservation is usually guaranteed by the use of topology-preserving thinning methods, based *e.g.* on simple point deletion [15,11]. The centering of the skeleton may be achieved to some extent by the use of parallel thinning methods [6] or by imposing geometrical constraints such as the preservation of the centers of maximal included balls [12]. However it is generally difficult, in discrete spaces, to satisfy conditions 1), 2) and 3) together. In particular, the methods evoked above fail to guarantee a thin result.

In order to overcome these limitations, we adopt in this article the framework of cubical complexes. Abstract (cubical) complexes have been promoted in particular by V. Kovalevsky [17] in order to provide a sound topological basis for image analysis. Intuitively, a cubical complex may be thought of as a set of elements having various dimensions (*e.g.* cubes, squares, edges, vertices) glued

---

* This work has been partially supported by the "ANR BLAN07–2_184378 MicroFiss" project.

together according to certain rules (see Fig. 3). In this framework, we can say that a complex is "thin" if it does not contain any 3-dimensional element (cube). The thinning methods that will be proposed in the sequel of this paper produce thin results in this sense.

In the framework of cubical complexes, the notion of critical kernel introduced by G. Bertrand constitutes a powerful tool to study parallel homotopic thinning in any dimension, which unifies and encompasses previous works on parallel thinning [4,7]. Indeed, the very notion of critical kernel may be seen as thinning scheme, which consists of iteratively computing the critical kernel of the result of the previous step. Critical kernels may also be used to design new algorithms, as well as to check the topological validity of existing ones [5,6]. However, thinning algorithms based on critical kernels do not guarantee that the result has a dimension strictly lower than the input. Early work on thinning in cubical complexes with the aim of reducing dimension can be found in [18].

In our work, topology preservation will be ensured by the use of the *collapse operation*, which is an elementary topology-preserving transformation that has been introduced by J.H.C. Whitehead [29] and plays an important role in combinatorial topology. It can be seen as a discrete analogue of a retraction, that is, a continuous deformation of an object onto itself.

The contributions of this article are the following. We introduce a parallel thinning algorithm with directional substeps based on collapse, which is guaranteed to preserve topology and to provide a thin result. Then, we propose two variants of a surface-preserving thinning scheme, based on this parallel directional thinning algorithm. Finally, we propose a methodology to produce filtered surface skeletons, based on the above thinning methods and the recently introduced discrete λ-medial axis [9].

2   Basic Notions

2.1   Cubical Complexes

Let $\mathbb{Z}$ be the set of integers, we consider the family of sets $\mathbb{F}_0^1$ and $\mathbb{F}_1^1$, such that $\mathbb{F}_0^1 = \{\{a\} \mid a \in \mathbb{Z}\}$ and $\mathbb{F}_1^1 = \{\{a, a+1\} \mid a \in \mathbb{Z}\}$. Any subset f of $\mathbb{Z}^n$ such that f is the cartesian product of m elements of $\mathbb{F}_1^1$ and $(n-m)$ elements of $\mathbb{F}_0^1$ is called a face or an *m-face* of $\mathbb{Z}^n$, m is the dimension of f, we write $dim(f) = m$. A 0-face is called a *vertex*, a 1-face is an *edge*, a 2-face is a *square*, and a 3-face is a *cube* (see Fig. 1).

We denote by $\mathbb{F}^n$ the set composed of all faces in $\mathbb{Z}^n$.

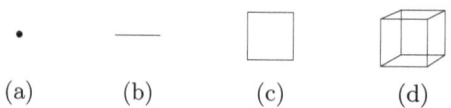

(a)   (b)   (c)   (d)

Fig. 1. Graphical representations of: (a) a 0-face, (b) a 1-face, (c) a 2-face, (d) a 3-face

Let $f \in \mathbb{F}^n$. We set $\hat{f} = \{g \in \mathbb{F}^n | g \subseteq f\}$, and $\hat{f}^* = \hat{f} \setminus \{f\}$. Any element of $\hat{f}$ is *a face of* f, and any element of $\hat{f}^*$ is *a proper face of* f. We call *star of* f the set $\check{f} = \{g \in \mathbb{F}^n | f \subseteq g\}$, and we write $\check{f}^* = \check{f} \setminus \{f\}$: any element of $\check{f}$ is a *coface of* f. It is plain that $g \in \hat{f}$ iff $f \in \check{g}$.

A set X of faces in $\mathbb{F}^n$ is a *cell*, or *m-cell*, if there exists an m-face $f \in X$ such that $X = \hat{f}$. The *closure* of a set of faces X is the set $X^- = \cup\{\hat{f} | f \in X\}$. The set $\overline{X}$ is $\mathbb{F}^n \setminus X$.

A finite set X of faces in $\mathbb{F}^n$ is a cubical *complex* if $X = X^-$, and we write $X \preceq \mathbb{F}^n$. Any subset Y of X which is also a complex is a *subcomplex of* X, and we write $Y \preceq X$.

A face $f \in X$ is *a facet of* X if f is not a proper face of any face of X. We denote by X^+ the set composed of all facets of X. A complex X is *pure* if all its facets have the same dimension. The *dimension of* X is $\dim(X) = \max\{\dim(f) | f \in X\}$. If $\dim(X) = d$, then we say that X is a d-complex.

2.2 From Binary Images to Complexes

Traditionally, a binary image (see Fig. 2a) is represented as a subset of $\mathbb{Z}^2$ (or $\mathbb{Z}^3$ for 3D images). Let $S \subseteq \mathbb{Z}^n$, the elements of S represent the pixels (or voxels) of the image, often called object pixels or black pixels. The set $\overline{S} = \mathbb{Z}^n \setminus S$ represents the background (set of white pixels or voxels).

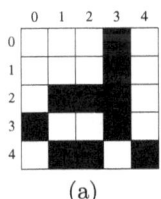
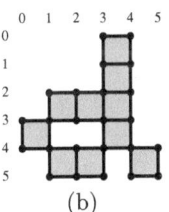

(a)   (b)

Fig. 2. (a): A binary image, represented by a set $S \subseteq \mathbb{Z}^2$ (black pixels). (b): The complex $X - \Phi(S)^-$.

To obtain a complex from a subset of $\mathbb{Z}^n$, we consider the following bijection which associates, to each point of $\mathbb{Z}^n$, a facet of $\mathbb{F}^n$. Let $x = (x_1, \ldots, x_n) \in \mathbb{Z}^n$, define $\Phi(x) = \{x_1, x_1 + 1\} \times \ldots \times \{x_n, x_n + 1\}$. See Fig. 2 an illustration in 2D, where the image of each pixel by Φ is a 2-face (a square). For example, the black pixel x in the first row has coordinates $(3, 0)$; the corresponding 2-face is $\Phi(x) = \{3, 4\} \times \{0, 1\} = \{(3, 0), (3, 1), (4, 0), (4, 1)\}$. The map Φ is straightforwardly extended to sets: $\Phi(S) = \{\Phi(x) | x \in S\}$. Then, the complex that we associate to a set S is $\Phi(S)^-$ (See Fig. 2b). Notice that the datum of an nD binary image is equivalent to the datum of a pure n-complex in $\mathbb{F}^n$.

2.3 Collapse

The collapse operation consists of removing two distinct elements (f, g) from a complex X under the condition that g is contained in f and is not contained

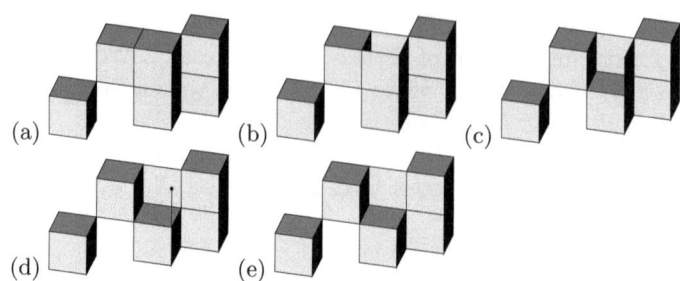

Fig. 3. (a-e): A series of complexes that illustrate a sequence of collapse operations

in any other element of X. This operation may be repeated several times (see Fig. 3). A more precise definition follows.

Let $X \preceq \mathbb{F}^n$, and let f, g be two faces of X. The face g is *free for X*, and the pair (f, g) is *a free pair for X* if f is the only face of X such that g is a proper face of f. In other terms, (f, g) is a free pair for X whenever $\check{g}^* \cap X = \{f\}$. In this case, we say that the complex $X \setminus \{f, g\}$ is an *elementary collapse of X*. It can be easily seen that if (f, g) is a free pair for X and $\dim(f) = m$, then f is a facet and $\dim(g) = (m - 1)$.

Let $X \preceq \mathbb{F}^n$ and $Y \preceq \mathbb{F}^n$ be two complexes. We say that X *collapses onto Y* if there exists a sequence of complexes $(X_0, ..., X_\ell)$ of $\mathbb{F}^n$ such that $X = X_0$, $Y = X_\ell$ and for all $i \in \{1, ..., \ell\}$, X_i is an elementary collapse of X_{i-1} (see Fig. 3).

Let f_0, f_ℓ be two n-faces of $\mathbb{F}^n$. An $(n-1)$-*path from f_0 to f_ℓ* is a sequence $\pi = (f_0, ..., f_\ell)$ of faces of $\mathbb{F}^n$ such that for all $i \in \{0, ..., \ell\}$, either i is even and f_i is an n-face, or i is odd and f_i is an $(n-1)$-face with $\check{f_i}^* = \{f_{i-1}, f_{i+1}\}$.

The following proposition will serve us to prove the thinness of our skeletons.

Proposition 1. *Let $X \preceq \mathbb{F}^n$ be an n-complex, with $n > 0$. Then X has at least one free $(n-1)$-face.*

Proof. Since X is an n-complex (hence X is finite) there exists an n-face a in X and an n-face b in $\overline{X}$. Obviously, there exists an $(n-1)$-path from a to b. Let h be the first n-face of π that is not in X, let k be the last n-face of π before h (thus k is in X), and let $e = k \cap h$ be the $(n-1)$-face of π between k and h. Since k and h are the only two n-faces of $\mathbb{F}^n$ that contain e, we see that the pair (k, e) is free for X. □

3   Parallel Directional Thinning Based on Collapse

The most "natural" way to thin an object consists of removing some of its border elements in parallel, in a symmetrical manner. However, parallel deletion of free pairs does not, in general, guarantee topology preservation: see for example Fig. 4, where simultaneously removing all free pairs would split the object into separate components.

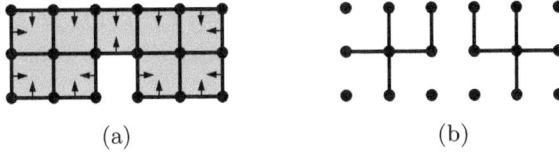

Fig. 4. (a): A 2-complex X, each free pair (f_i, g_i) is symbolized by an arrow from g_i to f_i. (b): The complex obtained by removing simultaneously from X all the pairs that are free for X.

In the framework of 2D digital topology, a popular method due to A. Rosenfeld [23] consists of dividing each thinning step into directional substeps. In each substep, only simple points that have no neighbor belonging to the object in one of the four main directions (north, south, east, west) are candidates for deletion.

However, this method cannot be straightfordly extended to 3D. In fact, the question of knowing whether this strategy has a "natural extension" to 3D was cited among three open questions relative to digital topology by Kong, Litherland and Rosenfeld in [16] (question 547). An answer to this question was recently given in [7], based on the critical kernels framework. Here, we show that the directional strategy can be directly adapted in the framework of cubical complexes, and provides interesting results.

First, we need to define the *direction* and the *orientation* of a free face.

Let $f \in \mathbb{F}^n$, the *center of* f is the center of mass of the points in f, that is, $c_f = \frac{1}{|f|} \sum_{a \in f} a$. The center of f is an element of $[\frac{\mathbb{Z}}{2}]^n$, where $\frac{\mathbb{Z}}{2}$ denotes the set of half integers. Let $X \preceq \mathbb{F}^n$, let (f, g) be a free pair for X, and let c_f and c_g be the respective centers of the faces f and g. We denote by $V(f, g)$ the vector $(c_f - c_g)$ of $[\frac{\mathbb{Z}}{2}]^n$.

Let (f, g) be a free pair, the vector $V(f, g)$ has only one non-null coordinate. We define $Dir(f, g)$ as the index of the non-null coordinate of $V(f, g)$. Thus, $Dir()$ is a surjective function from $\mathbb{F}^n \times \mathbb{F}^n$ to $\{1, \ldots, n\}$ such that, for all free pairs (f, g) and (i, j) for X, $Dir(f, g) = Dir(i, j)$ if and only if $V(f, g)$ and $V(i, j)$ are collinear. The number $Dir(f, g)$ is called the *direction* of the free pair (f, g). The free pair (f, g) has a *positive orientation*, and we write $Orient(f, g) = 1$, if the non-null coordinate of $V(f, g)$ is positive; otherwise (f, g) has a *negative orientation*, and we write $Orient(f, g) = 0$.

Let us now state an elementary property of collapse, which gives a necessary and sufficient condition under which two collapse operations may be performed in parallel while preserving topology.

Proposition 2. *Let $X \preceq \mathbb{F}^n$, and let (f, g) and (k, ℓ) be two distinct free pairs for X. The complex X collapses onto $X \setminus \{f, g, k, \ell\}$ if and only if $f \neq k$.*

Proof. If $f = k$, then it is plain that (k, ℓ) is not a free pair for $Y = X \setminus \{f, g\}$ as $k = f \notin Y$. Also, (f, g) is not free for $X \setminus \{k, \ell\}$. If $f \neq k$, then we have $g \neq \ell$, $\breve{g}^* \cap X = \{f\}$ (g is free for X) and $\breve{\ell}^* \cap X = \{k\}$ (ℓ is free for X). Thus, we have $\breve{\ell}^* \cap Y = \{k\}$ as $\ell \neq g$ and $k \neq f$. Therefore, (k, ℓ) is a free pair for Y. $\square$

From Prop. 2, the following corollary is immediate.

Corollary 1. *Let* $X \preceq \mathbb{F}^n$, *and let* $(f_1, g_1), \ldots, (f_m, g_m)$ *be* m *distinct free pairs for* X *such that, for all* $a, b \in \{1, \ldots, m\}$ *(with* $a \neq b$*),* $f_a \neq f_b$. *The complex* X *collapses onto* $X \setminus \{f_1, g_1, \ldots, f_m, g_m\}$.

Considering two distinct free pairs (f, g) and (i, j) for $X \preceq \mathbb{F}^n$ such that $Dir(f, g)$ $= Dir(i, j)$ and $Orient(f, g) = Orient(i, j)$, we have $f \neq i$. From this observation and Cor. 1, we deduce the following property.

Corollary 2. *Let* $X \preceq \mathbb{F}^n$, *and let* $(f_1, g_1), \ldots, (f_m, g_m)$ *be* m *distinct free pairs for* X *having all the same direction and the same orientation. The complex* X *collapses onto* $X \setminus \{f_1, g_1, \ldots, f_m, g_m\}$.

We say that a d-face of X is a *border face* if it contains a free $(d-1)$-face. Define $Border(X)$ as the set of all border faces of X. We are now ready to introduce our directional thinning algorithm (Alg. 1).

Algorithm 1. *ParDirCollapse*(X, W, ℓ)

Data: A cubical complex $X \preceq \mathbb{F}^n$, a subset $W \subseteq X$ that contains faces of X which should not be removed, and $\ell \in \mathbb{N}$, the number of layers of free faces which should be removed from X

1 **while** *there exists free pairs for* X *in* $X \setminus W$ *and* $\ell > 0$ **do**
2     $L = Border(X)^-$;
3   **for** $t = 1 \to n$ **do**
4   **for** $s = 0 \to 1$ **do**
5   **for** $d = n \to 1$ **do**
6              $E = \{(f, g)$ free for $X \mid g \notin W, f \notin W,$
                     $Dir(f, g) = t, Orient(f, g) = s, \dim(f) = d\}$;
7              $G = \{(f, g) \in E \mid f \in L$ and $g \in L\}$;
8              $X = X \setminus G$;

9     $l = l - 1$;
10 **return** X;

Let us first comment on the notion of "layer" and the role of the sets L and G (lines 2 and 7). Intuitively, we want a single execution of the loop line 1 to only affect facets that were on the border of X at the beginning of the loop, in order to ensure that the thinning will not reduce certain parts more quickly than others due to the scanning of directions. Thus, by line 7 we ensure that the layer which is removed in one iteration of the loop line 1 is included in L, which contains all the border facets of the complex X and all the faces included in those.

Remark that different definitions of the functions $Dir()$ and $Orient()$ could be given, corresponding to different orders in which directions and orientations are scanned in Alg. 1, and yielding different results. However in general some arbitrary choices must be made in order to thin a 3-complex into a 2-complex.

The choice of these two functions is the only arbitrary choice to be made, after which the results are uniquely defined.

For any complex $X \preceq \mathbb{F}^n$, any subset $W \subseteq X$ and for any $\ell \in \mathbb{N}$, it follows from Cor. 2 that X collapses onto $ParDirCollapse(X, W, \ell)$. Furthermore, if W does not contain any n-face nor any $(n-1)$-face that is not a facet of X, and if $\ell = +\infty$, then by Prop. 1 it can easily be deduced that $ParDirCollapse(X, W, \ell)$ contains no n-face.

Notice that checking whether a face is free or not, is quite easy to implement and can be done in constant time, whatever the dimension of the face, if n is considered as fixed. Indeed, Alg. 1 can be used to thin a cubical complex of any dimension. When a free pair (f, g) is removed from X, it is sufficient to scan the faces of f in order to check the appearance of new free faces, avoiding to search the whole complex for free faces. Thus, Alg. 1 may be easily implemented to run in linear time complexity (proportionally to the number of faces of the complex).

As shown in Fig.5, when the input complex has a nearly constant thickness, it is possible to obtain a surface skeleton of X by choosing a convenient value of ℓ as the last parameter of Alg. 1. However, in most cases, it is not possible to find a value of ℓ which is satisfying for the whole complex. In the next section, we explain how to use the directional thinning strategy to obtain a surface skeleton from a complex without having to tune any "thickness" parameter.

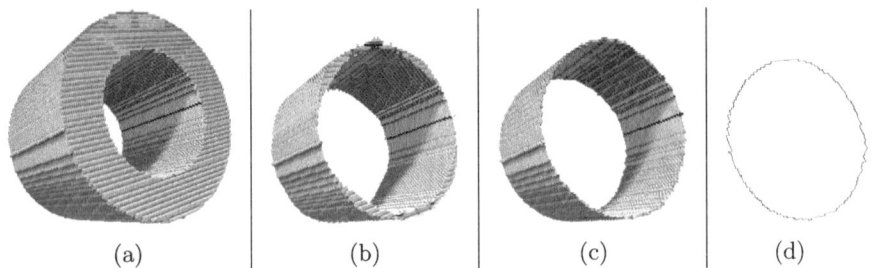

(a)   (b)   (c)   (d)

Fig. 5. (a): A 3-complex X having the shape of a thick tube with a skewed axis. (b): The result of $ParDirCollapse(X, \emptyset, 10)$ is not thin, some 3-faces remain in the object. (c): The result of $ParDirCollapse(X, \emptyset, 15)$ is a 2-complex. (d): The result of $ParDirCollapse(X, \emptyset, \infty)$ does not contain any cube nor any square, it is composed of edges and points.

4   Detection of Surfaces

In this section, two algorithms are proposed in order to perform a surface-preserving thinning of a cubical complex. Both algorithms are based on the same principle: a single layer of faces is removed using Alg. 1, then some 2-faces are detected and kept safe from any future removal. This process is repeated until stability.

Algorithm 2. *CollapseSurface*(X, W)

Data: A three-dimensional cubical complex $X \preceq \mathbb{F}^3$, a subset $W \subseteq X$ that
contains faces of X which should not be removed

1 **while** *there exists free pairs for X in $X \backslash W$* **do**
2   *ParDirCollapse*($X, W, 1$);
3     $W = W \cup \{f \in X^+ \mid \dim(f) = 2\}$;

4 **return** X;

For Alg. 2 (*CollapseSurface*), the detected 2-faces are simply those that are
not contained in any 3-face of the current complex, *i.e.*, the 2-facets of X.

Fig. 6a shows the result of *CollapseSurface*($X, \emptyset$), where X is the tube shown
in Fig. 5a. The resulting complex is a 2-complex containing some "branches"
and "surface patches" (as shown on the detailed view) which do not represent
significant "surfacic features" of the original object. This motivates the introduction of a variant of this method, which imposes a more restrictive condition
to preserve 2-facets from removal.

We say that a 2-facet $f \in X^+$ is a *2d-isthmus of X* if it contains no free edge,
in other words, if each edge in f is included in a facet of X distinct from f.
In Alg. 3 (*CollapseIsthmus*), the 2-facets that are detected and kept safe from
further removal are the 2d-isthmuses of X.

Algorithm 3. *CollapseIsthmus*(X, W)

Data: A three-dimensional cubical complex $X \preceq \mathbb{F}^3$, a subset $W \subseteq X$ that
contains faces of X which should not be removed

1 **while** *there exists free pairs for X in $X \backslash W$* **do**
2   *ParDirCollapse*($X, W, 1$);
3     $W = W \cup \{f \in X \mid f \text{ is a 2d-isthmus of } X\}$;

4 **return** X;

Notice that, following Prop. 1, the results of Alg. 2 and Alg. 3 are thin (they
do not contain any 3-face), provided that the input parameter W is set to $\emptyset$.
However, thanks to parameter W, it is possible to constrain these algorithms
to preserve selected parts of the original object, in addition to the surface parts
that are automatically detected.

Fig. 6b shows the result of *CollapseIsthmus*($X, \emptyset$), where X is the skewed
tube shown in Fig. 5a. The resulting complex contains less branches and spurious surfaces (as shown on the detailed view) than the complex obtained with
CollapseSurface($X, \emptyset$).

The results obtained show that both algorithms presented above allows one to
obtain a 2-dimensional skeleton from a 3-complex, containing important "shape
information" from the initial object. However, even if Alg. 3 produces better
results than Alg. 2, it fails in obtaining a completely satisfactory skeleton. In the

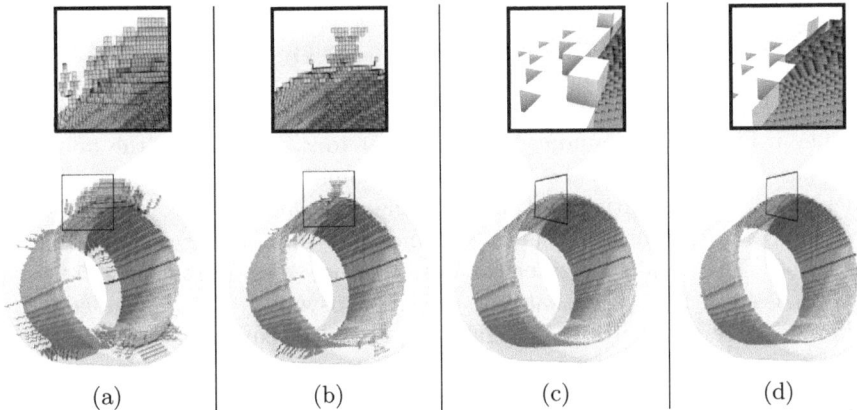

(a)   (b)   (c)   (d)

Fig. 6. Results of different algorithms performed on the complex X presented in Fig 5a. (a) : *CollapseSurface*$(X, \emptyset)$. (b) : *CollapseIsthmus*$(X, \emptyset)$. (c) : The discrete λ-medial axis (Sec. 5) of X, with $\lambda = 18$. (d) : Result of the method described in Sec. 5.3 performed on X, with $\lambda = 18$.

next section, we introduce a methodology which includes a filtering step based on the Euclidean distance, providing a better robustness to contour irregularities and a lower sensitivity to the orientation of shape features in the space.

5   Skeleton Filtering

A major difficulty when using the skeleton in applications (*e.g.*, shape recognition, shape analysis), is its sensitivity to small contour perturbations, in other words, its lack of stability. A recent survey [1] summarizes selected relevant studies dealing with this topic. This difficulty can be expressed mathematically: the transformation which associates a shape to its skeleton (or medial axis) is only semi-continuous. This fact, among others, explains why it is usually necessary to add a filtering step (or pruning step) to any method that aims at computing a skeleton. Hence, there is a rich literature devoted to skeleton pruning, in which different criteria were proposed in order to discard "spurious" skeleton points or branches.

In 2005, Chazal and Lieutier [10] introduced the λ-medial axis and studied its properties, in particular those related to stability. A major outcome of [10] is the following property: informally, for "regular" values of λ, the λ-medial axis of X remains stable under perturbations of $\overline{X}$ that are small with regard to the Hausdorff distance. Typical non-regular values are radii of locally largest maximal balls.

5.1   Intuitive Presentation of the λ-Medial Axis

Consider a bounded subset X of $\mathbb{R}^n$, as for example, for $n = 2$, the region enclosed by the solid curve depicted in Fig. 7 (left). The *medial axis of X* consists of the points of X that have several closest points on the boundary of X. For example in Fig. 7, the boundary points closest to x are a and b, the boundary points closest to x' are a' and b', and the only boundary point closest to x'' is a''.

Let λ be a non-negative real number, the *λ-medial axis of X* is the set of points x of X such that the smallest ball including all boundary points that are closest to x, has a radius greater than or equal to λ. Notice that the 0-medial axis of X is equal to X, and that any λ-medial axis with $\lambda > 0$ is included in the medial axis. We show in Fig. 7 (right) two λ-medial axes with different values of λ.

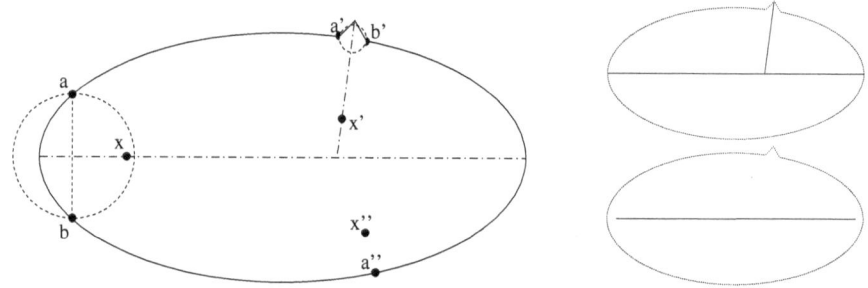

Fig. 7. Illustration of the λ-medial axis. Left: Points x, x' and x'' and their respective closest boundary points. Top right: λ-medial axis with $\lambda = \epsilon$, a very small positive real number. Bottom right: λ-medial axis with $\lambda = d(a', b') + \epsilon$, where $d(a', b')$ is the distance between a' and b'.

Intuitively, in 2D, every "bump" on the boundary of the object generates a medial axis branch. The goal of the filtering is to eliminate those branches which are due to unsignificant contour irregularities (noise). Roughly speaking, in the λ-medial axis, the branches that remain correspond to contour features having a "width" greater than the parameter value λ.

5.2   Definition of the Discrete λ-Medial Axis

The original definition of the λ-medial axis holds and make sense in the continuous Euclidean n-dimensional space. In [9], we introduced the definition of a discrete λ-medial axis (DLMA) in $\mathbb{Z}^n$. We evaluated experimentally its stability and rotation invariance. Furthermore, we introduced a variant of the DLMA which may be computed in linear time, for which the results are very close to those of the DLMA.

Notice that the DLMA applies on a binary image (*i.e.*, a set of voxels or a subset of $\mathbb{Z}^3$), not on a complex.

Let $x, y \in \mathbb{R}^n$, we denote by $d(x, y)$ the Euclidean distance between x and y, in other words, $d(x, y) = (\sum_{k=1}^{n}(y_k - x_k)^2)^{\frac{1}{2}}$. Let $S \subseteq \mathbb{R}^n$, we set $d(y, S) = \min_{x \in S}\{d(y, x)\}$.

Let $x \in \mathbb{R}^n, r \in \mathbb{R}^+$, we denote by $B_r(x)$ the *ball of radius r centered on x*, defined by $B_r(x) = \{y \in \mathbb{R}^n \mid d(x, y) \leq r\}$.

For each point $x \in \mathbb{Z}^n$, we define the *direct neighborhood of x* as $N(x) = \{y \in \mathbb{Z}^n \mid d(x, y) \leq 1\}$. The direct neighborhood comprises $2n + 1$ points.

Let S be a nonempty subset of $\mathbb{Z}^n$, and let $x \in \mathbb{Z}^n$. The *projection of x on S*, denoted by $\Pi_S(x)$, is the set of points y of S which are at minimal distance from x ; more precisely,

$$\Pi_S(x) = \{y \in S \mid \forall z \in S, d(y, x) \leq d(z, x)\}.$$

Let S be a finite subset of $\mathbb{Z}^n$, we denote by $R(S)$ the radius of the smallest ball enclosing S, that is, $R(S) = \min\{r \in \mathbb{R} \mid \exists y \in \mathbb{R}^n, B_r(y) \supseteq S\}$.

Transposing directly the definition of the λ-medial axis to the discrete grid $\mathbb{Z}^n$ would yield unsatisfactory results (see [9]), this is why we need the following notion. Let $S \subseteq \mathbb{Z}^n$, and let $x \in S$. The *extended projection of x on $\overline{S}$*, denoted by $\Pi_{\overline{S}}^e(x)$, is the union of the sets $\Pi_{\overline{S}}(y)$, for all y in $N(x)$ such that $d(y, \overline{S}) \leq d(x, \overline{S})$.

Let S be a finite subset of $\mathbb{Z}^n$, and let $\lambda \in \mathbb{R}^+$. We define the function $\mathcal{F}_S$ which associates, to each point x of S, the value $\mathcal{F}_S(x) = R(\Pi_{\overline{S}}^e(x))$. The *discrete λ-medial axis of S*, denoted by $DLMA(S, \lambda)$, is the set of points x in S such that $\mathcal{F}_S(x) \geq \lambda$.

Fig. 6c shows the DLMA of the object presented in Fig. 5a, with $\lambda = 18$. For more details, illustrations and performance analysis, see [9].

5.3   Surface Skeleton Based on the DLMA

Let us now describe our methodology to obtain a thin filtered surface skeleton from a binary image (a set of voxels). Let S denote our original set, a finite subset of $\mathbb{Z}^3$.

First, we compute the discrete λ-medial axis of S for a chosen value of λ. From here, we will consider the complex $X = \Phi(S)^-$ and the set $W = \Phi(DLMA(S, \lambda))$.

Notice that we have no guarantee that the complex W^- be topologically equivalent to the complex X (see counter-examples in [9]). Thus our next step consists of computing, using Alg. 1, the complex $Y = ParDirCollapse(X, W, +\infty)$ that contains W and that is topologically equivalent to X.

Now, as W is made of 3-faces (see the close-up of Fig. 6c), Y is a 3-complex and is not thin in the sense that we consider in this paper. Then to achieve our initial goal we compute $Z = CollapseIsthmus(Y, \emptyset)$ which is both thin (following Prop. 1 there is no 3-face in this complex, see the close-up of Fig. 6d for an illustration) and topologically equivalent to X (by Cor. 2). The centering of Z in X is achieved thanks to the use of the DLMA, based on the Euclidean distance. The parameter λ can be tuned in order to adjust the filtering to the characteristics (size, smoothness of contours . . .) of the input shape, and to the requirements of the user.

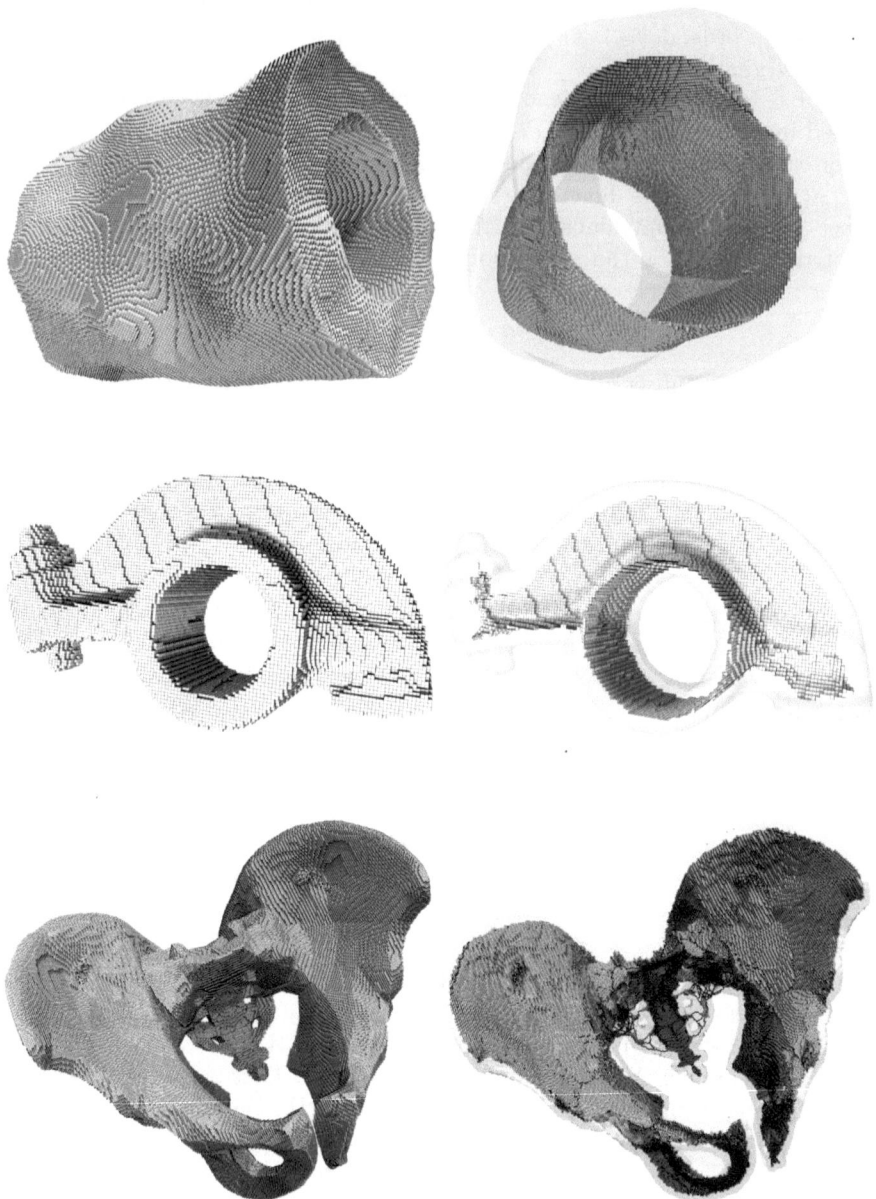

Fig. 8. Results of the methodology described in Sec. 5.3. On the left, the original object; on the right, the result obtained by our method. First row: A bumped and skewed tube, $\lambda = 20$. Second row: A rocker arm, $\lambda = 8$. Third row: A pelvis bone, $\lambda = 5$.

Figures 6d and 8 show various results using this methodology. It can be seen that the resulting 2-complexes indeed capture the main surfacic features of the original objects, without spurious branches or surface patches. A significant advantage of the 2D nature of the obtained skeletons, is to enable an easy analysis of important shape features such as intersections of surface parts.

6   Conclusion

We introduced in this work a methodology for obtaining filtered, thin surface skeletons of 3D objects. This methodology is based on a robust medial axis extraction method (DLMA), a parallel directional thinning algorithm based on collapse, and a surface detection strategy. The algorithms involved in this methodology can be implemented to run in linear time complexity.

References

1. Attali, D., Boissonnat, J.-D., Edelsbrunner, H.: Stability and computation of the medial axis — a state-of-the-art report. In: Möller, T., Hamann, B., Russell, B. (eds.) Mathematical Foundations of Scientific Visualization, Computer Graphics, and Massive Data Exploration. LNCS, pp. 1–19. Springer, Heidelberg (to appear, 2009)
2. Attali, D., Lachaud, J.O.: Delaunay conforming iso-surface, skeleton extraction and noise removal. Computational Geometry: Theory and Applications 19, 175–189 (2001)
3. Attali, D., Montanvert, A.: Modelling noise for a better simplification of skeletons. In: Proc. International Conference on Image Processing (ICIP), vol. 3, pp. 13–16 (1996)
4. Bertrand, G.: On critical kernels. Comptes Rendus de l'Académie des Sciences, Série Math. I(345), 363–367 (2007)
5. Bertrand, G., Couprie, M.: Two-dimensional parallel thinning algorithms based on critical kernels. Journal of Mathematical Imaging and Vision 31(1), 35–56 (2008)
6. Bertrand, G., Couprie, M.: A new 3D parallel thinning scheme based on critical kernels. In: Kuba, A., Nyúl, L.G., Palágyi, K. (eds.) DGCI 2006. LNCS, vol. 4245, pp. 580–591. Springer, Heidelberg (2006)
7. Bertrand, G., Couprie, M.: On parallel thinning algorithms: minimal non-simple sets, P-simple points and critical kernels. Journal of Mathematical Imaging and Vision 35(1), 23–35 (2009)
8. Borgefors, G., Ragnemalm, I., Sanniti di Baja, G.: The Euclidean distance transform: finding the local maxima and reconstructing the shape. In: Proc. of the 7th Scandinavian Conference on Image Analysis, vol. 2, pp. 974–981 (1991)
9. Chaussard, J., Couprie, M., Talbot, H.: A discrete lambda-medial axis. In: Brlek, S., Reutenauer, C., Provençal, X. (eds.) DGCI 2009. LNCS, vol. 5810, pp. 421–433. Springer, Heidelberg (2009)
10. Chazal, F., Lieutier, A.: The lambda medial axis. Graphical Models 67(4), 304–331 (2005)
11. Couprie, M., Bertrand, G.: New characterizations of simple points in 2D, 3D and 4D discrete spaces. IEEE Trans. on Pattern Analysis and Machine Intelligence 31(4), 637–648 (2009)

12. Davies, E.R., Plummer, A.P.N.: Thinning algorithms: a critique and a new methodology. Pattern Recognition 14, 53–63 (1981)
13. Ge, Y., Fitzpatrick, J.M.: On the generation of skeletons from discrete Euclidean distance maps. IEEE Trans. on Pattern Analysis and Machine Intelligence 18(11), 1055–1066 (1996)
14. Hesselink, W.H., Roerdink, J.B.T.M.: Euclidean skeletons of digital image and volume data in linear time by the integer medial axis transform. IEEE Trans. on Pattern Analysis and Machine Intelligence 30(12), 2204–2217 (2008)
15. Kong, T.Y., Rosenfeld, A.: Digital topology: introduction and survey. Computer Vision, Graphics and Image Processing 48, 357–393 (1989)
16. Kong, T.Y., Litherland, R., Rosenfeld, A.: Problems in the topology of binary digital images. In: Open problems in topology, pp. 376–385. Elsevier, Amsterdam (1990)
17. Kovalevsky, V.A.: Finite topology as applied to image analysis. Computer Vision, Graphics and Image Processing 46, 141–161 (1989)
18. Liu, L.: 3d thinning on cell complexes for computing curve and surface skeletons. Master's thesis, Washington University in Saint Louis (May 2009)
19. Malandain, G., Fernández-Vidal, S.: Euclidean skeletons. Image and Vision Computing 16, 317–327 (1998)
20. Ogniewicz, R.L., Kübler, O.: Hierarchic Voronoi skeletons. Pattern Recognition 28(33), 343–359 (1995)
21. Pudney, C.: Distance-ordered homotopic thinning: a skeletonization algorithm for 3D digital images. Computer Vision and Image Understanding 72(3), 404–413 (1998)
22. Rémy, E., Thiel, E.: Exact medial axis with Euclidean distance. Image and Vision Computing 23(2), 167–175 (2005)
23. Rosenfeld, A.: A characterization of parallel thinning algorithms. Information and Control 29, 286–291 (1975)
24. Serra, J.: Image analysis and mathematical morphology. Academic Press, London (1982)
25. Siddiqi, K., Bouix, S., Tannenbaum, A., Zucker, S.: The Hamilton-Jacobi skeleton. In: International Conference on Computer Vision (ICCV), pp. 828–834 (1999)
26. Soille, P.: Morphological image analysis. Springer, Heidelberg (1999)
27. Talbot, H., Vincent, L.: Euclidean skeletons and conditional bisectors. In: Proc. VCIP 1992, SPIE, vol. 1818, pp. 862–876 (1992)
28. Vincent, L.: Efficient computation of various types of skeletons. In: Proc. Medical Imaging V, SPIE, vol. 1445, pp. 297–311 (1991)
29. Whitehead, J.H.C.: Simplicial spaces, nuclei and m-groups. Proceedings of the London Mathematical Society 45(2), 243–327 (1939)

A General Bayesian Markov Random Field Model for Probabilistic Image Segmentation

Oscar Dalmau and Mariano Rivera

Centro de Investigación en Matemáticas (CIMAT),
A.C. Jalisco S/N, Colonia Valenciana, C.P. 36240, Guanajuato, Gto. México
{dalmau,mrivera}@cimat.mx

Abstract. We propose a general Bayesian model for image segmentation with spatial coherence through a Markov Random Field prior. We also study variants of the model and their relationship. In this work we use the Matusita Distance, although our formulation admits other metric-divergences. Our main contributions in this work are the following. We propose a general MRF-based model for image segmentation. We study a model based on the Matusita Distance, whose solution is found directly in the discrete space with the advantage of working in a continuous space. We show experimentally that this model is competitive with other models of the state of the art. We propose a novel way to deal with non-linearities (irrational) related with the Matusita Distance. Finally, we propose an optimization method that allows us to obtain a hard image segmentation almost in real time and also prove its convergence.

Keywords: segmentation, Markov random field, Matusita distance.

1 Introduction

Image segmentation is an important field in computer vision. It has been one of the most studied tasks in image processing and is considered to be a bridge between low and high level image processing tasks. The image segmentation problem consists of obtaining a partition of the image according to a homogeneous predicate. Depending on the image modeling viewpoint, many strategies have been proposed. If the image is modeled from a deterministic point of view, we can find different image segmentation approaches based on the Mumford-Shah functional [11,7]. When it is modeled as a graph [19] we find Graph Cut [3] and Normalized Cut [17]. In the context of data clustering, the fuzzy c-means (FCM) methods are widely used [1,5] and if the image is modeled as a Markov Random Field (MRF) several approaches have been reported [10,9,16]. Among them, Bayesian formulation, including spatial coherence through a MRF prior, has shown to be a powerful framework to design efficient and robust models for image segmentation.

In this work, we formulate the segmentation problem as the minimization of an energy functional in terms of the likelihood using a Bayesian regularization formulation based on MRF. Our strategy reduces the probabilistic image segmentation modeling to the choice of a convenient metric-divergence or a measure between distributions. In this work we study the Matusita Distance (MD) and the possibility of using it for both hard and soft segmentation. We also propose a way to deal with the non-linearities involved

P. Wiederhold and R.P. Barneva (Eds.): IWCIA 2009, LNCS 5852, pp. 149–161, 2009.

with MD. We propose an algorithm for solving a particular case of a quadratic optimization problem through a sequence of linear programming subproblems and prove its convergence.

2   Mathematical Formulation

2.1   Notation

In general, the segmentation problem can be written in terms of a set of sites and a set of labels. Let $\mathcal{L} = \{r = (n, m) : n = 1, 2, \ldots, N ; m = 1, 2, \ldots, M\}$ be a 2D regular lattice where $|\mathcal{L}| = NM$ represents the number of sites (or pixels) in the image. The neighborhood of the pixel r will be denoted as $\mathcal{N}_r$. Conveniently, we will also denote (or reindex) pixels (n, m) in the image as the subset of indices $\mathcal{I} = \{1, 2, \ldots, |\mathcal{L}|\}$ where $r_i, i \in \mathcal{I}$ represents a site in the image. The image is denoted by I, $I(r)$ is a scalar (the intensity in case of a gray level image) or a vector (color channels in the case of a color image). Let $\mathcal{K} = \{1, 2, \ldots, K\}$ be a subset of labels. A segmentation of an image is a mapping $\mathcal{R} : \mathcal{L} \to \mathcal{K}$ from the set of sites $\mathcal{L}$ to the set of labels $\mathcal{K}$. That is to say, it is a partition of $\mathcal{L}$ in regions $\mathcal{R}_k \subseteq \mathcal{L}$ in such a way that: $\cup_{k \in \mathcal{K}} \mathcal{R}_k = \mathcal{L}$ and $(i \neq j; i, j \in \mathcal{K}) \Rightarrow \mathcal{R}_i \cap \mathcal{R}_j = \emptyset$, where $\mathcal{R}_k = \{r \in \mathcal{L} : \mathcal{R}(r) = k\}$.

According to [15] the observation model at each pixel r of the image can be written as follows: $I(r) = \sum_{k \in \mathcal{K}} p_k(r) I_k(r) + \eta(r)$, where $p_k(r) \in [0, 1]$ is the component of a vector measure field p at a pixel r, $p(r) = (p_1(r), p_2(r), \ldots, p_K(r))$, and represents the probability that the pixel r belongs to the class k (or to the region $\mathcal{R}_k$). Then $p_k(r)$ must satisfy $\sum_{k \in \mathcal{K}} p_k(r) = 1$. Moreover, $\eta(r)$ is white noise and $I_k(r)$ is the intensity value of the layer I_k at the pixel r.

Definition 1. *The set of points $x = (x_1, x_2, \ldots, x_n) \in \mathbb{R}^n$ that satisfy $x_i \geq 0, \forall i \in \{1, 2, \ldots, n\}$ and*

$$\sum_{i=1}^{n} (x_i)^\gamma = 1, \tag{1}$$

will be called γ positive unit n-hypersurface and will be denoted as: S_γ^n.

For instance, if $\gamma = 1$ the positive unit n-hypersurface will be named *positive unit n-hyperplane* or a *simplex* (S^n) and if $\gamma = 2$ will be called *positive unit n-hypersphere*.

2.2   Review

Rivera et al. [15] defined a Consistence Condition Qualification (CCQ) that an Energy Functional for Gauss-Markov Measure Fields (GMMF) models should satisfy in an image segmentation problem.

Definition 2. (Consistence Condition Qualification) *If no a priori information about the vector measure field p is available, and also, its Maximum Likelihood (ML) estimator satisfies the condition: $\arg\max_k p_k(r) = \arg\max_k \hat{v}_k(r)$, $\forall r \in \mathcal{L}$, where $\hat{v}$ represents the normalized likelihood[1], then we say that the vector measure field p holds the Consistence Condition Qualification (CCQ).*

---

[1] The likelihood is defined as $v_k(r) = p(I(r)|p_k(r) = e_k, I_k)$.

In this work, we are interested in two particular CCQ cases. The first CCQ, allows us to obtain a soft segmentation of the image and from the second CCQ we can formulate a functional to obtain a hard segmentation. Now we introduce two definitions that will guide our study.

Definition 3. (Equality Consistence Condition) *If no a priori information about the vector measure field p is available, and also, its ML estimator satisfies the condition $p_k(r) = \hat{v}_k(r)$, $\forall r \in \mathcal{L}$, where $\hat{v}$ represents the normalized likelihood, we say that the vector measure field p holds the Equality Consistence Condition (ECC).*

Definition 4. (Hard Consistence Condition) *If no a priori information about the vector measure field p is available, and also, its ML estimator satisfies the following conditions*

$$k_r^* = \arg\max_k p_k(r) = \arg\max_k \hat{v}_k(r), \quad \forall r \in \mathcal{L}, \tag{2}$$

$$p_{k_r^*}(r) = 1, \quad \forall r \in \mathcal{L}, \tag{3}$$

where $\hat{v}$ represents the normalized likelihood, then we say that the vector measure field p holds the Hard Consistence Condition (HCC).

2.3 General Model

In the Bayesian regularization formulation, based on Markov Random Field, for instance: GMMF [10], HMMF [9] and QMMF [16], the modeling (observation model) begins with some assumptions about the generative model of the observation (the image). After some mathematical derivation, an energy functional that depends on the likelihood is obtained. In our formulation, different from the above models, we consider that the vector field v is the observation. Using a Bayesian formulation, the posterior distribution is

$$P(p|v) \propto P(v|p)P(p) = e^{-U(p;v)}, \tag{4}$$

the conditional probability $P(v|p)$ is obtained from

$$P(v|p) = \prod_{r \in \mathcal{L}} P(v(r)|p(r)) = e^{-D(p;v)},$$

where $P(v(r)|p(r))$ is the observation model and $P(p) \propto e^{-\lambda R(p)}$ is a Gibbsian distribution. In our case, we are not interested in the model that generates the data, but in finding energies that have particular properties, for instance those which satisfy CCQ, in particular HCC or ECC. Of course, each particular energy must have an underlaying observation model as in the cases mentioned above but we are not interested in this point. Therefore, we formulate the segmentation problem as the minimization of an energy functional (the MAP estimation of Eq. (4)). This problem has the general form:

$$\arg\min_{p \in S_\gamma^n} U(p;v) = D(p;v) + \lambda R(p), \tag{5}$$

where the parameter λ is positive and controls the granularity of regions to be segmented, $D(p;v)$ is the data term and can be related to the likelihood term in a Bayesian formulation, $R(p)$ is the regularization term and represents the prior knowledge about the vector measure field p. As we have said before, $D(p; v)$ must be chosen in such a way that enforces CCQ, and in particular we are interested in the HCC and ECC cases. In our formulation, we consider that the observations are given by the likelihood v, different from [10], [9] and [16] where the image is assumed to be the observation. So, we regularize directly on the likelihood. Then, the image segmentation problem can be solved by minimizing a metric-divergence between models associated to a certain feature vector [10,9]. Under this scheme many functionals can be obtained for segmentation purposes. A particular case of the functional in Eq. (5) is the following

$$U(p; \hat{v}) = D(p; \hat{v}) + \lambda R(p), \tag{6}$$

that depends on the normalized likelihood $\hat{v}$. A natural choice for the data term $D(p; \hat{v})$ in the functional (6) is to select a distance between distributions. On the other hand, in the data term of the potential (5) we can use other measures, for instance: divergence measures, inaccuracy measures, metrics and distances. The problem now is what measure to choose. In the literature we can find a lot of metric-divergences[2] between distributions [18]. Some of these metric-divergences have a symmetric version, for example, the Kullback and Leibler divergence, the main drawback is that they are highly non-linear. This leads us to a highly non-linear optimization problem. So, the algorithmic properties are an important issue in the selection of metric-divergences.

2.4   Model for ECC

In this section we study the Matusita Distance (MD)[3] [4]. Let f, h be discrete probability distributions, i.e. $\sum_{i=1}^{K} f_i = \sum_{i=1}^{K} g_i = 1$ and $f_i, g_i \geq 0$, then the MD is defined as:

$$d(f\|h) = \sqrt{2 - 2\sum_{i=1}^{K} \sqrt{f_i h_i}} . \tag{7}$$

This measure has a geometric interpretation and satisfies the axioms to be a metric. Hence, $d(f\|h) = 0 \iff f = h \iff f_i = g_i{}^4$ and therefore, MD can be used to obtain ECC models. Using (7) in the data term of (6) we obtain:

$$D(p; \hat{v}) = \sum_{r \in \mathcal{L}} \frac{1}{2} [d(p(r)\|\hat{v}(r))]^2 = \sum_{r \in \mathcal{L}} \left[1 - \sum_{i=1}^{K} \sqrt{p_i(r)\hat{v}_i(r)}\right] . \tag{8}$$

---

[2] From the mathematical point of view, they are not metrics because they do not satisfy the axioms to be a metric, most of them only satisfy the non-negativity and identity of indiscernible axioms, i.e.: $d(a, b) \geq 0$ and $d(a, b) = 0$ iff $a = b$, where a, b are elements of a metric space.
[3] MD is proportional to Hellinger Distance (HD), see [4].
[4] Observe that $d(f\|h) = 0 \iff f\|h$, i.e. the vectors $f, h \in \mathbb{R}^K$ are parallels, using the Cauchy-Bunyakovsky-Schwarz inequality. Then, there exists $\lambda \in \mathbb{R}$ such that $f = \lambda h$. Using the condition $\sum_{i=1}^{K} f_i = \sum_{i=1}^{K} g_i = 1$ we obtain that $\lambda = 1$ and finally $f_i = g_i$.

Similarly, the regularization potential is proposed in terms of the metric (7):

$$R(\boldsymbol{p}) = \sum_{r \in \mathcal{L}} \sum_{s \in \mathcal{N}_r} \frac{1}{2} \big[d(\boldsymbol{p}(r) \| \boldsymbol{p}(s)) \big]^2 = \sum_{r \in \mathcal{L}} \sum_{s \in \mathcal{N}_r} \Big[1 - \sum_{i=1}^{K} \sqrt{p_i(r) p_i(s)} \Big] . \quad (9)$$

As we can see, both terms (8) and (9) are non-rational and the minimization of (6) becomes a non-linear system. To deal with this problem, we propose to minimize the following functional that is a transformation of the original energy functional (6):

$$U(\tilde{\boldsymbol{p}}; \tilde{\boldsymbol{v}}) = U(\mathcal{T}_1 \boldsymbol{p}; \mathcal{T}_2 \hat{\boldsymbol{v}}) = D(\mathcal{T}_1 \boldsymbol{p}; \mathcal{T}_2 \hat{\boldsymbol{v}}) + \lambda R(\mathcal{T}_1 \boldsymbol{p}), \quad (10)$$

where $\mathcal{T}_j : [0, 1] \rightarrow [0, 1]$ with $j \in \{1, 2\}$ are increasing real functions.

So, instead of computing directly the vector measure field $\boldsymbol{p}$, we propose to compute the transformed $\tilde{\boldsymbol{p}} = \mathcal{T}_1 \boldsymbol{p}$ and $\tilde{\boldsymbol{v}} = \mathcal{T}_2 \hat{\boldsymbol{v}}$. To simplify the notation it is understood that when we apply $\mathcal{T}_j$ to a vector, we really mean to apply $\mathcal{T}_j$ on each component of the vector. Hence, substituting in Eq. (5) we have a new functional (10) in terms of $\tilde{\boldsymbol{p}}$ and $\tilde{\boldsymbol{v}}$.

A natural way for selecting $\mathcal{T}_j$ in the case of MD is choosing power functions $\mathcal{T}_j(.) = (.)^n$. We study the case $\mathcal{T}_1 = \mathcal{T}_2$ and $n = \frac{1}{2}$, that corresponds to a mapping of the simplex S^K onto the positive unit K-hypersphere S_2^K. Then, the original optimization problem is transformed into:

$$\min_{\tilde{\boldsymbol{p}} \in S_2^K} U(\tilde{\boldsymbol{p}}; \tilde{\boldsymbol{v}}), \ \tilde{\boldsymbol{v}} \in S_2^K \ . \quad (11)$$

The optimization problem (11) satisfies the following proposition:

Proposition 1. *The constraint quadratic optimization problem defined by Eq. (11) satisfies ECC when no a priori information is available (i.e. $\lambda = 0$).*

The proof is straightforward. We can see by using the Cauchy-Bunyakovsky-Schwarz inequality, that the vector measure field $\tilde{\boldsymbol{p}}$ under no prior knowledge (setting $\lambda = 0$) satisfies the ECC. Otherwise, the above optimization problem is a quadratic optimization problem with quadratic constraints that can be solved by using the Lagrange multiplier method. The Lagrangian for the constrained optimization problem, without including the non-negativity constraints, is

$$\mathcal{L}(\tilde{\boldsymbol{p}}, \boldsymbol{\pi}; \tilde{\boldsymbol{v}}) = U(\tilde{\boldsymbol{p}}; \tilde{\boldsymbol{v}}) + \sum_{r \in \mathcal{L}} \pi(r) \Big[\sum_{i=1}^{K} \tilde{p}_i^2(r) - 1 \Big], \quad (12)$$

where $\pi(r)$, $r \in \mathcal{L}$ are the Lagrange multipliers.

Now, we obtain the first-order necessary conditions for optimality, the Karush-Kuhn-Tucker (KKT) conditions, see [12]:

$$\frac{\partial \mathcal{L}(\tilde{\boldsymbol{p}}, \boldsymbol{\pi}; \tilde{\boldsymbol{v}})}{\partial \tilde{p}_k(r)} = 0, \forall r \in \mathcal{L}, k \in \mathcal{K}; \quad (13)$$

$$\sum_{k \in \mathcal{K}} p_k(r) = \sum_{k \in \mathcal{K}} \mathcal{T}_1^{-1}(\tilde{p}_k(r)) = 1, \forall r \in \mathcal{L} \ . \quad (14)$$

From the KKT condition in Eq. (13) we obtain

$$-\tilde{v}_k(r) - \lambda \sum_{s \in \mathcal{N}_r} \tilde{p}_k(s) + 2\pi(r)\tilde{p}_k(r) = 0 \; \forall r \in \mathcal{L} \; . \tag{15}$$

Defining $n_k(r) \stackrel{def}{=} \tilde{v}_k(r) + \lambda \sum_{s \in \mathcal{N}_r} \tilde{p}_k(s)$ and using a Gauss-Seidel scheme in the equation system (15), we can compute the vector measure field $\tilde{p}$ through an iterative process. By using the equality constraints (14) and substituting (15) we can compute the Lagrange multipliers $\pi(r)$. Substituting $\pi(r)$ in Eq. (15) we obtain the vector measure field $\tilde{p}$ whose components are computed with

$$\tilde{p}_k(r) = \frac{n_k(r)}{\sqrt{\sum_{i=1}^K n_i^2(r)}} \; . \tag{16}$$

Note that, if the vector measure field $\tilde{p}$ is initialized with non-negative values (i.e. $\tilde{p}^{(0)} \geq 0$) then the solution $\tilde{p}^*$ implicitly satisfies the non-negativity constraints and then $\tilde{p}^* \in S_2^K$. Finally, the optimum vector measure field p^* is computed by applying the inverse transformation of $\mathcal{T}_1$ to $\tilde{p}^*$: $p_k^*(r) = [\tilde{p}_k^*(r)]^2$.

2.5 Model for HCC

Based on the formulation defined by Eq. (11) we can think of changing the hypersurface on which $\tilde{p}$ is located. We study the case when $\tilde{p} \in S^K$ (i.e. we change the equality restrictions expressed in the L_2-Norm and use the L_1-Norm). Then, we have the following optimization problem:

$$\min_{\tilde{p} \in S^K} U(\tilde{p}; \tilde{v}) \; . \tag{17}$$

Under no prior knowledge, the above problem satisfies the following proposition:

Proposition 2. *If $\forall r \in \mathcal{L} \; \exists k_r \in K$ such that $\tilde{v}_{k_r}(r) > \tilde{v}_k(r) \; \forall k \neq k_r$ then the constraint quadratic optimization problem defined by Eq. (17) satisfies HCC when no a priori information is available (i.e. $\lambda = 0$).*

Observe that if we do not have prior knowledge, then the problem (17) is a linear programming (LP) problem. As is well-known in LP, under the conditions of the proposition the solution lies on a vertex of the polytope defined by the constraint and hence the solution at each pixel is $p(r) = e_{k_r}$, where e_{k_r} is a vector of the canonical base, i.e. $\tilde{p}_k(r) = \delta(k - k_r)$ and $k_r = \arg\max_{k \in K} \hat{v}_k(r), \; \forall r \in \mathcal{L}$.

On the other hand, if prior energy (i.e. $R(\tilde{p}) = \sum_r \sum_{s \in N_r} 1 - \tilde{p}(r)^T \tilde{p}(s)$) is given, then the optimization problem in (17) becomes a quadratic programming (QP) problem. In such a case, we still can use the idea explained above, but adapted to the QP problem (17).

We propose to solve the QP problem using a sequential linear programming (SLP) strategy. First, let us define the linear functional, at each step t

$$U^l(\tilde{p}; \tilde{V}^{(t)}) = \sum_{i=1}^{|\mathcal{L}|} -\tilde{p}^T(r_i)\tilde{V}^{(t)}(r_i), \tag{18}$$

where the components of $\tilde{\boldsymbol{V}}^{(t)}(r_i)$ are defined as $\tilde{V}_k^{(t)}(r_i) \overset{def}{=} \tilde{v}_k(r_i) + \lambda \sum_{r_j \in \mathcal{N}_{r_i}} \tilde{p}_k^{(t)}(r_j)$. Then, we need to solve the following LP subproblem at each step:

$$\min_{\tilde{\boldsymbol{p}} \in S^K} U^l(\tilde{\boldsymbol{p}}; \tilde{\boldsymbol{V}}^{(t)}) \ . \tag{19}$$

Observe that the functional defined in (18) is the linearization around $\tilde{\boldsymbol{p}}^{(t)}$ of the QP problem (17). In practice, we are not going to minimize the problem (19) as a whole, but instead we are going to reduce the energy minimizing pixel by pixel. So, we need to redefine $\tilde{V}_k^{(t)}(r_i)$ as follows:

$$\tilde{V}_k^{(t)}(r_i) \overset{def}{=} \tilde{v}_k(r_i) + \lambda \left(\sum_{r_j \in \mathcal{N}_{r_i}, j < i} \tilde{p}_k^{(t+1)}(r_j) + \sum_{r_j \in \mathcal{N}_{r_i}, j > i} \tilde{p}_k^{(t)}(r_j) \right) \ . \tag{20}$$

Therefore, in the current site $c \in \mathcal{I}$ we use the information of the previous step t and the updated sites in the current step $t+1$ (i.e. all the sites i such that $i < c$), see Eq. (20), as a Gauss-Seidel scheme, or *coordinate descent method*, see [12]. The strategy previously explained is summarized in the SLP algorithm.

Algorithm 1. SLP Algorithm

Given a starting point $\tilde{\boldsymbol{p}}^{(0)}$
repeat
    for $i \in \mathcal{I}$ **do**
        Evaluate $\tilde{\boldsymbol{V}}^{(t)}(r_i)$ using (20)
        Solve (19) to obtain $\tilde{\boldsymbol{p}}^{(t+1)}(r_i)$
    end for
until Convergence

Proposition 3. *The SLP algorithm converges, if at each step the vector $\tilde{\boldsymbol{V}}^{(t)}(r_i)$ has a unique maxima component for all r_i, $i \in \mathcal{I}$.*

For the proof see next Subsection.

2.6   Relation between ECC and HCC Proposed Models

The models studied in Subsections 2.4 and 2.5 can be enclosed in a more general model by changing the hypersurface on which $\tilde{\boldsymbol{p}}$ is defined. Now, we study the case in which $\tilde{\boldsymbol{p}}$ belongs to the γ unit positive hypersurface S_γ^K

$$\min_{\tilde{\boldsymbol{p}} \in S_\gamma^K} U(\tilde{\boldsymbol{p}}; \tilde{\boldsymbol{v}}) \ . \tag{21}$$

This model is a generalization of the models presented in the previous sections. It establishes the relation between ECC and HCC models and allows us to give a proof of the Algorithm 1 presented in Subsection 2.5. Following the same methodology used in Subsection 2.4 we obtain the following expression for $\tilde{p}_k(r)$:

$$\tilde{p}_k(r) = \frac{[n_k(r)]^{\frac{1}{\gamma-1}}}{\left\{\sum_{i=1}^{K} [n_i(r)]^{\frac{\gamma}{\gamma-1}}\right\}^{\frac{1}{\gamma}}} \ . \tag{22}$$

Observe that HCC model is the limit case of the problem (21) as γ approaches to 1 from the right side. If we make the assumption that for all $r \in \mathcal{L}$ there exists k_r such that $n_{k_r}(r) > n_i(r)$, $\forall i \neq k_r$ and we let γ tend to 1, then for $p_k(r) = \tilde{p}_k^\gamma(r)$ where $k \in \mathcal{K}$ we have:

$$\lim_{\gamma \to 1} \tilde{p}_k^\gamma(r) = \delta(k_r - k) = p_k(r) \tag{23}$$

where $k_r = \arg\max_{i \in \mathcal{K}} p_i(r)$. This means that for $\gamma = 1$ the update procedure for the vector $p(r)$ is simply to assign $p(r) = e_{k^*}$ where $k^* = \arg\max_k \tilde{p}_k(r)$. The relation obtained in Eq. (23) is a proof of Proposition 3.

3   Experiments

In this section, we make two quantitative comparisons between the proposed methods and others of the state of the art. Also, we present some experimental results of the ECC method proposed in Subsection 2.4 for interactive image segmentation and for color image segmentation.

As a comparison measure we use the Jaccard similarity coefficient (or Jaccard index).

$$J_{I_1,I_2}(k) = \frac{P_{I_1 \cap I_2}(k)}{P_{I_1 \cup I_2}(k)} \ . \tag{24}$$

The Jaccard coefficient (JC) measures the similarity between two sets. It is defined as the size of the intersection of the sets divided by the size of the union of the sets. In Eq. (24) $P_{I_1 \cap I_2}(k)$ denotes the number of pixels that belong to class k in both the original image (taken as the ground truth) and the segmented image. $P_{I_1 \cup I_2}(k)$ denotes the number of pixels that belong to class k in the original image or in the segmented image.

For evaluating the robustness, we compare the proposed methods with the following methods of the state of the art: Graph cut and QMMF. We use the binary image in Fig. 1 a). When we apply typical noise values to image Fig. 1 a), i.e. between 20 and

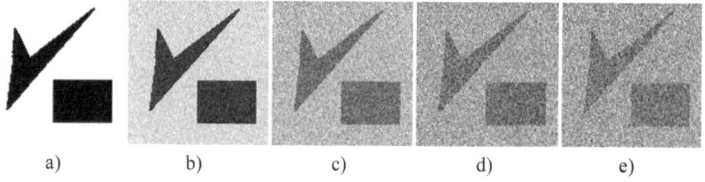

a)   b)   c)   d)   e)

Fig. 1. Binary image with different Peak Signal-to-Noise Ratio, a) Original image, b) 20 dB, c) 10 dB, d) 5 dB, e) 3 dB

Table 1. Jaccard index results using different segmentation methods. The segmented images are Fig. 1 c)-e) that are obtained from Fig. 1 a) after applying the noise levels 10, 5 and 3 dB PSNR respectively

Method	10 dB PSNR		5 dB PSNR		3 dB PSNR	
	white class	black class	white class	black class	white class	black class
QMMF	0.9907	0.9972	0.9707	0.9911	0.9618	0.9885
Graph cut	0.9936	0.9981	0.9759	0.9927	0.9720	0.9916
HCC	0.9907	0.9972	0.9274	0.9773	0.8743	0.9597
ECC	0.9923	0.9977	0.9772	0.9931	0.9631	0.9888

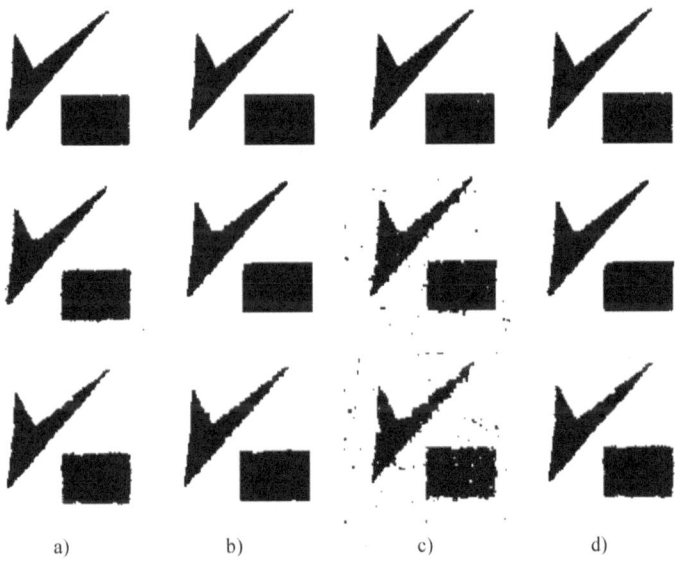

a)   b)   c)   d)

Fig. 2. Segmentation using different methods a) QMMF b) Graph cut c) Hard-segmentation proposed model, d) Soft-segmentation proposed model. First row corresponds to Fig. 1 c), second row corresponds to Fig. 1 d) and third row corresponds to Fig. 1 e).

40 dB of Peak Signal-to-Noise Ratio (PSNR) where a lower level of noise (in dB) is worse, all the methods obtained a JC equal to 1 for both classes. This means that all the methods have excellent performance when using typical values of PSNR.

In the next experiments we reduce the PSNR (i.e. we increase the noise level) up to 3 dB, see Fig. 1 c)-e). For the image with 10 dB PSNR, Fig. 1 c), the proposed methods presented as good results as those of the state of the art, see the JC in Table 1 and for segmentation results see Fig. 2. In the images with 5 and 3 dB PSNR, see Fig. 1 d)-e), the resulting quality of the hard-segmentation proposed method is reduced compared with the remainder methods, see Table 1 and Fig. 2 c). But experimentally, the HCC proposed model presents competitive results when the PSNR is greater than 10 dB, and

Table 2. Comparative performance between EC-QMMF and our ECC method in the context of interactive image segmentation. For the experiments we use Lasso's benchmark database available online in Ref. [8]. The results of EC-QMMF method are reported in Ref. [13]. The values in the table represent the classification percentage errors for each method.

Filename	EC-QMMF	ECC	Filename	QMPF+EC	ECC
21077	4.01	4.03	bush	7.86	5.93
20077	4.21	3.60	ceramic	1.73	1.66
37073	1.44	1.62	cross	1.75	1.60
65019	0.27	0.47	doll	1.03	0.52
69020	2.95	4.08	elefant	2.05	1.96
86016	1.99	2.28	flower	0.61	0.44
106024	7.55	7.19	fullmoon	0.27	0.00
124080	3.43	5.57	grave	1.27	1.65
153077	1.65	1.61	llama	4.32	3.82
153093	4.08	4.45	memorial	1.49	1.49
181079	7.41	7.70	music	2.26	2.03
189080	6.22	4.61	person1	1.16	0.25
208001	1.5	1.64	person2	0.71	0.30
209070	2.25	2.27	person3	0.87	0.75
227092	3.46	4.01	person4	3.27	3.59
271008	2.33	2.61	person5	2.48	2.22
304074	10.9	10.81	person6	5.19	4.92
326038	7.53	7.08	person7	0.96	0.59
376043	6.14	6.30	person8	0.93	0.92
388016	1.5	1.16	scissors	2.87	1.88
banana1	3.91	3.31	sheep	4.53	5.64
banana2	1.49	1.09	stone1	0.73	1.67
banana3	1.91	1.75	stone2	0.78	0.48
book	3.52	4.83	teddy	1.91	3.33
bool	1.74	1.65	tennis	7.31	5.60

is very fast. In an experiment with 1000 runs the reported average time was 0.00273 seconds, so it can be used in some real time image processing tasks. On the other hand, the soft-segmentation proposed method is very robust but it is slower, see the JC of the experiment in Table 1 and the segmentation in Fig. 2 d). For this experiment, the ECC proposed model presents competitive results according to the Jaccard index.

We also evaluate the performance of the ECC proposed method in the interactive color Image Segmentation task based on trimaps. For the experiments we use the Lasso benchmark database available online in Ref. [8]. According to the results reported in Refs. [14,13] our proposal demonstrates a better performance compared with methods of the state of the art. In Refs. [14,13] the authors compared the following methods: GraphCut, Gaussian Markov Measure Fields (GMMF), Random Walker (RW),

Table 3. Comparative summary of the results reported in Ref. [13] and our results. The values in the table are some statistics of the classification errors expressed in percentages.

	mean	median	stddev	parameters	AIC
GC	6.84	6.07	4.66	2	8.58
RW	5.47	4.66	3.58	2	6.50
GMMF	5.47	4.59	3.57	2	6.49
QMPF	5.08	3.87	3.46	2	6.04
QMPF+EC	3.03	2.15	2.40	3	3.58
ECC	2.98	2.12	2.34	2	3.42

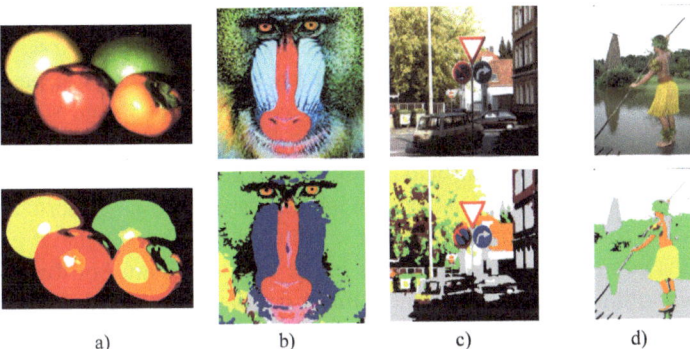

a)   b)   c)   d)

Fig. 3. Color image segmentation using the soft-segmentation proposed method based on MD and the perceptual color likelihood proposed by Alarcon and Marroquin in [2]

Quadratic Markov Measure Fields (QMMF or QMPF) and Entropy Controlled Quadratic Markov Measure Fields (EC-QMMF or EC-QMPF).

Table 2 shows, for each image of Lasso's benchmark database, a comparison between the errors obtained by the best method reported in Ref. [13] (EC-QMPF) and our ECC method. In Table 3 we summarize the experimental results reported in [13] and our results. In the last column of Table 3 we show Akaike's information criterion(AIC). As we can see, our method presents the best results. We notice that the improvement of ECC is marginal compared to the results obtained by EC-QMMF. The main advantage of our ECC method is that it has 2 free parameters.

Fig. 3 illustrates our method performance in the context of color image segmentation using the quasi-automatic segmentation method. In this experiment, we use the perceptual color likelihood obtained by Alarcon and Marroquin where the likelihood field is divided into 11 basic color categories, see [2] for details.

In Fig. 4 we show a possible application of the ECC proposed model for the colorization task. In this example we used the scheme for colorization proposed by Dalmau et al. in [6].

Fig. 4. Colorization experiment using the soft-segmentation proposed method based on MD. a) Original image b) Multimap c)-d) Colorizations.

4   Conclusions

We propose a general model for image segmentation in the context of Bayesian formulation with MRF prior. Under this formulation the modeling of image segmentation is reduced to select an appropriate metric-divergence. If the data term depends on the likelihood then a more general data term can be conveniently chosen. When the data term depends on the normalized likelihood, a metric or distance between distributions should be used. We study two particular cases of the general formulation. As metric-divergence we use the Matusita Distance. In the first case, HCC-based formulation, we propose a model for hard segmentation and we show experimentally that it gives competitive results compared to some methods of the state of the art. For this model, we give a fast algorithm based on a sequential linear programming and also we prove that under certain conditions the proposed method converges.

We study a second approach that satisfies the ECC condition. This model gives a soft segmentation method that is more robust to noise. We prove experimentally that this method gives excellent results and can be used for other image processing tasks.

References

1. Ahmed, M.N., Yamany, S.M., Mohamed, N., Farag, A.A., Moriarty, T.: A modified fuzzy c-means algorithm for bias field estimation and segmentation of mri data. IEEE Trans. Med. Imaging 21(3), 193–199 (2002)
2. Alarcón, T.E., Marroquín, J.L.: Linguistic color image segmentation using a hierarchical bayesian approach. Color Res. Appl. (to appear, 2009)
3. Boykov, Y., Jolly, M.P.: Interactive organ segmentation using graph cuts. In: Delp, S.L., DiGoia, A.M., Jaramaz, B. (eds.) MICCAI 2000. LNCS, vol. 1935, pp. 276–286. Springer, Heidelberg (2000)
4. Cha, S.H.: Comprehensive survey on distance/similarity measures between probability density functions. Int. J. of Math. Models and Methods in Applied Sciences, 300–307 (2007)
5. Chuang, K.S., Tzeng, H.L., Chen, S., Wu, J., Chen, T.J.: Fuzzy c-means clustering with spatial information for image segmentation. Computerized Medical Imaging and Graphics 30, 9–15 (2006)
6. Dalmau, O., Rivera, M., Mayorga, P.P.: Computing the alpha-channel with probabilistic segmentation for image colorization. In: IEEE Proc. Workshop in Interactive Computer Vision (ICV 2007), pp. 1–7 (2007)
7. Hewer, G.A., Kenney, C., Manjunath, B.S.: Variational image segmentation using boundary functions. IEEE Transactions on Image Processing 7, 1269–1282 (1998)

8. `http://research.microsoft.com/vision/cambridge/i31/ segmentation/GrabCut.htm`

9. Marroquin, J.L., Arce, E., Botello, S.: Hidden Markov measure field models for image segmentation. IEEE PAMI 25, 1380–1387 (2003)

10. Marroquin, J.L., Velazco, F., Rivera, M., Nakamura, M.: Gauss-Markov measure field models for low-level vision. IEEE Trans. PAMI 23 (2001)

11. Mumford, D., Shah, J.: Optimal approximation by piecewise smooth functions and associated variational problem. Commun. Pure Appl. Math., 577–685 (1989)

12. Nocedal, J., Wright, S.J.: Numerical Optimization. Springer Series in Operation Research (2000)

13. Rivera, M., Mayorga, P.P.: Comparative study on quadratic markovian probability fields for image binary segmentation. Technical Report I-07-15, Centro de Investigación en Matemáticas, A.C. (December 2007)

14. Rivera, M., Mayorga, P.: Quadratic Markovian probability fields for image binary segmentation. In: Interactive Computer Vision, ICV 2007 (2007)

15. Rivera, M., Ocegueda, O., Marroquin, J.L.: Entropy controlled Gauss-Markov random measure fields for early vision. In: Paragios, N., Faugeras, O., Chan, T., Schnörr, C. (eds.) VLSM 2005. LNCS, vol. 3752, pp. 137–148. Springer, Heidelberg (2005)

16. Rivera, M., Ocegueda, O., Marroquín, J.L.: Entropy-controlled quadratic markov measure field models for efficient image segmentation. IEEE Trans. on Image Processing 16, 3047–3057 (2007)

17. Shi, J., Malik, J.: Normalized cuts and image segmentation. IEEE Trans. PAMI 22(8), 888–905 (2000)

18. Taneja, I.: Generalized Information Measures and Their Applications, on-line book (2001)

19. Weiss, Y.: Segmentation using eigenvectors: A unifying view. In: ICCV, vol. (2), pp. 975–982 (1999)

An Order–Independent Sequential Thinning Algorithm

Péter Kardos, Gábor Németh, and Kálmán Palágyi

Department of Image Processing and Computer Graphics,
University of Szeged, Hungary
{pkardos,gnemeth,palagyi}@inf.u-szeged.hu

Abstract. Thinning is a widely used approach for skeletonization. Sequential thinning algorithms use contour tracking: they scan border points and remove the actual one if it is not designated a skeletal point. They may produce various skeletons for different visiting orders. In this paper, we present a new 2-dimensional sequential thinning algorithm, which produces the same result for arbitrary visiting orders and it is capable of extracting maximally thinned skeletons.

Keywords: shape representation, skeleton, thinning, topology preservation, order-independent thinning.

1   Introduction

Thinning algorithms [11] perform iterative object reductions. They delete some border points of binary objects that satisfy certain topological and geometrical constrains. Sequential thinning algorithms delete only one non-skeleton point at a time [1,7,8], hence it is easy to preserve the topology. However, they have the disadvantage that they can result various skeletons for different point–visiting orders. This can be explained with the fact that there can occur some pairs of points in the picture, from which any arbitrary point can be deleted without altering the topology, but removing the whole pair would split an object into two components, which is not allowed.

Although the idea of order–independence is already discussed, this area still yields questions for further research. Earlier, Ranwez and Soille [9], then Iwanowski and Soille [4,5] proposed order–independent sequential thinning algorithms. However, these algorithms simply preserve the pair of points, which fulfill the property mentioned above, so they cannot extract 1–point thin skeletons. In addition, these methods themselves can be used for reductive shrinking [3] (i.e., they can transform an object into a minimal structure, which is topologically equivalent to the original one, but some important geometrical information relative to the shape of the objects are vanished). In case of the existing order–independent algorithms, thinning can only be performed by a two–phase process: first, some endpoints are detected as anchors, then the branches relative to them are created by anchor–preserving shrinking.

P. Wiederhold and R.P. Barneva (Eds.): IWCIA 2009, LNCS 5852, pp. 162–175, 2009.

In this paper, we present a novel order–independent sequential algorithm. It preserves the topology and it has two advantageous properties over the previously published algorithms: our algorithm is maximal (i.e., it produces 1–point thin skeletons) and it is shape–preserving without an endpoint–detection phase.

2 Basic Notions and Results

The points of an image can be considered as a set of points in 2-dimensional digital space denoted by $\mathbb{Z}^2$. Each point p is represented as a pair $p = (p_x, p_y)$. Let $N_4(p) = \{q = (q_x, q_y) \in \mathbb{Z}^2 : |p_x - q_x| + |p_y - q_y| \leq 1\}$ and $N_8(p) = \{q = (q_x, q_y) \in \mathbb{Z}^2 : \max(|p_x - q_x|, |p_y - q_y|) \leq 1\}$. A point $q \in N_4(p)$ is called 4–adjacent to p, while a point $q \in N_8(p)$ is 8-adjacent to p. Further on, the notations $N_4^*(p) = N_4(p) \setminus \{p\}$ and $N_8^*(p) = N_8(p) \setminus \{p\}$ will be used to refer to the sets of the proper neighbors of p. The sequence $\langle s_0, s_1, \ldots, s_n \rangle$ of distinct points is a j-path if each point of the sequence is in S and for all i $(1 \leq i \leq n)$ s_i is j–adjacent to s_{i-1}. The point $s_1 \in S$ is j–connected in the set S, if there exists a j–path in S between s_1 and s_2. For any two sets S, S' for which $S' \supseteq S$, S is j–connected $(j = 4, 8)$ to S', if any two points of S are j–connected in S'. The notation $\overline{S}$ will be also used to refer to the complement of S.

A 2-dimensional $(8, 4)$ binary digital picture (in the following referred to as $(8, 4)$ picture or simply as picture) can be described with the quadruple $(\mathbb{Z}^2, 8, 4, \mathbf{B})$ [6], where $\mathbb{Z}^2$ is the set of picture points, $\mathbf{B} \subseteq \mathbb{Z}^2$ is the set of black points, for which we will assign the value "1"; its complement, $\overline{\mathbf{B}} = \mathbb{Z}^2 \setminus \mathbf{B}$ is the set of white points to which the value "0" is assigned. A black component is a maximal 8–connected set of black points, while a white component is defined as a maximal 4–connected set of white points.

A black point p in the picture $(\mathbb{Z}^2, 8, 4, \mathbf{B})$ is said to be a border point, if it is 4–adjacent to at least one white point. A black point p is called an interior point if it is not a border point. Let us denote by $B(p)$ the number of elements of the set $N_8^*(p) \cap \mathbf{B}$. Let $p \in \mathbf{B}$ be a black point in $(\mathbb{Z}^2, 8, 4, \mathbf{B})$. Let us denote by $C(p)$ and $A(p)$ the number of the 8-connected and 4-connected black components in the picture $(\mathbb{Z}^2, 8, 4, \mathbf{B} \cap N_8^*(p))$, respectively.

In order to preserve the shape of the objects, the so–called endpoints of an object must be retained in the picture during the whole thinning process. For our proposed algorithm, we will use the following endpoint–criterion [2]: the black point $p \in \mathbf{B}$ in $(\mathbb{Z}^2, 8, 4, \mathbf{B})$ is an endpoint if and only if $B(p) = 1$ or 2 and $A(p) = 1$.

A black point p is said to be a simple point, if its deletion (i.e., changing it to white) preserves the topology of the picture [6]. There are numerous characterizations of simple points. We make use of the following one.

Theorem 1. *[2] Black point p is simple in a (8,4) picture if and only if p is a border point and $C(p) = 1$.*

Furthermore, we say that the object point p is a corner point, if it suits any of the configurations in Fig. 1.

1	1	0
1	p	0
0	0	0

0	1	1
0	p	1
0	0	0

0	0	0
0	p	1
0	1	1

0	0	0
1	p	0
1	1	0

Fig. 1. Configurations of corner points

3   Decision Pairs and Their Properties

Let p and q be two 4–adjacent object points in $P = (Z^2, 8, 4, \mathbf{B})$. Let us introduce the notations $N_8(p,q) = N_8(p) \cup N_8(q)$ and $N_8^*(p,q) = N_8(p,q) \setminus \{p,q\}$. Henceforth, let $C(p,q)$ denote the number of black 8–components in $N_8^*(p,q)$.

Definition 1. *The pair of 4–adjacent border points $\{p,q\}$ is called a* decision pair, *if the following conditions hold:*

i) $C(p) = C(q) = 1$,
ii) neither p nor q is an endpoint,
iii) $C(p,q) = 2$.

It is easy to discover that a decision pair is a special minimal non-deletable set [10], but of course, these two notions do not coincide with each other (for example, consider a 2×2 square object). Actually, we could define a decision pair as a minimal non-deletable set of two 4-adjacent non-endpoints, however, Definition 1 is more precise and more useful for our purposes.

The fact that the set $\{p,q\}$ is a decision pair in a picture, is denoted with the term $\Gamma(p,q)$. We also define a relation $\prec$ between two 8–adjacent points $p = (p_x, p_y)$ and $q = (q_x, q_y)$ in the following way: $p \prec q$ holds if and only if $p_x + p_y < q_x + q_y$.

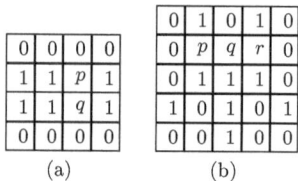

(a)   (b)

Fig. 2. Examples for comparing members of decision pairs with relation $\prec$: $p \prec q$ (a) and $p \prec q \prec r$ (b)

This relation is useful for order–independent rules in thinning. Let us consider, for example, the situation in Fig. 2(a), where $\{p,q\}$ is a decision pair, and neither p nor q is a member of any other decision pair. In such cases, it would be easy to define a proper rule for order–independent thinning: we could simply use the relation $\prec$ to decide which point to prefer, as exactly one of the conditions $p \prec q$ and $q \prec p$ holds. However, in some cases an object point can be a member of more than one decision pairs. In Fig. 2(b), for example, $p \prec q$ and $q \prec r$, and this makes clear, that if we would like to define correct rules for order–independent thinning, it is not enough to take only the relation $\prec$ into account. That is why we give the following classification of points, which will help to set up another kind of precedence between the points of a decision pair.

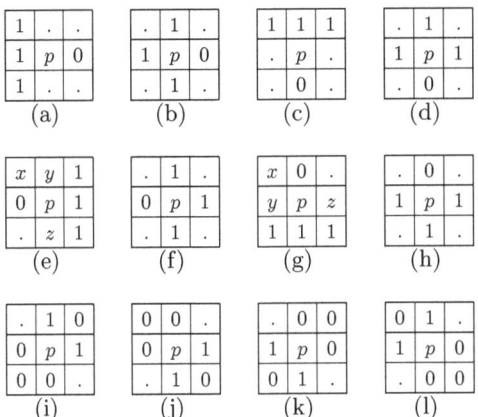

Fig. 3. In configurations (a)–(d), (e)–(h), and (i)–(l) p is an α–point, β–point, and γ–point, respectively. In configurations (e) and (g) $x \wedge y = 0$ or $z = 1$. A point marked "." may be either 0 or 1.

Let us consider the configurations shown in Fig. 3. The object point p in a picture will be called α–point, β–point, γ–point, and safe γ–point, if $N_8(p)$ matches any of the configurations in 3(a)–(d), (e)–(h), (i)–(l), and (i)–(j), respectively. The decision pairs will be called α–pair, β–pair, γ–pair, if both of their elements are α–, β–, γ–points, respectively, and we will talk about $\alpha\beta$–, $\alpha\gamma$–, $\beta\gamma$–pairs, if they contain exactly the denominated types of points. Here we discuss some important properties of decision pairs.

First, we show a necessary property of decision pairs, which is easier to check than the conditions in Definition 1, therefore we will make use of it in later proofs.

Proposition 1. If $\{p, q\}$ is a decision pair, then $N_8(p, q)$ matches at least one of the configurations of Fig. 4 or their rotations by 90, 180, or 270.

Proof. We give an indirect proof. Let us suppose that the decision pair $\{p, q\}$ does not fulfill the property stated in the proposition. Then, $N_8(p, q)$ must suit one of the configurations of Fig. 5 or their rotations by 90, 180 or 270. As $C(p, q) = 2$ holds, by Definition 1, it is easy to check that $\{x_2, x_4, x_5, x_7\} \cap \overline{\mathbf{B}} \neq \emptyset$. If $\{x_4, x_5\} \cap \overline{\mathbf{B}} \neq \emptyset$ held, then one of the conditions $i)$ and $ii)$ in Definition 1 would not be satisfied. Hence, $\{x_4, x_5\} \subset \mathbf{B}$.

.	1
p	q
.	1

1	.
p	q
1	.

1	.
p	q
.	1

.	1
p	q
1	.

Fig. 4. Possible configurations of the decision pair $\{p, q\}$, when q is the right neighbor of p. A point marked "." may be either 0 or 1.

Let us assume that $x_2 \in \overline{\mathbf{B}}$. From here, if $x_1 \in \overline{\mathbf{B}}$, then according to condition $iii)$ of Definition 1, $x_7 \in \overline{\mathbf{B}}$ and $x_8 \in \mathbf{B}$, which implies $C(q) = 2$. On the other hand, $x_1 \in \mathbf{B}$ implies $C(p) = 2$. This means, that both of these cases conflict with condition $i)$ of Definition 1, consequently, $x_2 \in \mathbf{B}$. Because of the symmetry of Fig. 5, we can deduce similar contradiction for the case $x_7 \in \overline{\mathbf{B}}$ as for $x_2 \in \overline{\mathbf{B}}$. □

x_1	0	0	x_8
x_2	p	q	x_7
x_3	x_4	x_5	x_6

x_3	x_4	x_5	x_6
x_2	p	q	x_7
x_1	0	0	x_8

Fig. 5. Configurations assigned to Proposition 1. $\{p, q\}$ cannot be a decision pair.

The following two propositions ensure that in α–pairs and β–pairs the relation $\prec$ can be used to unambiguously decide which point to prefer.

Proposition 2. *An object point may belong to at most one α–pair.*

Proof. Let $\{p, q\}$ be an α–pair. By examining the templates of α–points in Fig. 3(a)–(d), we can claim that the right or bottom 4-neighbor of an α–point must be a background point. Furthermore, according to Proposition 1, 4–neighbors of p and q from the same direction can not be both background points. Taking these observations into account it is easy to check that Fig. 6 shows all the possible 3×3 environments of p. If $x \in \mathbf{B}$, then x cannot be an α–point since none of its conceivable 3×3 environments match any of the configurations in Fig. 3(a)–(d). If $y_1 \in \mathbf{B}$ and y_1 is also an α–point, then, by necessity, $y_2 \in \overline{\mathbf{B}}$. But from this follows, according to Proposition 1, that the set $\{p, y_1\}$ cannot be a decision pair. □

1	q	0
y_1	p	x
y_2	0	

(a)

	x	1
y_1	p	q
y_2	0	1

(b)

1	y_1	y_2
q	p	0
0	x	

(c)

	y_1	y_2
x	p	0
1	q	1

(d)

Fig. 6. Configurations assigned to Proposition 2. $\{p, q\}$ is an α–pair.

Proposition 3. *An object point may belong to at most one β–pair.*

Proof. Note that if we rotate all the configurations of the β–points by 180, we get a subset of all the possible templates of α–points (see Fig. 3(a)–(h)). Based on this observation, the proof of this proposition may be similar to the previous one. (The reason why we can not get all α–point environments with the mentioned rotation, is the constraint "$x \wedge y = 0$ or $z = 1$" in Fig. 3(e) and (g), but this means only that this time there are even less cases to consider than in the previous proof.) □

Even if we take both the relation ≺ and the type of points into account in determining preference rules for our goal, it still can happen that a point q is a member of decision pairs $\{p, q\}$ and $\{q, r\}$, plus q is preferred to p, but not preferred to r. The observation below will help to formulate a suitable deleting condition for such situations as well.

Proposition 4. *Let $\{p, q\}$ be a β–pair where $p \prec q$ and let $\{r, s\}$ be an α–pair, where $r \prec s$. If $r \in N_4^*(p)$, then the sets $\{p, r\}$ and $\{q, s\}$ are αβ–pairs.*

Proof. The cases to consider are illustrated in Fig. 7. By examining the possible configurations of β–points, we can note that both members of a β–pair must suit Fig. 3(f) or 3(h). Hence, by necessity, $\{a, c, e, f\} \subset$ **B**. $r \neq a$, otherwise $r \prec s$ would imply $s = p$, which cannot happen. Therefore, $r = c$, and $s \in \{d, e\}$. $d \in \overline{\mathbf{B}}$, or else p would not be a border point, which contradicts Definition 1. Thus $s = e$. In addition, $b \in \mathbf{B}$, otherwise r would not be an α–point. Consequently, based on Proposition 1, the sets $\{p, r\}$ and $\{q, s\}$ are αβ–pairs. □

	0	b	
a	p	c	d
0	q	e	
	f		

	a	0	
0	p	q	f
b	c	e	
	d		

Fig. 7. Configurations assigned to Proposition 4. Both points of the β–pair $\{p, q\}$ must suit one of the templates in Fig. 3(f) and (h).

The next proposition tells us, that a decision pair must consist of points belonging to either of the earlier defined three classes.

Proposition 5. *If p is a member of a decision pair, then p must be an α–, β–, or γ– point.*

Proof. Let us suppose that $\{p, q\}$ is a decision pair. According to Proposition 1, $N_8(p, q)$ must match one of the templates in Fig. 4 or one of their proper rotated versions. If we complete these patterns to 3×3 environments of p in all the possible ways, then, by careful examination of the configurations in Fig. 3, we can observe that p must be an α–, β–, or γ– point. □

Finally, we point out to an important property of γ–pairs.

Proposition 6. *There is exactly one safe γ-point in a decision γ-pair.*

Proof. Let $\{p, q\}$ be a γ-pair. Without loss of generality we can assume that $p \prec q$. Fig. 8 shows the situations which can occur. In case (a),(b), and (d), only p is a safe γ-point in $\{p, q\}$, while in case (c), only q has this property in the considered pair.   □

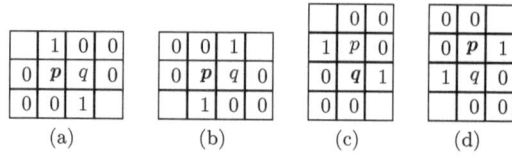

(a)   (b)   (c)   (d)

Fig. 8. Configurations assigned to Proposition 6. Only the highlighted point is safe γ-point in the γ-pair $\{p, q\}$.

4   The Proposed Algorithm KNP

Our algorithm called **KNP**[1] uses the subsets of black points denoted by S_{inner}, S_α, S_β, S_γ, S_{corner}, $S_{visited}$, and we need four additional Boolean functions. These functions have value of "1" (true) if the following conditions hold:

$\Delta_\alpha(p)$: $p \in S_\alpha$ and $\nexists q \in S_\alpha$ black point such that $\Gamma(p, q)$ and $p \prec q$.

$\Delta_\beta(p)$: $p \in S_\beta$, and the following two conditions hold:

    i) If $\exists q \in S_\beta$ black point such that $\Gamma(p, q)$ and $q \prec p$, then $\exists r \in S_\alpha$ such that $\Gamma(q, r)$.

    ii) For each $q \in S_\alpha$ black point, $\neg\Gamma(p, q)$ or $\exists r \in S_\alpha$ such that $\Gamma(q, r)$ and $q \prec r$.

$\Delta_\gamma(p)$: $p \in S_\gamma$, $\nexists q \in S_\alpha \cup S_\beta$ black point such that $\Gamma(p, q)$, and if p is not a safe γ-point, then $\nexists q \in S_\gamma$ black point such that $\Gamma(p, q)$.

$\Delta_{corner}(p)$: $p \in S_{corner}$ and $\exists q \in (S_{inner} \cap N_8^*(p))$ black point or $\exists r \in (S_{corner} \cap N_8^*(p))$ black point such that $r \prec p$.

Our preference rules for decision pairs are built in these Boolean functions. First of all, we have determined a priority order for the introduced classes of points, in which α-points are the most preferred, followed by β-points, which are preferred to γ-points, and finally, safe γ-points have higher priority than non-safe γ-points. If p has higher priority than q for every decision pair $\{p, q\}$, then p can be removed, but if q is preferred in such a pair, then p still has a chance to be removed, namely, if q is not preferred in another pair.

It can also occur that p and q have the same priority. (This can only happen in the case of α- and β-pairs, as we have already proved, that γ-pairs contain exactly one safe γ-point.) In such situations, we use the relation $\prec$ for the

---

[1] KNP stands for "Kardos-Németh-Palágyi", surnames of the authors.

decision: for α–points, we prefer p to q if $q \prec p$, however, in the case of β–pairs, $p \prec q$ must be fulfilled for deletion of p. (These opposite conditions for α– and β–pairs will help to ensure maximal thinning.)

Function $\Delta_\alpha(p)$ has only to check whether $q \prec p$ holds, as α–points have the highest priority. Function $\Delta_\beta(p)$ takes also care about the incidental α–points: criterion $i)$ deals with the case when q is a preferred β–point in $\{p, q\}$ ($q \prec p$), but not preferred in another decision pair, criterion $ii)$ stands for the situation when q is an α–point which is not preferred in an α–pair. As it is noticeable, function $\Delta_\gamma(p)$ does not examine, whether the points with higher priority in $\alpha\gamma$– and $\beta\gamma$–pairs are also preferred in other incidental decision pairs. This has only a technical reason: sometimes we would have to analyze a relatively large environment of p for this purpose. Instead of this, we drop this check, and simply postpone the decision to a subsequent iteration, when p will be visited again. (As we will see later, the absence of the mentioned examination does not affect the expected properties of the algorithm.) Finally, function $\Delta_{corner}(p)$ has the aim to prevent the fully removal of a 2×2 square object by preserving its upper–left point.

Using these conditions we define the term

$$\Delta(p) : \Delta_\alpha(p) \vee \Delta_\beta(p) \vee \Delta_\gamma(p) \vee \Delta_{corner}(p).$$

We are now ready to give the pseudo–code of our algorithm:

Order–Independent Sequential Thinning Algorithm KNP
Input: picture $(Z^2, 8, 4, X)$
Output: picture $(Z^2, 8, 4, Y)$

$S_\alpha = S_\beta = S_\gamma = S_{corner} = S_{visited} = \emptyset$
$S_{inner} = Y = X$
changed := true
repeat
       changed := false
       for each $p \in Y$ **do**   /* Phase 1 */
            if $p \in S_{corner}$ and $B(p) > 0$ **then**
                $Y := Y \setminus \{p\}$
                changed := true
            if p is a border point but not an endpoint **then**
                $S_{inner} := S_{inner} \setminus \{p\}$
                if p is an α–point **then**
                    $S_\alpha := S_\alpha \cup \{p\}$
                elseif p is a β–point **then**
                    $S_\beta := S_\beta \cup \{p\}$
                elseif p is a γ–point **then**
                    $S_\gamma := S_\gamma \cup \{p\}$
                elseif p is a corner point **then**
                    $S_{corner} := S_{corner} \cup \{p\}$

$$\textbf{for each } p \in \left(Y \cap \left(S_\alpha \cup S_\beta \cup S_\gamma \cup S_{corner}\right)\right) \textbf{ do} \qquad \text{/* Phase 2 */}$$
$$\quad \textbf{if } p \notin S_{visited} \textbf{ and } C(p) = 1 \textbf{ and } \Delta(p) = 1 \textbf{ then}$$
$$\quad\quad Y := Y \setminus \{p\}$$
$$\quad\quad \text{changed} := \text{true}$$
$$\quad \textbf{if } p \notin S_\gamma \textbf{ then}$$
$$\quad\quad S_{visited} := S_{visited} \cup \{p\}$$
$$\textbf{until } \text{changed=false}$$

An iteration step (i.e., the kernel of the **repeat** cycle) consists of two phases. In Phase 1, border points are assigned to the sets S_α, S_β, S_γ, and S_{corner} according to their classes. Furthermore, an earlier labeled corner point can be removed in this phase if it is not an isolated point. Then, in Phase 2 the algorithm visits the actual assigned border points, and removes point p, if it fulfills the deleting conditions ($p \notin S_{visited}$ and $C(p) = 1$ and $\Delta(p) = 1$). The set $S_{visited}$ has the role to memorize the already visited but not deleted α-, β-, and corner points.

5   Discussion

Now we will show that the proposed algorithm is order–independent, topology preserving, and maximal thinning. The latter property, however, needs to be concretely defined beforehand, as there can occur some special cases, where only non–simple object points and endpoints are present in the image, yet the object is not 1–point wide (see Fig. 9, based on the examples in [2] and [12]).

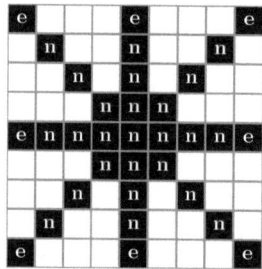 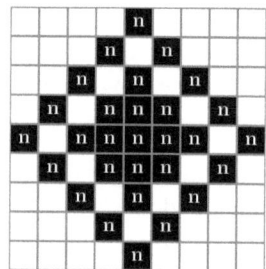

Fig. 9. Examples of objects which can not be thinned to 1-point thin line segments. Black points denoted by **e** are endpoints and points denoted by **n** are not simple.

Definition 2. *A thinning algorithm is* maximal, *if it produces a skeleton* $(Z^2, 8, 4, X)$ *such that for any* $x \in X$, x *is not a simple point or* x *is an endpoint.*

The following lemma is a preliminary result which shows that if $\{p, q\}$ is a decision pair, then we cannot have at the same time $\Delta(p)$ and $\Delta(q)$, implying that p or q must be retained in the picture.

Lemma 1. *In any iteration step of Algorithm* **KNP**, $\neg\left(\Gamma(p, q) \wedge \Delta(p) \wedge \Delta(q)\right)$ *holds for any object points* p *and* q.

Proof. Let p and $q \in Y$ be two object points in a given iteration. It is obvious that $\Delta(p)$ can be satisfied only if $p \in S_\alpha \cup S_\beta \cup S_\gamma \cup S_{corner}$. If $\Delta_\alpha(p)$ holds, then it is easy to verify, according to Definition 1 and to the conditions $\Delta_\alpha(q), \Delta_\beta(q), \Delta_\gamma(q)$, that $\neg(\Gamma(p,q) \wedge (\Delta_\alpha(q) \vee \Delta_\beta(q) \vee \Delta_\gamma(q)))$ holds. If $\Delta_\beta(p)$ is fulfilled, then it is obvious that $\neg\Delta_\gamma(q)$ also holds and from Proposition 4 follows that $\neg(\Gamma(p,q) \wedge (\Delta_\alpha(q) \vee \Delta_\beta(q)))$ is satisfied. If both $\Delta_\gamma(p)$ and $\Gamma(p,q)$ holds, then $q \notin S_\alpha \cup S_\beta$, which means $\neg(\Delta_\alpha(q) \vee \Delta_\beta(q))$ is fulfilled, and by Proposition 6, $\neg\Delta_\gamma(q)$ is also satisfied. Furthermore, due to Proposition 5, $\Delta_{corner}(p) \vee \Delta_{corner}(q)$ implies $\neg\Gamma(p,q)$. □

A necessary behavior of order–independent algorithms is that if a point p is the first visited point in an iteration and p is deletable in such a case, then p must fulfill the deleting conditions for any other visiting order as well. Of course, the same holds for the opposite case: if p could not be removed in the first iteration step, then p must be always preserved, no matter what the order of points is. Lemma 2 and Lemma 3 prepare the ground for the proof of order–independence by showing the above properties.

Lemma 2. *If $C(p) > 1$ or $\neg\Delta(p)$ holds for $p \in Y \cap \overline{S_{visited}} \cap \overline{S_\gamma}$ in the beginning of Phase 2, then the same holds when visiting p.*

Proof. Let us suppose that $C(p) > 1$ holds in either step of Phase 2 of a given iteration. Then, by examining the possible 8–neighbors of p, it can be shown that no one black component in $N_8^*(p)$ will be completely removed by the end of this iteration, because for at least one object point $q \in N_8^*(p)$ of any component, one of the following cases must occur:

- q was not a simple point or q was an endpoint in the beginning of Phase 2;
- q was a corner point for which $\neg\Delta(q)$ held in the beginning of Phase 2;
- there was an object point r in this component such that $\Gamma(q,r)$ and $\neg\Delta(q)$ held in the beginning of Phase 2.

In the first case, q is not visited in Phase 2. In the second case, it follows from the content of condition $\Delta_{corner}(q)$ that q remains in Phase 2. In the last case, if r was deleted, then according to the definition of decision pairs, $C(q)=2$ held right after the removal of r. By visiting q, the same can be stated for q as for p above, and this chain can be continued recursively until there will be an object point x for which one of the first two mentioned cases occur. As Y is finite, this chain is also finite. From this follows that q remains in the third case, too. Hence, the value $C(p)$ does not change in further steps of Phase 2. Now let us suppose that $\neg\Delta(p)$ holds in the beginning of Phase 2, and $\Delta(p)$ holds after some steps in this phase. By examining the deletion conditions, it is easy to see that in this case, there must have been an object point q for which $\Gamma(p,q)$ held in the beginning of Phase 2. But according to Definition 1, $C(p) = 2$ holds right after removing q, and we have already pointed to the fact that in such cases, $C(p)$ does not change anymore in Phase 2. □

Lemma 3. *If $C(p) = 1$ and $\Delta(p)$ holds for $p \in Y \cap \overline{S_{visited}}$ in the beginning of Phase 2 of a given iteration, then the same holds when visiting p.*

Proof. If $\Delta_{corner}(p)$ holds in the beginning of Phase 2, then based on the neighborhood of p and on the content of the mentioned condition, it is obvious that in any further step of this phase, $C(p) = 1$ and $\Delta_{corner}(p)$ holds. Let us suppose that $C(p) = 1$ and $\Delta_{\alpha}(p) \vee \Delta_{\beta}(p) \vee \Delta_{\gamma}(p)$ is satisfied in the beginning of Phase 2, but $C(p) > 1$ or $\neg\Delta(p)$ holds in either further step of this phase. By examining the conditions and Definition 1, it can be stated that this can only happen if $\exists q \in N_4^*(p)$ such that $\Gamma(p,q)$ in the beginning of Phase 2 and q must be deleted before visiting p. That means, $C(q) = 1$ and $\Delta(q)$ must be satisfied in the beginning of this phase, or else according to Lemma 2, $C(q) > 1$ or $\neg\Delta(q)$ would hold in any step of this phase. But taking into account Lemma 1, $\Delta(p)$ and $\Delta(q)$ cannot hold at the same time. So we came to a contradiction, which means, our lemma is satisfied.   □

Finally, Lemma 4 will be used to proove that our algorithm is maximal.

Lemma 4. *The output of Algorithm* **KNP** *does not contain any decision pair.*

Proof. Indirectly, let us suppose that there is at least one decision pair P in the output picture $(Z^2, 8, 4, Y)$. Y contains neither any α–points nor any β–points, or else based on Proposition 2 and Proposition 3, there would exist at least one object point $p \in Y$ satisfying condition $\Delta_{\alpha}(p)$ or $\Delta_{\beta}(p)$, and the algorithm would not have stopped. Therefore, according to Proposition 5, P must be a γ–pair. Due to Proposition 6, P must contain exactly one $p \in P$ safe γ–point, and as there is not any α–point or β–point in Y, $\Delta_{\gamma}(p)$ holds. This means, the algorithm continues with a further iteration, which conflicts with the initial assumption that the picture $(Z^2, 8, 4, Y)$ is the output of Algorithm **KNP**.   □

Theorem 2. *Algorithm* **KNP** *is order–independent, topology preserving, and maximal.*

Proof. According to Lemmas 2 and 3, we can say that except for the γ–points, the same points will be deleted for any visiting order in Phase 2 of a given iteration. It is guaranteed that an object point $p \in S_{\alpha} \cup S_{\beta}$ can be visited only in one iteration, because after that, p is added to $S_{visited}$. It is also easy to see that if an object point $p \in S_{\gamma}$ is not deleted at the first check, then p will be deleted in the succeeding iteration if and only if for all $q \in N_4^*(p)$, $C(q) > 1$ or $\neg\Delta(q)$ holds in the beginning of Phase 2 of the iteration in which p is first checked. That means, if there exists any visiting order, for which an object point $p \in S_{\gamma}$ will be deleted in a given iteration, then p will surely be deleted at the latest in the succeeding iteration for each visiting order. There is a chance that a corner point, which has been visited in a given iteration, can be checked again for deletion in Phase 1 of the succeeding iteration. It is easy to see, that this can only happen to some corner points of the horizontal or vertical 2–point thick line objects and to the upper–left corner points of the 2×2 square objects, from which always the square–corner points will be preserved according to the condition $B(p) > 0$. From the observations above, it follows that the proposed algorithm is order–independent.

It is guaranteed that in Phase 2 an object point p will be removed only if it is a border point and $C(p) = 1$ holds. It is also easy to show that if we remove some (but not all) black neighbors of a corner point p, then $C(p) = 1$ still holds in Phase 1 of the next iteration. Therefore, after Theorem 1 only simple points will be deleted, which means Algorithm **KNP** preserves the topology of the objects.

Let us suppose that our algorithm is not maximal, which means that in the last iteration of the algorithm, there is a simple point p in the picture, which is not an endpoint. Due to Lemma 4, there is not any decision pair in the skeleton. From this follows that $p \notin (S_\alpha \cup S_\beta \cup S_\gamma)$ in Phase 2 of the last iteration, otherwise $\Delta_\alpha(p) \vee \Delta_\beta(p) \vee \Delta_\gamma(p)$ would hold, which again would result in a contradiction. Hence, $p \in S_{corner}$ at the last visit of p. It can be easily verified that in this situation, p must be an upper–left corner point of a 2×2 square object $O = \{p, q, r, s\}$, or else p could be deleted. But in such an object, the conditions $\Delta_{corner}(q)$, $\Delta_{corner}(r)$, $\Delta_{corner}(s)$ hold, therefore, this situation cannot come into question. As there are no more possible cases to examine, we have indirectly proved that Algorithm **KNP** is maximal. □

6   Results

In experiments our Algorithm **KNP** and the previously published order–independent Algorithm **RS** proposed by Ranwez and Soille [9] were tested on numerous objects of different shapes and their results were compared. Here we present some illustrative examples below. In Figs. 10–13, "skeletons" produced by these two algorithms are superimposed on the original objects and numbers in parentheses mean the count of skeletal points. We can state that our algorithm (without an endpoint detection phase) does not produce spurious branches.

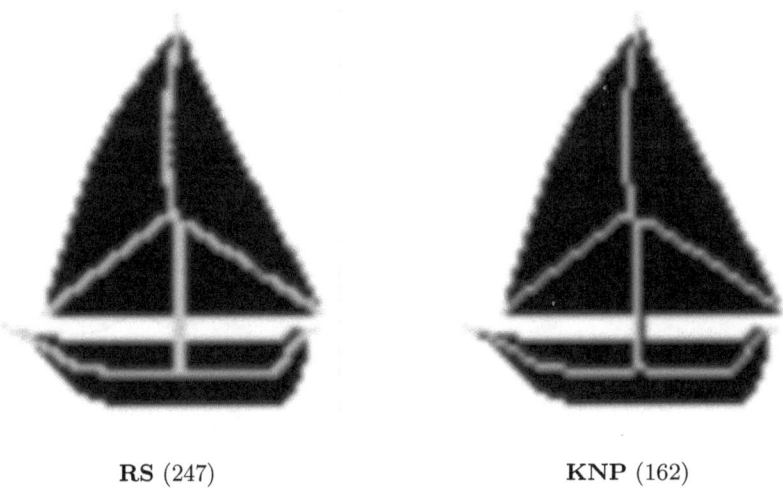

RS (247)   **KNP** (162)

Fig. 10. A 60×80 image of a ship and its "skeletons". Skeletal points are displayed in gray.

RS (331)   KNP (235)

Fig. 11. A 100×40 image and its "skeletons"

RS (2740)   KNP (1590)

Fig. 12. A 768×678 image of a duck and its "skeletons"

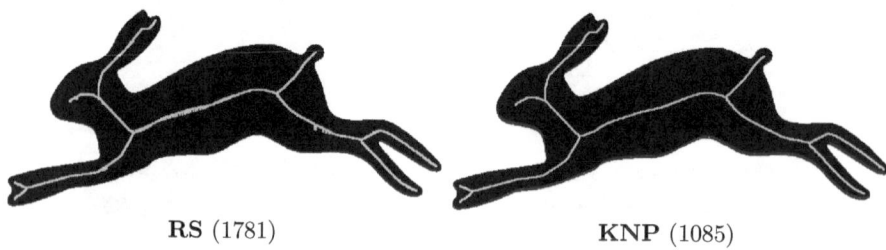

RS (1781)   KNP (1085)

Fig. 13. A 712×412 image of a hare and its "skeletons"

Acknowledgements

This research was partially supported by the TÁMOP-4.2.2/08/1/2008-0008 program of the Hungarian National Development Agency.

References

1. Arcelli, C.: Pattern thinning by contour tracking. Computer Graphics and Image Processing 17, 130–144 (1981)
2. Hall, R.W.: Parallel connectivity–preserving thinning algorithm. In: Kong, T.Y., Rosenfeld, A. (eds.) Topological Algorithms for Digital Image Processing, Machine Intelligence and Pattern Recognition, vol. 19, pp. 145–179. North–Holland, Amsterdam (1996)
3. Hall, R.W., Kong, T.Y., Rosenfeld, A.: Shrinking binary images. In: Kong, T.Y., Rosenfeld, A. (eds.) Topological Algorithms for Digital Image Processing, Machine Intelligence and Pattern Recognition, vol. 19, pp. 31–98. North–Holland, Amsterdam (1996)
4. Iwanowski, M., Soille, P.: Order independence in binary 2D homotopic thinning. In: Kuba, A., Nyúl, L.G., Palágyi, K. (eds.) DGCI 2006. LNCS, vol. 4245, pp. 592–604. Springer, Heidelberg (2006)
5. Iwanowski, M., Soille, P.: Fast algorithm for order independent binary homotopic thinning. In: Beliczynski, B., Dzielinski, A., Iwanowski, M., Ribeiro, B. (eds.) ICANNGA 2007. LNCS, vol. 4432, pp. 606–615. Springer, Heidelberg (2007)
6. Kong, T.Y., Rosenfeld, A.: Digital topology: Introduction and survey. Computer Vision, Graphics, and Image Processing 48, 357–393 (1989)
7. Kwok, P.C.K.: A thinning algorithm by contour generation. Communications of the ACM 31, 1314–1324 (1988)
8. Lam, L., Lee, S.-W., Suen, C.Y.: Thinning methodologies – A comprehensive survey. IEEE Trans. Pattern Analysis and Machine Intelligence 14, 869–885 (1992)
9. Ranwez, V., Soille, P.: Order independent homotopic thinning for binary and grey tone anchored skeletons. Pattern Recognition Letters 23, 687–702 (2002)
10. Ronse, C.: Minimal test patterns for connectivity preservation in parallel thinning algorithms for binary digital images. Discrete Applied Mathematics 21, 67–79 (1988)
11. Suen, C.Y., Wang, P.S.P. (eds.): Thinning Methodologies for Pattern Recognition. Series in Machine Perception and Artificial Intelligence, vol. 8. World Scientific, Singapore (1994)
12. Verwer, B.J.H., Van Vliet, L.J., Verbeek, P.W.: Binary and grey-value skeletons: Metrics and algorithms. Int. Journal of Pattern Recognition and Artificial Intelligence 7, 1287–1308 (1993)

Blurred Segments in Gray Level Images for Interactive Line Extraction*

Bertrand Kerautret and Philippe Even

LORIA, Nancy University - IUT de Saint Dié des Vosges
54506 Vandœuvre-lès-Nancy Cedex
{kerautre,even}@loria.fr

Abstract. The recognition of discrete straight segments is a significant topic in the field of discrete geometry and for many applications dealing with geometric feature extraction. It can be performed from noisy binary data using the concept of blurred segments [3,2]. However, to our best knowledge, these algorithms have never been defined to directly extract straight segments in gray level images. This article proposes a solution to extend the recognition by using gray level image information. Although initially intended to be implemented within a semi-automatic line selection tool used in an interactive 3D modeling application, it also meets more general parameter extraction requirements.

Keywords: Blurred segments, discrete geometry, line extraction.

1 Introduction

The recognition of discrete straight segments is widely used for geometric parameter extraction in already segmented digital objects. For instance, geometric features as perimeter, tangent, curvature [13,6,9] or polygonal approximation [1] can be estimated from the recognition of discrete straight segments. Many formulations have been proposed for the digital straight segments recognition (see [8] for a review). Some of them are based on the exact definition of arithmetic straight segments, and are not always well adapted to noisy digital contours.

To overcome this limitation Debled-Rennesson *et al.* [3] and Buzer [2] introduced two equivalent recognition algorithms allowing to set a parameter ν associated to the width of the blurred segment being recognized. The application of these algorithms allows to define robust to noise algorithms for extracting geometric parameters as for example curvature [7]. Blurred segment recognition was formerly designed for binary images, i.e all the input data of the algorithm is an ordered list of the pixel coordinates. This characteristics limits the use in many applications of image analysis and the contribution in this work is to extend the recognition algorithm to gray level images. As described in the following, the main idea is to directly analyze the pixels belonging to wide stripes

---

* This work was partially funded by ANR GeoDIB project, n° ANR-06-BLAN-0225. Bertrand Kerautret was partially funded by a BQR project of Nancy-Universités.

P. Wiederhold and R.P. Barneva (Eds.): IWCIA 2009, LNCS 5852, pp. 176–186, 2009.

defined by a start section and an initial direction. Geometric criteria are set so that a list of possible line candidates can be obtained according to the gray level information found.

We intend to apply this extension to a semi-automatic straight line selection tool intended for interactive 3D modeling from camera images. In this context, a close man-machine cooperation is still required to select relevant image features [10]. Extracted lines are mainly used to determine some orientation in space for camera calibration or 3D object matching tasks [4]. Because a large amount of segments must be extracted from images in order to build a complete 3D model, even slight improvements of this basic task may provide noticeable time reduction of a whole 3D modeling session [5].

In the following section we recall the definition of blurred segments introduced by Debled-Rennesson *et al.*. Afterward, we describe our approach extending the recognition of digital straight segments in gray level images. In section 4, we present the interactive process for line extraction. Finally, we conclude the paper with remarks and future works.

2   Blurred Segment Recognition

The notion of blurred segment is defined from the arithmetical definition of discrete lines. To describe briefly this notion we recall some definitions introduced by Debled *et al.* [3]. A set of integer points S belongs to the digital straight line $\mathcal{L}(a, b, \mu, \omega)$ if and only if all points verify $\mu \leq ax - by < \mu + \omega$. The real lines $ax - by = \mu$ and $ax - by = \mu + \omega - 1$ are denoted as the upper and lower leaning lines. By assuming S_b a set of 8-connected points, a discrete line $\mathcal{L}(a, b, \mu, \omega)$ is said bounding for S_b if all points of S_b belong to $\mathcal{L}$. Finally, a bounding line is said optimal if its vertical distance $\frac{\omega - 1}{max(|a|, |b|)}$ is minimal, i.e. if its vertical distance is equal to the vertical distance of the convex hull of the set S_b.

Definition 1 [3]. A set S is a blurred segment of width ν if and only if its optimal bounding line has a vertical distance less than or equal to ν.

From this definition, the authors proposed a linear time algorithm to recognize a blurred segment from a set of ordered discrete points. The algorithm is based on

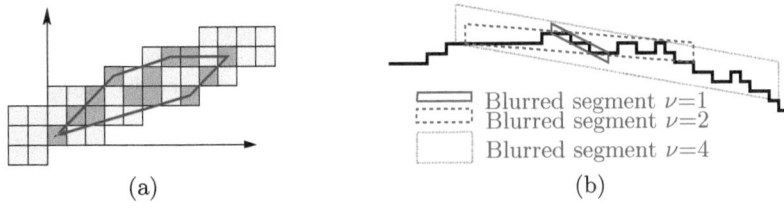

(a)   (b)

Fig. 1. Blurred segments with convex hull and bounding line of an initial set of pixels (dark gray) (a). Illustration of the recognition obtained with different values of the parameter ν (b).

the computation of the convex hull. More details about it can be found in [3] or [12] with different hypotheses on initial discrete points. A blurred segment and its associated convex hull is illustrated on Fig. 1 (a). The dark pixels represent the initial data while the light gray pixels represent the optimal bounding line.

Fig. 1 (b) illustrates the recognition of blurred segments obtained with several values of the parameter ν. The recognition was defined by the maximal blurred segments which can not be extended neither on the left nor on the right direction. As illustrated, the use of a large value of width ν allows to recognize segments without taking into account low contour variations.

3 Extension to Gray Level Images

In order to extend this binary blurred segment recognition to gray level images, we implement a directional scan of the image based on a coarse to fine strategy composed of three main steps. The first one consists in a preliminary recognition in order to obtain an approximation of the direction $\mathfrak{D}^{(1)}$ associated to a linear structure near the first direction $\mathfrak{D}^{(0)}$. The second (resp. third) step uses the directions $\mathfrak{D}^{(1)}$ (resp. $\mathfrak{D}^{(2)}$) in order to improve the recognition process.

The direction $\mathfrak{D}^{(0)}$ is defined by two points P_1 and P_2 which are selected on each side of the line to detect. They define a start scan segment $S_0^{(0)}$ associated to the initial direction $\mathfrak{D}^{(0)}$. Candidate line points are extracted along this scan segment. We then successively analyze parallel scan segments $S_i^{(0)}$ on each side of $S_0^{(0)}$ to detect next candidate pixels, that are used to feed the blurred segment recognition algorithm. The extension of the recognition algorithm relies on the use of the previous scans and on the exploitation of image gradient information, so that spatial consistency is ensured all along the scan. Once a possible blurred segment is found, a refined scan direction $\mathfrak{D}^{(1)}$ is set orthogonally to the detected segment in order to perform a new scan. This process is iterated once more with direction $\mathfrak{D}^{(2)}$.

In the following of this section, we first detail the directional scan computation. Then we explicitate how candidate pixels are selected and sorted. Finally the recognition algorithm is introduced.

3.1 Directional Scan

We perform a directional scan starting from the initial segment P_1P_2, and extending it in orthogonal directions. Let $P_1 = (x_0, y_0)$, $P_2 = (x'_0, y'_0)$, $b = x'_0 - x_0$ and $a = y'_0 - y_0$. The corresponding discrete line is defined by $\mathcal{L}_0(a, b, \mu_0, \omega)$ where $\omega = max(|a|, |b|)$.

In order to ensure that no pixel within the scan stripe is missed, we part the image space into parallel naive (8-connected) lines defined by vertical or horizontal 1-pixel shifts according to the segment slope configuration (Fig. 2 (a)). The start scan segment S_0 is computed using an arithmetical digitization algorithm, with remainder values $r_i = r_0 + i \cdot min(|a|, |b|) \mod (max(|a|, |b|))$, $i \in [0, max(|a|, |b|)]$, where the first remainder value $r_0 = 0$. Next scan segments

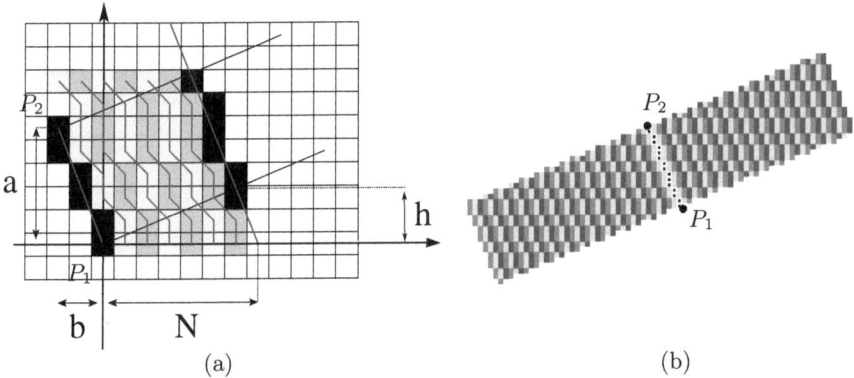

Fig. 2. Nth scan segment computation (a) and example of directional scan (b)

S_j are drawn using the same algorithm along the parallel lines, and clipped orthogonally to P_1 and P_2 using relevant start pixel (x_k, y_k) and initial remainder value r_k. Note that each scan segment configuration (Freeman code) may differ from the one of the start scan segment (Fig. 2 (a)).

Without loss of generality, in the present derivation we assume that the scan direction lies within the first octant (i.e. $a > -b > 0$). The Nth scan segment S_N belongs to a discrete line obtained by a horizontal N-pixels shift. Positive (resp. negative) values for N correspond to scan segments on the left (resp. right). The discrete line supporting S_N is defined by $\mathcal{L}_N(a, b, \mu_N, \omega)$ where $\mu_N = \mu_0 + N \cdot a$. Start pixel coordinates are $x_N = x_1 + N - \lfloor (N \cdot |b|)/a \rfloor$ and $y_N = y_1 + H$, where H is the nearest integer to $h = \frac{a \cdot b}{a^2 + b^2}$. Associated remainder value used in the digitization algorithm is set to $r_N = (N \cdot |b|) \mod a$.

3.2 Selection of the Candidates in the Scans

There exists different ways to select the candidate pixels for the recognition process. Since we search to extract straight segments contained in the image, it seems natural to use image gradient information. Another way could be to compare the image intensity profiles of the candidates with the profiles defined from the initial direction $\mathfrak{D}^{(0)}$. In this work we choose to focus on the first solution and we leave the other solution for future work.

In the following we denote by $g_x(p)$ and $g_y(p)$ the two components of the gradient vector $g(p)$ computed in the image and $\|g(p)\|$ represents the magnitude $\sqrt{g_x^2(p) + g_y^2(p)}$. Each component is computed by a simple Sobel filter with a fixed neighborhood size. The size of the filter can be adjusted to the level of noise present in the image. For all the tests in this paper a standard 3×3 pixels Sobel masks proved quite sufficient.

For the following we denote by p_m the mth point of a scan S_k. The first selection of the candidate points is defined by selecting only the point p_m for which its gradient norm $\|g(p_m)\|$ is a the local maxima. To retain only significant points we set a maximal angular deviation θ_{max} allowed between the direction $\mathfrak{D}$ and the direction of the gradient vector. Thus, for the step i of the algorithm, we remove from the list of candidates all points which angular deviation is greater than $\theta_{max}^{(i)}$. More formally, the pre-selection $\mathcal{C'}_k^i$ for a step i, is defined as follows:

$$\mathcal{C'}_k^i = \left\{ p_m \in S_k \mid (\exists\, a, b \ ; \ \forall j \in [a, b], \|g(p_j)\| \leq \|g(p_m)\|) \right.$$

$$\left. \wedge (|U_{\mathfrak{D}^{(i)}} \cdot U_{g(p_m)}| > \cos(\theta_{max}^{(i)})) \right\} ;$$

where $U_{\mathfrak{D}^{(i)}}$ and $U_{g(p_m)}$ are respectively the two unit vectors associated to the directions $\mathfrak{D}^{(i)}$ and $g(p_m)$ with $i \in \{0, 1, 2\}$. The final list $\mathcal{C}_k^i$ is obtained after a sort of the list $\mathcal{C'}_k^i$ by decreasing values of $\|g(p_m)\|$. Since in the first step of the algorithm the initial direction is not precise enough we remove this constraint by setting the value of $\theta_{max}^{(0)}$ to $\frac{\pi}{2}$. For the second and third steps, the parameter θ_{max} is set to $\theta_{max}^{(1)} = \frac{\pi}{8}$ and $\theta_{max}^{(2)} = \frac{\pi}{10}$. These empirical values needed not be adjusted to all the experiments we led on real images. However this parameter can be set in order to adjust the coarse to fine level depending on the application and the nature of the images.

3.3   Recognition Algorithm

For the definition of this algorithm we assume to have the two primitives used by the blurred segment algorithm: BlurredSegment(P_0, P_1, P_2, ν) which permits to obtain the initialization of a blurred segment of width ν and add(P_i) which is used to try to add points on it. From the previous definition of the candidate selection, we can define the algorithm to recognize a blurred segment in a gray level image. The input data are the two lists L_R, L_L of the selected candidates $\mathcal{C}_k$ of each scan S_k and the candidate list $\mathcal{C}_0$ of the central scan S_0. Since a list of candidates $\mathcal{C}_k$ can be empty and in order to simplify the initialization of the algorithm, the lists of candidates are constructed such that each first set of candidates $\mathcal{C}_1$ and $\mathcal{C}_0$ is not empty. The first parameter of the algorithm 1 is the width ν of the blurred segment. This parameter can be set according to the awaited precision of the straight segment being extracted. A small value of ν allows to obtain a precise segment but the recognition is sensible to small variations while with a larger value of ν the noise can be taken into account. In order to recognize straight segments containing missing parts, we add the parameter τ that allows to continue the recognition while the number of consecutive interrupted pixels is less than this parameter. Thus the value of τ permits to stop the algorithm 1.

The algorithm 2 is defined in order to initialize the first blurred segment from the list of candidates ($\mathcal{C}_0$, $\mathcal{C}_1^R$ and $\mathcal{C}_1^L$) and from a parameter associated to the number of the selected solution. This parameter is very useful when the initial segment ($P1P2$) intersects several solutions. The selection is only done in the first step of the recognition process since for the second and third steps we already have a precise estimation of the blurred segment.

Algorithm 1. Recognition algorithm

Input: $\mathcal{C}_0$:= the candidate list of the central scan S_0.
L_R := $\{\mathcal{C}_1^R, \ldots, \mathcal{C}_k^R, \ldots, \mathcal{C}_M^R\}$ the set of candidate lists $\mathcal{C}_k^R$ for a right scan S_k^R
L_L := $\{\mathcal{C}_1^L, \ldots \mathcal{C}_k^L, \ldots, \mathcal{C}_N^L\}$ the set of candidate lists $\mathcal{C}_k^L$ for a left scan S_k^L;
Parameters: real ν; integers τ, numSolution, algorithmStep;
Result: Maximal Blurred Segment
BlurredSeg := getInitialBS($\mathcal{C}_0$, $\mathcal{C}_1^L$, $\mathcal{C}_1^R$, ν , numSolution, algorithmStep);
integer skipLeft, skipRight:=0 ;
boolean rightScanStopped, leftScanStopped:=false; integer k:=1;
while *not rightScanStopped OR not leftScanStopped* **do**
    if *not scanRightStopped* **then**
        pointAdded:=false;
        while $\mathcal{C}_k^R$ *is not empty AND not pointAdded* **do**
            p:=first element of $\mathcal{C}_k^R$ and remove p from $\mathcal{C}_k^R$;
            pointAdded:=BlurredSeg.add(p) ;
        if *pointAdded* **then**
            skipRight:=0;
        else
            skipRight=skipRight+1;
    if *not scanLeftStopped* **then**
        pointAdded:=false;
        while $\mathcal{C}_k^L$ *is not empty AND not pointAdded* **do**
            p:=first element of $\mathcal{C}_k^L$ and remove p from $\mathcal{C}_k^L$;
            pointAdded:=BlurredSeg.add(p) ;
        if *pointAdded* **then**
            skipLeft:=0;
        else
            skipLeft:=skipLeft+1;
    rightScanStopped:=(skipRight=τ OR k=M) ;
    leftScanStopped:= (skipLeft=τ OR k=N) ;
    increment k;

Algorithm 2. Initialisation of initial Blurred Segment: getInitialBS

Input: Candidate lists $\mathcal{C}_0$, $\mathcal{C}_1^L$, $\mathcal{C}_1^R$; real ν; integer numSolution, algorithmStep;
Result: Maximal Blurred Segment
if *algorithmStep=0* **then**
    P_c:=the numSolutionth element of $\mathcal{C}_0$;
    P_1^R:=the nearest point to P_c extracted from all elements of $\mathcal{C}_1^R$;
    P_1^L:=the nearest point to P_c extracted from all elements of $\mathcal{C}_1^L$;
else
    P_c:=the first element of $\mathcal{C}_0$;
    P_1^R:=the first element of $\mathcal{C}_1^R$; P_1^L:=the first element of $\mathcal{C}_1^L$;
return BlurredSegment(P_c, P_1^R, P_1^L, ν);

The recognition algorithm can be summarized as follows:

1. Precompute image gradient $g_x(p)$ and $g_y(p)$ for all the image pixels p.
2. Compute all candidates L_R and L_L from the two points P_1 and P_2.
3. Determine the blurred segment BS by applying algorithm 1 with the parameter $algorithmStep$ equals to 0.
4. Change the position of P_1 and P_2 such that $\|P_1 P_2\| = 2\nu$, center of $(P_1 P_2)$ = first point of BS and the direction $(P_1 P_2)$ orthogonal to the direction of BS.
5. Apply again steps 2. ,3. and 4. with $algorithmStep$ set to 1.
6. Apply again steps 2. ,3. and 4. with $algorithmStep$ set to 2.

Note that the two repetitions of steps 5. and 6. appear experimentally sufficient and no significant improvement were detected with the use of more iterations. The application of the three steps is illustrated in the following section with Fig. 5.

4   Semi-automatic Segment Extraction

An single point could be judged sufficient to select a particular segment in the image. Local image gradient analysis may help to detect some line in the point vicinity. But this operating mode provides no valid information to orient the recognition process. It is then unreliable when other segments pass near the selected point. Therefore we better keep the control of the initial direction.

Abundant literature may be found about interactive edge detection, but much less is available if we restrict to straight segments extraction. Rodrigo *et al.* [11] introduced an energy minimization scheme to refine and extend an initial segment, but it requires an accurate initial solution and intensive computation. Even and Malavaud [5] proposed a sequential approach based on a two-step Hough transform, where the human operator first quickly draws a segment on the image and then lets it be attracted toward the best detected edge nearby. Noticeable time reduction is reported compared to standard manual modes, but it still requires a careful initialization to ensure a reliable attraction.

In the present approach, the human operator is just asked to coarsely strike out the desired segment on the image. Provided points P_1 and P_2 are immediately used to run the blurred segment extraction algorithm. Desired segments are here more quickly and more reliably extracted.

Different experimentations are presented on Fig. 3. For all the images (except image (b)), exactly the same parameters were used for the line detections. The red lines are the directions given by the user while the blue ones are the results of our algorithm. Image (d) illustrates the possibility to select multiple solutions when the line given by the user intersects several ones. The images (e-g) demonstrate the ability of our approach to extract straight segments made of very low gradient values. A cross was added in the image (e) of Fig. 3 with a gray level very close to the mean gray value of the local area. This cross is highlighted in blue color on image (f). Despite the low gradient values, the cross is well detected (g).

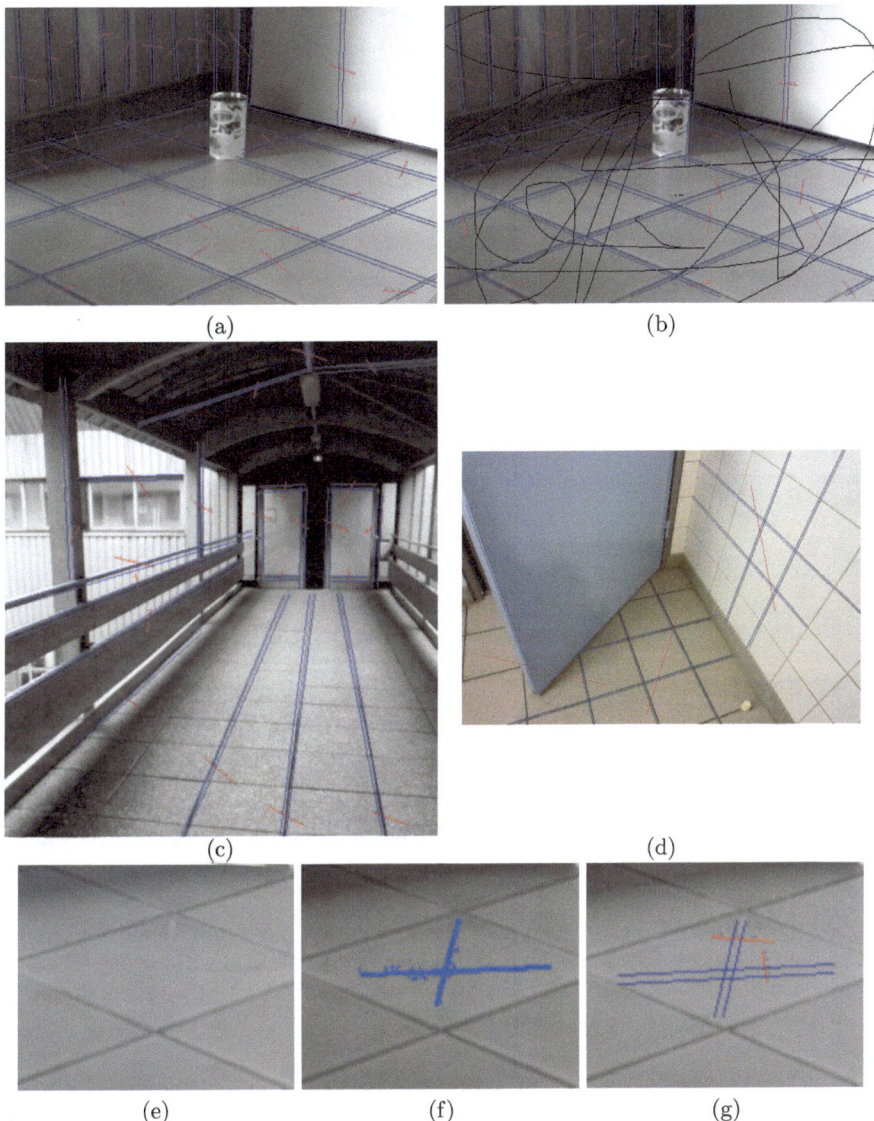

Fig. 3. Results obtained with the same parameters: ($\nu = 5, \tau = 5$) except for image (b) ($\nu = 5, \tau = 8$). Image (d) illustrates the selection of multiple straight segments intersecting the initialization. Low gradient detection is illustrated on images (e-g).

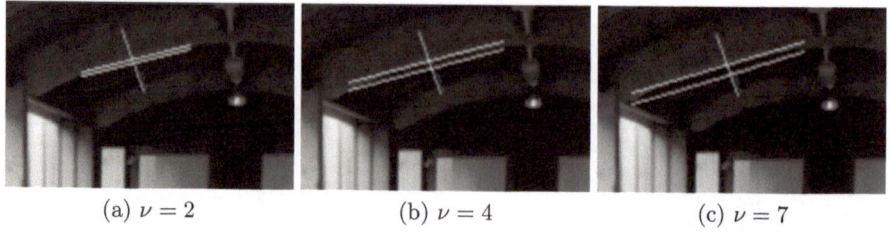

(a) $\nu = 2$ (b) $\nu = 4$ (c) $\nu = 7$

Fig. 4. Illustration of the recognition obtained with different values of ν

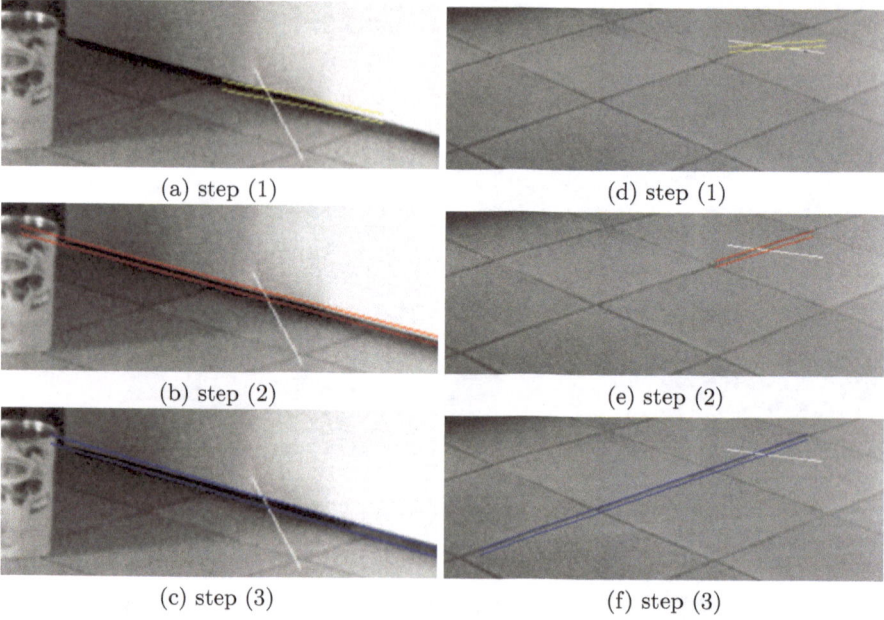

(a) step (1)   (d) step (1)

(b) step (2)   (e) step (2)

(c) step (3)   (f) step (3)

Fig. 5. Illustration of the three steps of the recognition algorithm applied on a noisy version of the image (Fig. 3 (a)). The source image was obtained with a Gaussian noise of standard deviation $\sigma = 10$. The recognition was obtained with the same parameters as the previous examples ($\nu = 5$ and $\tau = 5$).

Fig. 4 illustrates the recognition obtained with several values of ν on a curved area. We can see that a small value of ν gives a precise detection on a limited area while with a larger value of ν we obtain a coarser line detection of the initial shape.

To show the behaviour of the recognition process on noisy data we applied the algorithm on a noisy version of image Fig. 3 (a) obtained with a Gaussian noise with standard deviation $\sigma = 10$. The figure Fig. 5 shows the three steps of the recognition process obtained on two areas of the image (column left and right).

The algorithm parameters ($\nu = 5$ and $\tau = 5$) were not changed and the size mask was keep to 3×3. For the first example (a-c), we can see on the step (2) that the segment is too long, however on step (3) the recognition is well stopped (thanks to the value of $\theta_{max}^{(3)}$). The second example (d-f) shows some sensibilities to noise if we compare it to the previous results of Fig. 3 (a). More generally, the presence of noise induces more sensibility to the initial direction and induces less segments detected. The use of a larger mask size appears necessary to deal with noisy images.

5   Concluding Remarks

A simple and efficient method to extend the blurred segments to gray level images was proposed. This first extension allows to perform a fast straight segments detection on the fly and needs only few parameters which have a simple geometric interpretation (width ν of blurred segments and τ for the 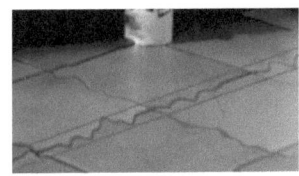 interruption length threshold). The next step of this work is to adapt the recognition in a multi-scale approach in order to be able to detect straight segments defined with numerous small variations as illustrated in the upper floating figure. Besides, ergonomy evaluations are planned in order to get a better estimation of the enhancements provided by the semi-automatic mode.

References

1. Bhowmick, P., Bhattacharya, B.B.: Fast Polygonal Approximation of Digital Curves Using Relaxed Straightness Properties. IEEE Trans. on PAMI 29(9), 1590–1602 (2007)
2. Buzer, L.: A simple algorithm for digital line recognition in the general case. Pattern Recognition 40(6), 1675–1684 (2007)
3. Debled-Rennesson, I., Feschet, F., Rouyer-Degli, J.: Optimal Blurred Segments Decomposition of Noisy Shapes in Linear Times. Comp. & Graphics 30, 30–36 (2006)
4. Even, P.: A generic procedure for interactive 3D model registration on images. The International Archives of the Photogrammetry, Remote Sensing and Spatial Information Sciences 35(B5), 204–209 (2004)
5. Even, P., Malavaud, A.: Semi-automated edge segment specification for an interactive modelling system of robot environments. International Archives of Photogrammetry and Remote Sensing 33(B5), 222–229 (2000)
6. Feschet, F., Tougne, L.: Optimal time computation of the tangent of a discrete curve: Application to the curvature. In: Bertrand, G., Couprie, M., Perroton, L. (eds.) DGCI 1999. LNCS, vol. 1568, pp. 31–40. Springer, Heidelberg (1999)
7. Kerautret, B., Lachaud, J.-O.: Curvature Estimation along Noisy Digital Contours by Approximate Global Optimization. Pattern Recognition 42(10), 2265–2278 (2009)
8. Klette, K., Rosenfeld, A.: Digital straightness–a review. Discrete Applied Mathematics 139(1-3), 197–230 (2004)

9. Lachaud, J.-O., Vialard, A., de Vieilleville, F.: Fast, Accurate and Convergent Tangent Estimation on Digital Contours. Image and Vision Computing 25, 1572–1587 (2007)
10. Ohlhof, T., Gülch, E., Müller, H., Wiedemann, C., Torre, M.: Semi-automatic extraction of line and area features from aerial and satellite images. The International Archives of the Photogrammetry, Remote Sensing and Spatial Information Sciences 34(B3), 471–476 (2004)
11. Rodrigo, R., Shi, W., Samarabandu, J.: Energy based line detection. In: Proc. of Canadian Conf. on Electrical and Comp. Engineering, Ottawa, Ontario, Canada, pp. 2061–2064 (2006)
12. Roussillon, T., Tougne, L., Sivignon, I.: Computation of Binary Objects Sides Number using Discrete Geometry Application to Automatic Pebbles Shape Analysis. In: Proc. 14th Int. Conf. on Image Analysis and Processing (ICIAP), pp. 763–768 (2007)
13. Vialard, A.: Geometrical parameters extraction from discrete paths. In: Miguet, S., Ubéda, S., Montanvert, A. (eds.) DGCI 1996. LNCS, vol. 1176, pp. 24–35. Springer, Heidelberg (1996)

Multi-scale Analysis of Discrete Contours for Unsupervised Noise Detection[*]

Bertrand Kerautret[1] and Jacques-Olivier Lachaud[2]

[1] LORIA, Nancy University - IUT de Saint Dié des Vosges
54506 Vandœuvre-lès-Nancy Cedex
kerautre@loria.fr
[2] LAMA, University of Savoie
73376 Le Bourget du Lac
jacques-olivier.lachaud@univ-savoie.fr

Abstract. Blurred segments [2] were introduced in discrete geometry to address possible noise along discrete contours. The noise is not really detected but is rather canceled out by thickening digital straight segments. The thickness is tuned by a user and set globally for the contour, which requires both supervision and non-adaptive contour processing. To overcome this issue, we propose an original strategy to detect locally both the amount of noise and the meaningful scales of each point of a digital contour. Based on the asymptotic properties of maximal segments, it also detects curved and flat parts of the contour. From a given maximal observation scale, the proposed approach does not require any parameter tuning and is easy to implement. We demonstrate its effectiveness on several datasets. Its potential applications are numerous, ranging from geometric estimators to contour reconstruction.

Keywords: Noise detection, discrete contour, maximal segments.

1 Introduction

Discrete or digital contours are natural outputs of image segmentation algorithms or digitisation processes like document scanning. Being constituted of horizontal and vertical steps, they are in essence non smooth. However their geometric analysis is of primary importance for later processing such as quantitative analysis, shape recognition or shape matching. Discrete geometry provides techniques to estimate correctly geometric characteristics on such contours, when the digital contour reflects perfect data. The most accurate techniques rely generally on the extraction of *maximal segments*, which are local affine reconstruction of contours.

In most cases, digital contours are not perfect digitizations of ideal shapes but present noise and perturbations. Rather recently, *blurred segments* were introduced to take into account both the discreteness and possible noise of data

---

[*] This work was partially funded by ANR GeoDIB project, n° ANR-06-BLAN-0225. Bertrand Kerautret was partially funded by a BQR project of Nancy-Universités.

P. Wiederhold and R.P. Barneva (Eds.): IWCIA 2009, LNCS 5852, pp. 187–200, 2009.

[2]. They are parameterized with a positive value related to the thickness of the perturbation. Based on this, discrete tangent and curvature estimators robust to noise have been developed [10,5]. Similarly, the curvature estimator of [9] requires a smoothing parameter related to the amount of noise.

Two factors limit the applicability of these techniques: first their parameterization requires a user supervision, secondly this parameter is global to the shape, while the amount of noise may be variable along the shape. This problem has been studied a lot in the image processing and edge detection community and led to the development of multi-scale analysis [11,6]. Further improvements lead to the automatic determination of a local scale [4,1]. This scale gives the minimum amount of local smoothing necessary to capture the local image geometry (i.e. edges). These techniques are thus able to determine the local amount of noise: adaptive filtering is then possible (generally edge detection).

Although very interesting for image processing, they cannot be used to process binary images or, equivalently, digital contours. They indeed rely on a local SNR analysis of the image, sometimes with a user-given global SNR parameter. Since we have only the discrete contour as input, the SNR is not calculable. We propose here a new method for estimating locally if the digital contour is damaged, what is the amount of perturbation, and what are the meaningful scales at which this part of the contour should be considered. Our method is similar in spirit to multi-scale analysis, but relies on specific properties of digital contours. The main idea is to look for the asymptotic properties of maximal segments in the multiresolution decomposition of the given contour. If they are present, then the scale is meaningful, otherwise the contour is still noisy at this scale and must be examined at a coarser scale. Our approach is local, requires no parameter tuning (from a given maximal observation scale), and is easy to implement. Its output can be used in many applications which requires a global or local noise parameterization. Among them, we may quote tangent or curvature estimators, dominant point and corner detection.

In Section 2 we recall standard notions of discrete geometry and known asymptotic results on maximal segments. We show that the length of maximal segments over scales can be used both to distinguish between flat and curved parts of a contour and to detect noise. In Section 3 we validate our technique on several datasets, containing different shape geometries, localized and variable noise, various resolutions. All datasets are processed without any specific parameterization. Noise is correctly determined in all cases. Section 4 presents some perspectives to this work.

2   Scale Properties of Maximal Segments

2.1   Definition and Known Asymptotic Results

A *standard digital straight line (DSL)* is some set $\{(x, y) \in \mathbb{Z}^2, \mu \leq ax - by < \mu + |a| + |b|\}$, where (a, b, μ) are also integers and $\gcd(a, b) = 1$. It is well known that a DSL is a 4-connected simple path in the digital plane. A *digital straight segment (DSS)* is a 4-connected piece of DSL. The interpixel contour of a simple digital shape is a 4-connected closed path without self-intersections. Given such a 4-connected path C, a *maximal segment* M is a subset of C that is a DSS and which is no more a DSS when adding any other point of $C \setminus M$.

We recall some asymptotic results related to maximal segments that lie on the boundary of some shape X digitized with step h. The digitization process is $\mathrm{Dig}_h(X) = X \cap h\mathbb{Z} \times h\mathbb{Z}$ (Gauss digitization). First, we assume the shape has smooth C^3-boundary and is strictly convex (no flat zones, no inflexion point). Theorem 5.26 of [7] states that the smallest discrete length of the maximal segments on the boundary of $\mathrm{Dig}_h(X)$ is some $\Omega(1/h^{1/3})$. The longest discrete length of the maximal segments on the boundary of $\mathrm{Dig}_h(X)$ is some $O(1/h^{1/2})$ (Lemma 15 of [8]).

Secondly, we observe maximal segments along the digitization of a flat zone of a shape. Since digital straight segments are digitization of straight line segments, there is at least one maximal segment that covers the straight line. It means that the discrete length of the longest maximal segment is some $\Theta(1/h)$.

As a corollary to the previous properties, we obtain:

Corollary 1. *Let S be a simply connected shape in R^2 with a piecewise C^3 boundary. Let p be a point of the boundary ∂S of S. Consider now an open connected neighborhood U of p on ∂S. Let (L_j^h) be the digital lengths of the maximal segments along the boundary of $\mathrm{Dig}_h(S)$ and which cover p. Then*

$$\text{if } U \text{ is strictly convex or concave, then } \Omega(1/h^{1/3}) \leq L_j^h \leq O(1/h^{1/2}) \quad (1)$$

$$\text{if } U \text{ has null curvature everywhere, then } \Omega(1/h) \leq L_j^h \leq O(1/h) \quad (2)$$

The first inequality expresses the asymptotic behaviour of the length of maximal segments in smooth curved parts of a shape boundary. The second one gives the analog properties in flat parts of a shape boundary.

2.2 From Asymptotic to Scale Analysis by Subsampling

In the context of image analysis and pattern recognition, we do not have access to asymptotic digitizations of shapes: we are not able to get finer and finer versions of the object. At first glance, it could mean that asymptotic properties are not useful to analyze shape boundaries. This is not true. We can use asymptotic properties in a reverse manner. We consider that our digital object O is the digitization of some Euclidean shape X at a grid step h, choosing for instance again the Gauss digitization. We then subsample the digital object O with covering "pixels" of increasing sizes $i \times i$, for $i = 2, 3, \ldots, n$. The subsampling process ϕ_i will be detailed in Section 2.3. The family of digital objects $\phi_n(O), \ldots, \phi_2(O), O$ is an approximation of the finer and finer digitized versions of X, $\mathrm{Dig}_{nh}(X), \ldots, \mathrm{Dig}_{2h}(X), \mathrm{Dig}_h(X)$. Corollary 1 holds for the latter family. Although this corollary does not formally hold for the former family, a similar behaviour is observed in practice.

When looking at lengths of maximal segments around some point P of the boundary of O, we should thus observe a decreasing sequence of lengths for the increasing sequence of digitization grid steps $h_i = ih$, whose slope is related to the fact that P was in a flat or curved region. More precisely, letting $(L_j^{h_i})_{j=1..l_i}$ be the discrete lengths of the maximal segments along the boundary of $\phi_i(O)$ and covering P, we can expect:

- If P is in a curved convex or concave zone, then the lengths $L_j^{h_i}$ follow (1).
- If P is in a flat zone, then the lengths $L_j^{h_i}$ follow (2).

The asymptotic bounds of these equations suggest:

Property 1 (Multi-scale). The plots of the lengths $L_j^{h_i}$ in log-scale should be approximately affine with negative slopes as specified below:

	expected slope	
plot	(curved part)	(flat part)
$(\log(i), \log(\max_{j=1...l_i} L_j^{h_i}))$	$\approx -\frac{1}{2}$	≈ -1
$(\log(i), \log(\min_{j=1...l_i} L_j^{h_i}))$	$\approx -\frac{1}{3}$	≈ -1

The plot is only approximately affine since the preceding properties are asymptotic. Given an object at a finite resolution, subsampling induces length variations that follow only approximately the asymptotic behaviour. Arithmetic artefacts play also a role in this. It is however clear that the approximation gets better when the initial shape is digitized with a finer resolution.

We can make several remarks about the preceding result. First, it allows to distinguish between flat parts and curved parts of an object boundary, provided the object was digitized with a reasonable precision. This distinction relies only on the classification of the plot slope between $[-1, -\frac{1}{2}[$ and $[-\frac{1}{2}, -\frac{1}{3}]$. Secondly, the preceding approach is not valid on (around) points that are (i) a transition between a flat and a curved part, (ii) corner points. Thirdly, this technique assumes smooth objects with perfect digitization: if the digital contour has been damaged by noise or digitization artefacts, these characterizations do not hold.

Although the two last remarks seem problematic for analyzing shapes, we will use them to detect *locally* the amount of noise and to extract *local* meaningful scales.

2.3   Subsampling of a Digital Contour

Our multi-scale analysis of digital contours will require several subsampling computation of the initial digital shape. A common way for subsampling a binary image by integer factors $i \times i$ is to cover the pixel space with a tiling of squares of size $i \times i$. Any big pixel of the subsampled object is lighted whenever any of the covered pixel are lighted. More formally,

$$\phi_i^{x_0, y_0}(O) = \left\{ (X, Y) \in \mathbb{Z}^2 | \exists (x, y) \in O, X = (x - x_0) \div i, Y = (y - y_0) \div i \right\},$$
(3)

where (x_0, y_0) defines the origin of the big pixels in $\mathbb{Z}^2$ and the integer division $\div$ is defined by truncation.

Also effective, this technique has two drawbacks. First, it does not give directly a correspondence between the digital contour of O and the digital contour of the subsampling of O (such a correspondence is illustrated in Fig. 1 (b)). We need it for computing the maximal segments at a lower resolution and assigning their lengths to the desired point. Secondly, the topology of O may change through the subsampling, for instance from a simply connected shape to a shape with holes or vice versa. A point P on the contour of O may have no point in the subsampling that is in its vicinity.

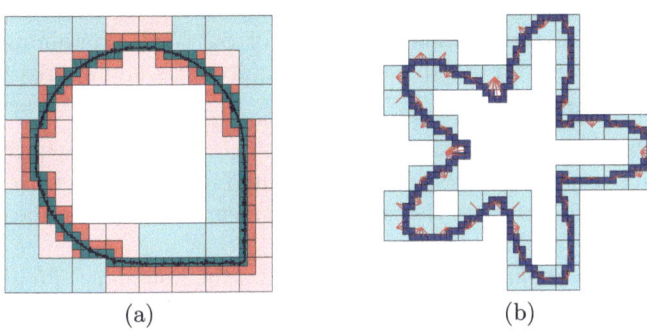

(a)   (b)

Fig. 1. (a) Subsampling $\phi_i^{0,0}(C)$ of a digital contour for $i = 64, 32, 16$ and 8. (b) The function $f_4^{0,0}$ is drawn with lines joining each pixel of C to it associated pixel of $\phi_4^{0,0}(C)$.

Although the preceding problems could be solved by *ad hoc* rules, we take another road. The subsampling is not spatial but operates on the digital contour. Its output is also a digital contour. We proceed in four steps, given a digital 4-connected contour C that is the boundary of the set of pixels O in the cellular model:

1. The interpixel contour C is shifted toward the inside so that it defines the 4-connected inner border of O. This 4-connected contour of pixels is denoted by C'. It is not necessarily simple and may contain some back-and-forth paths that are oriented toward the exterior of O (*outer spikes*).
2. The pixel contour C' is subsampled as the pixel contour C'', composed of the sequence of points $(X_j, Y_j) = ((x_j - x_0) \div i, (y_j - y_0) \div i)$, where C' is the sequence of points (x_j, y_j).
3. Consecutive identical pixels of C'' are merged to get a pixel contour C'''.
4. Outer and inner spikes of C''' are removed (by several linear scans). The obtained contour is shifted toward the outside so that it defines a 4-connected interpixel contour, that is the boundary of some digital shape.

The output subsampled contour is denoted by $\phi_i^{x_0, y_0}(C)$. Note that in the four preceding steps we keep the index correspondence between the contours. Along with $\phi_i^{x_0, y_0}$, there is thus a surjective map $f_i^{x_0, y_0}$ which associates any point P in

C to its image point in the subsampled contour $\phi_i^{x_0,y_0}(C)$. Several subsampled contours are illustrated on Fig. 1 (a) and a map $f_4^{0,0}$ for another contour is shown in (b). It is worth to note that this subsampling, similar to the common spatial one, can be done either locally around the point of interest or globally for the whole contour.

2.4   Local Geometric Evaluation with Multi-scale Criterion

We are now in position to analyze the local geometry of some point P on a digital contour C. For a resolution i and a shift (x_0, y_0), we compute the discrete lengths $L_j^{h_i,x_0,y_0}$ of the maximal segments of $\phi_i^{x_0,y_0}(C)$ containing $f_i^{x_0,y_0}(P)$. To take into account the possible digitization artefacts and approximations, we average these lengths as $\overline{L}^{h_i} = \frac{1}{i^2} \sum_{0 \leq x_0 < i, 0 \leq y_0 < i} \frac{1}{l_i^{x_0,y_0}} \sum_j L_j^{h_i,x_0,y_0}$, where $l_i^{x_0,y_0}$ represents the number of maximal segments containing $f_i^{x_0,y_0}(P)$. Note that all the maximal segments of $\phi_i^{x_0,y_0}$ can be computed in linear time with the number of points [8]. Fig. 2 illustrates the maximal segments obtained on the subsampled contour with several shift values.

The *multi-scale profile* $\mathcal{P}_n(P)$ of a point P on the boundary of a digital object O is the sequence of samples $(\log(i), \log(\overline{L}^{h_i}))_{i=1..n}$. According to Property 1, these samples should be correctly approached with an affine model. We thus define the *ideal multi-scale criterion* $\mu_n(P)$ of a point P on the boundary of a digital object O as the slope coefficient of the simple linear regression of $\mathcal{P}_n(P)$ (in the order regressor, regressand).

Property 1 indicates that $\mu_n(P)$ should be around -1 if P is in a flat zone, whereas it should be within $[-1/2, -1/3]$ if P is in a strictly convex or concave zone. Theorem 5.1 of [3] indicates that it is more likely to be around $-1/3$ in the latter case.

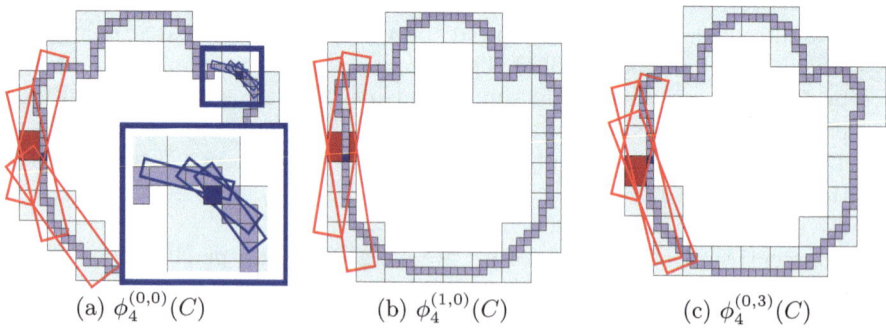

(a) $\phi_4^{(0,0)}(C)$ (b) $\phi_4^{(1,0)}(C)$ (c) $\phi_4^{(0,3)}(C)$

Fig. 2. Illustration of the maximal segments covering a point of the initial contour C (closeup view (a)). Several subsampled contours $\phi_4^{(x_0,y_0)}(C)$ obtained with different shifts with resolution 4 are given in light color (a-c). For each shift the maximal segments covering the considered point are drawn in dark.

Table 1. Distribution of the slopes of the multi-scale lengths of maximal segments (when plotted in log-space) for two different kinds of shapes generated with several digitization steps h. Two top groups: distribution for shapes with strictly positive curvature. Two bottom groups: distribution for shapes composed of flat sides. The last row gives the best choice of threshold $t_{f/c}$ to decide between flatness and curvedness for two shapes, assuming normal distribution. E and σ represent respectively the mean and the standard deviation of the slope distribution.

shape X	intervals	distribution of slopes $\mu_n(\cdot)$ according to digitization step h				
		$h=1$	$h=1/2$	$h=1/4$	$h=1/8$	$h=1/16$
circle $r=20$	$\in\,]-\infty,-\frac{2}{3}[$	0 %	0 %	0 %	0 %	0 %
	$\in[-\frac{2}{3},-\frac{1}{2}[$	0 %	0 %	0 %	0 %	0 %
	$\in[-\frac{1}{2},-\frac{1}{3}[$	100 %	54.0 %	62.1 %	67.0 %	58.5 %
	$\in\,]-\frac{1}{3},-\frac{1}{6}]$	0 %	46.0 %	37.9 %	33.0 %	41.5 %
	$\in\,]-\frac{1}{6},+\infty[$	0 %	0 %	0 %	0 %	0 %
	E,σ	$-0.391,0.028$	$-0.359,0.046$	$-0.356,0.041$	$-0.343,0.035$	$-0.342,0.055$
ellipse $a=20$ $b=14$ $\theta=0.2$	$\in\,]-\infty,-\frac{2}{3}[$	0 %	0 %	0 %	0 %	0 %
	$\in[-\frac{2}{3},-\frac{1}{2}[$	5.9 %	0 %	4.4 %	2.4 %	0 %
	$\in[-\frac{1}{2},-\frac{1}{3}[$	91.2 %	95.6 %	73.3 %	56.6 %	57.7 %
	$\in\,]-\frac{1}{3},-\frac{1}{6}]$	2.9 %	4.4 %	22.3 %	41.0 %	42.3 %
	$\in\,]-\frac{1}{6},+\infty[$	0 %	0 %	0 %	0 %	0 %
	E,σ	$-0.412,0.047$	$-0.392,0.039$	$-0.377,0.054$	$-0.353,0.056$	$-0.346,0.056$
triangle $r=20$ $\theta=0.3$	$\in\,]-\infty,-\frac{5}{4}[$	0 %	0 %	0 %	0 %	0 %
	$\in[-\frac{5}{4},-1[$	0 %	6.4 %	15.8 %	12.3 %	7.5 %
	$\in[-1,-\frac{3}{4}[$	81.2 %	89.8 %	83.8 %	87.6 %	92.4 %
	$\in[-\frac{3}{4},-\frac{1}{2}[$	18.8 %	3.8 %	0.4 %	0.1 %	0.1 %
	$\in[-\frac{1}{2},+\infty]$	0 %	0 %	0 %	0 %	0 %
	E,σ	$-0.860,0.100$	$-0.931,0.060$	$-0.956,0.052$	$-0.923,0.068$	$-0.920,0.047$
pentagon $r=20$ $\theta=0.2$	$\in\,]-\infty,-\frac{5}{4}[$	0 %	0 %	0 %	0 %	0 %
	$\in[-\frac{5}{4},-1[$	0 %	0 %	0.7 %	5.7 %	9.0 %
	$\in[-1,-\frac{3}{4}[$	37.3 %	89.3 %	95.8 %	91.2 %	89.6 %
	$\subset[-\frac{3}{4},-\frac{1}{2}[$	60.7 %	10.6 %	3.5 %	3.1 %	1.4 %
	$\in[-\frac{1}{2},+\infty]$	2.0 %	0 %	0 %	0 %	0 %
	E,σ	$-0.695,0.081$	$-0.815,0.070$	$-0.890,0.064$	$-0.914,0.066$	$-0.924,0.059$
$t_{f/c}$ ellipse/pentagon		-0.52	-0.55	-0.61	-0.61	-0.63

We have checked Property 1 on digitization of various smooth or polygonal shapes at different scales (from digitization step $h=1$ to $1/16$). For each experiment, we have chosen $n=10$. The results are summed up in Table 1. They clearly back up our claim on the validity of Property 1 as a reverse analog to asymptotic properties.

The preceding experiments suggest a flat/curved threshold $t_{f/c}$ between -0.6 and -0.5. Since the resolution of objects is generally low, setting $t_{f/c}=-0.52$ is a reasonable choice. It induces more than 98.3% of correct decision at $h=1$, and more than 99.9% of correct decision at finer grid steps.

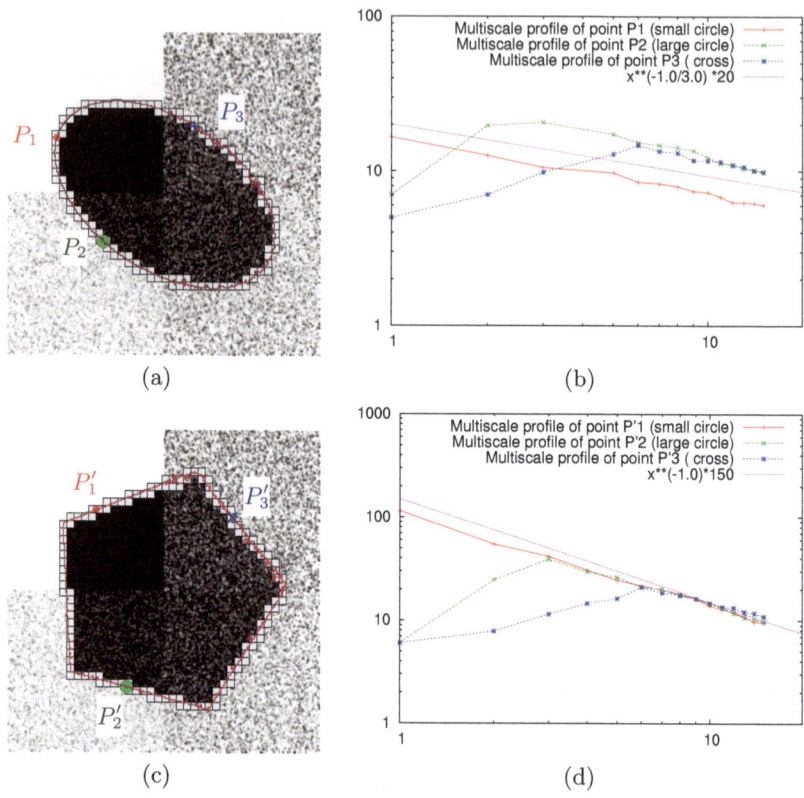

Fig. 3. Examples of multiscale profiles $\mathcal{P}_{15}(P)$ on ellipse (a,b): P_1 in a curved zone, P_2 in a slightly perturbated curved zone , P_3 in a strongly perturbated curved zone. Gaussian noise was added on each areas containing the point P_1, P_2 and P_3 with respectively the following standard deviation $\sigma_1 = 0$, $\sigma_2 = 75$, and $\sigma_2 = 175$. The same experimentation is applied on the polygon (c-d) with the points P'_1, P'_2, and P'_3.

2.5   Detecting Noise and Local Meaningful Scale

The multi-scale profile can be used to detect noisy digital contours. Indeed, if the multi-scale profile of some point P is not some approximation of an affine map with negative slope, it means that locally around P the shape geometry is neither a flat or curved zone. We display on Fig. 3 (a-b) the multi-scale profile of a point P_1 located on a perfectly digitized curved zone and the multi-scale profiles of the points P_2 and P_3 located in noisy zones. On the former profile, the decreasing affine relation is immediately visible. On the latter profiles, it is somewhat randomly increasing for fine resolution and then follow an expected decreasing affine profile after a given scale. A similar behaviour is observable for the multi-scale profiles on the pentagon shape with similar noisy regions. The only difference is the slope of the affine relation of the profiles (slopes near $-\frac{1}{3}$ for the plots of Fig. 3 (b) and near -1 for the plots of Fig. 3 (d)).

We therefore introduce a *noise threshold* t_m which discriminates between a curved zone and a noisy zone. This threshold should be somewhere between $]-\frac{1}{3}, 0[$. According to Table 1, setting $t_m = -1/6 \approx E + 3\sigma$ induces more than 99.7% of correct determination with some hypotheses (especially normal distribution). However after several experiments on noisy shapes it appears that the use of the upper threshold value $t_m = 0$ gives better results especially in curved zones.

A *meaningful scale* of a multi-scale profile $(X_i, Y_i)_{1 \le i \le n}$ is then a pair (i_1, i_2), $1 \le i_1 < i_2 \le n$, such that for all i, $i_1 \le i < i_2$, $\frac{Y_{i+1} - Y_i}{X_{i+1} - X_i} \le t_m$, and the preceding property is not true for $i_1 - 1$ and i_2.

If (i_1, i_2) is a meaningful scale of the profile $\mathcal{P}_n(P)$, the (i_1, i_2)-*multi-scale criterion* $\mu_{i_1, i_2}(P)$ of point P is then the slope coefficient of the simple linear regression of $\mathcal{P}_n(P)$ restricted to its samples from i_1 to i_2.

Obviously meaningful scales of $\mathcal{P}_n(P)$ do not overlap and are thus naturally ordered. If the first meaningful scale of $\mathcal{P}_n(P)$ is (k_1, k_2), then the integer $k_1 - 1$ is called the *noise level* at point P and we denote it by $\nu(P)$.

We will show in the experiment section that both definitions of meaningful scales and noise level have a clear intuitive interpretation. They determine precisely where the contour is perturbated and how it should be interpreted to be meaningful.

3   Experiments

Noise Detection. A robust noise detector should not detect noise on perfectly digitized data and should not be sensitive to the object initial resolution. To experiment this properties, different shapes have been generated with a manual addition of noise on some specifics areas (Fig. 4). These noisy areas are highlighted by red boxes on subfigures (a-c) and (g-i). Note that for certain areas (on bottom left on the shape) only one or three pixels were changed (highlighted in red). For each pixel P of the initial contour, the result of the noise detection is illustrated by drawing its associated pixel $f_K^{0,0}(P)$ (of size K) where K is $\nu(P) + 1$, i.e. its first meaningful scale.

The obtained noise detection displayed on Fig. 4 (d-f,j-l) shows a good precision. Even with low resolution shapes and with a one pixel change, the noise is well detected. Only a few false positive noise detections can be seen on some small areas on the flower (near corners): however these errors are limited to one noise level. Note that for all of these experiments no parameter was changed for the detection. The variable t_m associated to the noise threshold was set to 0 for all the experiments as suggested in section 2.5. The maximal resolution n used in the definition of the *multi-scale profile* $\mathcal{P}_n(P)$ has only an influence on the level of scale of the detected noise. Indeed, for example the use of the minimal value of $n = 2$ induces a noise detection only at scales 1 and 2. This value was set to 15 in all the presented experiments.

After experimenting the noise detection on shapes containing some local noisy parts, we apply this detection on a shape globally damaged with different noise

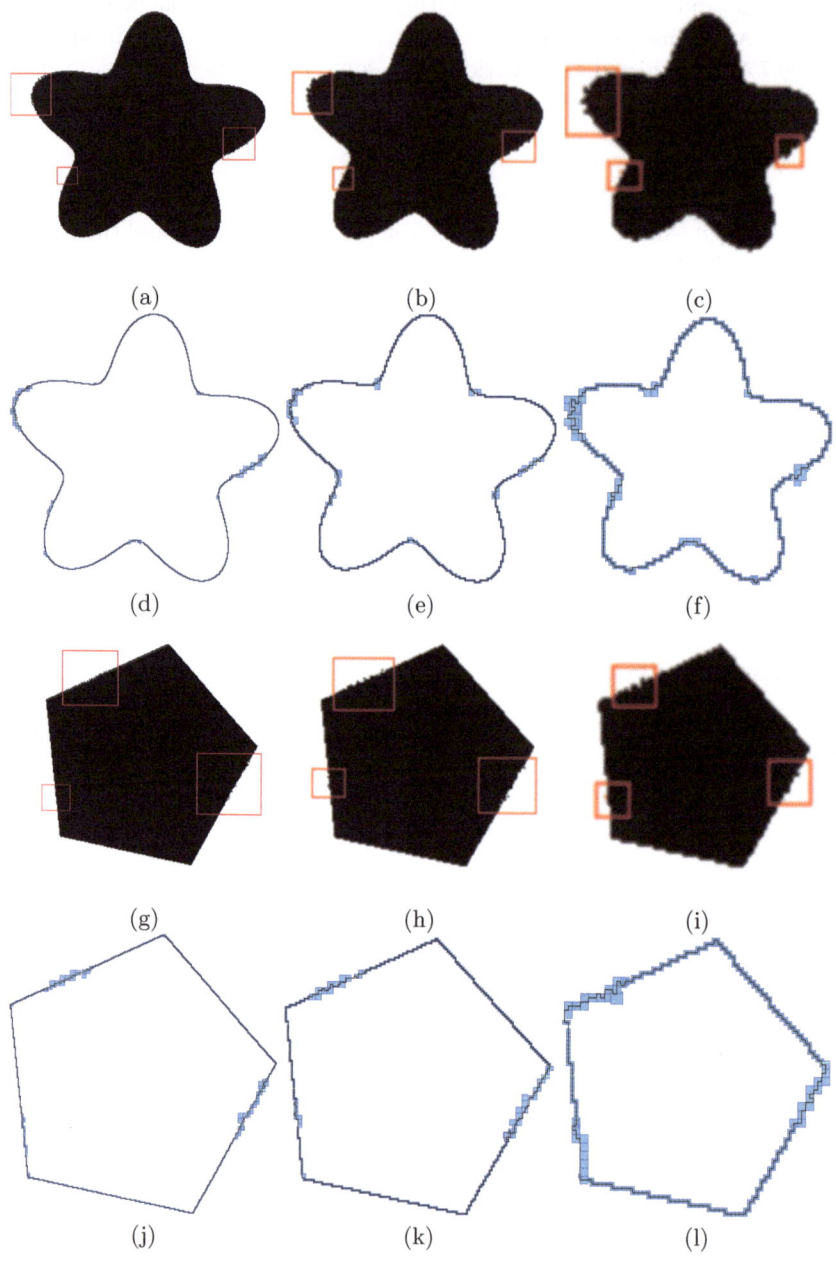

Fig. 4. Noise detection on flower-like and polygonal shapes defined with several grid sizes. Noisy areas are highlighted by red box (images (a-c,g-i)). Images (d-f,j-l) display for each pixel P its associated pixel $f_K^{0,0}(P)$ with K equal to $\nu(P) + 1$.

intensities. The initial image of the shape is thus divided into 4 parts on which a gaussian noise is applied with four standard deviations $\sigma_0 = 0$, $\sigma_1 = 75$, $\sigma_2 = 125$ and $\sigma_3 = 175$. The discrete contour is then extracted from the set of the biggest connected component selected with the threshold value 128 (Fig. 5). Subimages (a-d) shows the points whose noise level is below some scale K. We can see that the original local repartition of the noise per image quadrant is well visible.

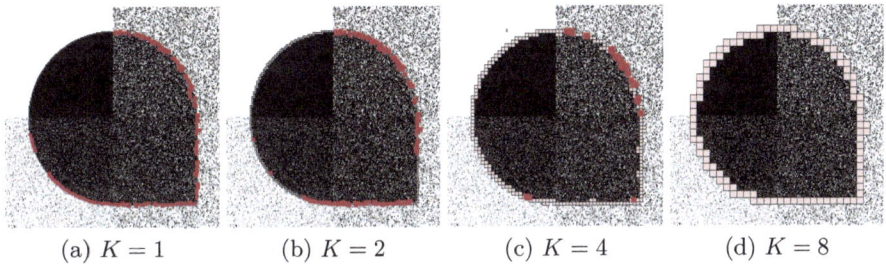

(a) $K = 1$ (b) $K = 2$ (c) $K = 4$ (d) $K = 8$

Fig. 5. Illustration of the contour parts which are considered as valid on the scale less or equals to K (light grey). Pixels with a higher noise level are drawn in dark red.

Detection of Flat and Curved Contour Areas. As described in the previous section the analysis of the *multi-scale profile* $\mathcal{P}_n(P)$ can discriminate the curved and flat areas of the contour. This discrimination relies on a constant threshold value $t_{f/c}$ which is set to -0.52 to maximize the good detection as suggested in Section 2.4. Exactly as in previous experimentations, the maximal resolution n of the *multi-scale profiles* was not changed (set to 15) and has only a limited influence since the estimation of the slope is only done within meaningful scales.

In the same way than as the noise detection, a shape containing curved and flat areas was generated with several resolutions and with gaussian noise (see Fig. 6 (a-c)). For the noisy version of the shapes, the profile analysis was initiated from the scale 4. This value was deduced from the global level of noise of the image. We can see on these examples that the curved and flat areas are well detected (subimages (a,b)) and the noisy shape (subimage (c)) presents very few pixels incorrectly classified. Fig. 6 (d-g) shows also other good detections.

Experiments on Real Images. Our method was applied on two real images with unchanged parameters (Fig. 7 and Fig. 8). The contours were obtained by extraction of the connected components defined with a simple threshold value. The image of Fig. 8 (a) was directly extracted without subsampling from a digital camera picture obtained at resolution 4000×2672 with a sensibility of 250 ISO. As for the previous experiments, the detected noise level K is illustrated by drawing for each pixel P its associated pixel $f_K^{0,0}(P)$ (Fig. 7 (a)). For the contours of Fig. 8 we change the noise visualisation by using a box of size K centered on each point. These results are comparable to the ones obtained on synthetic images and even for the detection of flat areas which are well detected (Fig. 7 (b)).

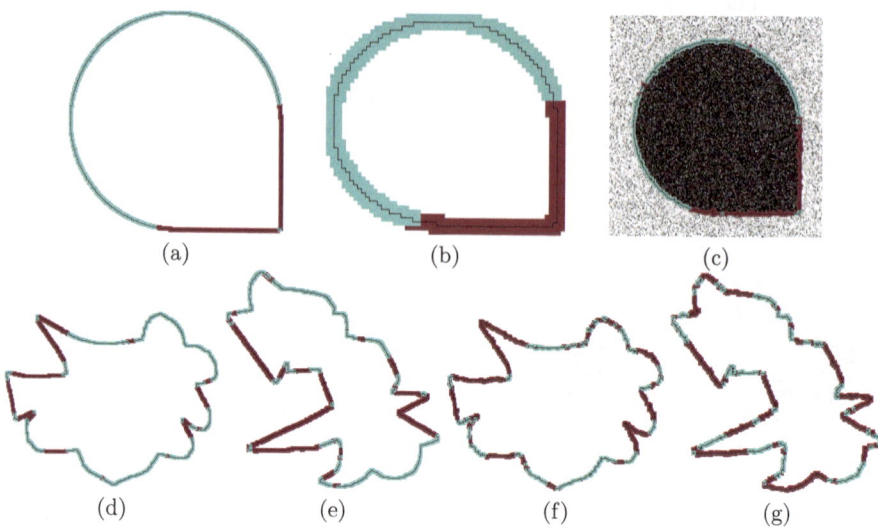

Fig. 6. Detection of flat/curved areas on a fine (a) and coarse (b) grid size, and with Gaussian noise with standard deviation 100 (c). Images (f,g) are the noisy version of (d,e) obtained after adding noise . For the shapes (c,f,g) the scale analysis was initiated from the scale 4.

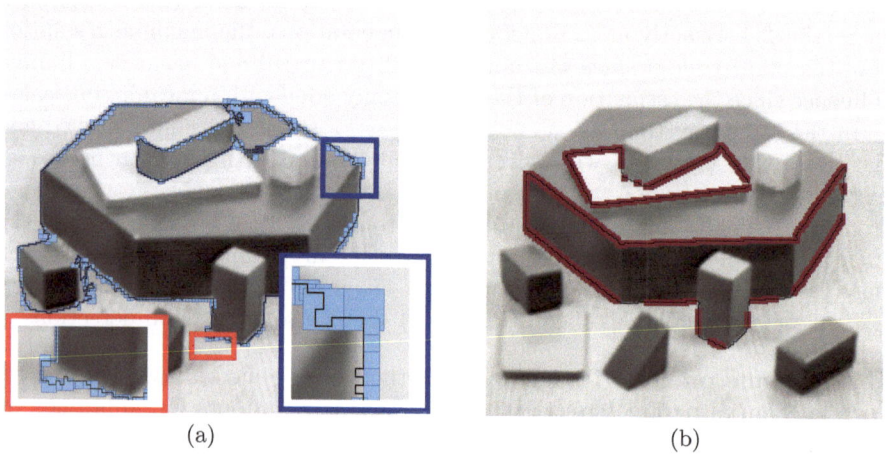

Fig. 7. (a) Example of meaningful scales obtained on contours in a real image (a). (b) Detection of the flat areas of selected contours.

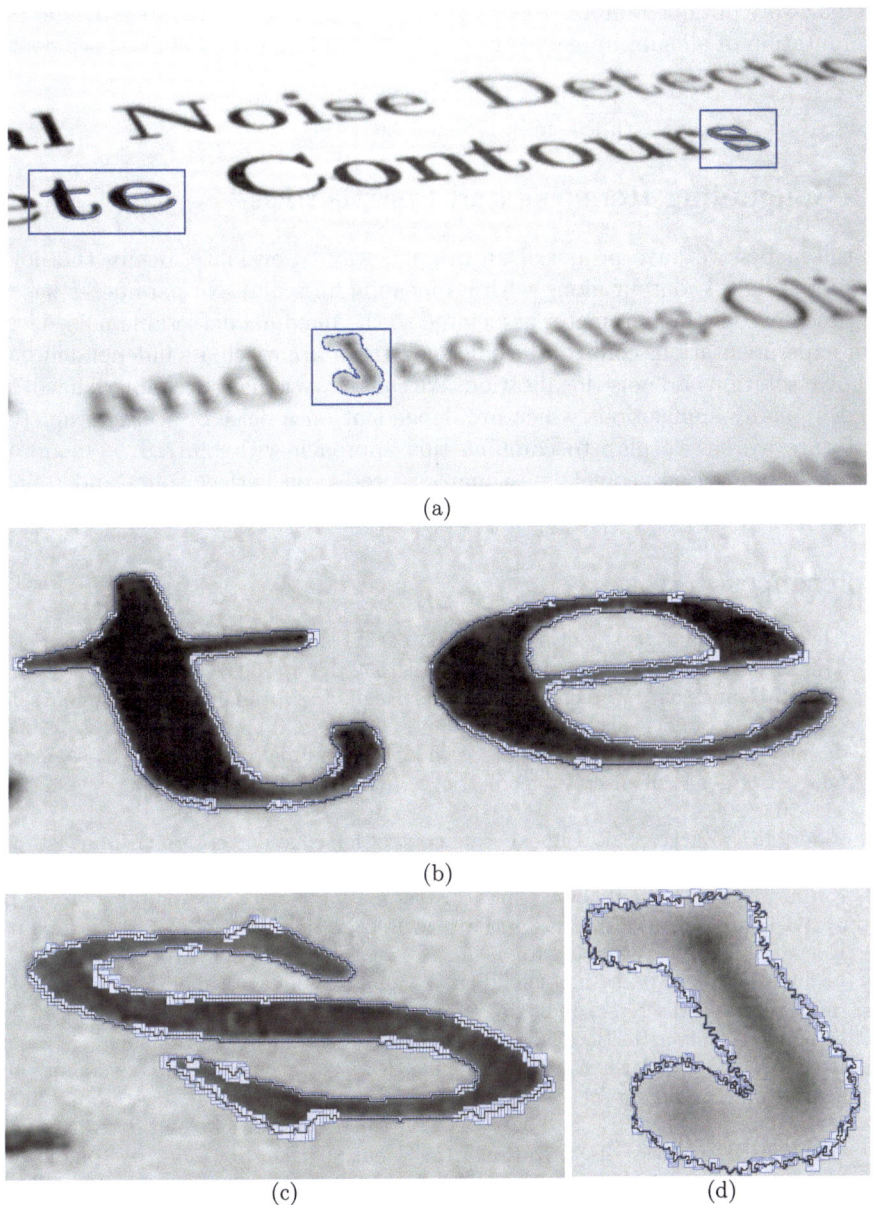

Fig. 8. Noise detection applied on a real text photography (a). The contours of the characters (b-d) were extracted by a simple threshold.

Timing Measures. To conclude the experimentation part some runtime measures were performed on the contours of Fig. 8. The times given on the following tabular were obtained on a 2.4 GHz *Intel Core Duo*. The measures include the computation of all subsampled contours $\phi_i^{x_0,y_0}(C)$ and their maximal segments.

nb points	226	702	788	874	1450
times (ms)	178	363	387	411	513

4   Concluding Remarks and Perspectives

In this paper we have proposed an original way to evaluate locally the noise level of a digital contour along with its meaningful scales. No parameter has to be tuned except the parameter associated to the maximal detectable noise level. Our experimentations confirm the efficiency of our approach, its independence to shape resolution and noise localisation. We thus expect that this approach will be used by many applications which are dependent on a noise or scale parameter. In future works, we plan to combine this approach with blurred segments to obtain unsupervised geometric estimators, precise on perfect zones and robust elsewhere.

References

1. Chen, K.: Adaptive smoothing via contextual and local discontinuities. IEEE Trans. Pattern Anal. Mach. Intell. 27(10), 1552–1566 (2005)
2. Debled-Rennesson, I., Feschet, F., Rouyer-Degli, J.: Optimal blurred segments decomposition of noisy shapes in linear times. Comp. and Graphics 30, 30–36 (2006)
3. de Vieilleville, F., Lachaud, J.O., Feschet, F.: Maximal digital straight segments and convergence of discrete geometric estimators. J. Math. Imaging Vis. 27(2), 471–502 (2007)
4. Elder, J.H., Zucker, S.W.: Local scale control for edge detection and blur estimation. IEEE Trans. Pattern Anal. Mach. Intell. 20(7), 669–716 (1998)
5. Kerautret, B., Lachaud, J.O.: Curvature estimation along noisy digital contours by approximate global optimization. Pattern Recognition 42(10), 2265–2278 (2009)
6. Koenderink, J.J.: The structure of images. Biol. Cyb. 50, 363–370 (1984)
7. Lachaud, J.O.: Espaces non-euclidiens et analyse d'image : modèles déformables riemanniens et discrets, topologie et géométrie discrète. Habilitation à diriger des recherches, Université Bordeaux 1, Talence, France (2006) (en français)
8. Lachaud, J.O., Vialard, A., de Vieilleville, F.: Fast, accurate and convergent tangent estimation on digital contours. Image Vision Comp. 25(10), 1572–1587 (2007)
9. Malgouyres, R., Brunet, F., Fourey, S.: Binomial convolutions and derivatives estimations from noisy discretizations. In: Coeurjolly, D., Sivignon, I., Tougne, L., Dupont, F. (eds.) DGCI 2008. LNCS, vol. 4992, pp. 370–379. Springer, Heidelberg (2008)
10. Nguyen, T., Debled-Rennesson, I.: Curvature estimation in noisy curves. In: Kropatsch, W.G., Kampel, M., Hanbury, A. (eds.) CAIP 2007. LNCS, vol. 4673, pp. 474–481. Springer, Heidelberg (2007)
11. Witkin, A.P.: Scale-space filtering. In: Proc. 8th Int. Joint Conf. Artificial Intelligence, pp. 1019–1022 (1983)

Sub-pixel Segmentation with the Image Foresting Transform

Filip Malmberg, Joakim Lindblad, and Ingela Nyström

Centre for Image Analysis
Uppsala University and Swedish University of Agricultural Sciences
Sweden
http://www.cb.uu.se

Abstract. The Image Foresting Transform (IFT) is a framework for image partitioning, commonly used for interactive segmentation. Given an image where a subset of the image elements (seed-points) have been assigned user-defined labels, the IFT completes the labeling by computing minimal cost paths from all image elements to the seed-points. Each image element is then given the same label as the closest seed-point. In its original form, the IFT produces crisp segmentations, i.e., each image element is assigned the label of exactly one seed-point. Here, we propose a modified version of the IFT that computes region boundaries with sub-pixel precision by allowing mixed labels at region boundaries. We demonstrate that the proposed sub-pixel IFT allows properties of the segmented object to be measured with higher precision.

Keywords: Image foresting transform, Interactive image segmentation, Sub-pixel precision.

1   Introduction

Image segmentation, i.e., the partitioning of an image into relevant regions, is a fundamental problem in image analysis. Accurate segmentation of objects of interest is often required before further analysis can be performed. Despite years of active research, fully automatic segmentation of arbitrary images is still seen as an unsolved problem. Semi-automatic, interactive segmentation methods [12] use human expert knowledge as additional input, thereby making the segmentation problem more tractable.

The segmentation process can be divided into two tasks: *recognition* and *delineation* [6]. Recognition is the task of roughly determining where in the image an object is located, while delineation consists of determining the exact extent of the object. Human users outperform computers in most recognition tasks, while computers are often better at delineation. A successful semi-automatic method combines these abilities to minimize user interaction time, while maintaining tight user control to guarantee the correctness of the result.

One popular paradigm for interactive segmentation is *seeded region segmentation*, where the user assigns labels (e.g., object and background) to a small

P. Wiederhold and R.P. Barneva (Eds.): IWCIA 2009, LNCS 5852, pp. 201–211, 2009.

subset of the image elements (known as *seed-points*). An automatic algorithm then completes the labeling for all image elements. If the result is not satisfactory, the user can add or remove seed-points until a desired segmentation has been obtained. Many different algorithms have been proposed for performing the label completion, see, e.g, [1,7]. Here, we will focus on one such algorithm, the Image Foresting Transform (IFT) [4,5].

The IFT belongs to a family of graph-based methods, where the image is interpreted as a graph. Each image element corresponds to a node in the graph, and adjacent image elements are connected by edges. For each node in the graph, the minimum cost path from the node to the set of seed-points is computed. The cost of a path typically depends on local image features. By choosing an appropriate path cost function, popular image segmentation methods such as relative fuzzy-connectedness [3] and watersheds [11] can be implemented.

The IFT can be computed efficiently using Dijkstra's algorithm, slightly modified to allow multiple seed-points [5]. In interactive segmentation applications, a user often adds or removes seed-points to refine an existing segmentation. In [4], it was shown that seed-points can be added to, or removed from, an existing IFT solution, without recomputing the entire solution. This modified algorithm is called the *differential* IFT, and has been shown to give a significant reduction of the total time required for interactive segmentation.

In the original IFT, the resulting labels are *crisp*, i.e., each image element is assigned the label of exactly one seed-point. However, due to the finite resolution of digital images, an image element may be partially covered by more than one (continuous) object. By allowing mixed labels, it is possible to obtain segmentations with sub-pixel precision. Numerous studies have confirmed that *pixel coverage segmentation* [14] outperforms crisp segmentation for subsequent measuring of object properties such as length and area/volume, see, e.g., [13,15]. In [9], it is shown that consequently misplacing the tissue borders, in a brain volume having voxels of size 1 mm^3, by one voxel resulted in volume errors of approximately 30%, 40% and 60% for white matter, grey matter and cerebrospinal fluid, respectively. Segmentation methods with sub-pixel precision can also produce more visually pleasing results than their crisp counterparts. Surface extraction algorithms such as Marching Cubes [10] can utilize sub-pixel precision to produce visually smoother surfaces. In the context of image compositing, sub-pixel segmentation is necessary to avoid aliasing artifacts [2].

Here we propose a modified version of the IFT, that computes labels with sub-pixel precision. In the following, we will refer to the original IFT method as *crisp IFT*, and the proposed method as *sub-pixel IFT*. Like the crisp IFT, the proposed method is defined on general graphs. Therefore, the method can be applied to higher-dimensional data without modification. Our method does not rely on any assumptions about the *shape* of the image elements. This makes the method more general, but also means that the output of the method is not strictly a pixel coverage segmentation. Instead, we see the previously demonstrated advantages of pixel coverage segmentation as a motivation for including sub-pixel information in the IFT. We demonstrate that similar improvements in feature estimation can be achieved with the proposed sub-pixel IFT.

2   Notation and Definitions

2.1   Images and Graphs

An *image* $\mathbf{I}$ is a pair $(\mathcal{I}, I)$ consisting of a set $\mathcal{I}$ of image elements and a mapping I that assigns to each image element $p \in \mathcal{I}$ a value in some arbitrary set (e.g., $\mathcal{I} \subset \mathbb{Z}^n$ and $I : \mathcal{I} \to [0, 255]$).

Over the image elements, we define an *adjacency* function $\mathcal{N}$ that maps each image element $p \in \mathcal{I}$ to a set $\mathcal{N}(p) \subset \mathcal{I}$ of *adjacent* nodes. Once the adjacency function has been fixed, the image $\mathbf{I}$ can be interpreted as a directed graph, whose nodes are the image elements and whose edges are all ordered pairs of image elements $p, q \in \mathcal{I}$ such that $q \in \mathcal{N}(p)$.

For each ordered pair of adjacent nodes p and q, we assign a real valued *edge weight* $w(p, q)$. This weight typically depends on local image features such as intensity or gradient magnitude. A thorough discussion on how the choice of adjacency function and edge weights affect the segmentation results can be found in [8].

2.2   Paths and Path Costs

A *path* $\pi = \langle p_1, p_2, \ldots, p_k \rangle$ of length $|\pi| = k$ is a sequence $p_1, p_2, \ldots, p_k$ such that $p_{i+1} \in \mathcal{N}(p_i)$. We denote the *origin* p_1 and the *destination* p_k of π by $org(\pi)$ and $dst(\pi)$, respectively. If π and τ are paths such that $dst(\pi) = org(\tau)$, we denote by $\pi \cdot \tau$ the concatenation of the two paths. Given a set Π of paths such that $dst(\pi) = p$ for all $\pi \in \Pi$, and a path τ such that $org(\tau) = p$, we define the set $\Pi \cdot \tau$ as

$$\{\pi \cdot \tau \,|\, \pi \in \Pi\} . \tag{1}$$

The *cost* of a path is denoted $f(\pi)$. This cost is typically a function of the edge weights along the path, e.g., the sum of all the edge weights along the path or the maximum edge weight along the path. We require $f(\pi)$ to be strictly increasing with respect to $|\pi|$, i.e., for two non-empty paths π, τ such that $dst(\pi) = org(\tau)$

$$f(\pi) < f(\pi \cdot \tau) \text{ if } |\pi| < |\pi \cdot \tau| . \tag{2}$$

This requirement is slightly more strict than the corresponding requirement in [5], where the path cost function was only required to be monotonically increasing. The stricter requirement is necessary to guarantee the existence of the sub-pixel region boundaries in Section 3.2. In practice, this additional restriction is not very limiting. For a given monotonically increasing path cost function f, a corresponding strictly increasing function g can be defined as

$$g(\pi) = f(\pi) + \epsilon |\pi| , \tag{3}$$

where ϵ is a small positive number.

Given three paths π, τ and υ such that $dst(\pi) = dst(\tau) = org(\upsilon)$ and $f(\pi) = f(\tau)$, we also require of the path cost function that

$$f(\pi \cdot \upsilon) = f(\tau \cdot \upsilon) \, . \tag{4}$$

A path π is a *minimum cost path* if $f(\pi) \leq f(\tau)$ for any other path τ with $org(\tau) = org(\pi)$ and $dst(\tau) = dst(\pi)$. Note that in general, the minimum cost path is not unique. The set of minimum cost paths between two nodes p and q is denoted $\pi_{min}(p, q)$. For a node p and a set $A \subset \mathcal{I}$, the set of minimal cost paths between p and A is defined as

$$\pi_{min}(p, A) = \bigcup_q \pi_{min}(p, q) \, , \tag{5}$$

for all $q \in \underset{r \in A}{\operatorname{argmin}}(f(\pi_{min}(p, r)))$.

2.3   Spanning Forests

A *predecessor map* is a function P that assigns to each image element $p \in \mathcal{I}$ a (possibly empty) subset of $\mathcal{N}(p)$. Note that in contrast to [5], we here allow $|P(p)|$ to be greater than one, i.e. a node can have more than on predecessor. For any node $p \in \mathcal{I}$, a predecessor map P defines a set $P^*(p)$ of one or more paths recursively as

$$P^*(p) = \begin{cases} \{\langle p \rangle\} & \text{if } P(p) = \emptyset \\ \bigcup_{q \in P(p)}(P^*(q) \cdot \langle q, p \rangle) & \text{otherwise} \end{cases} . \tag{6}$$

A *spanning forest* is a predecessor map that contains no cycles, i.e., $|\pi|$ is finite for all paths $\pi \in P^*(p)$.

2.4   The Image Foresting Transform

The IFT takes an image $\mathbf{I}$, a path cost function f, an adjacency function $\mathcal{N}$ and a set of seed-points $S \in \mathcal{I}$, and returns a spanning forest P such that $P^*(p) = \pi_{min}(p, S)$ for all nodes $p \in \mathcal{I}$.

During this process, a *cost map* C and a label map $\mathcal{L}$ is built. The cost map contains the cost of the minimum cost path from each pixel to S, i.e., $C(p) = f(\pi_{min}(p, S))$. The label map assigns to each node a *label vector* $\mathcal{L}(p) = (l_1, l_2, ..., l_k)$ where $l_i \in [0, 1]$ (For the original, crisp IFT, $l_i \in \{0, 1\}$). Each element in the label vector indicates the *belongingness* of the node to a certain class (such as object or background). The labels l_i sum up to 1, i.e.

$$\sum_{i=0}^{k} l_i = 1 \, . \tag{7}$$

For all nodes $p \in \mathcal{I}$, $\mathcal{L}(p)$ should represent the label of the seed-point *closest* to p. As discussed in previous sections, assigning a single element of the set $\mathcal{L}_S =$

$\{L(q) \,|\, q \in S\}$ to $\mathcal{L}(p)$ is problematic, due to both the ambiguity of minimal cost paths in the graph and the limited resolution of the image. We therefore calculate $\mathcal{L}(p)$ as a weighted average of the label vectors in $\mathcal{L}_S$. The exact procedure for calculating this average is described in Section 3. To differentiate between crisp and sub-pixel labels, label maps that have been computed with sub-pixel precision are denoted $\mathcal{L}_{sub}$.

3   Method

In this section, we present the proposed sub-pixel IFT method. The method consists of three steps. First, the IFT is computed, using a new policy for handling cases where the minimum cost path is not unique. Pseudo-code for this modified IFT is given in Section 3.1. Secondly, region boundaries between the nodes are estimated with sub-pixel precision using a linear model. Finally, the sub-pixel label of each node is computed by integrating the labels over the graph edges connected to the node. In Section 3.2, we show that this integral can be evaluated analytically.

3.1   Handling Ties

As observed in Section 2.2, the minimum cost path between a node and the set of seed-points is not necessarily unique. Therefore, a strategy is needed for assigning labels in cases where the minimal cost path is ambiguous. In previous literature on the IFT, such a strategy is usually referred to as a *tie-breaking policy*. In [5], two strategies were suggested: the *first-in-first-out* (FIFO) strategy and *last-in-first-out* (LIFO) strategy. With these strategies, each node is assigned the label corresponding to the minimum cost path found first and last, respectively. Both these strategies are somewhat ad-hoc and have the effect that the output of the algorithm depends on the order in which we process the seed-points. A better strategy is therefore desirable.

The problem with handling ties in the crisp IFT framework is that a single label must be determined for each node. For the sub-pixel IFT this requirement is lifted, and we therefore have more flexibility for handling ties. Here, we have used a *mean* tie-breaking scheme, where each node p is assigned the mean of the labels of the predecessors $P(p)$ along the minimal cost paths from p to S. Pseudo-code for computing the IFT with mean tie-breaking is given in Algorithm 3.1. Note that due to Equation (4), $f(P^*(p) \cdot \langle p, q \rangle)$ is well defined even if $|P^*(p)| > 1$.

3.2   Sub-pixel Estimation of Region Boundaries

In the crisp IFT, labels are defined for all nodes. To obtain sub-pixel precision, we define the labels over the graph edges as a piecewise constant function. Given two adjacent nodes p and q with corresponding labels, the label along the edge p, q is given by

$$\mathcal{L}_{p,q}(t) = \begin{cases} \mathcal{L}(p) \text{ if } t \leq t'_{p,q} \\ \mathcal{L}(q) \text{ if } t > t'_{p,q} \end{cases}, \tag{8}$$

Algorithm 1. Computing the IFT with mean tie-breaking

Input: An image $\mathbf{I}$, a path-cost function f, an adjacency function $\mathcal{N}$, and a set of seed-points $S \subset \mathcal{I}$.
Output: A spanning forest P such that $P(p) = \pi_{min}(p, S)$ for all $p \in S$, a cost map C such that $C(p) = f(\pi_{min}(p, S))$ for all $p \in S$, and a label map $\mathcal{L}$.
Auxiliary data structures: A list Q of active nodes.
Initialization: Set $C(p) \leftarrow 0$ for source nodes $p \in S$ and $C(p) \leftarrow \infty$ for remaining nodes. Assign appropriate labels $\mathcal{L}(p)$ to all source nodes. Insert all source nodes in Q. Set $P(p) = \emptyset$ for all nodes.

```
while Q is not empty
    Remove from Q a node p such that C(p) is minimal.
    forall nodes q ∈ N(p)
        if f(P*(p) · ⟨p,q⟩) < C(q)
            Set C(q) ← f(P*(p) · ⟨p,q⟩)
            Set L(q) ← L(p)
            Set P(q) ← p
            Insert q in Q
        elseif f(P*(p) · ⟨p,q⟩) = C(q)
            Set L(q) ← |P|L(q)+L(p)/(|P|+1)
            Set P(q) ← P(p) ∪ {p}
        endif
    endfor
endwhile
```

---

where the parameter $t \in \{0, 1\}$ determines the position along the edge and $t'_{p,q} \in \{0, 1\}$ is the point along the edge where the label changes.

To approximate $t'_{p,q}$, we compare the cost of a minimum path to p with the cost of a minimal path to q appended by the path $\langle q, p \rangle$, and vice versa. We denote these four scalar values a, b, c and d:

$$
\begin{aligned}
a &= f(\pi_{min}(p, S)) \\
b &= f(\pi_{min}(p, S) \cdot \langle p, q \rangle) \\
c &= f(\pi_{min}(q, S)) \\
d &= f(\pi_{min}(q, S) \cdot \langle q, p \rangle) \, .
\end{aligned}
\tag{9}
$$

We assume that the path costs vary linearly between nodes, a natural assumption if we consider the edge weights to be constant along each edge. Thus we can solve a linear equation to find $t'_{p,q}$. See Figure 1. For the intersection point $t'_{p,q}$, we obtain the equation

$$
t'_{p,q} a + (1 - t'_{p,q})b = (1 - t'_{p,q})c + t'_{p,q} d \, .
\tag{10}
$$

Solving Equation (10) for $t'_{p,q}$, we obtain

$$
t'_{p,q} = \frac{b - c}{(b - a) + (d - c)} \, .
\tag{11}
$$

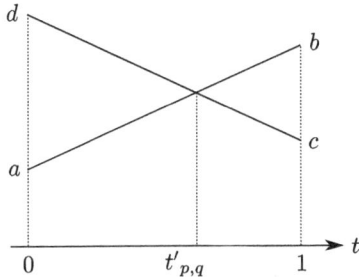

Fig. 1. Finding the intersection point $t'_{p,q}$ between two adjacent nodes. The scalars a,b,c and d are defined in the text.

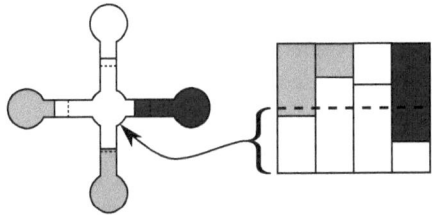

Fig. 2. Determining the sub-pixel label for a node with four neighbors. The sub-pixel label for the middle node is calculated by integrating over the closest half of all edges connected to the node.

Since the path cost function f is required to be strictly increasing with respect to path length, $(b-a) > 0$ and $(d-c) > 0$. Thus, the denominator of Equation (11) is non-zero.

Once the intersection points are determined, we calculate a sub-pixel label for each node by integrating the labels over edges connected to the node. For each edge, the domain of integration is the half of the edge that is closest to the node. This concept is illustrated in Figure 2. Formally, for each node $p \in G$, the sub-pixel label $\mathcal{L}_{sub}(p)$ is determined as

$$\mathcal{L}_{sub}(p) = \frac{2 \sum_{q \in \mathcal{N}(p)} \int_0^{\frac{1}{2}} \mathcal{L}_{p,q}(t) \, dt}{|\mathcal{N}(p)|}. \tag{12}$$

The integral in the numerator of Equation (12) can be written in closed form as

$$2 \int_0^{\frac{1}{2}} \mathcal{L}_{p,q}(t) \, dt = \begin{cases} s\mathcal{L}(p) + (1-s)\mathcal{L}(q) & \text{if } s < 1 \\ \mathcal{L}(p) & \text{otherwise} \end{cases}, \tag{13}$$

where $s = 2t'_{p,q}$.

3.3 Implementation Details

We have implemented the sub-pixel IFT in an interactive segmentation application. In our implementation, the sub-pixel boundaries are computed in a post-processing step. For efficiency, sub-pixel labels are only calculated for nodes that are adjacent to at least one node having different label. The time required for computing the sub-pixel boundaries is small compared to the computation time of the IFT itself.

The proposed sub-pixel model can also be implemented differentially. For each node p whose label is changed due to insertion or removal of a seed-point, only the sub-pixel labels of p and $\mathcal{N}(p)$ need to be updated, since the sub-pixel label only depends on the labels at adjacent nodes.

4 Evaluation

All interactive segmentation methods are subject to variations in user input. A labeling method used for seeded region segmentation should therefore be insensitive to small variations in seed placement. To compare the crisp IFT and sub-pixel IFT with respect to this property, we perform an experiment where we segment the spleen in a slice from a CT volume of a human abdomen. The spleen is selected because it is a non-trivial object from a real application, yet it can be accurately segmented using a limited number of seed-points.

For this experiment, an additive path cost functions is used, i.e.,

$$f(\pi) = \sum_{i=2}^{|\pi|} w(\pi_i - 1, \pi_i) \,. \tag{14}$$

The edge cost is

$$w(p, q) = |I(q) - I(p)| + \epsilon \,, \tag{15}$$

where ϵ is a small positive number.

To simulate realistic variations in user input, we select regions that are determined to be inside and outside the spleen, respectively. We then compute both the crisp IFT and the sub-pixel IFT, using a single pixel from the inside region and the complete outside region as seed-points. See Figure 3. We thus obtain 41 crisp and 41 sub-pixel segmentations, one for each pixel in the inside region. Visually, all (crisp and sub-pixel) segmentations correctly delineate the spleen. Each segmentation was computed in less than a second on a standard PC (3.6 GHz, 3 GB RAM). The area of each segmented object is measured by summing all pixel values of the segmented image. The results are plotted in Table 1. The difference between the maximum and the minimum area as well as the standard deviation of the area measurements is smaller for the sub-pixel IFT than for the crisp IFT. The measured area for all individual segmentations are shown in Figure 4. For all segmentations, the area of the sub-pixel segmentation deviates less from the mean area than the area of the crisp segmentation. This is true regardless of whether the area is larger or smaller than the mean area.

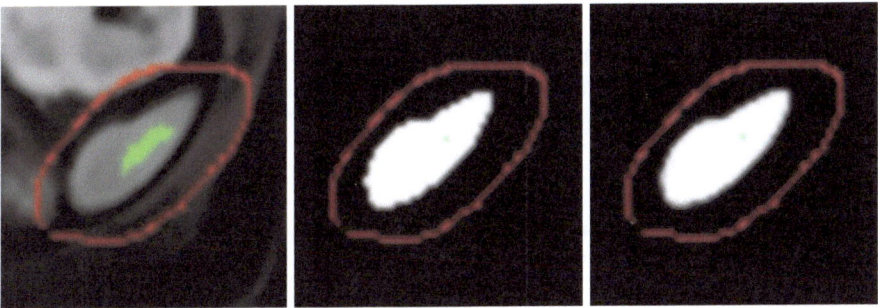

Fig. 3. Segmentation of the spleen in a slice from a CT volume.(Left) Seed-point regions used in the experiment. The green pixels define all object seeds, while the red pixels define background seeds. Single pixels from the green region were used to define object seeds. (Middle) Example result of crisp IFT. (Right) Example result of the proposed sub-pixel IFT.

Table 1. Statistics on the measured area for the 41 segmentations in the experiment. (Areas are given in number of pixels.)

Method	Mean area	Min area	Max area	σ
Crisp IFT	266.5	265	269	0.98
Sub-Pixel IFT	266.2	265.4	266.7	0.40

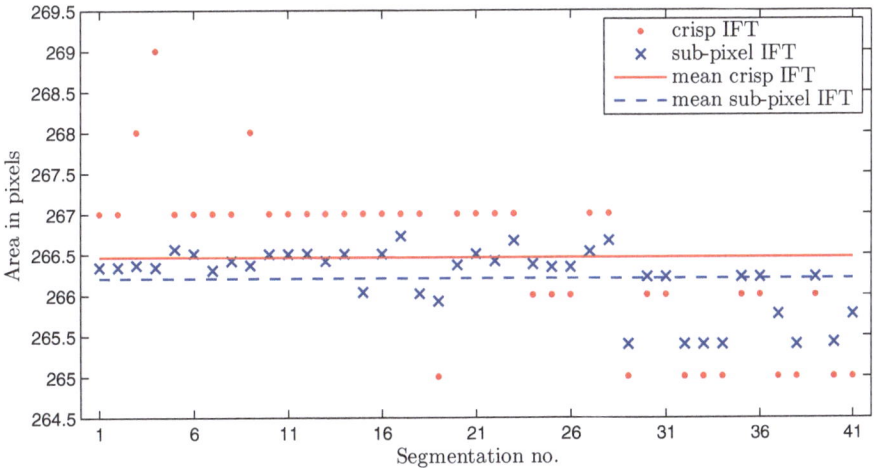

Fig. 4. Area of the segmented object for the 41 different segmentations in the experiment, sorted by the position of the seed-point inside the object. For all segmentations, the area of the sub-pixel segmentation deviates less from the mean area than the area of the crisp segmentation.

The results of the experiment indicate that the sub-pixel IFT is less sensitive to variations in seed placement than the crisp IFT, for the purpose of estimating area/volume of a segmented object.

5   Conclusions

We have presented a modified version of the IFT algorithm, that computes labels with sub-pixel precision. The sub-pixel IFT is straightforward to implement in an existing IFT implementation, and preserves the advantages of the crisp IFT. It can be computed efficiently, and can be implemented differentially to allow fast editing. Like the crisp IFT, the sub-pixel IFT is defined for general graphs, and can therefore be applied to images of any dimension. In addition to 2D segmentation, an example of volume image segmentation with the sub-pixel IFT is shown in Figure 5.

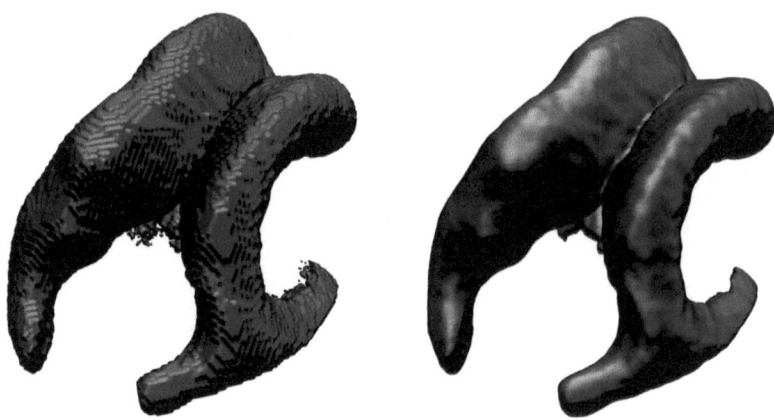

Fig. 5. Lateral ventricles of a human brain, segmented from an MR volume image using 20 single-voxel seed-points. A polygonal surface was extracted from the segmented volume using the Marching Cubes algorithm, which takes sub-pixel information into account. Both segmentations were produced using the same seed-points and path-cost function. (Left) Surface extracted from crisp IFT segmentation. (Right) Surface extracted from sub-pixel IFT segmentation.

Here, we have assumed that all edges that connect to a node affect the sub-pixel label of the node equally. For some graphs, this assumption may be invalid. In such cases, the uniform average in Equation (12) could be replaced by a weighted average that better reflects the influence of each edge.

References

1. Boykov, Y., Funka-Lea, G.: Graph cuts and efficient N-D image segmentation. International Journal of Computer Vision 70(2), 109–131 (2006)
2. Braquelaire, J.-P., Vialard, A.: A new antialiasing approach for image compositing. Visual Computer 13(5), 218–227 (1997)
3. Ciesielski, K., Udupa, J., Saha, P., Zhuge, Y.: Iterative relative fuzzy connectedness for multiple objects with multiple seeds. Computer Vision and Image Understanding 107(3), 160–182 (2007)
4. Falcão, A.X., Bergo, F.P.G.: Interactive volume segmentation with differential image foresting transforms. IEEE Transactions on Medical Imaging 23(9), 1100–1108 (2004)
5. Falcão, A.X., Stolfi, J., Lotufo, R.A.: The image foresting transform: Theory, algorithms, and applications. IEEE Transactions on Pattern Analysis and Machine Intelligence 26(1), 19–29 (2004)
6. Falcão, A.X., Udupa, J.K., Samarasekera, S., Sharma, S., Hirsch, B.E., Lotufo, R.A.: User-steered image segmentation paradigms: Live wire and Live lane. Graphical Models and Image Processing 60(4), 233–260 (1998)
7. Grady, L.: Random walks for image segmentation. IEEE Transactions on Pattern Analysis and Machine Intelligence 28(11), 1768–1783 (2006)
8. Grady, L., Jolly, M.-P.: Weights and topology: A study of the effects of graph construction on 3D image segmentation. In: Metaxas, D., Axel, L., Fichtinger, G., Székely, G. (eds.) MICCAI 2008, Part I. LNCS, vol. 5241, pp. 153–161. Springer, Heidelberg (2008)
9. Leemput, K.V., Maes, F., Vandermeulen, D., Suetens, P.: A unifying framework for partial volume segmentation of brain MR images. IEEE Transactions on Medical Imaging 22(1), 105–119 (2003)
10. Lorensen, W.E., Cline, H.E.: Marching cubes: A high resolution 3D surface construction algorithm. Computer Graphics 21(4), 163–169 (1987)
11. Lotufo, R.A., Falcão, A.X., Zampirolli, F.: IFT–Watershed from gray-scale marker. In: XV Brazilian Symposium on Computer Graphics and Image Processing, pp. 146–152. IEEE, Los Alamitos (2002)
12. Olabarriaga, S.D., Smeulders, A.M.: Interaction in the segmentation of medical images: A survey. Medical Image Analysis 5(2), 127–142 (2001)
13. Sladoje, N., Lindblad, J.: High-precision boundary length estimation by utilizing gray-level information. IEEE Transactions on Pattern Analysis and Machine Intelligence 31(2), 357–363 (2009)
14. Sladoje, N., Lindblad, J.: Pixel coverage segmentation for improved feature estimation. In: Foggia, P., et al. (eds.) Designing Privacy Enhancing Technologies. LNCS, vol. 5716. Springer, Heidelberg (2009) (in press)
15. Sladoje, N., Nyström, I., Saha, P.K.: Measurements of digitized objects with fuzzy borders in 2D and 3D. Image and Vision Computing 23(2), 123–132 (2005)

Phase-Correlation Guided Search
for Realtime Stereo Vision

Alfonso Alba* and Edgar Arce-Santana**

Facultad de Ciencias, Universidad Autónoma de San Luis Potosí
fac@fc.uaslp.mx, arce@fciencias.uaslp.mx

Abstract. In this paper, we propose a new theoretical framework, which is based on phase-correlation, for efficiently solving the correspondence problem. The proposed method allows area matching algorithms to perform at high frame rates, and can be applied to various problems in computer vision. In particular, we demonstrate the advantages of this method in the estimation of dense disparity maps in real time. A fairly optimized version of the proposed algorithm, implemented on a dual-core PC architecture, is capable of running at 100 frames per second with an image size of 256×256.

Keywords: stereo vision, real time, phase correlation, area matching.

1 Introduction

Disparity estimation is a fundamental task in 3D stereo vision, whose goal is to determine the distance from the camera array to each of the objects in the scene. Disparity is defined as the difference in position between a given pixel in the reference image, and the corresponding pixel in the matching image. This means that, in order to obtain the disparity for each pixel, one must find a pixel-to-pixel correspondence between both images. This is a seriously ill-posed problem due to various reasons, such as: (1) occlusions, which are regions visible only to one of the cameras and occluded by other objects for the other camera; (2) homogeneous regions and periodic textures, where a pixel may be matched with many other pixels in the region; and (3) noise, which generates artificial differences between the images. State-of-the-art stereo vision techniques are designed to be robust to noise and provide disparity values for regions where the matching process fails, typically by introducing regularization constraints which enforce (piecewise) smoothness of the disparity map. Also, one can usually make certain assumptions which may drastically reduce the search space, for example, when the reference and matching images are adequately aligned, the epipolar lines are perfectly horizontal, thus corresponding points are always on the same row.

---

* A. Alba was supported by Grant PROMEP/103.5/09/573.
** E. Arce-Santana was supported by Grant PROMEP/103.5/04/1387.

P. Wiederhold and R.P. Barneva (Eds.): IWCIA 2009, LNCS 5852, pp. 212–223, 2009.

Many algorithms have been proposed to solve the correspondence problem, and most of them can be classified as global or local optimization methods. Global methods typically attempt to find the disparity field that minimizes a global energy function involving a term that measures the number and degree of mismatches, and a term which enforces spatial coherence by penalizing abrupt disparity changes. Several methods have been used to estimate the optimal disparity field, including: Markov random fields [18], hidden Markov measure fields (HMMF) [1], and belief propagation [24]. In contrast, local methods, which are typically less accurate than global methods, take a pixel or region in the reference image and perform a search along the epipolar line in the matching image to find the pixel or region which minimizes a local error measure, or maximizes a correlation measure between both areas. Among these methods, the most common approach relies on the estimation of a similarity measure between a window centered at pixel (x, y) in one of the images and a window centered at pixel $(x + d, y)$ in the other image, for all d in a given range. In a winner-takes-all (WTA) approach, the disparity assigned to pixel (x, y) is the value of d which maximizes the similarity measure. Variuos similarity measures have been proposed, including the sum of absolute differences (SAD), sum of square differences (SSD), statistical correlation, and rank measures. It is also possible to use multiple windows [11], or adaptive windows [12] to improve the estimations within homogeneous regions. An extensive review of various of these methods is provided by Scharstein et al. [23].

In recent years, state-of-the-art algorithms for stereo depth estimation have advanced significatively in terms of quality; however, some of these algorithms may take a significant amount of time (from a few seconds up to a few minutes) to process a single pair of stereo images, which seriously limits their applications in problems which require real-time processing, such as: robot and vehicle navigation, surveillance, human-computer interfaces, and augmented reality. This has led researchers to develop faster methods by taking advantage of parallel and/or reconfigurable architectures, and by implementing clever techniques to reduce artifacts at low computational expense. Various recently proposed algorithms and architectures are already capable of computing dense disparity estimations at video-rate. Most of these algorithms have some degree of parallelization; for example, on a modern PC computer it is possible to use single-instruction multiple-data (SIMD) extensions such as MMX [11] or SSE2 [4] [5], multi-core processing [26], or GPU implementations [6] [2]. Algorithms designed for reconfigurable architectures such as FPGA's are also relatively common [3] [10]. Other proposals assume that the location of the cameras is fixed with respect to the scene, so that they can perform a background/foreground segmentation, and compute the disparity only for those pixels that correspond to new or moving objects [15]. These methods achieve framerates which vary between 5 and 60 frames per second for relatively small images (e.g., 320 × 240 pixels), and between 3 and 30 fps for larger images (see [25] for a good comparison).

One technique that has not been fully exploited is the phase-correlation method, which relies on the fact that a displacement in the spatial domain is observed as a phase change in the frequency domain. Specifically, if $F(\omega)$ is the Fourier transform of $f(x)$, and $g(x) = f(x+d)$, then the Fourier transform of $g(x)$ is given by $G(\omega) = e^{j\omega d} F(\omega)$. One can then compute the normalized cross-spectrum $R(\omega)$ between F and G, which in this case is given by

$$R(\omega) = \frac{F(\omega)G^*(\omega)}{|F(\omega)G^*(\omega)|} = e^{-j\omega d}. \tag{1}$$

The inverse Fourier transform $r(x)$ of $R(\omega)$ is called the phase-only correlation (POC) function. In this case, one obtains $r(x) = \delta(x - d)$ (i.e., an impulse centered at $x = d$); therefore, one can easily recover the displacement by finding the position of the peak of $r(x)$. This technique is commonly used for image registration [21] [7] [13] [14], but it has also been applied in stereo vision [20].

In this paper we introduce a technique to reduce the search space in local methods, which is based on the estimation of multiple phase-correlation peaks. This results in a very efficient disparity estimation algorithm that is between 2 and 3 times faster than methods that use a full-search approach. The paper is divided as follows: Section 2 presents the basic algorithm and some quality refinements, in Section 3 we present some test results with both static and moving images, and finally, our conclusions are presented in Section 4.

2   Methodology

Consider two images $f(x, y)$ and $g(x, y)$ that represent the same scene in slightly different situations (e.g., taken at different times or from two distinct points of view). In general, the same objects, although possibly in different positions, will appear in both images. If the displacement is the same for all objects (including the background), the phase correlation function would approximate a delta function located at the displacement. If, however, different displacements are observed for different objects, the POC function will show various peaks, contaminated and distorted by noise due to borders, occlusions, and homogeneous regions. Intuitively, each of the peaks observed in the POC function may be related to the displacement of a specific object. The goal of this work is to experimentally show that the POC maxima provide useful information that can be used to implement more efficient area matching algorithms.

We are particularly interested in estimating the dense disparity map from a pair of stereo images with horizontal epipolar lines. In this case, corresponding pixels always lie in the same row, and thus the problem can be solved with a row-by-row scheme. In the rest of this paper we will assume that $f(x, y)$ and $g(x, y)$ are, respectively, the left and right images obtained from an horizontal stereo camera arrangement; also, the observation model is $g(x, y) = f(x + d(x, y), y)$, where $d(x, y)$ is the piece-wise smooth disparity field. Note that, under this model, the reference image is the right image $g(x, y)$; however, our methodology can be easily adapted to the case where $f(x, y)$ is the reference image. A typical winner-takes-all (WTA) full-search approach finds, for each pixel (x, y), the disparity value $d \in \{0, \ldots, D\}$ which minimizes a local error function $S(x, y, d)$ that measures the difference between a neighborhood of $g(x,y)$ and a similar-shaped neighborhood of $f(x+d,y)$. Therefore, the disparity $d(x,y)$ for each pixel is given as

$$d(x,y) = \arg\min_{d' \in \{0,...,D\}} \{S(x,y,d')\}, \qquad (2)$$

where D is a given disparity range (typically between 32 and 64). Two common error measures are the SAD and SSD, respectively defined as

$$S_{SAD}(x,y,d') = \sum_{j=-w}^{w} \sum_{i=-w}^{w} \|f(x+d'+i,y+j) - g(x+i,y+j)\|, \qquad (3)$$

and

$$S_{SSD}(x,y,d') = \sum_{j=-w}^{w} \sum_{i=-w}^{w} \|f(x+d'+i,y+j) - g(x+i,y+j)\|^2, \qquad (4)$$

where w determines the size of the correlation window, and it directly affects the granularity of the results: small windows will produce noisy disparity fields, mainly due to ambiguities in homogeneous regions, whereas large windows may blur the borders and are more sensitive to projective distortions between the images. Larger windows may also significatively increase the computational cost, however, for fixed-size rectangular windows, it is possible to use cost-aggregation techniques to efficiently compute the error measure for a large number of pixels.

2.1   Phase-Correlation Guided Stereo Matching

Computation of the scores $S(x,y,d')$ is the most expensive step in a WTA-based stereo matching algorithm. The full-search approach, described above, requires the computation of D score values per pixel. To reduce the search space, we compute the phase-correlation function for each row, and estimate the most likely disparity values (for the pixels in that row) by finding the highest peaks of the POC function. Only those peaks whose value exceeds a given threshold (which, in our case, is zero) are considered.

The basic approach is summarized as follows: given a stereo pair of images $f(x,y)$ and $g(x,y)$, representing the left and right images, respectively, we estimate the disparity field $d(x,y)$ by performing these steps:

1. For each row y, do the following:
   (a) Compute the 1D FFT's $F_y(\omega)$ and $G_y(\omega)$ of the y-th row of $f(x,y)$ and $g(x,y)$, respectively.
   (b) Compute the normalized cross-spectrum $R_y(\omega)$, given by

$$R_y(\omega) = \frac{F_y(\omega)G_y^*(\omega)}{|F_y(\omega)G_y^*(\omega)|}. \qquad (5)$$

   (c) Compute the POC function $r_y(x)$ as the 1D IFFT of $R_y(\omega)$.

(d) Find the M most significative positive maxima of $r_y(x)$ in the interval $x \in [0, D]$ where D is the disparity range. Let $d_1, d_2, \ldots, d_M$ be the positions of these maxima; that is, the most likely disparity values for row y. In particular, we have obtained good results with M between $D/4$ and $D/3$.

(e) For each pixel x in the row, find the disparity $d(x, y)$ as

$$d(x, y) = \arg \min_{k \in \{1, \ldots, M\}} S(x, y, d_k). \tag{6}$$

Note that, in step (e) of the algorithm, the search is performed only over the M candidates, and not across the full disparity range as in other methods.

2.2 Regularization and Collision Detection

Reduction of the displacement search space induces a certain quantization of the disparity values. Under certain smoothness assumptions, this quantization enforces regularization constraints in the disparity field, but only along the horizontal direction. Because of this, the basic algorithm may suppress discontinuities at the borders of the objects, causing some artifacts in the form of streaks of incorrect disparity values along the scanline. One way to reduce these artifacts is to enforce inter-line consistence in the set of likely disparities. To achieve this, we first compute the POC function $r_y(x)$ for all rows and reorganize it as an image $r(x, y)$, and then we apply a 1D low-pass filter to $r(x, y)$ along the Y-axis. One can then find the maxima corresponding to each row of the filtered POC image. Note that this regularization filter is applied not to the final disparity map, but to the phase correlation function that determines the disparity candidates for each scanline, in order to induce inter-line consistence. We have obtained good results using a Gaussian low-pass filter with standard deviation between 3 and 16. On the other hand, applying a median filter to the estimated disparity map may improve the quality of the results with little computational cost.

We have also implemented a simple detection stage for invalid disparities due to collision, based on the same idea used by Di Stefano et al. [4]. This idea relies on the uniqueness constraint, which states that each point in the 3D scene will be projected to at most one point in each image of the stereo pair. Therefore, if pixel $g(x, y)$ is matched with pixel $f(x+d, y)$, and later, pixel $g(x', y)$ is matched with $f(x' + d', y)$, so that $x + d = x' + d'$, then at least one of those matchings is wrong. We simply mark both matchings as invalid; however, in the future we expect to implement a mechanism to decide which is the correct matching, and possible recompute invalid matchings. In any case, due to the quantization induced in the disparity field by the search space reduction, the number of invalid matchings produced by our algorithm is relatively low.

2.3 Implementation Details

We have implemented the proposed method in a PC platform with several optimizations to achieve high frame rates. The most relevant algorithmic optimizations are:

- Since the images f and g are real, half of the coefficients of their Fourier transforms and the normalized cross spectrum are redundant, thus one can only compute the coefficients corresponding to non-negative frequencies. This reduces the POC computation time roughly by half.
- Computation of error scores (SAD and SSD measures) is performed by cost accumulation. Specifically, for a given row y, we first compute the partial errors $C(x, y, d)$ corresponding to the columns of the correlation window; for example, when using the SAD measure, the column errors are given by

$$C(x, y, d) = \sum_{j=-w}^{w} |f(x+d, y+j) - g(x, y+j)|. \tag{7}$$

Error scores $S(x, y, d)$ for all x are then simply computed as

$$S(x, y, d) = S(x-1, y, d) - C(x-1-w, y, d) + C(x+w, y, d). \tag{8}$$

- Median filter computation is performed efficiently using the algorithm described in [17].

Our implementation also takes advantage of parallelization provided by the Intel Core 2 Duo platform. Specifically, the most notable platform-dependant optimizations are:

- FFT's are performed using the FFTW library [8], which includes SSE2 optimizations and efficient algorithms for real-valued input data.
- Computation of error scores and matching is performed in parallel under a multi-core processor, where each core processes a given range of rows.
- SSE2 extensions used for column error estimation and POC image filtering, allowing the algorithm to process up to 8 contiguous pixels in parallel.

A faster algorithm was also implemented for input images whose width is a power of 2. In this case, the FFTW algorithms are particularly efficient, and any multiplication by the image width can be replaced by a binary shift.

It is worth mentioning that our implementation could be further optimized by using SSE2 and multi-core processing during the estimation of the normalized cross-spectrum, and the median filter stages, both of which are implemented using plain C++ code.

3   Results

To determine the quality and computational efficiency of our method under different circumstances, we have performed a series of tests using both static and moving images. In the case of static images, the ground truth is also known, which permits us to obtain quantitative quality measurements. All tests were performed under an Intel Pentium Core 2 Duo workstation running at 2.4 GHz with 2 Gb of RAM.

Table 1. Percentages of disparity values with error above 1 pixel, measured over different image regions (non-occluded, all regions, and near discontinuities). These scores were computed by the Middlebury evaluation system.

	Tsukuba	Venus	Teddy	Cones
Non-occluded	7.86	6.06	37.0	22.5
All regions	9.78	7.65	43.3	31.0
Discontinuities	29.1	45.2	50.1	42.3

Table 2. Percentages of correct, incorrect, and invalid matchings for various real-time algorithms

Tsukuba scene			
Method	% Correct	% Incorrect	% Invalid
SAD5 [11]	85.12	4.56	10.32
SMP [4]	60.06	30.62	9.32
Hierarchical [16]	95.93	4.07	—
Hardware LWPC [3]	75.10	24.90	—
Our method	86.08	8.79	5.13

Venus scene			
Method	% Correct	% Incorrect	% Invalid
SMP [4]	93.79	4.19	2.02
Hardware LWPC [3]	89.49	10.51	—
Our method	87.40	8.67	3.93

3.1   Results with Static Images

We have applied our algorithm to four stereo pairs (Tsukuba, Venus, Teddy, and Cones) obtained from the Middlebury database [23] [22], and submitted the resulting maps to the Middlebury evaluation system. In all cases, the search space is reduced to only $M = 16$ disparity values per row. Results from this evaluation are shown in Table 1. Note that our scores are comparable to some of the worst-performing methods reported in the Middlebury database; however, most of those methods are designed for quality instead of speed, and only a few are able to perform in real-time. In any case, this means that, in terms of quality, our method is roughly comparable to other recently published algorithms. To perform a more fair assessment, we have compared our algorithm against other real-time oriented methods using the Tsukuba and Venus scenes. In this case, we compute the percentage of correct disparity estimations, incorrect disparity estimations (error above 1), and invalid disparities (due to collisions), and compare them with the results reported for the other algorithms. These scores are shown in Table 2. Note that, in the case of the Tsukuba scene, only the Hierarchical algorithm by Kim et al. performs significantly better than our method; however, this algorithm is based on a background/foreground segmentation, which requires the camera to be in a fixed position, and re-computation of the background disparities takes up to a few seconds. On the other hand, our method clearly does not perform so well

Left image   Right image   Ground truth

Full search   POC search (unfiltered)   POC search (filtered)

5 x 5

11 x 11

19 x 19

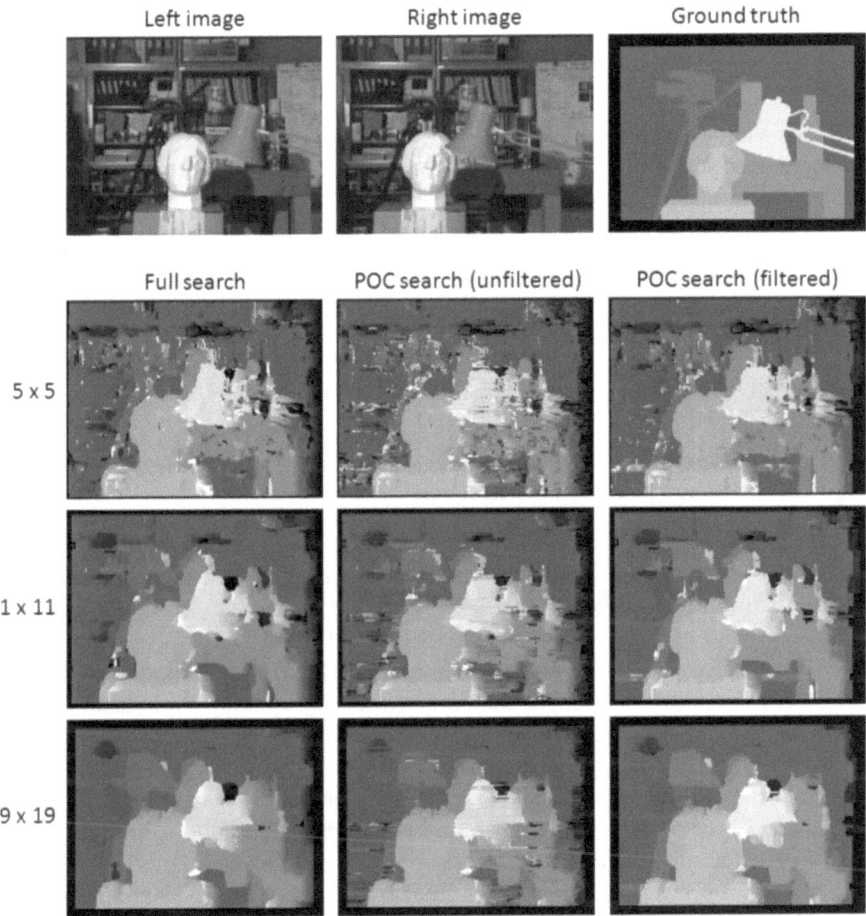

Fig. 1. Disparity estimation for the Tsukuba scene. The top row shows the left and right images, and the true disparity map. Rows 2-4 show the estimated disparity maps using an area-based WTA approach with the SAD measure computed for different window sizes (5×5, 11×11, and 19×19). The left-most column shows the result using a full-search approach over a disparity range of 48 pixels. The middle column shows the results using the basic POC-driven search with $M = 12$ maxima. The right-most column shows the results when a Gaussian filter with a standard deviation of 10 pixels is applied to the POC along the Y-axis.

Table 3. Average computation times for the estimated disparity maps shown in Figure 1. The image size is 256×192.

	Full search	POC search	POC Filter	POC + median filters
5 × 5	14 ms	5 ms	6 ms	8 ms
11 × 11	17 ms	6.5 ms	7 ms	9 ms
19 × 19	22 ms	7.5 ms	8 ms	10 ms

with the Venus scene; this is because the objects in the Venus scene form oblique planes whose disparity varies smoothly and cannot be correctly estimated using only 16 values.

We have also measured the computation time for the Tsukuba scene varying the size of the correlation window and the type of regularization used. In this case we used a 256×192 sub-sampled version of the Tsukuba images, which allows us to use our more efficient algorithm. Figure 1 shows the results obtained for the 256×192 Tsukuba scene with varying correlation window sizes, using both full and POC-guided search (with $M = 12$). Performance results for these tests are shown in Table 3.

3.2   Results with Video Sequences

We have also tested our method with some of the real video sequences used by Di Stefano et al. in their 2004 paper [4] [19]; in particular, we tested the rectified 640×480 "Indoor" and "Outdoor" sequences. Figure 2 shows the results obtained for frame 0050 of the Outdoor sequence using both Di Stefano's SMP method (where white areas represent invalid matchings), and our algorithm. The true disparities for these sequences are unknown, thus quality evaluation can only be performed from a qualitative perspective. In our opinion, both algorithms perform similarly: the main objects in the scene and their relative distance to the cameras can be distinguished, but there are some clear mismatches in homogeneous such as the wood board, and the person's shoulders, and in occluded regions such as object borders.

Finally, we present in Table 4, a comparison of the computational performance of our method against various state-of-the-art real-time algorithms. For this comparison to be meaningful, we have included a column which describes the platform where each algorithm was implemented, as well as the size and disparity range of the test images. Still, this comparison is not completely fair, since some of the algorithms include extra steps such as image rectification, and sub-pixel interpolation; however, the computational cost of these refinements is relatively small. For example, the S^3E algorithm proposed by Zinner et al. [26], which also runs in an Intel Core 2 Duo platform and takes advantage of dual-core and SSE2 optimizations, includes various quality refinement steps and a 3D reconstruction stage; however, according to their results, these extra steps account only for 11% of the total computation time (for an image size of 450×375). Therefore, even if one removed the extra steps from the S^3E system, the POC-driven search method would still be roughly twice as fast.

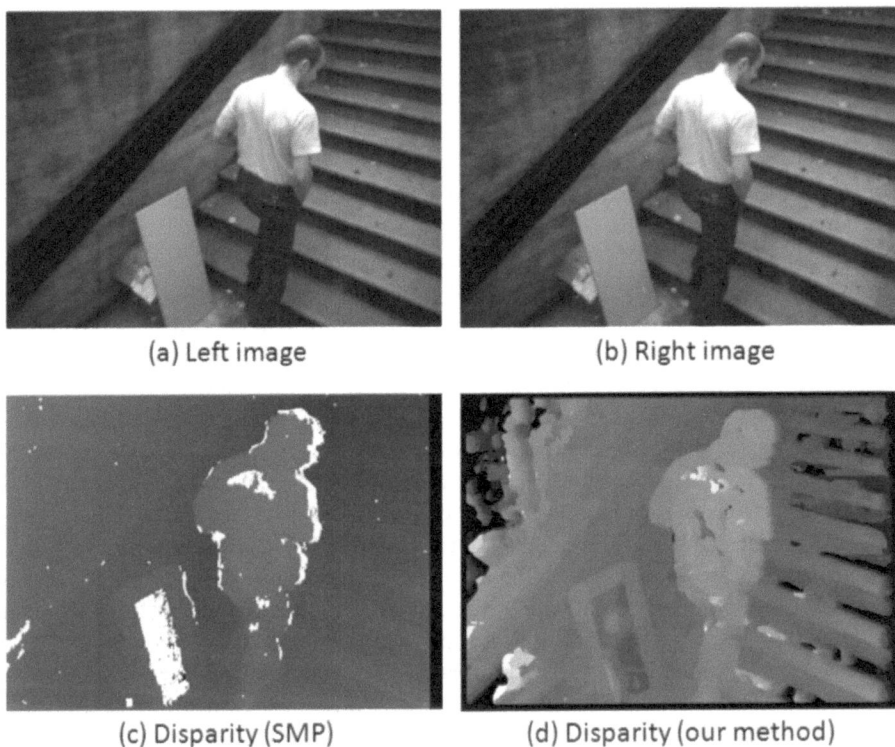

(a) Left image

(b) Right image

(c) Disparity (SMP)

(d) Disparity (our method)

Fig. 2. Results on frame 0050 of the "Outdoor" sequence: (a) rectified left image, (b) rectified right image, (c) disparity map obtained with the SMP method [4], (d) disparity map obtained with our method. In both cases, the disparity range is 48 pixels and the size of the correlation window is 15×15.

Table 4. Performance comparison between our POC-based implementation and other recent methods. Note that some of these methods may include additional processes such as frame grabbing from the video cameras, and image rectification.

Method	Platform	Image size	Range	Rate
SMP [4]	Intel Pentium III, 800 MHz	320×240	32	31.25 fps
SMP [4]	Intel Pentium III, 800 MHz	640×480	48	5.77 fps
LWPC [3]	Xilinx Virtex2000E FPGAs	256×360	20	30 fps
[9]	ATI Radeon X800 GPU	288×216	40	17.8 fps
[6]	GeForce 8800 Ultra GPU	320×240	64	13 fps
[6]	GeForce 8800 Ultra GPU	640×480	128	4.2 fps
S^3E [26]	Intel Core 2 Duo, 2.0 Ghz	320×240	30	42 fps
S^3E [26]	Intel Core 2 Duo, 2.0 Ghz	480×360	50	13 fps
Our method	Intel Core 2 Duo, 2.4 Ghz	256×256	48	110 fps
Our method	Intel Core 2 Duo, 2.4 Ghz	512×360	48	38 fps

4   Conclusions

The phase-correlation technique is commonly used to determine the geometrical translation that may exist between two images. In general, however, each object in the scene may suffer a different displacement in one image with respect to the other. In this case, the phase-correlation function does not approximate a delta function, but shows instead several peaks, some of which correspond to the displacements corresponding to the different objects. This information can be used in a stereo vision algorithm to reduce the search space in a local area matching approach, simply by finding a set of likely disparity candidates for each row of the images. This induces a quantization of the disparity values in the image, which leads to undesirable artifacts; however, these artifacts can be significatively reduced by smoothing the phase-correlation function along the vertical axis.

Preliminary results show that a well-optimized implementation of this technique may yield disparity estimations which are at least as accurate as some of the recently published methods, but in only a fraction of the time. On an Intel Core 2 Duo workstation running at 2.4 GHz, our implementation is able to process 100 frames of size 256×256, or 35 frames of size 512×512 per second. This leaves enough processing power for additional steps which may be necessary in a real-time computer vision application, such as: image rectification, sub-pixel refinements, object segmentation, and motion estimation.

Future research will focus on: (1) improving the quality of the disparity estimations, (2) using these techniques in an augmented reality application, and (3) applying the POC-search matching approach to other problems in computer vision, such as motion estimation. In particular, we are currently investigating various ways to refine the disparity measurements to obtain sub-pixel accuracy and recover bad matchings due to collisions. We are also developing a fast multi-channel segmentation method for real-time object detection in stereo video sequences.

References

1. Arce, E., Marroquin, J.L.: High-precision stereo disparity estimation using HMMF models. Image and Vision Computing 25, 623–636 (2007)
2. Bartczak, B., Jung, D., Koch, R.: Real-time neighborhood based disparity estimation incorporating temporal evidence. In: Rigoll, G. (ed.) DAGM 2008. LNCS, vol. 5096, pp. 153–162. Springer, Heidelberg (2008)
3. Darabiha, A., MacLean, W.J., Rose, J.: Reconfigurable hardware implementation of a phase-correlation stereo algorithm. Machine Vision and Applications 17(2), 116–132 (2006)
4. Di Stefano, L., Marchionni, M., Mattoccia, S.: A fast area-based stereo matching algorithm. Image and Vision Computing 22, 983–1005 (2004)
5. Di Stefano, L., Marchionni, M., Mattoccia, S.: A pc-based real-time stereo vision system. Machine Graphics & Vision 13(3), 197–220 (2004)
6. Ernst, I., Heiko, H.: Mutual information based semi-global stereo matching on the GPU. In: Bebis, G., Boyle, R., Parvin, B., Koracin, D., Remagnino, P., Porikli, F., Peters, J., Klosowski, J., Arns, L., Chun, Y.K., Rhyne, T.-M., Monroe, L. (eds.) ISVC 2008, Part I. LNCS, vol. 5358, pp. 228–239. Springer, Heidelberg (2008)

7. Foroosh, H., Zerubia, J.B., Berthod, M.: Extension of phase correlation to subpixel registration. IEEE Trans. PAMI 11(3), 188–200 (2002)
8. Frigo, M., Johnson, S.G.: FFTW Web Page (2009), http://www.fftw.org
9. Gong, M.: Enforcing temporal consistency in real-time stereo estimation. In: Leonardis, A., Bischof, H., Pinz, A. (eds.) ECCV 2006. LNCS, vol. 3953, pp. 564–577. Springer, Heidelberg (2006)
10. Han, S.K., Jeong, M.H., Woo, S., You, B.J.: Architecture and implementation of real-time stereo vision with bilateral background subtraction. In: Huang, D.-S., Heutte, L., Loog, M. (eds.) ICIC 2007. LNCS, vol. 4681, pp. 906–912. Springer, Heidelberg (2007)
11. Hirschmüller, H., Innocent, P.R., Garibaldi, J.: Real-time correlation-based stereo vision with reduced border errors. Int. J. Comput. Vision 47(1-3), 229–246 (2002)
12. Kanade, T., Okutomi, M.: A stereo matching algorithm with an adaptive window: Theory and experiment. IEEE Trans. PAMI 16(9), 920–932 (1994)
13. Keller, Y., Averbuch, A., Moshe, I.: Pseudopolar-based estimation of large translations, rotations, and scalings in images. IEEE Trans. on Image Processing 14(1), 12–22 (2005)
14. Keller, Y., Shkolnisky, Y., Averbuch, A.: The angular difference function and its application to image registration. IEEE Trans. PAMI 27(6), 969–976 (2005)
15. Kim, H., Min, D.B., Choi, S., Sohn, K.: Real-time disparity estimation using foreground segmentation for stereo sequences. Optical Engineering 45(3), 037402-1–037402-102 (2006)
16. Kim, H., Min, D.B., Sohn, K.: Real-Time Stereo Using Foreground Segmentation and Hierarchical Disparity Estimation. In: Ho, Y.-S., Kim, H.-J. (eds.) PCM 2005. LNCS, vol. 3767, pp. 384–395. Springer, Heidelberg (2005)
17. Kopp, M., Purgathofer, W.: Efficient 3x3 median filter computations. Technical University, Vienna (1994)
18. Lorenzo, J.T., Portillo, J., Alberola-López, C.: A novel markovian formulation of the correspondence problem in stereo vision. IEEE Trans. on Systems, Man and Cybernetics - Part A: Systems and Humans 34(3), 428–436 (2004)
19. Mattoccia, S.: Real-time stereo vision based on the uniqueness constraint: experimental results and applications (2004),
    http://www.vision.deis.unibo.it/smatt/stereo.htm
20. Muquit, M.A., Shibahara, T., Aoki, T.: A High-Accuracy Passive 3D Measurement System Using Phase-Based Image Matching. IEICE Trans. Fundam. Electron. Commun. Comput. Sci. E89-A(3), 686–697 (2006)
21. Reddy, B.S., Chatterji, B.N.: An FFT-Based Technique for Translation, Rotation, and Scale-Invariant Image Registration. IEEE Trans. on Image Processing 5(8), 1266–1271 (1996)
22. Scharstein, D., Szeliski, R.: Middlebury stereo vision page (2001),
    http://vision.middlebury.edu/stereo/
23. Scharstein, D., Szeliski, R.: A taxonomy and evaluation of dense two-frame stereo correspondence algorithms. International Journal of Computer Vision 47, 7–42 (2002)
24. Sun, J., Zheng, N.N., Shum, H.Y.: Stereo matching using belief propagation. IEEE Trans. PAMI 25(7), 787–800 (2003)
25. van der Mark, W., Gavrila, D.M.: Real-time dense stereo for intelligent vehicles. IEEE Trans. on Intelligent Transportation Systems 7(1), 38–50 (2006)
26. Zinner, C., Humenberger, M., Ambrosch, K., Kubinger, W.: An optimized software-based implementation of a census-based stereo matching algorithm. In: Bebis, G., Boyle, R., Parvin, B., Koracin, D., Remagnino, P., Porikli, F., Peters, J., Klosowski, J., Arns, L., Chun, Y.K., Rhyne, T.-M., Monroe, L. (eds.) ISVC 2008, Part I. LNCS, vol. 5358, pp. 216–227. Springer, Heidelberg (2008)

PCIF: An Algorithm for Lossless True Color Image Compression

Elena Barcucci[1], Srecko Brlek[2], and Stefano Brocchi[1,*]

[1] Dipartimento di Sistemi e Informatica, Università di Firenze
Viale Morgagni, 65 - 50134 Firenze - Italy
{barcucci,brocchi}@dsi.unifi.it
[2] LaCIM, Université du Québec à Montréal,
CP 8888 Succ. Centre Ville, Montréal (QC) Canada H3C3P8
brlek.srecko@uqam.ca

Abstract. An efficient algorithm for compressing true color images is proposed. The technique uses a combination of simple and computationally cheap operations. The three main steps consist of predictive image filtering, decomposition of data, and data compression through the use of run length encoding, Huffman coding and grouping the values into polyominoes. The result is a practical scheme that achieves good compression while providing fast decompression. The approach has performance comparable to, and often better than, competing standards such JPEG 2000 and JPEG-LS.

Keywords: Lossless compression, predictive coding, Huffman codes.

1   Introduction

Much recent effort has been spent on developing new image compression techniques. The most widely used algorithms are lossy. However, in many cases, this is undesirable. For example, in biomedical imaging, any loss of image quality may be considered to be unacceptable. It is therefore important to have efficient ways for achieving lossless compression of images, and clearly, it is very useful in cases where an image has to be modified many times. Indeed, the image quality could suffer from cumulative distortion as the image is decoded, modified and re-encoded several times.

Many lossless compression techniques have been developed, some of which got the label of ISO standard. The widely used Portable Network Graphics (PNG) [6] standard, was introduced as an alternative to the patented Graphics Interchange Format (GIF) (which can be freely used since the patents has expired). PNG is fast and produces files that are significantly smaller than the files obtained with GIF. PNG supports true color RGB images, but does not support other color spaces. However it does not have a very good compression ratio, and is not intended for professional use. Several algorithms, originating from recent

---

* Corresponding author.

P. Wiederhold and R.P. Barneva (Eds.): IWCIA 2009, LNCS 5852, pp. 224–237, 2009.

research, achieve very high compression rates at the cost of a memory and time tradeoff: these include MRP [9], CALICA-A [13] and GLICBAWLS [10]. On the other hand, some compression schemes such as FELICS [5] and its successor SFALIC [11], yield very fast executions that maintain a good compression ratio. Nevertheless, these algorithms often do not provide an implementation for true color images, making a relevant comparison impossible. Two very popular standards achieving lossless compression are JPEG 2000[7] and JPEG-LS [12]. Well suited for handling true color (RGB based model) photographic images, they are fast and achieve good compression ratio. They are used hereafter for comparison purpose.

Our goal is to develop an efficient algorithm, also suitable for parallel computations, that performs a lossless image compression. For sake of simplicity, the description and its actual implementation is limited to RGB images with 8-bit depth, but can be easily adapted to other color spaces and depths. We propose a compression scheme, consisting of three phases, *filtering*, *decomposition* and *compression*, always having in mind the performance. For that purpose, we focused our attention on the decompression efficiency, and when suitable, a slower compressing process is preferred if the gain in the decompressing one is worth.

The algorithm PCIF (standing for Polyomino Compressed Image Format), is fast, requires low memory, and produces challenging compression ratios as confirmed by the benchmarks performed against PNG[6], JPEG 2000[7] and LOCO [12] (the algorithm at the core of JPEG-LS). It offers a good alternative to these standards. Indeed, on the average, the compression ratio of PCIF is similar to LOCO but better than PNG (20%) and JPEG 2000 (10%). PCIF is slightly inferior on photographic images, but is significantly better on computer generated images or images having edges. Furthermore, PCIF is asymmetrical: contrarily to LOCO and JPEG 2000 it requires noticeably less time and resources during decoding than while encoding. This is clearly an advantage if we suppose that for an image the decompression process is much more frequent than compression. This is the case for a large part of end users of image compression: each time a user visualizes an image stored in a compressed format, as in JPG or PNG, then it must be decompressed before being sent to a display; the same happens each time a webpage containing images is opened. On the other hand, only users that actually modify or create images will perform the encoding.

Finally, we wish to underline the fact that PCIF is released as free and open source, and represents an interesting alternative to patented software.

2   Overview

In a *truecolor* image, each pixel is encoded on three bytes, each one corresponding to a channel (R)ed, (G)reen, or (B)lue. For our purpose we use 8 bits per channel in order to allow $256(= 2^8)$ intensities for each primary color. Each pixel takes therefore a value in the set of 2^{24} colors roughly approximated to 16 million. An image is represented conveniently by a three dimensional matrix $M(width, height, color)$ where *width* and *height* are integer values determined by the size of the image, and *color* ranges in the color space $\{R, G, B\}$.

We propose an algorithm consisting in three phases; the first phase applies filtering to the image and changes values of the matrix that represent it. After its completion, it is expected to obtain lots of near zero values that have consequently a highly-compressible distribution. This phase is formed by three substeps; in the first of them from each value the algorithm subtracts a prediction made using some neighbouring values representing the same color. In the second substep we make an analogous operation using instead values representing different colors of the same pixel to predict each other. In the third step each possible value is exchanged with another according to a permutation array built-in function of the frequencies of the symbols; this determines the behaviour of the second phase explained hereafter. Still in the first phase, an analysis is performed on every zone of the image in order to determine which one of the available filtering functions is more effective; this requires to store some extra data with the image. The choice to do this analysis during the filtering phase instead of using a backward-adaptive technique has been done to speed up the decompression process and to guarantee the stability of the method at the cost of a slower compression speed.

The second phase consists in decomposing the matrix representing the image in 24 bitplanes. These binary matrices are compressed independently; this should not affect noticeably compression efficiency, as the major part of the correlation between them should have been removed in the filtering phase, but it allows a highly parallelable execution of the next phase, namely the image compression.

In the third phase we proceed with the compression of the layers. According to information gathered in the previous elaboration each layer is processed in three possible ways; a first technique groups the binary values representing 1's in hv-convex polyominoes, for which an efficient coding is provided. Another choice is to proceed with a run-length encoding followed by an Huffman coding. To speed up the process, some layers, reported as being particularly chaotic, are stored as they are; this is also useful to isolate uncompressible parts of data that could otherwise result in efficiency loss.

A flowchart of the compression process is depicted in figure 1.

Even if the method is actually focused on images with an 8 bit depth for every color channel, it could be applied without any substantial change to images with different bit depth; the filtering phase could be applied exactly in the same way, and once the matrix is decomposed in binary layers, again the compression step would not be affected in any way by the original bit depth of the image. Major changes would be necessary only for binary images (represented with 1 bit per pixel), where the filtering phase would have no sense as actually defined, and for images that make use of palettes.

Main features. The proposed algorithm rests on a combination of techniques to obtain the described results.

Polyomino compressed layers. This compression of polyominoes in certain layers is fast and efficient for bitplanes with sparse ones. It is applied, on average, on $11 - 15$ of the 24 bitplanes for photographical images, and this number may be

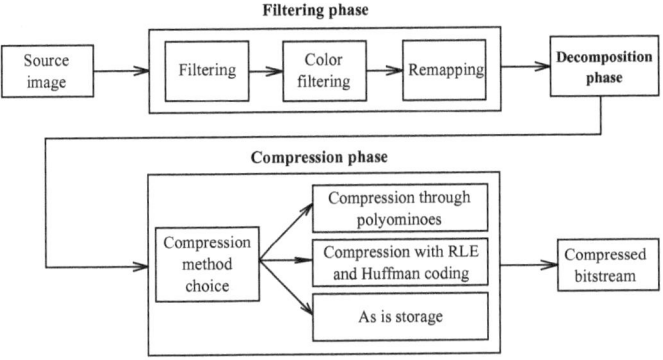

Fig. 1. Compression scheme

higher in artificial ones with smaller prediction errors. Compared to the alternative technique consisting of RLE followed by Huffman coding, it allows for most of the benchmark images, a gain in the global filesize ranging from 0.5% to 2%, and a 5% gain in the polyomino compressed layers.

Color filtering. As block adaptive filters clearly deliver better results than the use of a single filter for the entire image, this suggests the use of local-adaptive color transforms. Unfortunately, this would lead to using different color spaces in the boundaries between blocks, breaking the continuity assumption which is necessary in the filtering phase. For this reason we apply the color filtering phase *after* the filtering phase, so that prediction errors are predicted through others relative to the other bitplanes. Their correlation is expected as an edge or a sudden change of color in an image will usually yield high prediction errors in all color planes. Our assumption is confirmed by experimental results: with this order, the algorithm achieves a global filesize 2% smaller than with the reverse order. The result is also better than the standard approach, which consists in applying any of the color transforms described hereafter to the entire images.

Using a double filtering phase, determining an optimal combination of filters is difficult, and in our algorithm we settle for an approximation. Sacrificing time efficiency to determine the best pair of filters for a given block could be an interesting challenge for future research.

Total forward adaptiveness. The proposed algorithm is totally forward adaptive, and therefore, no model is dynamically built nor modified during the decompression phase. This is likely the best approach to achieve a faster decompression process than more complex algorithms as [13] and [7] offer. Usually, even other algorithms based on forward adaptiveness contain some backward adaptiveness during decoding; for example in [9] the prediction errors are encoded through arithmetic coding, and the relative model must be dynamically updated during decompression as well. The decompression phase could also be competitive with the one in [12]: the latter uses Golomb codes for fast encoding and decoding, but to determine which family to use for a given sample it has to perform a computation based on the previous ones. Making a comparison, even if LOCO uses Golomb codes (faster than the Huffman codes we use), the time required for such adaptation could widely obscure this advantage. Additionally, our algorithm is better suited for parallel computation as the bitplanes are coded independently.

This technique requires some additional side data to be stored in the file, but benchmarks reveal that, in spite of this, the compression ratios are very similar to LOCO and better than JPEG 2000. To the authors' knowledge, there is no other totally forward adaptive algorithm that achieves such results.

Some minor adjustment also contribute to the efficiency: using different compression methods for different bitplanes, optimizing filters for 8×8 square blocks (a good compromise between smaller and bigger blocks), and encoding some bitplanes through RLE and Huffman codes obtained by a zigzag scan order.

3   Filtering

The image is divided into a grid consisting of chunks of 8×8 pixels, and different filters are used on each chunk. The 8×8 size of the chunk is a good compromise between using larger zones, that would reduce the efficiency of the filtering phase, and smaller zones, that would require a greater amount of auxiliary data that could reduce the compression rate. Even if using 4×4 or 16×16 zones gave slightly better results for some images, the adopted choice obtained globally the best results for the image set used during the development of the algorithm[1].

In this phase we consider both, local filtering that takes into account the local dependency of pixels, and color filtering which benefits from highly correlated colors. For both steps, it is clear that computing the output using all filters and then selecting the one that minimizes the output, outperforms on the average any single choice. This selection is performed on each chunk, so that chunks may have different filters, each one being locally optimal.

The pixels of the image are processed column by column from right to left, each column being scanned from top to bottom. To make the operation reversible, for every sample only the values that follow it in the defined order are used for its prediction; the filtering operation will be reversed by scanning the image in opposite order during decoding. As for the PNG standard for instance [6], we take into consideration the local dependencies, that is, only the points in Figure 2 (a) are actually used in order to reduce the computation time. The resulting predicted value for every point is then subtracted from the original value mod 256 to form the *predictor error* $e(x)$. In Figure 2 (b), there is an example of filtering of a small matrix where the applied prediction function subtracts from every sample the average of its left and lower neighbour. Among the possible filters that are commonly used we implemented the following ones, where the points a, b, c and d represent the neighbour samples of the value to be estimated as in Figure 2 (a), and $f(x)$ is the predictor function for x. These

---

[1] To avoid biased results, this image set is not the same as the well-known sets used for benchmarks.

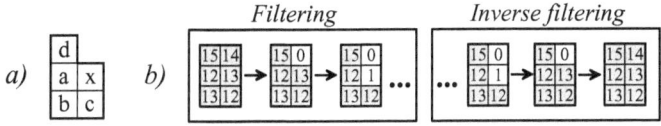

Fig. 2. Neighborhood filtering points (a), and filtering and its reversal on a small matrix (b). Here the prediction function is the average between a and c.

prediction functions are well known in literature and have been used in other compression algorithms such as [6] and [12]. These filters include the following:

$f(x) = 0$	$f(x) = a$	$f(x) = b$	$f(x) = c$
$f(x) = d$	$f(x) = a + (b - c)/2$	$f(x) = b + (a - c)/2$	$f(x) = (a + b)/2$
$f(x) = (a + b + c + d + 1)/4$	$f(x) = (a + d)/2$		

$$f(x) = \begin{cases} a + b - c & \text{if } 0 \le a + b - c < 256, \\ 0 & \text{if } a + b - c < 0, \\ 255 & \text{if } a + b - c > 255. \end{cases}$$

$f(x) = \text{Paeth}(a, b, c)$ is the Paeth filter, as defined in [6]

$$f(x) = \begin{cases} a + b - c & \text{if } a < c < b, \\ \text{Paeth}(a, b, c) & \text{otherwise.} \end{cases}$$

All divisions are assumed to be integer divisions; hence their computation can be achieved using bit shifts, that are essentially costless operations. The definition of the special behaviors for border points of the image is omitted.

In the algorithm, an estimated encoding cost is defined for each possible value of the prediction error; these costs have been experimentally determined. The exact values are omitted in this work, as the achieved results are only slightly better that those obtained by minimizing the sum of the absolute values of the predicion errors. The algorithm elects for a given zone the function yielding the minimal estimated coding cost; in the compressed bitstream the algorithm records which filter has been chosen for every block.

The second substep consists in filtering by colors, and is aimed at reducing the correlation between the different color planes. Again, every 8×8 chunk of the image may be processed by a different color filtering function. More precisely, for a given pixel, let (R, G, B) be the components relative to the three primary colors; the three available filtering functions are the following:

1. $R = R - G; G = G - B$;
2. $R = R - B; G = G - B$;
3. No operations.

It is easy to see how this operations can be reversed by scanning the colors in the order (R, G, B) during encoding and in reverse order when decoding. While the first step uses continuity assumptions for computing an estimate of a sample, this second step takes into account the correlation between colors. It is expected that, in parts of the image containing edges or discontinuities, the filtering phase produces high predictor errors in more than one color plane; for this reason a filtered sample relative to a color can be a good estimate for the filtered samples of the others. As discussed previously, this step can be considered as a zone dependant color transform applied on the result of the spatial filtering step.

The third substep consists in mapping every possible value into another number. According to the frequency of symbols, each one is mapped as follows:

the ith most frequent symbol is replaced with the value $i - 1$.

Following this operation the image is decomposed in layers, and the 1's relative to the least significative bits are the most chaotic, i.e. where the frequencies of 0's and 1's are close to 50%. In higher order layers instead, the number of 1's is more limited. This separation allows the application of different compression techniques accordingly to the proper characteristic of each layer.

4   Compression

With the decomposition phase the algorithm splits the filtered samples in 24 distinct layers, each one relative to one of the 8 bits of a sample and to one of the 3 base colors. During this operation the frequency f of 0's in each layer is computed by dividing the numer of zeros for the total number of bits in the layer. Let b be the index of the bit relative to a given layer, considering 7 to be the most significant bit and 0 the least. The compression algorithm used for a given layer is determined as follows:

- if $f * 100 + b > 90$ the layer is compressed through RLE and polyomino encoding;
- otherwise, if $f * 100 + b > 52$ the layer is compressed through RLE and Huffman coding
- otherwise the layer is not compressed and is stored 'as is'.

Notice that, thanks to the remapping phase, we have $f \geq 0.5$; thus, we cannot have cases of too regular layers due to the low density of 0's.

Different techniques are used for different cases; for very chaotic layers the algorithm does nothing in order to avoid a costly computation time in exchange of a very little spatial gain. Otherwise one of the two available compression techniques is selected, considering that the polyomino encoding based algorithm gives optimal results for layers having lots of 0's and few clusters of 1's, while the simple RLE followed by Huffman coding gives better results in more chaotic layers. The bit index is also taken into account as usually layers relative to higher bits appear to have the 1's often appearing in clusters, and this situation favours the polyomino compression method.

The parameters 52 and 90 have been determined experimentally by searching the parameter configuration yielding globally the best results on an image set selected in-house, different from the sets used for benchmarking in the next sections. Efficiency has been also taken into consideration: for the first rule, parameters ranging from 87 to 90 gave approximately the same results, but for 90 the time performances were a bit better. Lowering the second parameter 52 to 49 would also allow a gain of few bits, but we considered this wasn't worth the speedup allowed by storing some layers with no compression.

Even being globally optimal for this set, this setting does not give the best result for every possible layer; searching for other methods to determine the compression method could be an interesting subject for future research, especially since even a more complex criteria would require more time only on the compression phase while leaving unaltered the decompression performances.

4.1 Encoding Layers Using Polyominoes Extraction

The polyomino compression method aims at compressing quickly binary matrices having a large number of 0's and few clusters of 1's. We start by recalling a few standard definitions. A *polyomino* is a 4 connected set of cells (here 1's), and it is called *hv*-convex if each column (respectively line) is connected. These objects have been extensively studied and the reader is referred to [4,1,3] for a basic introduction and some work relevant to this paper. For a given polyomino P, area(P) is the number of it cells, width(P) and height(P) are the dimensions of the smallest rectangle (called *bounding rectangle*) having sides parallel to the horizontal and vertical axes, and which contains all of its cells. The *distance* between a sequence of 1's and a side of a bounding rectangle is the number of 0's between them. Finally, Topf(P) and Botf(P) are respectively the *top foot* and the *bottom foot* of P, that is the first cell, from left to right, of the bottom line (resp. top line).

In this step, two different Huffman trees are used, one for the polyominoes' dimensions and the other for the distances between them. A proper Huffman code is determined for each layer, and the related trees are stored in the compressed bitstream. Again, in comparison to forward-adaptive techniques, this approach requires some extra resources during compression but allows a more efficient decompression. The Extraction proceeds as follows:

E1. The matrix is scanned sequentially from left to right and from bottom to top, counting the number d of 0's (a cyclic run-length counting). When reaching a 1, the algorithm checks if this is contained in an already encoded polyomino. If so, the bit is skipped and the scan continues from the next bit; otherwise a polyomino P that contains it is chosen, then d and P are encoded in the bitstream as described in the next steps.

E2. The number d is encoded with an Huffman tree containing the first 63 integers and an escape value; if a value exceeding 62, say k, has to be coded, the escape symbol is written in the bitstream followed by $63 - k$, encoded in a variable bit coding that uses a number of bits proportional to $\log_2(k)$.

E3. When a bit x containing 1 is encountered, an increasing sequence $P_1 \subseteq P_2, \subseteq \cdots \subseteq P_n$ of *hv*-convex polyominoes (ending with the largest one) is generated as follows:

$$P_1 := x; \quad P_i := P_{i-1} \cup y \quad \text{if } y \text{ is adjacent to } P_{i-1} \text{ and } P_i \text{ is } hv\text{-convex.}$$

This is obtained by a filling procedure: once a line L is completed, the next line is obtained by finding a 1 above L, and filling right and left, while maintaining the polyomino v-convex. The gain of encoding a cluster of 1's in a polyomino P is estimated by

$$c(P) = p - 2 \cdot \lfloor \log_2 m \rfloor - 2 \cdot (m - 1) \cdot \lfloor \log_2 M \rfloor, \tag{1}$$

where $M = \max\{\text{width}(P), \text{height}(P)\}$, $m = \min\{\text{width}(P), \text{height}(P)\}$, and $p = \text{area}(P)$. Then one selects P such that $c(P) = \max\{c(P_i) : i = 1..n\}$ is selected and the Extraction resumes with step E1, starting from the first bit following the run containing $\text{Botf}(P)$.

In Equation (1), the integer part is applied to the logarithm to allow its computation through a series of bit shifts, and to execute all operations in integer arithmetic; the second term represents the approximate number of bits necessary for encoding the feet, and the third is a predicted cost of the remaining information, as described below. A small example of the extraction is shown in Figure 3.

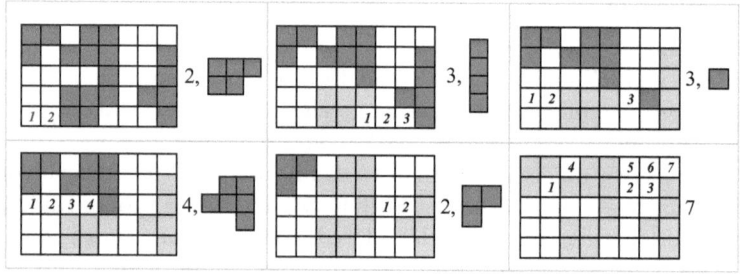

Fig. 3. Extraction of polyominoes in a layer, for each step a distance and a polyomino to be encoded are displayed. We assume that the algorithm estimated this decomposition to be the best possible according to equation 1. An alternative could have been obtained, for example, by merging the second and third polyominoes, but this would have required more bits for encoding the layer.

The coding of the polyomino P is achieved in three steps:

P1. First, the pair $(\text{width}(P), \text{height}(P))$ is coded with an Huffman tree. If P is such that $\text{width}(P) = 1$ or $\text{height}(P) = 1$ the next three steps are skipped.

P2. If $\text{width}(P) > \text{height}(P)$, the polyomino is rotated 90 degrees clockwise.

P3. The feet of P: the indexes p_s and p_n of the columns of respectively $\text{Botf}(P)$ and $\text{Topf}(P)$ are written in the bitstream. Since $1 \leq p_s, p_n \leq \text{width}(P)$, storing these values requires $\lceil \log_2(\text{width}(P)) \rceil$ bits.

P4. Every column of P is encoded in the bitstream. This is done by writing the distances between the run of 1's in the column and the bottom and top sides of the bounding rectangle. Thanks to the convexity constraint, this information is sufficient for restoring all cells of P (see Figure 4a).

Prior to recording each distance, an upper and lower bound is computed for reducing the range of its possible values. They are derived from the following properties of hv-convex polyominoes, where d_k and u_k are, for some column of index k, the distances from the sequence of 1's to the bottom and top side of the bounding rectangle.

(D1) $k < p_s \implies d_{k-1} \geq d_k$;   (D2) $k > p_s \implies d_{k-1} \leq d_k$;
(D3) $k < p_n \implies u_{k-1} \geq u_k$;   (D4) $k > p_n \implies u_{k-1} \leq u_k$.

An example of these limitations is shown in Figure 4b.

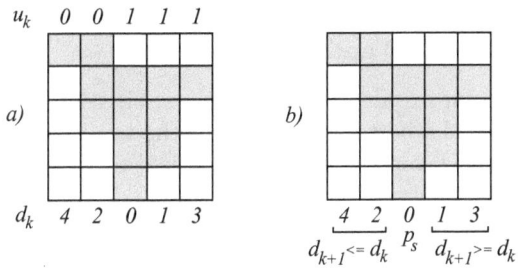

Fig. 4. Limitations on values of d_x for a convex polyomino

This range determines the number of bits used for encoding the distance; during the decompression phase, the decoder rebuilds this interval and correctly determines how many bits have to be read for obtaining the encoded distance. To maximize the information given by these conditions, the columns are encoded in an order defined by the following pseudo code, where the first call to Visit is done with indexes of the first and last columns as arguments.

```
Visit(l, r) =  if (l < r) { med = (l + r) / 2;
                            encode(med);
                            Visit(l , med - 1);
                            Visit(med + 1, r); }
```

The encoding of a polyomino is a computationally expensive step of the compression, but in any case, it is performed only a few times for every image. Nevertheless, we have experimentally observed that this technique yields a better compression ratio for more regular layers of an image, as it takes into account in a more flexible way the correlation of data in both horizontal and vertical directions. This is also the reason why we exploit at great extent the 4-topology of the image during encoding; also, the zig zag visiting order described in the next section is developed following the same motivations. Pixels that are adjacent in the original image, and hence likely correlated, are represented by bits that are neighbours in the binary layers; hence by always keeping in consideration the bidimensional nature of the original data we take advantage at most of the correlation between the original pixels.

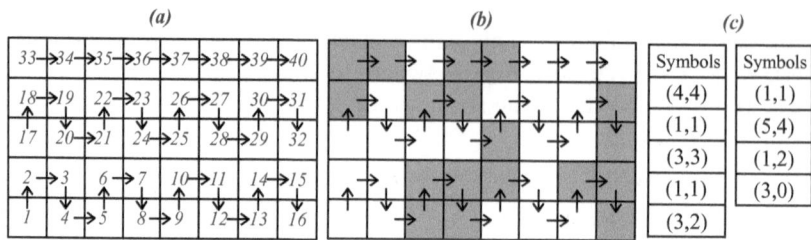

Fig. 5. Encoding layers with the zig zag visiting order

4.2   Encoding with RLE and Huffman Codes

The other selected compression scheme uses a combination of RLE and Huffman coding. Starting from the bottom line towards the top line, lines are scanned in pairs. The bits of the layer are visited, from left to right, in a zig zag order consisting in four steps in the directions up, right, down and right as shown in Figure 5 (a). When two lines have been fully scanned the algorithm proceeds to the next two; if the image contains an odd number of rows the last one is scanned simply from left to right. During this scan the number of consecutive 1's and 0's is counted and grouped in pairs; this operation is shown on a small example in Figure 5 (b) and (c). Finally, the obtained symbols are encoded with an Huffman tree, built to generate optimal codes for the specific layer. The choice to use a coding with an integer number of bits per symbol, instead of adopting arithmetic coding, does not compromise the compression ratio as the packed size of a layer results experimentally to be greater than its first order entropy of only a $0.5\% - 1\%$. The described combination of techniques is also interesting because, even if at the last stage a first order model is used, the average number of considered bits represented by a symbol can be quite high, allowing with a moderate use of resources to consider correlation between several elements.

To determine dynamically the optimal Huffman tree for each layer, two passes are required in the encoding phase, and the trees have to be stored in the compressed bitstream. On the other hand, this saves a substantial amount of computation during the decompression phase and avoids unstable results. To save some space and accelerate the tree reconstruction during decompression, the nodes with frequency zero are removed from the tree before encoding.

5   Results

The algorithm has been tested on two popular true color image sets, the Kodak image set[2] and the Waterloo color image set[3]. The current implementation is realized in Java to allow a simple and efficient web integration of images compressed with this technique; through a developed applet it is actually possible

---

[2] Available at http://r0k.us/graphics/kodak/
[3] Available at http://links.uwaterloo.ca/colorset.base.html

to visualize the compressed images in a Java supporting browser with no other type of client side compatibility. On the other hand we could expect that the optimization of the algorithm in a totally compilable language would improve the time performances. Further, since the layers are compressed independently, a multithreaded implementation on a multi processor environment would allow a substantial gain as well.

The various parameters used for the compression of these images have been determined and tuned on another image set. Nevertheless, the numerous experimentations confirmed them to be globally the best for this benchmarking set; this suggests that this setting could be fairly good in the general case. On the other hand, since these parameters do not always constitute the optimal choice for every single image, we think that finding a method that optimizes the parameters for every single image could further improve the compression performances.

The algorithm PCIF has been tested on an Athlon 2600 processor. The program actually takes, for all the 32 images, 93.78 seconds for compression and 29.21 seconds for decompression. The results obtained fulfill adequately our efficiency goals as we obtained, even with a non compiled language on an old machine, an average of less than one second for the decompression and execution times of the order of few seconds for the compression of each image.

In order to compare the decompression time performances with the other algorithms, especially for what regards LOCO, a fair comparison is impossible at this stage since we use a Java implementation while competing algorithms are written in compilable languages and compiled in machine language. However, the actual decompression performances of the *compiled* implementation of LOCO are about 3.5 times faster, but we believe that translating our algorithm in a low level language would beat it. Indeed, even if the difference of performances between Java and other languages as C is the object of vivid discussions, there is a consensus on the fact that this gap depends of the type of involved application. Hence we refer to the study [8], where several codecs for the JPEG 2000 standard are compared, and it is shown how time performances of the codec realized in Java (JJ2000) result to be from three to ten times slower than an optimized compiled codec (Kakadu), and the average results tend to be much closer to the latter case[4]. Moreover, since our study case involves image compression codecs, we expect a similar speedup in our algorithm after the translation and a low level optimization; this definitely would prove the superiority of our algorithm's decompression performances. We plan in future work to translate the algorithm in C in order to perform a direct comparison with the most popular compression standards.

The algorithm PCIF has been compared with different types of losslessly compressed image formats; we have considered the actually most popular, the PNG

---

[4] Recently, more recent versions of the kakadu codec have been announced on the website (http://www.kakadusoftware.com/) and the author affirms that a noticeable speedup has been obtained; this would make the gap in performances even bigger.

Table 1. Compression results: the uncompressed file sizes (column 'bmp') are compared with the size of compressed files relatively to algorithms PCIF, PNG, JPEG 2000 and LOCO. All values are expressed in KBytes.

Filename	bmp	pcif	png	jp2	loco	Filename	bmp	pcif	png	jp2	loco
Kodak 01	1152	516	719	498	511	Kodak 17	1152	454	587	435	437
Kodak 02	1152	467	603	439	450	Kodak 18	1152	565	762	516	562
Kodak 03	1152	398	491	388	374	Kodak 19	1152	490	655	463	482
Kodak 04	1152	463	622	437	456	Kodak 20	1152	375	480	387	362
Kodak 05	1152	559	767	519	547	Kodak 21	1152	495	622	468	481
Kodak 06	1152	477	604	460	465	Kodak 22	1152	524	685	483	517
Kodak 07	1152	426	553	408	404	Kodak 23	1152	434	544	407	417
Kodak 08	1152	553	769	534	554	Kodak 24	1152	508	689	488	493
Kodak 09	1152	443	569	427	437	clegg	2100	571	490	1369	646
Kodak 10	1152	450	579	434	441	frymire	3621	641	361	1560	806
Kodak 11	1152	464	606	446	448	lena	768	431	500	434	442
Kodak 12	1152	420	518	415	402	monarch	1152	468	621	431	448
Kodak 13	1152	604	803	569	601	peppers	768	343	441	327	327
Kodak 14	1152	520	675	487	501	sail	1152	554	777	511	544
Kodak 15	1152	443	598	431	426	serrano	1463	242	147	623	282
Kodak 16	1152	432	521	421	416	tulips	1152	527	691	477	508
Total	39827	15274	19064	16706	15201						

format[5] (see [6]), the JPEG 2000 standard[6] (see [7]) and the LOCO algorithm[7] (see [12]). We can see in table 1 the size in KB of the resulting files, where the column labelled "bmp" contains the uncompressed file sizes.

The algorithm PCIF gives very good results compared to the mainly used formats. The size of the PCIF compressed files is very close to those of the JPEG 2000 standard, that achieves generally slightly better results for photographic images but generates much larger files for artistic or high edged images. Compared to the LOCO algorithm, again compression ratios are very similar but our algorithm is more stable: the LOCO algorithm compresses at most a 5% better (Kodak image 7) but for other images it produces up to 25% bigger files (Frymire image).

In conclusion, the PCIF algorithm gives good results in lossless compression both in execution time and file size. Further refinements can probably be done; since the technique is composed of independent phases it is very flexible towards possible changes. We could think of using a more complex filtering phase or a higher order coding in the compression phase, trading time efficiency for higher compression ratios.

A file format for the described algorithm has been also developed; to encourage the use and study of this technique the program, source code of the implementation, and some detailed examples about the algorithm's execution, are available online (see [2]).

---

[5] PNG files were created with a program optimizing their size; used options are maximum compression and not interlaced images.

[6] As implemented on http://www.kakadusoftware.com

[7] As available at http://www.hpl.hp.com/loco/; we were unable to find an implementation that executes a color transform, so we implemented the one suggested in the original paper [12] to allow the LOCO algorithm to be competitive.

References

1. Barcucci, E., Del Lungo, A., Nivat, M., Pinzani, R.: Reconstructing convex polyominoes from horizontal and vertical projections. Theoretical Computer Science 155, 321–347 (1996)
2. Brocchi, S.: Polyomino Compressed Format (2008),
   http://www.researchandtechnology.net/pcif/
3. Brocchi, S.: Un algoritmo per la compressione di immagini senza perdita. Thesis, University of Florence (2006)
4. Golomb, S.W.: Polyominoes. Scribner, New York (1965)
5. Howard, P.G., Vitter, J.S.: Fast and efficient lossless image compression. In: Proceedings DCC 1993 Data Compression Conference, pp. 351–360. IEEE Comput. Soc. Press, Los Alamitos (1993)
6. ISO/IEC 15948:2003, Portable Network Graphics (PNG) Specification. W3C Recommendation
7. ISO/IEC JTC 1/SC 29/WG 1, ISO/IEC FCD 15444-1, Information Technology - JPEG 2000 Image Coding System (March 2000)
8. Man, H., Docef, A., Kossentini, F.: Performance Analysis of the JPEG 2000 Image Coding Standard. Multimedia Tools and Applications 26, 27–57 (2005)
9. Matsuda, I., Ozaki, N., Umezu, Y., Itoh, S.: Lossless coding using variable block-size adaptive prediction optimized for each image. In: Proceedings of 13th European Signal Processing Conference, WedAmPO3 (September 2005)
10. Meyer, B., Tischer, P.: Glicbawls - Grey Level Image Compression by Adaptive Weighted Least Squares. In: Proc. of 2001 Data Compression Conf., March 2001, p. 503 (2001)
11. Starosolski, R.: Simple Fast and Adaptive Lossless Image Compression Algorithm. Software: Practice and Experience 37, 65–91 (2006)
12. Weinberger, M.J., Seroussi, G.: The LOCO-I Lossless Image Compression Algorithm: Principles and Standardization into JPEG-LS. IEEE Trans. of Image Processing 9(8), 1309 (2000)
13. Wu, X., Memon, N.: Context-Based, Adaptive, Lossless Image Coding. IEEE Transactions on Communications 45(4) (April 1997)

Ego-Vehicle Corridors for Vision-Based Driver Assistance

Ruyi Jiang[1], Reinhard Klette[2], Tobi Vaudrey[2], and Shigang Wang[1]

[1] Shanghai Jiao Tong University, Shanghai, China
jruy001@aucklanduni.ac.nz
[2] The University of Auckland, Auckland, New Zealand
{r.klette,t.vaudrey}@auckland.ac.nz

Abstract. Improving or generalizing lane detection solutions on curved roads with possibly broken lane marks is still a challenging task. This paper proposes a concept of a (virtual) corridor for modeling the space an ego-vehicle is able to drive through, using available (but often incomplete, e.g., due to occlusion, road conditions, or road intersections) information about the lane marks but also about the motion and relative position (with respect to the road) of the ego-vehicle. A corridor is defined in this paper by special features, such as two fixed starting points, a constant width, and a unique relationship with visible lane marks. Robust corridor detection is possible by hypothesis testing based on maximum a posterior (MAP) estimation, followed by boundary selection, and road patch extension. Obstacles are explicitly considered. A corridor tracking method is also discussed. Experimental results are provided.

Keywords: driver assistance, lane detection, corridor detection.

1 Introduction

A lane is an area on the road surface that is completely confined by road boundaries or lane marks. For example, [4] defines a lane by criteria or hypotheses all depending on such features of the real road. This common approach for identifying a lane critically depends on the detection of physical features. Another way to model the road in front of the *ego-vehicle* (i.e., "our" car) for driver assistance or navigation is given by *free space*, which is the road region in front of the ego-vehicle where navigation without collision is guaranteed [2,3,6]; Similarly, *terrain* is usually analyzed in navigation in order to find a way to drive through the environment [11]. The term "corridor" has been used in [3] to address the driving space based on free space computation; however, this paper specifies the term "corridor" a bit further, for modeling the space the ego-vehicle is expected (!) to drive through based on the ego-vehicle's state (velocity, yaw rate, etc.) and the lane marks. We define a *corridor* (see Fig. 1) as a road patch in front of the ego-vehicle that will be driven through shortly, with a width which is slightly wider than the known width of the ego-vehicle.

The corridor starts at the current position of the ego-vehicle (defined by lateral position on the road, driving direction, and width of ego-vehicle). When this

P. Wiederhold and R.P. Barneva (Eds.): IWCIA 2009, LNCS 5852, pp. 238–251, 2009.

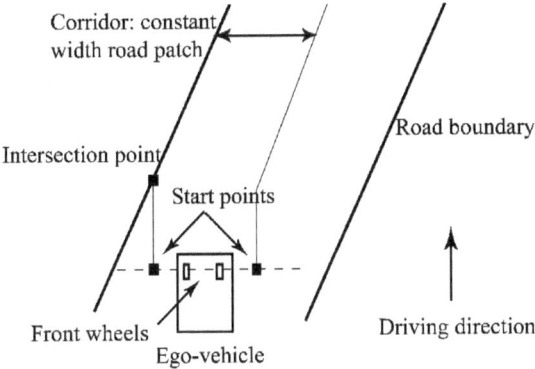

Fig. 1. Illustration for our definition of a corridor

initial road patch of constant width hits a road boundary or lane mark at an intersection point (see *intersection point* in Fig. 1), then it will smoothly bend accordingly, to follow those physical features defined by the minimum deviation from the original direction. In this way, the corridor is partly decided by physical road features, and partially by the state of the ego-vehicle. Altogether, our corridor is characterized as follows:

1. It has two fixed starting points.
2. It has a constant width that is a bit larger than that of the ego-vehicle.
3. It has smooth boundaries (note: 'smooth' remains vague at this point).
4. It will proceed in driving direction if is not curved by lane boundaries.
5. If constrained by lane boundaries then it will comply with those.

Two examples of detected corridors are shown on the right in Fig. 2.

Challenges for lane detection algorithms occur if lane boundaries cannot be properly detected or hypothesized. For example, when the ego-vehicle is changing lanes (see top of Fig. 2), there are two lanes in front of the vehicle, and the ego-vehicle is driving partially on each of them. Another example is when there is none, or only one lane boundary, and it is difficult to tell where the lane is in such a case based only on visible road features (see bottom of Fig. 2). Also, a corridor will give the driver a more reasonable presentation of the road area than a lane, even when assuming a 100% accurate lane detection algorithm; see Fig. 3. The limited value of lane detection in a *driver assistance system* (DAS) is due to its single source of information, as only physical lane marks are utilized without considering the ego-vehicle's state. For the detection of a corridor we combine methods as already available for detecting a lane and for analyzing the ego-vehicle motion. Compared with stereo-based free space computation, as in [2,3], a corridor here pays attention to the lane-based free space which will be important for driver assistance and navigation on the marked road; but we do not (yet) utilize stereo analysis in this paper. The detected corridor will indicate

Fig. 2. Comparison of lanes and corridors. Left: lanes are manually marked by lines. Right: corridors are marked by patches using our program.

Fig. 3. Comparison of lane detection and corridor detection in a sequence with broken lane marks. Top: actual lane on the road, manually marked. Bottom: corridor detection results, in real-time, using the algorithm described below.

to the the driver where he/she is expected to go shortly according to style of driving and the given lane marks.

The proposed concept of a corridor allows one to deal with various road situations, such as variable road width, non-parallel lane boundaries, non-existing lane marks, lane crossings, and some difficulties caused by illumination. The problem of detecting a corridor is identical to the detection of its boundaries.

Fig. 4. Corridor detected with obstacle in front

Instead of modeling both left and right boundaries (so far, in lane detection they are commonly assumed to be parallel), that boundary, either left or right, which is detected best (with respect to some optimality characteristics) is chosen in our corridor detection method as a guide.

If there are some obstacles in the path pre-detected as a corridor, then the corridor will be stopped in front of the obstacle. Generally, corridor detection is based on a monocular image; while considering obstacle detection, there are plenty of methods using monocular [7] or stereo [5] images. A complete example of the corridor detected with the obstacle is shown in Fig. 4

This paper is structured as follows: Section 2 explains a robust corridor detection method using hypotheses testing based on MAP (maximum *a posteriori*) estimation, combined with a modified obstacle detection method from [5] using a pair of stereo images. Section 3 explains the tracking method. Experimental results are presented in Section 4. Finally, our conclusions are stated in Section 5.

2   Robust Corridor Detection

Figure 5 illustrates the overall flow of the proposed algorithm by means of an example. The algorithm starts with mapping the input image into a bird's-eye perspective view. A low-level image processing method, as introduced in [5], is then adopted to detect edges in the bird's-eye image. Next, MAP-based hypotheses testing is conducted to detect points on left and right corridor boundaries separately, applying constraints based on information about the edges in the bird's-eye image as well as about the car's state (direction, speed, etc.). After that, a comparison between these two boundaries will select the one with the better characteristics. Points on the selected corridor boundary are then smoothed by a sliding mean, and then back-projected from the bird's-eye image into the input image, as the input image is more suitable for presentation to the driver. Finally, using a road patch extension based on the identified boundary, a corridor is produced which is in front of the ego-vehicle, with controllable patch width defined by the width of the ego-vehicle.

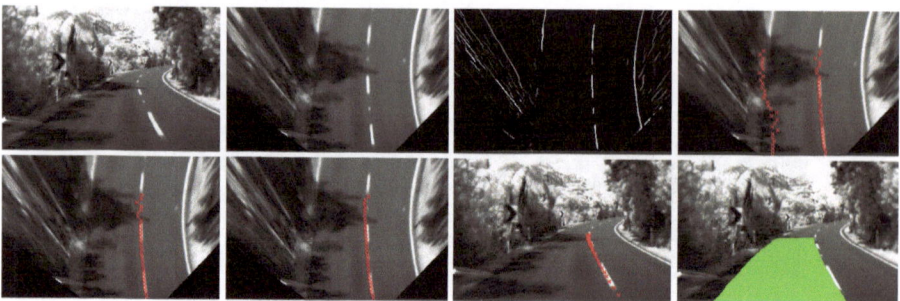

Fig. 5. Top, left to right: input image, bird's-eye image, edge map of the bird's-eye image, and detected left and right boundaries. Bottom, left to right: boundary selection, smoothing of corridor boundary using a sliding mean, projection of the boundary into the input image, and the identified corridor (in green).

2.1   Bird's-Eye Mapping

As in [8,9], a four-point correspondence is used for the mapping from the input image into the bird's-eye image. We use the locally planar ground plane assumption for this mapping. An important reason for using a bird's-eye mapping is that the driving direction is vertical in the bird's-eye image.

The mapping is achieved by selecting four points when calibrating the ego-vehicle's camera(s), and by using the locally planar ground plane assumption. The four points from the input image are in the driving direction such that they would form corners of a rectangle in the bird's-eye image (see Fig. 6) and make sure that the driving direction is vertical. In this way, the bird's-eye image provides a clear indication of the driving direction of the ego-vehicle. Another benefit of the bird's-eye image is that a used distance scale can be adjusted by selecting different sets of four correspondence points (i.e., by scaling the "length" of the rectangle); this proved to be useful for detecting dashed lane markers, as gaps between theses dashed lane marks can be adjusted by the used scaling, which also then adjusts further forward-looking situations. Also, lane marks in

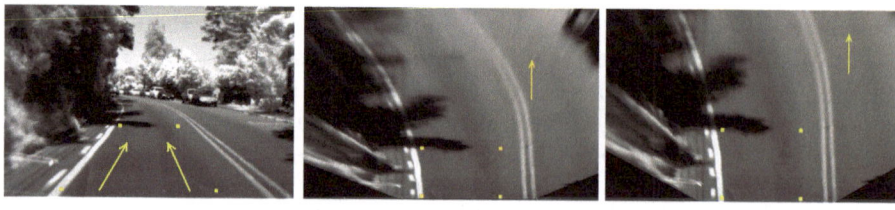

Fig. 6. Left: input image. Middle and Right: bird's-eye images based on different distance definitions. Four-point correspondence (points shown in yellow) is established by calibration; the driving direction is always vertical in the generated bird's-eye images.

the bird's-eye image have a constant width, which may be used for for edge detection in low-level image processing [5].

2.2   Corridor Detection Using MAP-Based Hypotheses Testing

Based on our corridor definition in Section 1, the procedure for detecting a corridor is composed of three stages: initialization, prediction, and hypotheses testing. After initialization (at the selected start points), we will not continue with searching scanline by scanline in the original image (as in [14]) or with an inverse-perspective-mapped image (as in [10]); we search for potential corridor boundary points in the initialized driving direction, using a fixed distance interval in the bird's-eye image. Doing so proved to be convenient, also due to the fact that the lateral position and the distance of points from the ego-vehicle in the bird's-eye image are already known due to calibration.

Individual steps of the procedure predict three points, using the previously detected points. Then, a *search region* $\mathbb{S}$ is used with fixed width, centered at a predicted point; hypotheses testing of pixels in the search region $\mathbb{S}$ uses MAP (maximum *a posteriori*) estimation. In this way, each predicted point leads to a detected point at the corridor boundary, with a MAP probability in driving direction. The distance threshold in front of the ego-vehicle (for corridor definition) can easily be controlled, and is regarded as the *forward looking distance limit*.

Initialization. The selection of the first point on a boundary is a difficult task when initializing a lane detection process. As in [10], a particle filter is applied to search for this first point. Note that one of the main difference between corridor and lane detection is that start points are predefined in corridor detection according to the width of the ego-vehicle.

In the initialization stage, lateral positions of potential boundary points are assigned to the defined two start points (see Fig. 1), which are determined based on calibration results. The distance between these two start points is assumed to be larger than that between both front wheels. This initializes the search for the corridor, but not for the lane. Furthermore, based on calibration, a few more initial points (in driving direction from the start points) are predefined (at constant distance increments $\triangle$) to ensure that the following prediction may work.

Prediction of Corridor Boundary Points. In our definition of a corridor we assumed a smooth boundary. We will ensure this by using the sliding mean (see below). Based on the smooth boundary, the following procedure can be used to predict potential boundary points by using previously detected ones. For robustness, three points are predicted in each step using different previously detected ones. $X_n(u_n, v_n)$ denotes a detected lane boundary point at the nth interval in driving direction, where v_n (longitudinal pixel position) increases with assumed step size $\triangle$, and only u_n (lateral pixel position) needs to be determined,

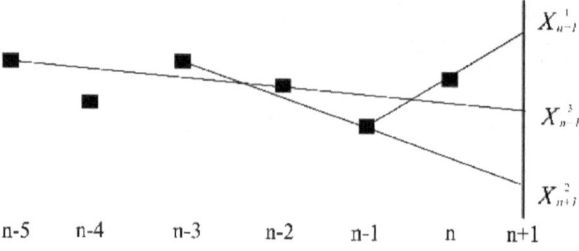

Fig. 7. Points X^1, X^2, and X^3 as generated at the prediction stage

for every X_n. In order to obtain the three boundary points at X_{n+1}, we use three predictions $X^i_{n+1} = (u^i_{n+1}, v_{n+1})$, for $i = 1, 2, 3$, as follows (see Fig. 7):

$$u^1_{n+1} = 2u_n - u_{n-1}$$
$$u^2_{n+1} = 2u_{n-1} - u_{n-3}$$
$$u^3_{n+1} = 2u_{n-2} - u_{n-5}$$

Global distance scaling (based on the used rectangle) in the bird's-eye can be achieved by modifying the parameter *triangle*. The prediction method has the same effect, and can be used to detect broken road features. Experiments showed that this prediction method may also generate irrelevant outliers, and thus we constrain predicted points to the range:

$$u^i_{n+1} = \begin{cases} 2u_{n-i+1} - u_{n-2i+1} & \text{if } (u_{n-i+1} - u_{n-2i+1}) < T \\ u_{n-i+1} + T & \text{else} \end{cases}$$

where T is the a threshold parameter. Each of these three predictions will have its own search region, and then undergo hypotheses testing independently to obtain the corresponding MAP points. These three points are then compared to produce just one estimated point at distance v_{n+1}, namely the point with the largest MAP.

Hypotheses Testing Based on MAP. A 1D search region $\mathbb{S}$ (with fixed width in the row of the edge map of the bird's-eye image) is attached to every predicted point X^i_{n+1}. Let $x(u, v)$ denote pixels in the search region of the edge map of the bird's-eye image. A likelihood function $p(z|x)$, with z for observed features, denotes the probability of observing a lane boundary edge at pixel location $x = (u, v)$. Then, the MAP estimation can be written as follows:

$$x^* = arg \max_{x \in \mathbb{S}} p(x|z)$$

Using Bayes' theorem, we obtain that

$$x^* = arg \max_{x \in \mathbb{S}} p(z|x)p(x)$$

Here, $p(x)$ is a prior probability density function. Assuming smoothness between neighboring boundary points, $p(x)$ is defined as (a_1, b_1 are constants):

$$p(x) = \frac{1}{a_1} \exp \left[-b_1(u - u_{n+1}^i)^2\right]$$

The determination of the likelihood function $p(z|x)$ uses edge information. [14] states that "edge-based methods often fail to locate the lane boundaries in images with strong distracting edges". However, edges are still useful as a source of information to discriminate a lane from its surroundings; it is reasonable to assume that the stronger an edge, the more likely it is that it is part of a lane boundary; see [10]. Let S be the edge strength at $x(u, v)$, and let $x_{max}(u_{max}, v)$ be the pixel with the greatest edge strength S_{max} in $\mathbb{S}$. Then, we use the following (a_2, b_2, b_3 are constants):

$$p(z|x) = \frac{1}{a_2} \exp \left[-\frac{1}{b_2} \cdot (S - S_{max})^2 - \frac{1}{b_3} \cdot (u - u_{max})^2\right]$$

Experiments show that the detected lane boundaries are (typically) distracted by other strong edges such as at bright areas or shadows on the road, other objects on the road or on the roadside – if only edge information is used. The smoothness assumption (for corridor boundaries) and the proposed prediction method can relieve the distractions (noise) caused by edges on the road.

2.3 Post Processing for Robust Detection

Based on the detected corridor boundaries in Section 2.2, post processing will produce a reasonable and smooth corridor. Optimal boundary selection will select a better boundary, points of which will then be smoothed by a sliding mean. After back-projecting to the input image, a road patch defined by these boundaries will be extended as being the corridor.

Optimal Corridor Boundary Selection. Until now, both a left and a right corridor boundary have been detected. Considering different deterioration features in the left and right part of the road as well as if those boundaries are parallel or not, it can be expected that both boundaries may not define a patch of constant width on the road.

In this situation, this paper suggests that a better corridor boundary is selected for robustness and practicality reasons, according to the following criteria. For a straight road, the criterion may be something like "less lateral variation" to pay more attention to the straight boundary. Actually, as we have made no assumption of a straight road, two criteria are actually useful for selecting the corridor boundary: first, the preference of "stronger edge strength" (as edge information is used for corridor detection, a boundary with points showing stronger edge strength will be selected); second, the "minimization of variation in boundary direction" (due to requested smoothness of corridor boundaries, a boundary with less change in direction will be selected with higher probability.)

Sliding Mean. As no specific road geometry is assumed, no curve model is used to fit the detected boundary points. A simple sliding mean is applied for ensuring smoothness of corridor boundaries. We obtain smoothed points X'_n from the point sequence $\{X_n\}$ by applying the following:

$$X'_n = X'_{n-1} + (X_n - X_{n-s})/s$$

The constant s determines the step size.

Road Patch Extension. Following the corridor definition, once a "dominant" corridor boundary (left or right) is selected and mapped back to the input image, a road patch will be calculated, having one of its sides identical to the selected boundary, and by calculating the other side with the pre-defined width. The width is adaptable as it should provide enough space for the car to drive through. Compared with a constant road width assumption, this method will provide a similar result if driving on a lane of constant width, and a fixed width identical to the lane's width. However, at other occasions, when the constant road width assumption is not applicable, the provided method still detects a reasonable corridor using the road patch extension.

2.4   Obstacle Detection

Generally, various obstacle detection can be used in corridor detection. A benefit from corridor detection in detecting obstacle is that the search area can be confined to a small patch. In [5], a method based on inverse perspective mapping using stereo images was introduced to detect obstacle. However, there's a problem when left and right images have a different brightness scale, which is common in real situation for DAS. The method is modified here, with left and right images after sobel edge detection as input. Furthermore, bird's-eye mapping as in Section 2.1 is used instead of inverse perspective mapping. For more details of the implementation, see [5].

3   Corridor Tracking

This section presents a corridor tracking approach which is not utilizing a time-consuming particle filter, as in [8,10,12,13], nor a model-based Kalman filter, as in [1]; both techniques are commonly used for lane tracking. After a corridor is detected (as discussed in Section 2), a practical way to represent a corridor is by using points on the central line of the corridor (in the bird's-eye image) and its width. Such a point sequence $\{C_n, n = 0, 1, \ldots, N\}$ (N is defined by the lookahead distance) of the center line can be calculated from some of the corridor's boundary points (in the bird's-eye image), also using the constant corridor width. Tracking of a corridor is composed of two modules: non-interrupted (i.e., by updating the previous corridor) corridor tracking, and interrupted corridor tracking.

3.1 Non-interrupted Corridor Tracking

Note that a corridor estimates a road patch that will be driven through shortly by the ego-vehicle. This means that a corridor detected at time t will have been partly driven through at time $t+1$. The ratio of the already driven part depends on the cycle time between two frames as well as the ego-vehicle's speed. If the ego-vehicle does not change much the driving direction (i.e., the yaw angle), and is also not in the process of a lane crossing, then there will be a non-interrupted corridor update between subsequent frames. Note that the point sequence $\{C_n\}$ is in the bird's-eye image. Corridor tracking is then easy, defined by tracking of the ego-vehicle's motion state for an adjustment of sequence $\{C_n\}$, and composed of two steps: adjustment caused by the driven distance and the variation in driving direction; possibly also by detecting new points. For points $\{C_n\}$ in frame $t+1$ and $\{C'_n\}$ in frame t, because of the driven distance, it follows that

$$C_i = C'_{i+m}, \qquad i = 0, 1, \ldots, N - m$$

Here, m is determined by the driven distance, and usually it is small. Furthermore, points $\{C_n, n = 0, 1, \ldots, N - m\}$ will all have an added shift in lateral position (according to n) caused by the variation of driving direction between these two frames. For the detection of center line points $\{C_n, n = N - m + 1, \ldots, N\}$, the same method is used as introduced in Section 2.2. The only difference is that left and right boundary points are calculated starting at points on the center line. By combining points from the last frame and points detected in the current frame, a corridor can be efficiently updated using the given sequence of frames.

3.2 Interrupted Corridor Tracking

However, a corridor will not always change non-interrupted between subsequent frames, which is obvious from its definition. If the change in driving direction is above some threshold, then the corridor may differ greatly compared to the corridor of the previous frame. Another situation is when an ego-vehicle is in the process of lane changing. The example in Fig. 8 gives an illustration of this case of interruption. In order to deal with such situations, a simple re-initialization by corridor detection is applied. The change of driving direction can be calculated

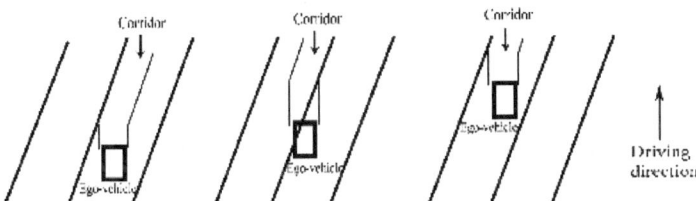

Fig. 8. Illustration of corridor interruption between frames. Left: before lane crossing. Middle: during lane crossing. Right: after lane crossing.

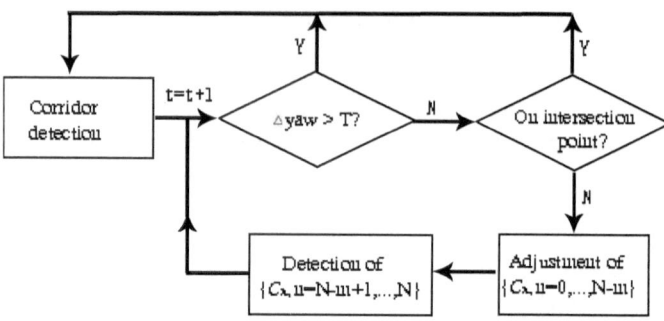

Fig. 9. The proposed scheme for corridor tracking. "On intersection point" means that the ego-vehicle is at the intersection point defined by the previous frame.

from ego-vehicle's motion model. Any occurrence of a lane change, or of any other boundary variation, can be identified by tracking the intersection points (see Fig. 1). A pass through this intersection point means a change of a corridor. Then, corridor detection will be applied as re-initialization, to restart a new process of tracking. Actually, as corridor detection is really time-efficient (see time measurements in Section 4), re-initialization will not harm non-interrupted tracking of the corridor. Furthermore, driving direction and intersection point allow an easy way to re-initiate using backtracking. – Figure 9 summarizes the proposed scheme of corridor tracking.

4   Experimental Results

Experiments were conducted on images and sequences recorded with our test vehicle "HAKA1" (High Awareness Kinematic Automobile No. 1). A pair of stereo cameras was installed in HAKA1 for those test drives. Corridor detection results are illustrated (only in left image) in Fig. 10 for a few selected (e.g., challenging) road situations. For better understanding of the shown situations, intermediate results of boundary detections (after smoothing) are also shown. Note that other vehicles detected as obstacles in the corridor are indicated by a white line (see the second column of Fig. 10), which is slightly different from the exact start position of the obstacle, but is relatively accurate enough for corridor detected in the input image.

Corridor detection only takes less than 0.1 seconds for a 640×480 image, using an off-the-shelf laptop (Pentium 2.1 GHz processor), implemented in OpenCV without runtime optimization. As no time-consuming computation is needed for detection, it is very "reasonable" in its computational efficiency.

Results for corridor tracking are presented in Fig. 11, for a few frames of some sequences. For interrupted tracking, corridor detection is commonly used for re-initialization. Non-interrupted corridor tracking is much faster than the steps at interruption. A whole procedure of lane crossing is illustrated in Fig. 12 with the detected corridor shown in green.

Fig. 10. Experimental results. Left: original image. Middle: both raw corridor boundaries (in red). Right: detected corridor (in green).

Fig. 11. Experimental results for two sequences (top/bottom row)

Fig. 12. Example lane change with detected corridor (from left to right, top to bottom)

Obtained experimental results (see Fig.4) show that corridor detection and tracking provides a good indication of the road patch that the ego-vehicle is expected to drive through shortly, even with difficult road situations.

5   Conclusions

A new concept of a corridor was introduced in this paper, and a possible corridor detection and tracking method is proposed. Compared with lane or free space detection, our corridor also pays attention to the driver's intention, which is indicated by the car's lateral position on the road and the driving direction. Still, a corridor is partly constrained by the physical lane marks or boundaries, and it will follow those if suitable. A corridor can be utilized in a DAS to indicate to the driver the road area where they will drive through shortly, if there is no rapid change in driving (i.e., by breaking or making a sharp turn).

Future corridor detection work will also incorporate stereo analysis.

Acknowledgments. This work is supported by the National Natural Science Foundation of China under Grant 50875169.

References

1. Aufrére, R., Chapuis, R., Chausse, F.: A model-driven approach for real-time road recognition. Machine Vision and Application 13, 95–107 (2001)
2. Badino, H., Franke, U., Mester, R.: Free space computation using stochastic occupancy grids and dynamic programming. In: Proc. ICCV Workshop Dynamical Vision (2007)

3. Badino, H., Mester, R., Vaudrey, T., Franke, U.: Stereo-based free space computation in complex traffic scenarios. In: Proc. IEEE Southwest Symp. Image Analysis Interpretation, pp. 189–192 (2008)
4. Bellino, M., de Meneses, Y.L., Ryser, P., Jacot, J.: Lane detection algorithm for an onboard camera. In: Proc. SPIE Photonics in the Automobile, vol. 5663, pp. 102–111 (2005)
5. Bertozzi, M., Bróggi, A.: GOLD:A parallel real-time stereo vision system for generic obstacle and lane detection. IEEE Trans. Image Processing 7, 62–81 (1998)
6. Cerri, P., Grisleri, P.: Free space detection on highways using time correlation between stabilized sub-pixel precision IPM images. In: Proc. IEEE Robotics and Automation, pp. 2223–2228 (2005)
7. Chen, Z., Pears, N., Liang, B.: Monocular obstacle detection using reciprocal-polar rectification. Image and Vision Computing 24, 1301–1312 (2006)
8. Kim, Z.: Realtime lane tracking of curved local road. In: Proc. IEEE Intelligent Transportation Systems, pp. 1149–1155 (2006)
9. Kim, Z.: Robust lane detection and tracking in challenging scenarios. IEEE Trans. Intelligent Transportation System 9, 16–26 (2008)
10. Sehestedt, S., Kodagoda, S., Alempijevic, A., Dissanayake, G.: Efficient lane detection and tracking in urban environments. In: Proc. European Conf. Mobile Robots, pp. 126–131 (2007)
11. Thrun, S., Montemerlo, M., Aron, A.: Probabilistic terrain analysis for high-speed desert driving. In: Proc. Robotics Science and Systems Conference, Philadelphia, PA (2006)
12. Wang, Y., Teoh, E.K., Shen, D.: Lane detection and tracking using B-Snake. Image and Vision Computing 22, 269–280 (2004)
13. Wang, Y., Bai, L., Fairhurst, M.: Robust road modeling and tracking using condensation. IEEE Trans. Intelligent Transportation Systems 9, 570–579 (2008)
14. Zhou, Y., Xu, R., Hu, X., Ye, Q.: A robust lane detection and tracking method based on computer vision. Measurement Science Technology 17, 736–745 (2006)

Adaptive Pixel Resizing for Multiscale Recognition and Reconstruction

Marc Rodríguez, Gaëlle Largeteau-Skapin, and Eric Andres

Laboratory XLIM, SIC Department,
University of Poitiers BP 30179, UMR CNRS 6712
86962 Futuroscope Chasseneuil Cedex, France
rodriguez@sic.sp2mi.univ-poitiers.fr, glargeteau@yahoo.fr,
andres@sic.sp2mi.univ-poitiers.fr

Abstract. This paper present an adaptive pixel resizing method based on a parameter space approach. The pixel resizing is designed for both multiscale recognition and reconstruction purposes. The general idea is valid in any dimension. In this paper we present an illustration of our method in 2D. Pixels are resized according to the local curvature of the curve to control the local error margin of the reconstructed Euclidean object. An efficient 2D algorithm is proposed.

Keywords: Reconstruction, recognition, parameter space, multiscale.

1 Introduction

For several years now, the discrete geometry community works on an invertible reconstruction of a discrete object. A reconstruction is a transformation from the discrete space to the Euclidean space that associates a Euclidean object to a discrete one. The reconstruction is said to be invertible if the digitization of the reconstructed object is equal to the original discrete object. Recently, Isabelle Debled-Renesson et al. have started to consider reconstruction methods that allow a margin of error in the reconstruction [8,9]. These methods are not invertible but allow a reconstruction of, for instance a discrete curve, where a pixel might be missing or out of place. This brought them to define the notion of blurred line segments and consider applications such as discrete curve denoising. Their method is based on thick arithmetical Reveillès discrete straight lines. Contrary to an invertible reconstruction algorithm where all the pixels that verify the Reveillès straight line equation $0 \leq ax + by + c < max(|a|, |b|)$ belong to the discrete line, for blurred lines of thickness ω a pixel has to verify $0 \leq ax + by + c < \omega$. The value ω defines in some way an error margin. A set of pixels might not, strictly, be a discrete straight line but as long as its pixels are not too far away from it, it will form a blurred line. The only problem with this approach is that the arithmetical thickness of those lines, and thus the error margin, is uniform for a given curve. This was the starting point of our research.

Our goal in this paper is to propose an adaptive voxel resizing method based on a parameter space approach and give an illustration of this method with a

P. Wiederhold and R.P. Barneva (Eds.): IWCIA 2009, LNCS 5852, pp. 252–265, 2009.

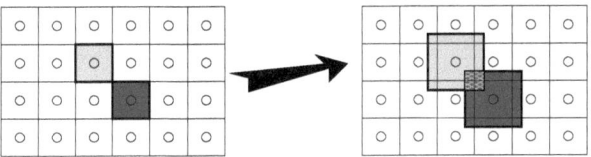

Fig. 1. After resizing the pixels may overlap

2D pixel resizing that fits the local curvature of a discrete curve. This example is a direct extension of the work of Debled-Renesson et al. and shows how our approach can generalize their work. Our long term goal however is to build a theoretical and practical framework for multiscale pattern recognition and object reconstruction. A resized voxel can be seen as a voxel at a different scale and the parameter space approach allows to work with voxels of various sizes, at various scales.

The approach is based on a parameter space approach (see the J.Vittone approach [19,20]). A Euclidean point, in nD, in the image space is associated to a Euclidean hyperplane, in nD, in the parameter space. The dual of a voxel (in the parameter space) will therefore be a sort of a butterfly figure that represents all the hyperplanes that cross the voxel. This is called the generalized preimage of the voxel. The intersection of all the generalized preimages of the voxels of a set is the generalized preimage of the voxel set. The preimage approach we propose here allows us to locally mix voxels of different scales. Having voxels of different scales means that voxels can overlap (see Fig.1). As an illustration of the voxel resizing approach, we propose a 2D application where pixels of a discrete curve are resized according to the maximal symmetric tangent [13,18] (we can also use other curvature estimators such as the estimator defined by J.W. Bullard [5] which is defined in any dimension). This provides an adaptive pixel size according to the smoothness of the curve. Indeed, if the maximal symmetric tangent around a pixel is *long* it means that locally, around this pixel, the curve is flat and thus smooth. The curve does not need to be denoised much (in the sense defined by Debled-Renesson et al. [8,9]) and the pixel size will be left unchanged. On the contrary, a *short* maximal symmetric tangent around a pixel means that locally the curve is not flat (high curvature) and thus can be simplified by increasing the margin of error in the reconstruction by increasing the size of the pixel. The pixel resizing involves several modifications of the reconstruction algorithm proposed by M.Dexet [10]. Those modifications come directly from the fact that all pixel of a set do not share the same size. We state in this paper that the computation of the two-dimensional polytopes collecting all the straight lines crossing a pixel set can be reduced to the linear time computation of half-plane intersections. We use it to compute the generalized preimage of a resized pixel set. Further applications in discrete curve denoising and Euclidean curve simplification are considered.

The paper is organized as follow: section two presents basic notions and notations, it describes the generalized preimage and its use. Section three presents our version of the reconstruction with resized pixels. We also propose in this section an iterative half-plane intersection algorithm in linear time for the preimage computation. Finally, some applications examples in simplification and denoising are presented.

2   Preliminaries

2.1   Notation

Let $\mathbb{Z}^n$ be the subset of the nD Euclidean space $\mathbb{R}^n$ containing all the integer coordinate points. A *discrete* (resp. Euclidean) *point* is an element of $\mathbb{Z}^n$ (resp. $\mathbb{R}^n$). A *discrete* (resp. Euclidean) *object* is a set of discrete (resp. Euclidean) points.

Discrete-Euclidean operations [14] are operations that operate partially in the discrete space $\mathbb{Z}^n$ and partially in the continuous space $\mathbb{R}^n$. Appropriate transforms are used to travel between both spaces. A transformation from the Euclidean to the Discrete space is called *discretization*. The transformation from the discrete space to the Euclidean space is classically called *reconstruction*.

We denote p_i the ith coordinate of a point or vector p. The *voxel* $\mathbb{V}(p) \subset \mathbb{R}^n$ of a discrete nD point p is defined by $\mathbb{V}(p) = [p_1 - \frac{1}{2}, p_1 + \frac{1}{2}] \times ... \times [p_n - \frac{1}{2}, p_n + \frac{1}{2}]$.

Standard hyperplanes [1] are a particular case of *analytical discrete hyperplanes* defined by J.-P. Reveilles [15,3]. A set of successive points $C_{i,j}$ of a digital curve C is called a *Digital Straight Segment* (DSS) if there exists a standard discrete line (a, b, μ) containing them. The predicate $C_{i,j}$ is a DSS is denoted $S(i, j)$. Around any point C_k of a digital curve C, the DSS $C_{k-l,k+l}$ with $S(k-l, k+l)$ and not $S(k-l-1, k+l+1)$ is called *Symmetric Tangent* [13,18].

Let h be the direct *Hausdorff distance*: $A \subset \mathbb{R}^n$, $B \subset \mathbb{R}^n$, $h(A, B) = max_{a \in A}(min_{b \in B}(d_2(a, b)))$, where d_2 is the Euclidean distance. The Hausdorff distance between A and B is $H(A, B) = max(h(A, B), h(B, A))$.

An *homeomorphism* or *topological isomorphism* is a bicontinuous function between two topological spaces. Two spaces are called *homeomorphic* if they are the same from a topological point of view.

2.2   Recognition and Invertible Reconstruction Based on a Preimage

In computer imagery, *parameter spaces* are often used to recognize shapes in a picture. Those spaces are defined with a transformation that associates a point in the image space to a geometrical object in the parameter space. In this paper, we use the generalized preimage definition [10]: any Euclidean point $p_\mathcal{E} = (x_1, ..., x_n) \in \mathbb{R}^n$ (the image space) is associated to a hyperplane (in the parameter space $\mathbb{R}^n$):

$$D_\mathcal{E}(p_\mathcal{E}) = \left\{ (y_1, ..., y_n) \in \mathbb{R}^n \mid y_n = x_n - \sum_{i=1}^{n-1}(x_i y_i) \right\}.$$

Contrary to the Hough transform [12], the parameter space here is not digitized and the transform is analytical. The parametric image of a voxel will therefore be a sort of a butterfly figure (see Fig.2 for examples of pixel preimages) that

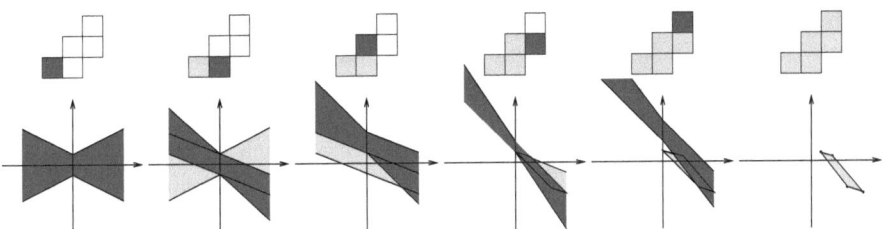

Fig. 2. Straight line recognition process: In dark, the current pixel and below the preimage of the current pixel. In grey the pixels already processed and the intersection of the preimages of those pixels. At the end, the intersection is not empty: the pixels belong to a discrete straight line segment.

represents all the hyperplanes that cross the voxel. This is called the generalized preimage of the voxel. Each point of the generalized preimage in the parameter space corresponds to one hyperplane that cuts the voxel in the image space. The generalized preimage of a set of voxels is defined as the intersection of all the preimages of each voxel of the set. Each point of the generalized preimage of a set of voxels in the parameter space corresponds to a hyperplane that cuts all the voxels of the set in the image space. This corresponds to an invertible reconstruction if the standard [1] or the supercover discrete analytical model is considered [2].

The analytical reconstruction consists of two steps: firstly a recognition step is performed that divides the discrete object into discrete analytical hyperplanes pieces [15,1] (subsets of discrete analytical hyperplanes) and computes their generalized preimage in the parameter space. A hyperplane recognition algorithm in dimension n based on the general preimage definition has been proposed by M.Dexet [10]. Secondly comes the reconstruction itself. It consists in picking a Euclidean solution from each generalized preimage of each hyperplane piece and group them to form the final Euclidean reconstructed object. The main difficulty here is not so much the recognition step than the reconstruction step. A largely unsolved problem is the problem of reconstructing a closed Euclidean surface from the different Euclidean hyperplane pieces. Invertible reconstructions have been proposed in dimension 2 by R.Breton et al. [4] for the standard analytical model for discrete objects [1] and in dimension 3 by I. Sivignon et al. [17]. The problem is still partly open in dimension 3 and largely open in dimension higher than three.

In two dimensions, the preimage of a point is a Euclidean straight line. The preimage of a pixel is an unbounded polygon with edges that are the preimages of the four vertices of the pixel (see example of pixel preimages in Fig.2).

The invertible reconstruction of a discrete curve goes basically as follows (see [4] for more details): the preimage of consecutive pixels are intersected as long as the result is not empty. This process is illustrated in fig.2. At the end, we obtain a polygon in the parameter space: for figure 2, the generalized preimage of the five pixels. Each point of this polygon corresponds to a Euclidean line that crosses

all the five pixels of the pixel set. The standard or supercover discretization
of each of these Euclidean straight lines contains the five pixels. At this point
we have recognized a DSS. When handling a discrete curve, we start over with
the last pixel of the previous DSS recognition in order to recognize a new DSS
until we have dealt with all the pixels of the curve. Each DSS is then replaced
by a Euclidean straight line segment corresponding to a point chosen inside
the preimage of the DSS (the intersection of the preimage of its pixels). Some
attention needs to be given to the intersections of the different Euclidean straight
line segments in order to have an invertible reconstruction.

There are two important advantages to this method. Firstly, it computes the
exhaustive set of all the Euclidean straight lines (resp. hyperplanes) crossing all
the pixels (resp. voxels) of a given pixel set. Secondly, nothing in the method
imposes that the pixels or voxels we are considering are of the same size (i.e. at
the same scale). The grid elements we are considering do not even have to be
squares, rectangles, hypercubes, etc. as long as the preimage is known. Heterogeneous grids can therefore be handled with the generalized preimage approach.
It is this second property that will be used in what follows with pixels and voxels
of different sizes scales.

3   Reconstruction with Resized Pixels

The aim of this section is to propose a new reconstruction method with resized
pixels. The basic idea is to reproduce the principle described in section 2 with
pixels that have different sizes (that come from grids at different scales). This
generalizes the approach with blurred lines proposed by I. Debled-Rennesson et
al. [8,9] where the thickness of the blurred line is a fixed parameter. In our case
the pixels may, independently, have different sizes. The general principle of the
reconstruction method based on generalized preimages does indeed not change
with pixels of different sizes, even if they overlap. This general idea is valid in
dimension n as well. The algorithm proposed by M. Dexet [10] in dimension
n needs however some adaptation which we present here in dimension 2 in a
specific application: Adaptive Pixel Resizing according to the local curvature of
the discrete curve. First however, let us show how these ideas are well adapted
for recognition and reconstruction within multiscale and hereterogeneous grids.

3.1   Multiscale Framework

In what follows (starting in section 3.2), we consider a curve in a grid where
the individual pixels are resized according to a given criteria (in our case a local
discrete curvature value). This is the basically the same point of view than the
one proposed by I. Debled-Rennesson et al. [8,9]. There is however another way
of looking at this, not as pixels that are resized but as pixels of different grids
at different scales. As we can see in Fig. 3, one can consider that the preimage method allows recognition and reconstruction in a multiscale setting where,
contrary to many applications, the different scales (i.e. corresponding grids) do

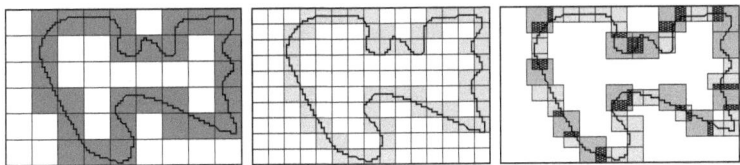

Fig. 3. Pixel are given at different scales

not necessarily have to match each other. The pixels from the different grids may overlap or not. Recognition and reconstruction within heterogeneous grids is another possible application field of general preimage based approaches. These approaches work in every dimension as long as the grid cells are convexes, that we know what their preimages are (and know how to intersect them with each other) and that we define an order with which one adds cells into the recognition process. If the grid cells are not convex then it requires that we know how to divide the cell into convex pieces or, if we do not consider invertible reconstruction, that we know how to compute the cells convex hulls. As we can see, preimage methods are very flexible and allow recognition in a wide range of multiscale frameworks. Note that, contrary to classical line stabbing/transversal problems [11], we are not interested in simply knowing if there is a hyperplane that cuts a set of objects or proposing one hyperplane as solution. In order to tackle the much more complicated "reconstruction" problem in discrete geometry, we are looking for all the solutions (all the hyperplanes that cut a set of grid cells represented by the generalized preimage).

3.2 Adaptive Pixel Resizing According to the Curvature

From here on we consider an illustration of the preimage approach where pixels of a given discrete curve are resized according to a local criteria. In this particular case, with resized pixels, their center does not change (see Fig.1). This is different from what happens when the grid scale is simply changed since then, usually, the center of the pixels change. The preimage approach allows pixel resizing. One can imagine many ways of considering resized pixels/voxels or pixels/voxels with different scales. We are going to focus here on one such idea where we allow a bigger margin of error for the reconstruction in high curvature parts of a discrete curve. It is meant as a direct extension of the work of Debled-Rennesson et al. [8,9]. In this paper however, in order to simplify the computations, we do not compute the actual curvature at a given discrete point of the curve but only a first order approximation. An ongoing work with an other curvature estimator (the estimator defined by J.W. Bullard [5] which is defined in any dimensions) is briefly illustrated in 3D in Fig. 5. Here, we consider the length of the longest discrete straight line included in the curve and centered on the discrete point (i.e. the length of the maximal discrete symmetric tangent). That means, for instance, that discrete cusps [4] will have their size increased much more than pixels in the middle of long straight lines (flat parts of the curve).

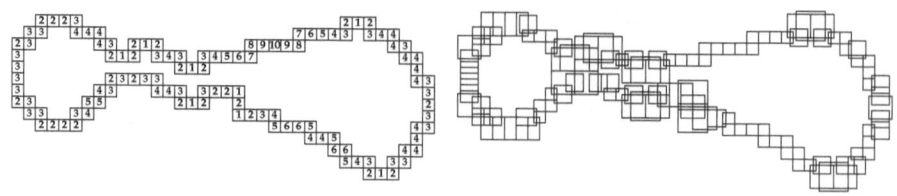

Fig. 4. Pixel resizing using the maximal symmetric tangent

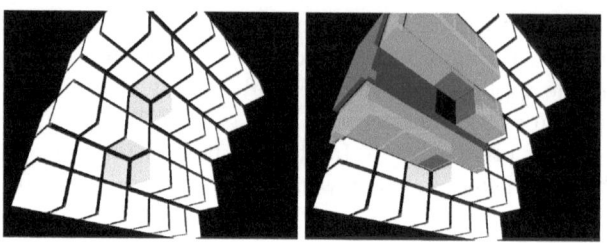

Fig. 5. Voxel resizing using the bullard curvature estimator in dimension three

We want to focus the simplification on *uneven* parts of curve where the curve is not flat. An increase in the size of the pixels there means that we will be able to smooth/denoise the discrete curve (see Fig.4). In other words, the greater the curvature of the discrete curve centered on the considered pixel, the greater the new size of this pixel. Our recognition method will smooth the curve and (small) oscillations will disappear. Of course, this is not meant as a realistic simplification method as it smoothes out points of interest in curves. It illustrates however well how the recognition and reconstruction with pixels of various sizes/scales works. To evaluate the curvature at a pixel, the length of the maximal symmetric tangent [13,18] centered on this pixel is considered. The maximal symmetric tangent centered on a discrete point p for a discrete curve, is the longest discrete line centered on p such that there are the same number of points on the discrete line on each side of p. It is a very simple criteria that gives a measure of the *flatness* of the curve around this pixel. It can be computed in linear time according to the number of discrete points of the curve. On the left of the figure 4, each pixel is labeled with $p \in \mathbb{N}$ where $2p + 1$ is the size of the maximal discrete symmetric tangent. On the right of the figure 4, pixels are resized according to the resizing function $r = max(1, 3 - \frac{1}{2}p)$.

The pixel resizing is function of the maximal symmetric tangent length. To increase the pixel size, the resizing function has to be greater than one at any point. This function can be modified to control where the error margin is located.

3.3   Error Measure

In order to evaluate the reconstruction transform, we propose an upper bound on the Hausdorff distance between a Euclidean object and the reconstruction of its digitization. It is indeed not very easy to measure otherwise the impact of a reconstruction transform. The reconstruction method certifies that every resized voxel is crossed by at least one edge of the boundary of the reconstructed Euclidean object. The error is therefore directly proportional to the size of the biggest resized voxel.

Theorem 1. *The Hausdorff distance between a Euclidean object A and the simplified object B (the reconstruction of its digitization) depends on the maximal voxel resizing factor $R = max_i(r(i))$: $H(A,B) \leq \frac{(R+1)\sqrt{n}}{2}$ where n is the dimension.*

Proof. The Hausdorff distance is controlled by the voxel resizing. We know that, for each reconstruction, a Euclidean hyperplane crosses the original voxel and the resized voxel. The biggest distance between those two hyperplanes for a resized voxel corresponds to two hyperplanes crossing the voxels at opposite corners. The biggest distance over all the voxels is obtained for the biggest resized voxel. This leads to the result.   □

If many such operations are chained, the Hausdorff distance is not equal to the sum of the resizing factors because additional errors are added by the discretizations between each reconstruction. The error bound can diverge when chaining such operations. This reflects the fact that we can have geometrical deformations [14]. One way to avoid this problem is to apply every successive operation (with different resizing function for instance) on the original object.

3.4   Half-Plane Intersection Algorithm for a 2D Reconstruction

Let us now take a look at the algorithmic aspects of the preimage computation. In general, we can determine in $\theta(n)$ that there is a hyperplane intersecting all the elements of a set of n voxels independently of the dimension [11]. However, as already mentioned, in our case we are not looking for a single solution but for all the Euclidean straight lines that cut the set of all the (resized) pixels. For an orthonormal grid with pixels that all have the same size, the preimage of a set of pixels has at most four edges [6]. With pixels of various sizes, the number of edges of the preimage polytope can be infinite if the number of different pixel sizes is not bounded. Without an hypothesis on the number of vertices of the preimage, the iterative preimage construction can be computed in $\Theta(n^2)$ with n the number of pixels. However, we know that the center of the resized pixels are discrete points in $\mathbb{Z}^2$ which means that the number of vertices of the preimage depends only on the number of different possible pixel sizes. Here we propose a limited number of possible pixel sizes which leads to a generalized preimage construction algorithm in $\Theta(n)$ where n is the number of (resized) pixels.

During the reconstruction process, the first important step is the analytical primitive recognition in the parameter space. For orthonormal grids, a very efficient algorithm was proposed by M.Dexet in [10] but with resized pixels it is no longer working since pixels may now overlap.

The preimage of a pixel with vertices $(v_i)_{i \in [1,4]} = ((x_i, y_i))_{i \in [1,4]}$ is an unbounded polygon with edges defined by the preimage of the four vertices: the straight lines of equation $d_i : a_i x + b_i = 0$, $i \in [1,4]$.

A point $p = (x, y)$ is inside the preimage if:

$$(a_1 x + b_1 \leq 0 \ \cup \ a_2 x + b_2 \leq 0) \ \cap \ (a_3 x + b_3 \geq 0 \ \cup \ a_4 x + b_4 \geq 0).$$

This is equivalent to:

$$(a_1 x + b_1 \leq 0 \ \cap \ a_3 x + b_3 \geq 0) \ \cup \ (a_2 x + b_2 \leq 0 \ \cap \ a_4 x + b_4 \geq 0).$$

The problem can be divided into two groups: the straight line solutions with positive slopes corresponding to any point $p = (x_p, y_p)$ verifying $(a_1 x + b_1 \leq 0 \ \cap \ a_4 x + b_2 \geq 0)$, $(x, y) \in \mathbb{R}^+ \times \mathbb{R}$ and the solutions with negative slopes corresponding to any point $q = (x_q, y_q)$ verifying $(a_2 x + b_2 \leq 0 \ \cap \ a_4 x + b_4 \geq 0)$, $(x, y) \in \mathbb{R}^- \times \mathbb{R}$.

The preimage of a pixel (α, β) can therefore be described by four constraints:

(1) $-(\alpha - \frac{1}{2})x - y + (\beta + \frac{1}{2}) \geq 0$, $(x, y) \in \mathbb{R}^+ \times \mathbb{R}$.
(2) $(\alpha + \frac{1}{2})x + y - (\beta - \frac{1}{2}) \geq 0$, $(x, y) \in \mathbb{R}^+ \times \mathbb{R}$.
(3) $-(\alpha + \frac{1}{2})x - y + (\beta + \frac{1}{2}) \geq 0$, $(x, y) \in \mathbb{R}^- \times \mathbb{R}$.
(4) $(\alpha - \frac{1}{2})x + y - (\beta - \frac{1}{2}) \geq 0$, $(x, y) \in \mathbb{R}^- \times \mathbb{R}$.

Proposition 1.   *The solution polytope computation problem can be reduced to a half-plane intersection problem. The solution polytope is the union of two convexes, one in $\mathbb{R}^+ \times \mathbb{R}$ and one in $\mathbb{R}^- \times \mathbb{R}$.*

Let us now describe our half-plane intersection algorithm. Without any hypothesis on the half-planes, the algorithms suppose that the half-planes are ordered according to, for example, the slope. The complexity of these algorithms is therefore in $O(n \ln(n))$ [7]. In our case however, there are several hypothesis that allow us to compute the half-space intersections, and thus construct the preimage, in linear time. Firstly, the intersections of the half-planes are performed respectively in the two halves of the space $x \leq 0$ and $x \geq 0$. The first half-space to be considered in our intersection algorithm is therefore $x \leq 0$ and $x \geq 0$ respectively. Secondly, the parameter space definition ensures that no constraint (half-plane) can be vertical. This hypothesis is important because we know that the first constraint intersects the vertical axis $x = 0$. Lastly, we know that the number of edges of the preimage in each half of the space is bounded because the center of the resized pixels are integer coordinate points and the number of possible different pixel sizes is bounded.

Algorithm main idea (see p.262):
Four constraints (half planes) are associated to each pixel. Therefore $4n$ constraints are associated to a set of n pixels. Each constraint is a half-space in the parameter space bounded by a straight line. We add a first constraint which is $x \leq 0$ and $x \geq 0$ respectively. For each half-space constraint, we have two Boolean markers that are set to false at start. A marker indicates if a vertex of the preimage is on the half-plane border. Two markers at true means that we

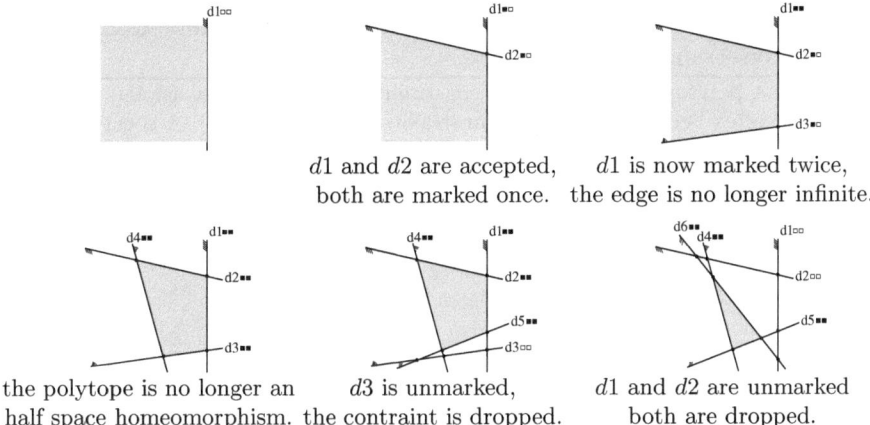

$d1$ and $d2$ are accepted, $d1$ is now marked twice,
both are marked once.   the edge is no longer infinite.

the polytope is no longer an $d3$ is unmarked, $d1$ and $d2$ are unmarked
half space homeomorphism.   the contraint is dropped.   both are dropped.

Fig. 6. Example of the iterative half plane intersection algorithm

have a straight line segment. When we add a new constraint, we check all the vertices that we have already identified and determine like that if the constraint cuts the polytope into two pieces (and then the intersected lines are immediately identified), or if the vertices all verify or do not verify the new constraint. There is a slight difference in treatment here if the preimage at that point is still unbounded. When all the half-spaces have two markers at true then the preimage is bounded (it is not a half-space homeomorphism anymore). When a constraint is not marked (both at false) anymore then the constraint is dropped. A constraint with only one marker at true corresponds to a half line. The algorithm works in linear time because we know that the number of vertices of the preimage that needs to be checked is bounded. Figure 6 shows an example of the half-space intersection algorithm at work on the half-space $x \leq 0$.

Theorem 2. *The algorithm that computes iteratively the intersection of half-planes works in linear time in the number of pixels of the border in the case of an Adaptive Pixel Size Reconstruction where the number of different possible new pixel sizes is bounded and if the center of each pixel has integer coordinates.*

Proof. The algorithm complexity is $\mathcal{O}(k \times n)$ where n is the number of pixels of the digital curve and k the number of vertices of the solution polytope. If the center of pixels are discrete points of $\mathbb{Z}^n$ and the number of different pixel sizes is bounded, then the number of vertices of the preimage is bounded. Indeed, the number of support points and the number of support lines is bounded (see [8,9] for more details on support lines and support points). Each support point corresponds to an edge of the solution polytope in the parameter space. The number of vertices k of the preimage only depends on the number of different pixel sizes and not on the length of the curve.   □

Algorithm 1. Iterative computation of a preimage polytope using half-plane intersections

---

Data: A polytope $\mathcal{P}$ defined by a constraint list CL, a vertex list VL where a vertex is defined with two constraints $v(C1)$ and $v(C2)$. A constraint $\mathcal{C}$: $\alpha X + \beta Y + \gamma \geq 0$

Result: a polytope $\mathcal{P}'$ defined by a constraint list CL' and a vertex list VL (if $\mathcal{P} = \mathcal{P}'$ the constraint $\mathcal{C}$ is redundant).

begin

  if $CL.size() = 0$ **then**
      add the edge $\mathcal{C}$ to CL

  else if $CL.size() = 1$ **then**
      add the vertex $\mathcal{C} \cap CL(0)$ to VL;
      add the edge $\mathcal{C}$ to CL ;
      mark $\mathcal{C}$ and $CL(0)$ one time

  else

    for all $v \in VL$ **do**
      if v *do not verify* $\mathcal{C}$ **then**
         unmark v_{C1} and v_{C2} one time ;
         remove v from VL

    if *The polytope is an half-space homeomorphism* **then**
      for all $c \in CL$ **do**
        if c *is not marked* **then**
          if c *is not an infinite edge* **then**
            remove c from CL
          else
            add the vertex $\mathcal{C} \cap c$ to VL ;
            mark $\mathcal{C}$ and c one time

        else if c *is marked one time* **and** c *is not an infinite edge* **then**
          add the vertex $\mathcal{C} \cap c$ to VL ;
          mark $\mathcal{C}$ and c one time
        if *the intersection between* $\mathcal{C}$ *and an infinite edge* e *verify all the* CL *constraints* **then**
          add the vertex $\mathcal{C} \cap e$ to VL ;
          mark e and $\mathcal{C}$ one time ;
          //if $\mathcal{C}$ *has two marks, the polytope is no longer an half-space homeomorphism ;*
          //else, $\mathcal{C}$ *become an infinite edge*

    else
      for all $c \in CL$ **do**
        if c *is not marked* **then**
          remove c from CL
        else if c *is marked one time* **then**
          add the vertex $\mathcal{C} \cap c$ to VL ;
          mark $\mathcal{C}$ and c one time

end

---

4   Example of Applications

Let us now present two short applications of our Adaptive pixel resizing reconstruction method: discrete curve denoising and Euclidean curve simplification.

The *discrete curve denoising* is a discrete-Euclidean transform (see [14] for details). A discrete-Euclidean transform is a transform in the discrete world that operates partially in the Euclidean world. In this case, the starting point is a discrete curve that we reconstruct with our new Adaptive pixel resizing reconstruction method. The reconstructed Euclidean curve is then discretized to obtain a new discrete curve that is *smoother*. Missing information or misplaced pixels can also be replaced with this method as long as the pixels are ordered in a sequence on the discrete curve (see Fig. 7).

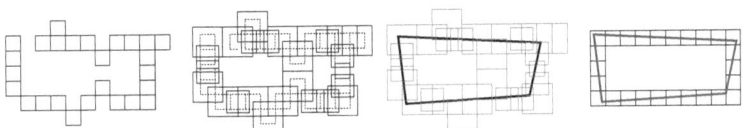

Fig. 7. Example of discrete curve denoising

In the case of a *Euclidean-discrete curve simplification*, the original object is a Euclidean curve. The adaptive pixel size reconstruction is applied to the discretization of the Euclidean curve. This transform is therefore a Euclidean-discrete transform (see [14] for details) that transforms a Euclidean curve into a Euclidean curve. The transform operates partially in the discrete world. On

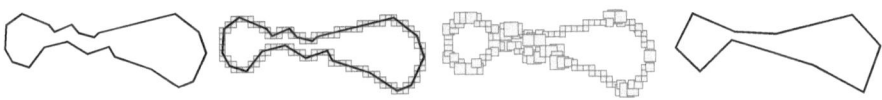

Fig. 8. Discrete-Euclidean simplification

figure 8, we can see the discretization and the reconstruction with resized pixels. Those operations transform the original Euclidean object into an other one with less vertices depending on the resizing function.

5   Conclusion and Perspectives

In this paper we have presented a new two-dimensional reconstruction based on the computation of the generalized preimage with an adaptive pixel size. This extends the recognition and reconstruction based on blurred lines proposed by Isabelle Debled-Rennesson et al. [8,9]. In our method, each pixel can be individually resized according to a given criteria. We have proposed to resize each pixel according to the length of the maximal discrete symmetric tangent at that pixel.

Extensions to more precise curve estimators is one of the immediate perspectives we are working on. We have already started to work with the Bullard curvature estimator that is defined for all dimensions. First voxel resizing according to the curvature have been done in 3D (See Fig. 5). An efficient algorithm was implemented that provides a two-dimensional reconstruction in linear time if the number of different possible pixel sizes is bounded and if the centers of the pixels belong to a predefined regular grid.

One of the major perspectives of this work is its extension to higher dimensions. The Bullard curvature estimator [5] can be computed very quickly at the fly in any dimension. High dimensional primitive recognition in a multiscale, heterogeneous grid setting seems now to be achievable in a very general framework. Questions about algorithmic complexity of the generic recognition algorithms and heuristics behind the way voxels are added into the recognition algorithm (there is no natural order anymore on surfaces) is however still an open and difficult problem. Reconstruction in dimensions higher than two is still quite open, even on regular grids, if you consider global techniques such as the ones we discussed in this paper rather than local techniques such as, for instance, the Marching Cubes technique. This is one of the long term goals we have. A more immediate perspective is the topological control of the reconstruction in 2D. Right now, the method does not guaranty that the reconstructed curve does not modify the topology of the object when pixel sizes are increased. Features of the simplified curve might disappear and the different edges might intersect each other elsewhere than on the common vertices. This represents right now the short term focus of our research. A first look at these topological problems has been proposed at IWCIA2009 [16]. A last perspective comes from the comparison with the work of Isabelle Debled-Rennesson et al. There are obvious links between both approaches on a theoretical level that need to be explored. This is certainly an interesting perspective especially when considering higher dimensions.

References

1. Andres, E.: Discrete linear objects in dimension n: the standard model. Graphical Models 65, 92–111 (2003)
2. Andres, E.: The supercover of an m-flat is a discrete analytical object. Theoretical Computer Science 406, 8–14 (2008); Discrete Tomography and Digital Geometry: In memory of Attila Kuba
3. Andres, E., Acharya, R., Sibata, C.: Discrete analytical hyperplanes. Graphical Models and Image Processing 59, 302–309 (1997)
4. Breton, R., Sivignon, I., Dexet, M., Andres, E.: Towards an invertible euclidean reconstruction of a discrete object. In: Nyström, I., Sanniti di Baja, G., Svensson, S. (eds.) DGCI 2003. LNCS, vol. 2886, pp. 246–256. Springer, Heidelberg (2003)
5. Bullard, J.W., Garboczi, E.J., Carter, W.C., Fuller Jr., E.R.: Numerical methods for computing interfacial mean curvature. Computational Materials Science 4, 103–116 (1995)
6. Coeurjolly, D., Sivignon, I., Dupont, F., Feschet, F., Chassery, J.M.: On digital plane preimage structure. Discrete Applied Mathematics, 78–92 (2005)

7. De Berg, M., Cheong, O., Van Kreveld, M., Overmars, M.: Computational Geometry: Algorithms and Applications, 3rd edn. Springer, Heidelberg (2008)
8. Debled-Rennesson, I., Feschet, F., Rouyer-Degli, J.: Optimal blurred segments decomposition of noisy shapes in linear time. Computer & Graphics 30 (2006)
9. Debled Rennesson, I., Tabbone, S., Wendling, L.: Multiorder polygonal approximation of digital curves. Electonic Letters on Computer Vision and Image Analysis 5, 98–110 (2005)
10. Dexet, M., Andres, E.: A generalized preimage for the digital analytical hyperplane recognition. Discrete Applied Mathematics 157, 476–489 (2009)
11. Edelsbrunner, H.: Finding transversals for sets of simple geometric figures. Theoretical Computer Science 35, 55–69 (1985)
12. Hough, P.V.C.: Method and means for recognizing complex patterns. In: United States Pattent 3069654, pp. 47–64 (1962)
13. Lachaud, J.O., Vialard, A., De Vieilleville, F.: Analysis and comparative evaluation of discrete tangent estimators. In: Andrès, É., Damiand, G., Lienhardt, P. (eds.) DGCI 2005. LNCS, vol. 3429, pp. 240–251. Springer, Heidelberg (2005)
14. Largeteau-Skapin, G., Andres, E.: Discrete-euclidean operations. Discrete Applied Mathematics 157, 510–523 (2009)
15. Reveillés, J.P.: Géometrie Discrete, calcul en nombre entiers et algorithmique. PhD thesis, Université Louis Pasteur (1991)
16. Rodríguez, M., Largeteau-Skapin, G., Andres, E.: Adaptive pixel size reconstruction with topological control. In: Wiederhold, P., Barneva, R.P. (eds.) Progress in Combinatorial Image Analysis. Research Publishing Services, Singapore (to appear 2009)
17. Sivignon, I., Dupont, F., Chassery, J.: Reversible vectorisation of 3d digital planar curves and applications. Image Vision and Computing 25, 1644–1656 (2007)
18. Vialard, A.: Geometrical parameters extraction from discrete paths. In: Miguet, S., Ubéda, S., Montanvert, A. (eds.) DGCI 1996. LNCS, vol. 1176, pp. 24–35. Springer, Heidelberg (1996)
19. Vittone, J., Chassery, J.M.: (n,m)-cubes and farey nets for naive planes understanding. In: Bertrand, G., Couprie, M., Perroton, L. (eds.) DGCI 1999. LNCS, vol. 1568, pp. 76–90. Springer, Heidelberg (1999)
20. Vittone, J., Chassery, J.M.: Recognition of digital naive planes and polyhedrization. In: Nyström, I., Sanniti di Baja, G., Borgefors, G. (eds.) DGCI 2000. LNCS, vol. 1953, pp. 296–307. Springer, Heidelberg (2000)

Contour Reconstruction for Multiple 2D Regions Based on Adaptive Boundary Samples

Peer Stelldinger and Leonid Tcherniavski

University of Hamburg
Vogt-Koelln-Str. 30
{stelldin,tchernia}@informatik.uni-hamburg.de

Abstract. There exist a lot of algorithms for 2D contour reconstruction from sampling points which guarantee a correct result if certain sampling criteria are fulfilled. Nevertheless nearly none of these algorithms can deal with non-manifold boundaries of multiple regions. We discuss, which problems occur in this case and present a boundary reconstruction algorithm, which can deal with partitions of multiple regions, and non-smooth boundaries (e.g. corners or edges). In comparison to well-known contour reconstruction algorithms, our method requires a lower sampling density and the sampling points can be noisy.

Keywords: Contour Reconstruction, Gabriel graph, Curve Reconstruction, Boundary Reconstruction, Topology.

1 Introduction

Contour reconstruction from boundary sampling points is a well known problem and many algorithms for solving it have been proposed. Typically, they give guarantees on the topological correctness of the result if certain sampling conditions are fulfilled, e.g. the sampled contour has to be smooth and the ratio of sampling density and local curvature has to exceed a certain value [1,2,3]. In [9] an algorithm is given, which outperforms previous ones experimentally and which works well even in case of non-smooth boundaries. Unfortunately the algorithm is only a heuristic without any proven guarantees.

In contrast to the problem of curve reconstruction the related problem of reconstructing multiple region boundaries has gained very few attention. This problem is much more complicated, since the boundary is no longer a manifold, which is a common assumption for curve reconstruction algorithms. Recently, an algorithm has been proposed which can deal with non-smooth contours (i.e. having corners and edges), with noisy samplings (i.e. when the sampling points lie not on but only near the contour) and with multiple region boundaries [11], but the algorithm requires a known global bound on the sampling density as parameter, which is more restrictive than local sampling approaches.

In this paper we will discuss the difficulties of reconstructing multiple region boundaries given locally adaptive sampling points and derive a parameter-free new reconstruction algorithm, which is also able to deal with noise and local

P. Wiederhold and R.P. Barneva (Eds.): IWCIA 2009, LNCS 5852, pp. 266–279, 2009.
© Springer-Verlag Berlin Heidelberg 2009

violations of the sampling density. We show, that our algorithm generates a boundary reconstruction, which is an oversegmentation, i.e. it contains a correct boundary reconstruction as subset but may also contain additional boundary parts. We discuss, under which circumstances these additional boundary parts occur and why such cases are unavoidable.

2   Preliminaries

Since we want to reconstruct multiple regions by sampling their boundary, we have to define what we mean by a partition of the plane into different regions. The following definition is adapted from [11] (using usual Euclidean topology in $\mathbb{R}^2$ for the definition of connectivity, closure, open sets, set boundary, etc.):

Definition 1. *In 2D, a* plane partition $\mathcal{R}$ *is defined by a finite set of pairwise disjoint regions* $\mathcal{R} = \{R_i \subset \mathbb{R}^2\}$, *such that each region* $R_i \in \mathcal{R}$ *is a connected open set and the union of the closures of the regions covers the whole plane,* $\bigcup_i \overline{R_i} = \mathbb{R}^2$. *The* boundary *of the partition is* $\partial \mathcal{R} := \bigcup_i \partial R_i$.

Now, if each region of the plane partition represents an object of the real world, the reconstruction task is to construct a second plane partition from a discrete set of sampling points which lie on or near the boundary of the first partition. The two partitions should share as much as possible properties of the objects and the relations between them. Thus, we have to distinguish between the original plane partition and the resulting approximation. The approximation is based on a complete simplicial complex of the sampling points.

Definition 2. *Let* $S \subset \mathbb{R}^2$ *be a finite set of points. Then the convex hull of* $n \in \{1, 2, 3\}$ *points* $s_1, ..., s_n \in S$ *is called an* $(n-1)$-simplex. *Any simplex* σ_1 *based on the convex hull of a subset of the points defining a second simplex* σ_2, *is called a* face *of* σ_2, *and* σ_2 *is called a* coface *of* σ_1. *The 2-simplices are* triangles, *the 1-simplices are* edges, *and the 0-simplices are the points of* S. *Now a (simplicial)* complex $\mathcal{K}$ *is a set of simplices such that any face of a simplex in* $\mathcal{K}$ *is also a simplex in* K, *and the intersection of any two simplices in* $\mathcal{K}$ *is also a simplex in* $\mathcal{K}$. *A simplicial complex* $\mathcal{K}$ *is called* complete, *if the union of the simplices* $|\mathcal{K}| := \bigcup \sigma \in \mathcal{K}$ *equals the convex hull of* S. *A* subcomplex *of a complex is a subset, which itself is also a complex.*

Let σ *be an* $(n-1)$-simplex *with corner points* $s_1, ..., s_n \subset S$ *for a finite point set* $S \in \mathbb{R}^2$. *Then the smallest circle containing all corner points* $s_1, ..., s_n$ *is called the* circumcircle *of the simplex. The interior, the center and the radius of the circle are called the* circumdisc, *the* circumcenter *and the* circumradius *of the simplex. When talking of the size of a simplex we relate to the radius of its circumdisc, i.e. we call a simplex* σ *of circumradius* r_σ *to be smaller than a simplex* τ *with circumradius* r_τ, *if* $r_\sigma < r_\tau$. *A triangle is called* acute, *if it contains its circumcenter. An edge is called* Gabriel, *if its circumdisc contains no other point of* S.

Note, that the size of an edge is equal to half of its length and not the length itself. One example of a complete simplicial complex is the *Delaunay triangulation,* where the open circumdisc of any triangle is free of points of S. Now the result of a reconstruction can be defined in the following way:

Definition 3. *Let $\mathcal{K}$ be a complete simplicial complex based on a set of points S being embedded in $\mathbb{R}^2$. Then a simplicial complex partition $\mathcal{D}$ is a set of disjoint subsets D_i of $\mathcal{K}$ including the infinite region $D_0 := \mathbb{R}^2 \setminus |\mathcal{K}|$, such that the regions $|D_i|$ covered by the sets D_i define a plane partition $|\mathcal{D}| := \{|D_i|\}$. In case of $\mathcal{K}$ being a Delaunay triangulation, the subcomplex $\partial\mathcal{D} \subset \mathcal{D}$, $\partial\mathcal{D} := \mathcal{K} \setminus \bigcup_i D_i$ is called the* result of a reconstruction. *Then, $|\partial\mathcal{D}|$ is called the* reconstructed boundary, *and the pairwise disjoint components D_i* interiors of reconstructed regions. *For each D_i, the underlying area $|D_i|$ is the* reconstructed region. *An edge σ is called a* boundary edge *if at least two of its cofaces lie in different interiors of reconstructed regions. Given two regions $|D_i|$ and $|D_j|$, we call $|D_i|$ to be smaller than $|D_j|$, $|D_i| < |D_j|$, if the greatest simplex in D_i is smaller than the greatest simplex in D_j. Analogously we define $D_i < D_j$.*

In order to avoid degenerate cases of the Delaunay triangulation, the points in S are assumed to be *in general position*, which normally means that no three points are collinear and no four points are cocircular. In addition to that we assume that no two edges of the Delaunay triangulation have the same length. Note, that any set of points, which is not in general position, i.e. which does not fulfill all of these properties, can be brought into general position by displacing the points independently by at most some arbitrarily small $\varepsilon > 0$.

3   Feature Sizes

Some contour reconstruction algorithms base their proof of correctness on the so-called *local feature size (lfs)*, e.g. [1,5,6], by making demands on the local sampling density based on the lfs. The local feature size lfs(b) of a boundary point b is simply its shortest distance to the medial axis transform (MAT), which itself is defined as the locus of all circumcenters of maximal discs being empty of all boundary points. Since the local feature size is zero at non-smooth boundary points (like e.g. corners), all reconstruction algorithms which require a sampling density based on the lfs can only be applied to smooth contours, since they would need an infinite number of sampling points at non-smooth contour parts. Instead we introduce the *local region size (lrs)*, which directly relates on the *distance transform*:

Definition 4. *The* distance transform dist_B *of a set $B \subset \mathbb{R}^2$ is defined as $\text{dist}_B(x) = \min_{y \in B} ||x - y||$. The distance transform is called* continuous *if B is infinite and* discrete *otherwise.*

Definition 5. *Let $B \subset \mathbb{R}^2$ be a set of boundary points. Then the* local region size (lrs) *of a boundary point $b \in B$, $\text{lrs}(b)$ is defined as the smallest value $\text{dist}_B(x_i)$ for all local maxima x_i of the distance transform, which can be reached from b by a simple path on which the distance transform is strictly monotonic and which goes from b straight to the circumcenter of a maximal disc being empty of* *all boundary points and touching b, and then follows the MAT until reaching a maximum x_i.*

Note, that the lrs is strongly related to the watershed transform (see e.g. [10]): Taking the positions of the local maxima of $-\text{dist}_B$ as seeds, the combined boundary of the resulting regions covers B and the locally adjacent region with smallest corresponding maximum defines the lrs in a boundary point. Loosely speaking, the plane is partitioned into regions containing each exactly one maximum of the distance transform, the size of the regions is defined by the distance value of the corresponding maximum, and the lrs of a boundary point is defined as the size of the smallest adjacent region. Obviously, for any boundary point, the lrs is greater than or equal to the lfs, since at least one simple path passes the nearest point of the MAT.

The definition of the lrs solves several disadvantages of the lfs when being used for defining the required sampling density: it allows corners without becoming zero (which would imply infinitely dense sampling points). This implies, that sampling using the lfs can also not be applied to multi-region boundaries, since at any junction of three or more regions at least one region must have a corner. In addition to that, the distance to the nearest point of the MAT is strongly influenced by local distortions (e.g. by noise), while the value of the distance transform at its maxima can robustly be approximated from boundary samples - even if the sampling point positions are distorted by noise.

The lrs is not the first approach to define an alternative to the lfs which is not zero at corners. E.g. in [11] so-called *r-stable plane partitions* are used. A plane partition is *r*-stable, if the *weak feature size (wfs)* of the complement of the boundary is at least equal to r. The wfs is has been introduced in [4], and is defined as the smallest distance between a point of a given set and a critical point (i.e. a saddle or a maximum) of the distance transform outside the set. Obviously, the weak feature size does never exceed the lrs, since the smallest value of the lrs relates to the smallest local maximum of the distance transform, which is greater or equal to the smallest critical point. Moreover the wfs is globally defined and thus is useless if one wants to allow a locally adaptive sampling density. The wfs is always at least equal to the smallest value of the lfs, which is often called *reach* [7]. The reach of the boundary of a set strongly relates to the definition of *r-regular sets* as used in [2]: a set is *r*-regular, if the reach is at least r.

To sum up, we get the following hierarchy:

$$\forall b \in B : \text{reach}(B) = \min_{b' \in B} \text{lfs}(b') \leq \frac{\text{lfs}(b)}{\text{wfs}(\mathbb{R}^2 \setminus B)} \leq \text{lrs}(b).$$

Thus the local region size is always greater than or equal to the other feature size definitions.

4   The Sampling Criterion

If a region contains more than one maximum of the distance transform, our idea is to separate this region at first into different subregions, each containing exactly one maximum. In a second step we try to merge reconstructed subregions by using the fact, that the separating boundaries are free of sampling points. The separating contours between two such subregions directly correspond to saddles of the original distance transform (which all lie on the MAT), thus they occur at narrowings of the region like waists.

In literature one can find different approaches for defining a suitable sampling for reconstruction. E.g. in [1,5,6], an ε-sampling is defined as a set of sampling points on the boundary, such that every boundary point b has a sampling point in a distance of at most $\varepsilon \cdot \mathrm{lfs}(b)$. In [11], a (p, q)-sampling is defined as a set of sampling points on or near a boundary, such that every boundary point b has a sampling point in a distance of at most p and every sampling point is not more than q away from the boundary. Here we use an alternative sampling criterion, which is based on the local region size:

Definition 6. *Let ∂R be the boundary of a plane partion $\mathcal{R} = \{R_i \subset \mathbb{R}^2\}$ and $S \subset \mathbb{R}^2$ be a finite set of points, not necessarily lying on dSP. Then S is said to be a stable sampling of ∂R, if*

$$- \ \forall b \in \partial R : \exists s \in S : ||b - s|| < \tfrac{1}{2}\,\mathrm{lrs}(b),$$
$$- \ \forall s \in S : \exists b \in \partial R : ||b - s|| < \tfrac{1}{2}\,\mathrm{lrs}(b).$$

The lrs-dilation $\partial R^{\oplus}$ of the boundary ∂R is defined as the union of the open discs of radius $\tfrac{1}{2}\,\mathrm{lrs}(b)$ centered in b for all $b \in \partial R$.

Note, that according to the results of the last section, an ε-sampling for $\varepsilon \leq \tfrac{1}{2}$ is also a stable sampling, while the opposite is not necessarily true. Analogously, a (p, q)-sampling of an r-regular set or an r-stable partition with $p \leq \tfrac{r}{2}$ and $q \leq \tfrac{r}{2}$ is also a stable sampling, while the opposite is in general not true.

5 Acute Triangle Condition

Assuming the sampling of a plane partition to be *stable* (see definition 6) we ensure the correspondence between the regions of the original partition and the reconstructed regions. First, we show that the interior of each region contains at least one acute Delaunay triangle.

Lemma 1. *Let $R \in \mathcal{R}$ be a region. Let S be a stable sampling of the boundary ∂R, and let $\mathcal{K}$ be the Delaunay triangulation of S. Then there exists at least one acute triangle of $\mathcal{K}$ with its center lying inside R and its circumradius being greater than $\tfrac{1}{2}\max_{b \in \partial R} \mathrm{lrs}(b)$.*

Proof. Due to the definition of the local region size, there exists a point x inside R, where the continuous distance transform built on ∂R is greater than or equal to $\mathrm{lrs}(b)$ for all $b \in \partial R$. Let r be the value of the continuous distance transform in x. Thus, the distance of x to ∂R is r. Since $\forall b \in \partial R : r \geq \mathrm{lrs}(b)$, and since $\forall b \in \partial R : \exists s \in S : ||b - s|| < \tfrac{1}{2}\,\mathrm{lrs}(b)$, the value of the discrete distance transform built on S at the point x is strictly greater than $r - \tfrac{1}{2}\,\mathrm{lrs}(b) \geq \tfrac{1}{2}\,\mathrm{lrs}(b)$ for all $b \in \partial R$. By ascending the discrete distance transform starting in x one finally arives in a local maximum of the discrete distance transform with distance value greater than $\frac{1}{2}$ lrs(b). The traversed path can not intersect ∂R, since at any point $b' \in \partial R$, the discrete distance transform is smaller than $\frac{1}{2}$ lrs(b'). Thus the discrete distance transform admits a local maximum m inside R. This maximum is the circumcenter of an acute Delaunay triangle τ.

Definition 7. *Let x be point, which is a local maximum in the continuous distance transform. Further let y be the local maximum of the discrete distance transform, which will be reached from x by steepest ascend regarding the discrete distance transform. Then y is called the* associated discrete maximum *of x.*

Note, that for each continuous maximum there exists by definition exactly one discrete maximum, but this mapping is neither injective nor surjective: there may exist discrete maxima being associated to zero or more than one continuous maximum. The discrete maxima are exactly the circumcenters of the acute triangles of the Delaunay triangulation $\mathcal{K}$ based on a stable sampling of a plane partition. Thus by detecting them, we hit each region at least once. Lemma 1 guarantees that if two or more continuous maxima are associated to the same discrete maximum, all continuous maxima lie inside the same continuous region. Thus we can use the acute triangles as starting points for growing the discrete regions which at the end correspond to an oversegmentation of the continuous regions.

At the end we want to have a thin reconstructed boundary consisting of small edges, i.e. edges which fulfill everywhere the locally required sampling density. E.g. consider a point b on the original boundary having two sampling points as its nearest sampling points in exactly the same distance. Then the corresponding edge is a Delaunay edge. If the sampling is stable, the circumradius of this edge is smaller than $\frac{1}{2}$ lrs(b) (and thus its length is smaller than lrs(b)). The union of all such edges defines a thin boundary reconstruction separating the different regions from each other (but possibly containing additional small regions), which is called the *normalized mesh*, see [2]. Unfortunately in practice we do not know the exact position of the boundary, and thus we can not derive the normalized mesh by just knowing the sampling points. Note, that in case of smooth curves this is possible, as shown in [2], but their solution fails in case of non-smooth boundaries having e.g. corners.

Therefore our aim is to get a boundary reconstruction with its edges being as small as possible. In addition to that, given a plane partition $\mathcal{R}$ with regions R_i, the simplicial complex partition $\mathcal{D}$ with regions D_i should fulfill the following property: every two local maxima of the continuous distance transform, which lie inside different continuous regions R_{i_1}, R_{i_2}, are associated to discrete maxima which lie also inside different reconstructed regions $D_{i'_1}, D_{i'_2}$. We call such a simplicial complex partition a *refinement* of $\mathcal{R}$:

Definition 8. *Given a plane partition $\mathcal{R}$ and a stable sampling S. Then a discrete complex partition $\mathcal{D}$ is called a refinement of $\mathcal{R}$, if it fulfills the following property:*

– Let x_1, x_2 be two local maxima of the continuous distance transform lying inside different regions R_{i_1}, R_{i_2} of $\mathcal{R}$. Then the discrete maxima y_1, y_2 being associated to x_1, x_2 lie in different reconstructed regions $D_{i'_1}, D_{i'_2}$ of $\mathcal{D}$.

Two refinements $\mathcal{D}, \mathcal{D}'$ are called compatible, if there exists a one-to-one mapping of their reconstructed regions, such that every reconstructed region is mapped onto a reconstructed region containing both local maxima of the discrete distance transform which are associated to exactly the same local maxima of the continuous distance transform. A refinement is called elementary if every reconstructed region contains exactly one local maximum of the discrete distance transform.

For compatible refinements $\mathcal{D}, \mathcal{D}'$ we define an ordering relation $<$ in the following way: Let $T := (\tau_1, \tau_2, \ldots \tau_m)$ and $T' := (\tau'_1, \tau'_2, \ldots \tau'_n)$ be the lists of all boundary triangles of $\mathcal{D}$, resp. $\mathcal{D}'$, ordered descending by their size. Then we say $\mathcal{D}$ is smaller than $\mathcal{D}'$, $\mathcal{D} < \mathcal{D}'$ if $T < T'$ regarding lexicographic order. A refinement $\mathcal{D}$ is called minimal, if there exists no compatible refinement $\mathcal{D}'$, such that $\mathcal{D}' < \mathcal{D}$.

The idea of our algorithm is as follows: first we construct a minimal elementary refinement. Due to the general position assumption, different minimal elementary refinements ony differ in triangles which are no boundary triangles, i.e. which lie inside a reconstructed region and not on its boundary. In a second step, we merge neighbouring reconstructed regions if we can be sure that their maxima lie in the same continuous region. For merging neighbouring reconstructed regions we will use a reduction algorithm which guarantees that the result still remains minimal.

The result of our first step has to be a minimal elementary refinement. To achieve this we simply compute the Gabriel graph, which is the subcomplex of the Delaunay triangulation, which is built of the Gabriel edges. It was originally introduced in [8].

Definition 9. *Given a Delaunay triangulation $\mathcal{K}$, the* Gabriel graph *is the subcomplex consisting of all Gabriel edges of $\mathcal{K}$ and non further edge or triangle of $\mathcal{K}$. A* Gabriel path *is a sequence of triangles τ_i and edges ν_i, $(\tau_0, \nu_1, \tau_1, \ldots, \nu_n, \tau_n)$, such that τ_0 is an acute triangle, and each ν_i is face of both τ_i and τ_{i-1} and ν_i is not a Gabriel edge.*

Obviously any Gabriel path lies completely inside one of the regions being separated by the Gabriel graph, and the triangles on a Gabriel path are ordered by their size. As we show now, the Gabriel graph gives the minimum of all elementary refinements:

Lemma 2. *The Gabriel graph $\mathcal{K}G$ is the boundary of a minimal elementary refinement.*

Proof. Any local maximum of the discrete distance transform must be the circumcenter of an acute triangle. Due to the definition of the Gabriel graph, any edge, which is not Gabriel, is the longest edge of a not acute triangle. Thus by removing a not-Gabriel edge, one always connects a region (possibly covering already more than one triangle) with a not-acute triangle. Thus by induction, any resulting region covers at most one acute triangle. Since a not-Gabriel edge can not be the longest edge of two triangles at the same time due to the definition of the Delaunay triangulation, every reconstructed region covers exactly one acute triangle and thus contains exactly one local maximum of the discrete distance transform. Now let x_1, x_2 be two maxima of the continuous distance transform lying inside different continuous regions $R_1 \in \mathcal{R}$ and $R_2 \in \mathcal{R}$ and let y_1, y_2 be their associated maxima of the discrete distance transform. Due to Lemma 1 it follows that $y_1 \neq y_2$. Thus y_1 and y_2 are circumcenters of different acute triangles, which implies that they lie inside different reconstructed regions. Thus $\mathcal{K}_G$ is a refinement. Since each acute triangle defines a separate region, $\mathcal{K}_G$ is an elementary refinement.

Thus it remains to show that $\mathcal{K}_G$ is minimal. Let's now assume the opposite. Then it follows that there is a compatible refinement $\mathcal{K}' < \mathcal{K}_G$ which is minimal, ie.$T' < T$, where T, T' are the corresponding lists of boundary triangles σ_i' respectively σ_i ordered descending by their size. Let i be the position on which the first difference between T' and T appears: $\sigma_i' < \sigma_i$. Without loss of generality let σ_i be the boundary edge between two reconstructed regions $|D_1|', |D_2|'$ with their acute triangles τ_1, τ_2. Then σ_i is smaller than all edges lying on a Gabriel path from τ_1 or τ_2 to a coface of σ_1. Since every elementary refinement separates the acute triangles from each other, the edge σ_i' must lie on one such Gabriel path. Since the triangles of a Gabriel path are ordered by their size, this is in contradiction to the assumption $\sigma_i' < \sigma_i$. Thus $\mathcal{K}_A$ is a minimal elementary refinement.

6   Undersampled Merge

In this section we are going to discuss the possibility of merging the Gabriel regions. The Gabriel graph gives a minimal elementary refinement. However, this refinement is an oversegmentation of the original plane partition $\mathcal{R}$, since the maxima of several reconstructed regions may lie inside the same original region. We now want to reduce the amount of oversegmentation without destroying the property of being a minimal refinement. For detecting boundary edges which can be removed with the partition still remaining a minimal refinement, we can use the fact that the sampling is stable. Therefore we at first show that the normalized mesh (see [2]) gives a refinement:

Theorem 1. *Given a stable sampling S of a plane partition $\mathcal{R}$, the normalized mesh is the boundary of a refinement of $\mathcal{R}$.*

Proof. The normalized mesh consists of all edges of the Delaunay triangulation, whose dual Voronoi edge intersects the original boundary (see [2]). Thus there exists for any of its edges a boundary point $b \in \partial\mathcal{R}$ having the edge end points as its nearest sampling points in exactly the same distance. It follows that the size of the edge is smaller than $\frac{1}{2}$ lrs(b) (and its length is smaller than lrs(b)), which implies that any point on this edge has a distance of strictly less than lrs(b) to the boundary $\partial \mathcal{R}$. Thus the normalized mesh is covered by the lrs-dilation $\partial \mathcal{R}^\oplus$ of $\partial \mathcal{R}$. Now let p_1, p_2 be two intersection points of $\partial \mathcal{R}$ with Voronoi edges, such that there exists a path in dSP between them which goes through no other such intersection point. Then they lie on the boundary of a common Voronoi region and thus the two Delaunay edges being dual to the Voronoi edges meet in a common sampling point. It follows that any two points being separated from each other by $\partial \mathcal{R}^\oplus$ are also separated from each other by the edges of the normalized mesh. Thus, since the local maxima of the continuous distance transform are not covered by $\partial \mathcal{R}^\oplus$, any path from a continuous maximum to another must intersect the normalized mesh. Thus it separates continuous maxima lying in different regions of $\mathcal{R}$. For each local maximum of the continuous distance transform x with continuous distance value r, the associated local maximum y of the discrete distance transform can be reached by a path with increasing discrete distance value. Since the value of the discrete distance transform at x is at least $\frac{1}{2}r$, the path can not intersect R. Thus y lies inside the same reconstructed region as x.

Now let R_i be a region of a plane partition $\mathcal{R}$ with stable sampling S and let r be the distance value of the greatest local maximum of the continuous distance transform inside R_i (i.e. the maximal value of the local region size of all points on the boundary of R_i). Then due to Theorem 1, we know that there exists a refinement such that the length of every boundary edge of R_i is smaller than r (i.e. its size is smaller than $\frac{1}{2}r$). Obviously, this must also be true for a compatible minimal refinement. So, edges with circumradius greater than $\frac{1}{2}r$ (i.e. edges being longer than r) can not contribute to separation of two different ground truth regions, i.e. they are undersampled.

Obviously, the lrs can not be known during the reconstruction process. But the Gabriel graph speparates the plane into regions, which have a special property: the interior of each region contains only one acute triangle of the Delaunay triangulation and its circumradius is the greatest of all contained triangles. We use this circumradius to estimate the lrs and then remove undersampled edges.

Definition 10. *Let σ be a boundary edge between two neighbouring reconstructed regions D_i, D_j, and let r_σ denote its circumradius. Further let $\tau_i \in D_i$ and $\tau_j \in D_j$ be the acute triangles with greatest circumradius r_{τ_i} and r_{τ_j} respectively in the interiors of the reconstructed regions. Then σ is called an* undersampled *simplex if $2r_\sigma \geq r_{\tau_i}$ or $2r_\sigma \geq r_{\tau_j}$.*

In the following we prove that the undersampled edges can be removed and the resulting simplicial complex partition is still a minimal refinement.

Definition 11. *The* undersampled merge *is a processing step which deletes a simplex τ fulfilling the following properties: there exists a reconstructed region $|D|'$ with at least one undersampled simplex in its boundary, and τ is the greatest boundary simplex of $|D|'$. The* top undersampled merge *is an undersampled merge of the greatest of all simplices fulfilling the properties.*

Applying the undersampled merge iteratively, always terminates after a finite number of steps, since only a finite number of simplices is given.

Theorem 2. *Let $\mathcal{D}$ be a minimal refinement of the plane partition $\mathcal{R}$ being built by applying a finite number of top undersampled merge to the minimal elementary refinement given by the Gabriel graph based on a stable sampling. Then top undersampled merge on $\mathcal{D}$ results again in a minimal refinement $\mathcal{D}'$.*

Proof. Let $\sigma \in \partial\mathcal{D}$ be the edge to be deleted due to the top undersampled merge and let $|D|, |D|'$ be the adjacent reconstructed regions, with $|D|$ being the one for which σ is undersampled. Then we have to show that all local maxima of the continuous distance transform which are associated to maxima of the discrete distance transform inside $|D|$ or $|D|'$, lie inside the same continuous region.

If one of the reconstructed regions does not contain any local maximum of the discrete distance transform which is associated to a local maximum of the continuous distance transform, the top undersampled merge obviously results in a refinement. Thus suppose there exist two local maxima x, x' of the continuous distance transform having distance values r, r' and having corresponding local maxima of the discrete distance transform y, y' inside $|D|$ respectively $|D|'$. Without loss of generality let x, x' be the greatest of such local maxima regarding $|D|$ respectively $|D|'$. Then the value of the discrete distance transform at x and x' is at least $\frac{1}{2}r$ respectively $\frac{1}{2}r'$. By applying the construction used in Lemma 1 we can go from x respectively x' on an ascending path regarding the discrete distance transform to y respectively y'. By construction, the value of the discrete distance transform at y and y' is greater than $\frac{1}{2}r$ respectively $\frac{1}{2}r'$. Since all previously deleted edges are greater than σ, there exists a path between y and y' through the circumcenter c of σ, which is everywhere greater of equal to the discrete distance value in c. Putting together we get a path from x to x' such that at any point of the path the discrete distance transform is at least equal to the minimum of the discrete distance value in x, x' and c. Since c is undersampled, the discrete distance value at c is at least r'. This implies that the values of the continuous distance transform along the path are always greater than zero. This implies, that this path does not hit the boundary $\partial\mathcal{R}$ and all maxima lie inside the same continuous region. Thus the result $\mathcal{D}'$ is still a refinement. This refinement is even minimal, because if otherwise D' would not be minimal, then $\mathcal{D}$ could also not be minimal.

Unfortunately top undersampled merge deletes not always all of the inappropriate simplices, since there is always a bias due to the overestimation. But the following is valid: the denser is the sampling, the less is the amount of overestimation and consequently, the more accurate is the reconstruction. It follows the refinement reduction algorithm, which iteratively applies the top undersampled merge step:

The refinement reduction *algorithm*

1. *Given a locally stable sampling S of the boundary of some plane partition, compute the Gabriel graph and its corresponding simplicial complex partition.*

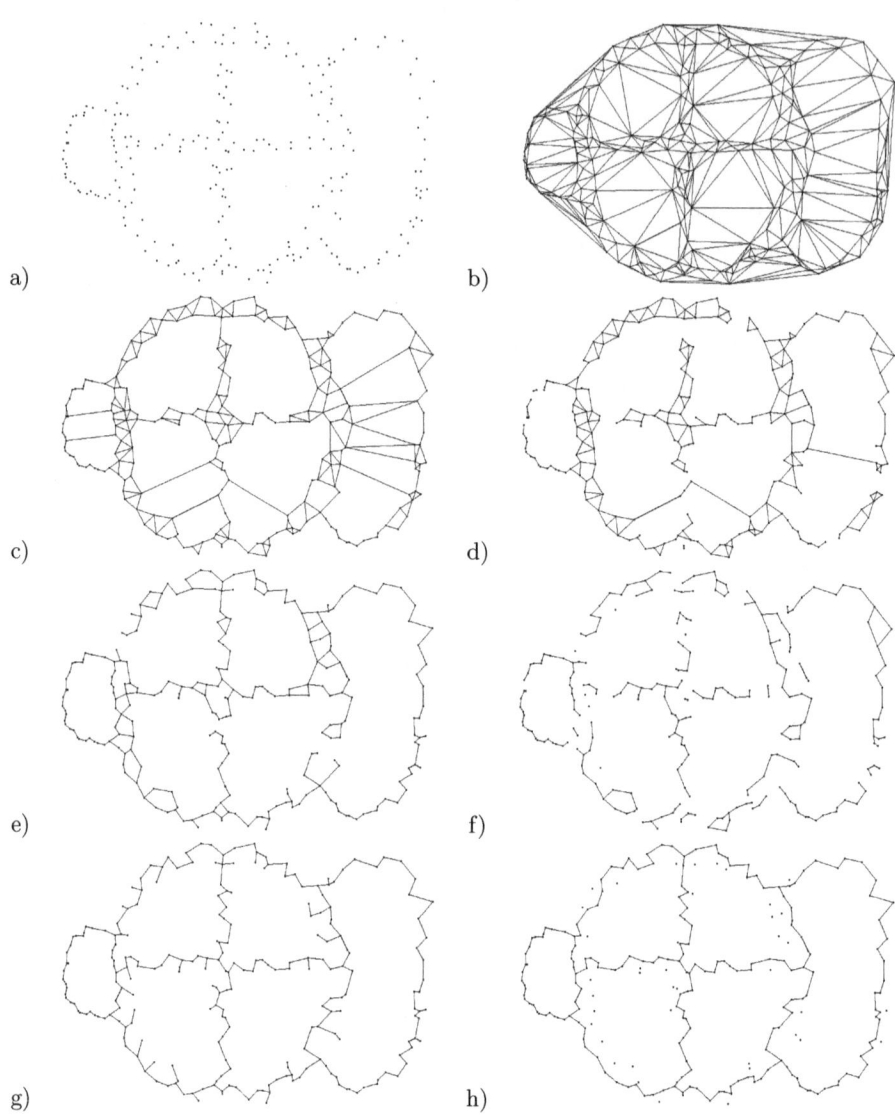

Fig. 1. Comparison of different reconstruction algorithms. a) original noisy point set. b) Delaunay triangulation. c) Gabriel graph. d) Reconstruction based on [2]. e) Reconstruction based on [9]. f) Crust algorithm [1]. g) Our refinement reduction algorithm. h) Our algorithm after additional removal of all edges which do not separate two different regions. Our algorithm is the only one, which reconstructs the correct number of regions.

2. *Find for all reconstructed regions all undersampled edges and put them in a priority queue (Sorted lexicographically according to pairs $r_\tau, -r_\sigma$ in increasing order, where r_σ is the cirumradius of the undersampled edge σ, τ is the triangle with the greatest cirumcenter r_τ in the reconstructed region.*
3. *As long as the queue is not empty:*
   (a) Pop the first edge σ from the queue
   (b) If σ is still undersampled, remove σ from $\mathcal{K}_A$.

The refinement reduction algorithm starts with the smallest of all reconstructed regions having an undersampled edge and deletes its greatest boundary simplex. So in each step a region is merged with a greater one. In such a way all undersampled edges of the greater region stay undersampled, while the previously undersampled simplices of the smaller region not necessarily remain undersampled.

The algorithm complexity is dominated by the Gabriel graph construction, which takes $O(n \log n)$ time in the number of sampling points. The greatest circumcenters in the regions can be found on the fly when constructing the Gabriel graph. Sorting the queue due to the priority values also takes $O(n \log n)$ time. Thus the overall complexity of the refinement reduction algorithm is $O(n \log n)$. Interestingly, the restriction to *top* undersampled merge is not necessary; we can apply undersampled merge in any order - the result is always exactly the same:

Lemma 3. *Let $\hat{D}_C(S)$ be a refinement of the plane partition. Then (under the assumption of general position) applying iteratively an arbitrary undersampled merge on $\hat{D}_C(S)$ until convergence results in exactly the same plane partition for any order of undersampled merge steps.*

Proof. Let τ_1, τ_2 be two different arbitrary simplices fulfilling the properties for undersampled merge, i.e. for both there exists a reconstructed region R'_1, resp. R'_2 with at least one undersampled edge in its boundary, and τ_1 resp. τ_2 is the greatest boundary edge of R'_1 resp. R'_2, and there exists a second region R''_1, resp. R''_2, which lies on the other side of the regarded edge. Obviously, R'_1 can not be equal R'_2. Now we remove τ_2.

If $R'_1 = R''_2$, τ_2 (and all other boundary edges of R'_2) must be smaller than τ_1, thus τ_1 is still the greatest edge of the new region $R'_1 \cup R'_2$. In addition to that, τ_1 is still undersampled in this new region, since it is greater than τ_2. It follows that τ_1 still fulfills the properties for undersampled merge.

Thus, assume $R'_1 \neq R''_2$. Then the undersampled merge of τ_2 does not affect the region R'_1 at all and τ_1 still fulfills the properties for undersampled merge.

Thus every edge which fulfills the properties for undersampled merge will never loose them during the iteration and eventually it will be removed.

The only remaining case are edges, which do not fulfill the properties at the beginning, but will get these properties after some undersampled merge. This can only be the case if two regions R'_1 and R''_1 have both the same greatest undersampled edge in their border. Then the undersampling merge removes the greatest edge from both regions and it has to be tested, if the greatest of the remaining edges is also undersampled. So let again τ_1, τ_2 be two different edges fulfilling the properties for undersampled merge, i.e. for both there exists a reconstructed region R_1', resp. R_2' with at least one undersampled edge in its boundary, and τ_1 resp. τ_2 is the greatest boundary edge of R_1' resp. R_2', and there exists a second region R_1'', resp. R_2'', which lies on the other side of the regarded edge. But now let τ_1 be also the greatest edge of R_1'' and τ_3 the second greatest edge of R_1' and R_1''. Without loss of generality let τ_3 be inside R_1'. Now we remove τ_2.

If $R_1' = R_2''$, τ_2 (and all other boundary edge of R_2') must be smaller than τ_3, thus τ_3 is still the second greatest edge of the new region $R_1' \cup R_2'$. In addition to that, τ_3 is still undersampled in this new region, since it is greater than the undersampled τ_2. It follows that τ_3 still will get the properties after undersampled merge of τ_1.

Thus, lets assume, $R_1'' = R_2''$. Then τ_2 (and all other boundary edges of R_2') must be smaller than τ_3, thus τ_3 is still greater than every edge in the new region $R_1' \cup R_2'$ except of τ_1. It again follows that τ_3 still will get the properties after undersampled merge of τ_1.

Thus, lets assume, $R_1' \neq R_2''$ and $R_1'' \neq R_2''$. Then the undersampled merge of τ_2 does not affect the regions R_1' and R_1'' at all and τ_3 again will get the properties after undersampled merge of τ_1.

Thus the set of edges which will be removed by applying undersampled merge iteratively until convergence is not affected by the order of the undersampled merge steps. Consequently, the algorithm terminates always with the same result.

Corollary 1. *Let $\mathcal{D}$ be a minimal refinement of the plane partition $\mathcal{R}$ being built by applying a finite number of undersampled merge to the minimal elementary refinement given by the Gabriel graph based on a stable sampling. Then undersampled merge on $\mathcal{D}$ results again in a minimal refinement $\mathcal{D}'$.*

This makes it possible to increase the speed of the algorithm further by using parallelization, if the Delaunay triangulation is already given. We compared the output of our algorithm experimentally with different algorithms for boundary reconstruction, namely the reconstruction algorithm of Attali given in [2], the Crust algorithm [1] and the heuristic approach given in [9]. Our experiments showed that the algorithm of Attali is very sensitive even to small noise, while the other two algorithms perform quite well in case of moderade noise. But the Crust algorithm fails in case of a significant amount of noise. All tested algorithms perform worse than our algorithm in case of multiple regions and a significant amount of noise, as illustrated in Figure 1.

7   Conclusion

We introduced an alternative to the concepts of local feature size and weak feature size. We used this to derive a new algorithm being able to reconstruct noisy sampled non-smooth 2D multi-region partitions. Our algorithm is parameter-free and can thus be applied to applications, where nothing is known about the real plane partition. We proved that the Gabriel graph is minimal under all graphs which separate the acute triangles from each other and showed that our algorithm also results in a minimal separation of the detected regions. As far as we know, this is the first attempt to solve the contour reconstruction problem for multiple-region-boundaries by using locally adaptive sampling criteria.

References

1. Amenta, N., Bern, M., Eppstein, D.: The crust and the β-skeleton: Combinatorial curve reconstruction. Graph. Models and Image Proc. 60(2), 125–135 (1998)
2. Attali, D.: r-Regular shape reconstruction from unorganized points. In: Proceedings of the 13th annual ACM Symposium on Comput. Geom., pp. 248–253 (1997)
3. Bernardini, F., Bajaj, C.L.: Sampling and reconstructing manifolds using alpha-shapes. In: Proc. 9th Canad. Conf. Comput. Geom. (1997)
4. Chazal, F., Lieutier, A.: Weak feature size and persistent homology: computing homology of solids in Rn from noisy data samples. In: Proc. of the 21st annual Symposium on Computational Geometry, pp. 255–262 (2005)
5. Dey, T.K., Kumar, P.: A simple provable algorithm for curve reconstruction. In: Proceedings of the 10th annual ACM-SIAM Symposium on Discr. Algorithms, Society for Industrial and Applied Mathematics, Philadelphia, PA, USA, pp. 893–894 (1999)
6. Dey, T.K., Mehlhorn, K., Ramos, E.A.: Curve reconstruction: Connecting dots with good reason. In: Proceedings of the 15th annual symposium on Computational geometry, pp. 197–206. ACM, New York (1999)
7. Federer, H.: Curvature measures. Transactions of the American Mathematical Society 93, 418–491 (1959)
8. Gabriel, K.R., Sokal, R.R.: A new statistical approach to geographic variation analysis. Systematic Zoology, 259–278 (1969)
9. Mukhopadhyay, A., Das, A.: An RNG-based heuristic for curve reconstruction. In: Voronoi Diagrams in Science and Engineering, 3rd International Symposium ISVD 2006, pp. 246–251 (2006)
10. Roerdink, J.B.T.M., Meijster, A.: The Watershed Transform: Definitions, Algorithms and Parallelization Strategies. Fundam. Inform. 41(1-2), 187–228 (2000)
11. Stelldinger, P., Köthe, U., Meine, H.: Topologically Correct Image Segmentation Using Alpha Shapes. In: Kuba, A., Nyúl, L.G., Palágyi, K. (eds.) DGCI 2006. LNCS, vol. 4245, pp. 542–554. Springer, Heidelberg (2006)

Reconstruction of Canonical hv-Convex Discrete Sets from Horizontal and Vertical Projections

Péter Balázs*

Department of Image Processing and Computer Graphics
University of Szeged
Árpád tér 2, H-6720 Szeged, Hungary
pbalazs@inf.u-szeged.hu

Abstract. The problem of reconstructing some special hv-convex discrete sets from their two orthogonal projections is considered. In general, the problem is known to be NP-hard, but it is solvable in polynomial time if the discrete set to be reconstructed is also 8-connected. In this paper, we define an intermediate class – the class of hv-convex canonical discrete sets – and give a constructive proof that the above problem remains computationally tractable for this class, too. We also discuss some further theoretical consequences and present experimental results as well.

Keywords: discrete tomography; hv-convex discrete set; reconstruction from projections; polynomial-time reconstruction.

1 Introduction

One of the main tasks of *Discrete Tomography* (DT) is to reconstruct discrete sets (finite subsets of the two-dimensional integer lattice) from few projections. Opposite to methods of Computerized Tomography – like filtered backprojection and algebraic reconstruction – which use several hundreds of projections [10], in DT just a few (typically less than ten) projections are available. For the basic algorithms and the wide area of applications of DT the reader is referred to [8,9]. The main problem arising from the very limited number of projections is that the reconstruction task is usually extremely underdetermined, i.e. there may be many different discrete sets with the same projections.

One way to reduce the number of possible solutions is to restrict the reconstruction to a class of discrete sets which satisfy some geometrical properties, such as connectivity and convexity. In the past 20-25 years many reconstruction algorithms have been developed for different classes of discrete sets, and also some strong theoretical results concerning the complexity of the reconstruction in those classes have been presented. Concerning the reconstruction from the horizontal and vertical projections with additional geometrical constrains the first

---

* This research was partially supported by the TÁMOP-4.2.2/08/1/2008-0008 program of the Hungarian National Development Agency and by the János Bolyai Research Scholarship of the Hungarian Academy of Sciences.

P. Wiederhold and R.P. Barneva (Eds.): IWCIA 2009, LNCS 5852, pp. 280–288, 2009.

approach was presented in [11] where the author gave a reconstruction heuristic for the class of horizontally and vertically convex (shortly, *hv*-convex) discrete sets using only two projections. Later, it was shown that this reconstruction task is NP-hard [13]. However, by this time it was known that assuming that the set to be reconstructed is also connected makes polynomial-time reconstruction possible [4,5,6,12]. The aim of this paper is to introduce an intermediate class between the classes of general and connected *hv*-convex discrete sets, and study the computational complexity of the reconstruction in that class.

The structure of the contribution is the following. First, the necessary definitions are introduced in Section 2. In Section 3 we supply a polynomial time algorithm to reconstruct canonical *hv*-convex discrete sets and do some further theoretical observations. In Section 4 we present experimental results concerning the average running time of the algorithm. Finally, Section 5 is for the conclusion.

2   Preliminaries

An arbitrary finite subset of the two-dimensional integer lattice defined up to translation is called *discrete set* and it can be represented by a binary image or binary matrix, too (see Fig. 1). In the following – depending on technical convenience – we will use both terms discrete set and binary matrix in the same sense. To avoid confusion, without loss of generality, we will assume that the vertical axis of the 2D integer lattice is directed top-down and the upper left corner of the smallest containing rectangle of a discrete set is the position $(1, 1)$.

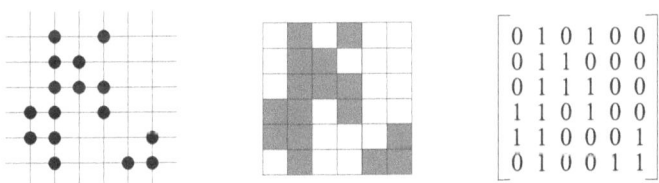

Fig. 1. A discrete set represented by its elements (*left*), a binary picture (*center*), and a binary matrix (*right*)

A discrete set F is *4-connected (8-connected)* if for any two distinct positions $P, Q \in F$ there exists a sequence of distinct positions $(i_0, j_0) = P, \ldots, (i_k, j_k) = Q$ such that $(i_l, j_l) \in F$ and $|i_l - i_{l+1}| + |j_l - j_{l+1}| = 1$ $(|i_l - i_{l+1}| + |j_l - j_{l+1}| \leq 2)$ for each $l = 0, \ldots, k - 1$. The 4-connected sets are also called *polyominoes*. If a discrete set is not 4-connected then it can be partitioned (in a uniquely determined way) into maximal 4-connected subsets which are called *components* of the discrete set. We say that the discrete set is *canonical (anticanonical)* if consists of just one component or the smallest containing rectangles of the components are connected to each other with their bottom-right and upper-left (bottom-left and upper-right) corners (see Fig. 2). A discrete set is called *hv-convex* if all the rows and columns of the set are 4-connected, i.e., the 1s of the corresponding representing matrix are consecutive in each row and column. For example, the discrete sets in Fig. 2 is *hv*-convex.

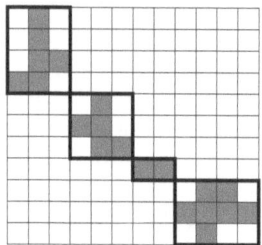 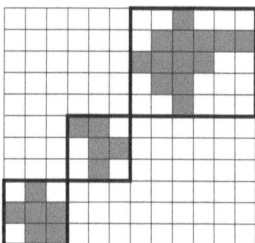

Fig. 2. A canonical *hv*-convex discrete set with 4 components (*left*) and an anticanonical *hv*-convex discrete set with 3 components (*right*). The smallest containing rectangles of the components are drawn bold.

The *size* of the discrete set is defined by the size of its minimal bounding rectangle. Given a discrete set F of size $m \times n$, its *horizontal* and *vertical projections* are defined by the vectors $\mathcal{H}(F) = H = (h_1, \ldots, h_m)$, and $\mathcal{V}(F) = V = (v_1, \ldots, v_n)$, respectively, where

$$h_i = \sum_{j=1}^{n} f_{ij}, \qquad i = 1, \ldots, m \ , \tag{1}$$

and

$$v_j = \sum_{i=1}^{m} f_{ij}, \qquad j = 1, \ldots, n \ . \tag{2}$$

For example, the horizontal and vertical projections of the discrete set in Fig. 1 are $H = (2, 2, 3, 3, 3, 3)$ and $V = (2, 6, 2, 3, 1, 2)$, respectively. Discrete sets having zero row or column sums are not of interest in our work. In this paper we are going to study the reconstruction of certain discrete sets from two projections. Given a class $\mathcal{G}$ of discrete sets and two vectors $H \in \mathbb{N}^m$, $V \in \mathbb{N}^n$ (for arbitrary m and n) this problem is the following

RECONSTRUCTION($\mathcal{G}, H, V$)
Instance Two non-negative vectors $H \in \mathbb{N}^m$ and $V \in \mathbb{N}^n$.
Task Construct a discrete set $F \in \mathcal{G}$ such that $\mathcal{H}(F) = H$ and $\mathcal{V}(F) = V$.

3   Reconstruction of Canonical *hv*-Convex Discrete Sets

We first recall a result which describes the special structure of *hv*-convex 8-connected discrete sets.

Lemma 1. *An hv-convex 8-connected discrete set is either canonical or anti-canonical.*

Proof. See Theorem 2 of [2]. □

Now, let us denote the class of hv-convex 4-connected, hv-convex 8-connected canonical, hv-convex canonical, and general hv-convex discrete sets by $\mathcal{HV}_4$, $\mathcal{HV}_8^c$, $\mathcal{HV}^c$, and $\mathcal{HV}$, respectively. On the basis of Lemma 1 the class $\mathcal{HV}_8^c$ is non-empty. Moreover, based on the definitions it is obvious that $\mathcal{HV}_4 \subset \mathcal{HV}_8^c \subset \mathcal{HV}^c \subset \mathcal{HV}$ (see Fig. 3 for the proper inclusions).

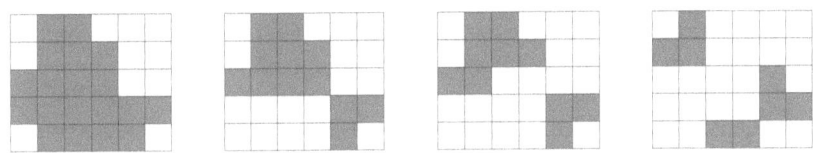

Fig. 3. An hv-convex 4-connected, an hv-convex 8-connected canonical, an hv-convex canonical, and a general hv-convex discrete set (from left to right, respectively)

Several algorithms have been proposed for solving the reconstruction problem in the class $\mathcal{HV}_4$ of hv-convex polyominoes in polynomial time using the horizontal and vertical projections, among them the fastest one has a worst case time complexity of $O(mn \cdot \min\{m^2, n^2\})$ [3,4,6]. Later, this result was extended to the class of (canonical) hv-convex 8-connected discrete sets, too [5,12]. Surprisingly, it also turned out that the reconstruction in the class of hv-convex 8- but not 4-connected discrete sets (with our notation in the class $\mathcal{HV}_8^c \setminus \mathcal{HV}_4$) can be performed faster, namely in $O(mn \cdot \min\{m, n\})$ time [2]. On the other hand, it was proven that the general problem RECONSTRUCTION($\mathcal{HV}, H, V$) is NP-hard [13]. Since the class $\mathcal{HV}^c$ of hv-convex canonical discrete sets forms an extension of the class $\mathcal{HV}_8^c$ and – in the same time – a restriction of the class $\mathcal{HV}$, it is an important and fascinating question whether the reconstruction complexity in this intermediate class is polynomial. In the followings we give a positive answer to this question.

Consider that the two vectors $H \in \mathbb{N}^m$ and $V \in \mathbb{N}^n$ are given and we want to reconstruct a canonical hv-convex discrete set with those projections. For each $1 \leq i_1 \leq i_2 \leq m$ and $1 \leq j_1 \leq j_2 \leq n$ let us define a boolean variable A_{i_1,i_2,j_1,j_2} such that $A_{i_1,i_2,j_1,j_2} = true$ if and only if there exists an hv-convex polyomino P with the minimal bounding rectangle $[i_1, i_2] \times [j_1, j_2]$ such that $\mathcal{H}(P) = (h_{i_1}, \ldots, h_{i_2})$ and $\mathcal{V}(P) = (v_{j_1}, \ldots, v_{j_2})$. In addition, for each $1 \leq i \leq m$ and $1 \leq j \leq n$ let $B_{i,j}$ be a boolean variable such that $B_{i,j} = true$ if and only if there exists a canonical hv-convex discrete set S with the minimal bounding rectangle $[1, i] \times [1, j]$ such that $\mathcal{H}(S) = (h_1, \ldots, h_i)$ and $\mathcal{V}(S) = (v_1, \ldots, v_j)$. In order to decide whether there exists a discrete set of $\mathcal{HV}^c$ with the given projections we have to calculate the value of $B_{m,n}$. For this we need the following

Lemma 2. *The variables $B_{i,j}$ satisfy the following recurrence*

(i) $B_{1,j} = A_{1,1,1,j}$ *for each* $1 \leq j \leq n$,

(ii) $B_{i,1} = A_{1,i,1,1}$ *for each* $1 \leq i \leq m$, *and*
(iii) $B_{i,j} = A_{1,i,1,j} \vee \bigvee_{1<i'<i,1<j'<j} A_{i',i,j',j} \wedge B_{i'-1,j'-1}$ $(1 < i \leq m; 1 < j \leq n)$.

Proof. The components of a (canonical) hv-convex discrete set are hv-convex polyominoes. If the minimal bounding rectangle of an hv-convex discrete set F consists of just one row or column then F necessarily has just a single component, i.e. it is an hv-convex polyomino. Thus, (i) and (ii) hold. Now, suppose that $1 < i \leq m$ and $1 < j \leq n$ and we have to evaluate $B_{i,j}$. The structure of a canonical hv-convex discrete set S in the minimal bounding rectangle $[1,i] \times [1,j]$ having projections $\mathcal{H}(S) = (h_1, \ldots, h_i)$ and $\mathcal{V}(S) = (v_1, \ldots, v_j)$ can only be of two types. It may happen that it consists of a single hv-convex component with the minimal bounding rectangle $[1,i] \times [1,j]$ and the proper projections (first term of the right hand side of (iii)). If not so, then it has an hv-convex polyomino in the bottom-right corner having a minimal bounding rectangle $[i',i] \times [j',j]$ for certain $1 < i' < i$ and $1 < j' < j$ and the rest of the set is a canonical hv-convex discrete set in the minimal bounding rectangle $[1,i'-1] \times [1,j'-1]$ with the corresponding projections, which proves the second term of the right hand side of (iii)) (see also Fig. 2, again). □

Based on this observation we now can outline the reconstruction algorithm for the class $\mathcal{HV}^c$. The algorithm uses an additional two-dimensional array L for storing links to one of the *true* conjunction terms of formula (iii) (if exists) for each $B_{i,j}$ in order to make it possible to identify the smallest bounding rectangles of the components.

Algorithm REC-HVC for reconstructing canonical hv-convex discrete sets.
Input: Two vectors $H \in \mathbb{N}^m$ and $V \in \mathbb{N}^n$.
Output: A discrete set $F \in \mathcal{HV}^c$ such that $\mathcal{H}(F) = H$ and $\mathcal{V}(F) = V$ or the message "NO SOLUTION".

Step 1 for each $1 \leq i_1 \leq i_2 \leq m$ and $1 \leq j_1 \leq j_2 \leq n$ calculate A_{i_1,i_2,j_1,j_2};
Step 2 for each $1 \leq i \leq m$ and $1 \leq j \leq n$
          { calculate $B_{i,j}$;
            if $(B_{i,j} = true)$ **then** link $L_{i,j}$ to one of $B_{i,j}$'s *true* conjunction terms;
                          else link $L_{i,j}$ to *null*; }
Step 3 if $(B_{m,n} = true)$ **then** output the solution using the link array L;
                          else output "NO SOLUTION";

The main result of the paper is the following

Theorem 1. *Algorithm   REC-HVC   solves   problem   RECONSTRUCTION* $(\mathcal{HV}^c, H, V)$ *in* $O(m^3 n^3 \cdot \min\{m^2, n^2\})$ *time with* $O(m^2 n^2)$ *memory require-ments.*

Proof. In Step 1 the algorithm checks for each pair $1 \leq i_1 \leq i_2 \leq m$ and $1 \leq j_1 \leq j_2 \leq n$ whether an hv-convex polyomino exists with the correspond-ing horizontal and vertical projections. The pairs (i_i, i_2) and (j_i, j_2) identify the smallest bounding rectangle of the polyomino. The number of such pairs is

$O(m^2n^2)$ and the values of all A's can be stored on $O(m^2n^2)$ bytes. For a certain pair it takes $O(mn \cdot \min\{m^2, n^2\})$ time to reconstruct an hv-convex polyomino from the given projections (if exist) (see for example [3] for the reconstruction algorithm). Thus, performing Step 1 takes $O(m^3n^3 \cdot \min\{m^2, n^2\})$ time. The recursive formula given in Lemma 2 for all $B_{i,j}$ can be calculated by a dynamic programming in $O(m^2n^2)$ time and stored on $O(mn)$ bytes. So the total space complexity of the algorithm follows. Based on the definition of the boolean variables used and the recursive relations of Lemma 2, there exist a solution of the given task if and only if $B_{m,n} = true$, and in this case – using the link array L – it is also possible to locate the smallest bounding rectangles of the components (otherwise the algorithm terminates with "NO SOLUTION"). Knowing those rectangles one can reconstruct each component independently. Since those components are hv-convex polyominoes this can be achieved in $O(mn \cdot \min\{m^2, n^2\})$ time (see again [3]). This does not affects the total execution time which is of $O(m^3n^3 \cdot \min\{m^2, n^2\})$. □

Some more restrictive forms of the above reconstruction problem can also be solved in polynomial time.

Theorem 2. *Given two vectors $H \in \mathbb{N}^m$ and $V \in \mathbb{N}^n$ it is possible to reconstruct a discrete set $F \in \mathcal{HV}^c$ (if exists) in polynomial time such that $\mathcal{H}(F) = H$ and $\mathcal{V}(F) = V$ and F has exactly k components for a prescribed $k \in \mathbb{N}$.*

Proof. We alter algorithm REC-HVC in the following way. For each $1 \le i \le m$, $1 \le j \le n$, and $1 \le c \le \min\{m, n\}$ let $B_{i,j}^c$ be a boolean variable such that $B_{i,j}^c = true$ if and only if there exists a canonical hv-convex discrete set S with the minimal bounding rectangle $[1, i] \times [1, j]$ such that $\mathcal{H}(S) = (h_1, \ldots, h_i)$ and $\mathcal{V}(S) = (v_1, \ldots, v_j)$ and S has exactly c components. Note, that a (canonical) hv-convex discrete set of size $m \times n$ can have at most $\min\{m, n\}$ components. Then, the task is to calculate $B_{i,j}^k$ for which – in a similar way to Lemma 2 the following recursion can be applied

(i) $B_{i,j}^1 = A_{1,i,1,j}$ for each $1 \le j \le n$ and $1 \le i \le m$,
(ii) for each $1 < i \le m$, $1 < j \le n$, and $1 < c \le \min\{m, n\}$

$$B_{i,j}^c = \begin{cases} \bigvee_{1 < i' < i, 1 < j' < j} A_{i',i,j',j} \wedge B_{i'-1,j'-1}^{c-1}, & \text{if } c \le \min\{i, j\} , \\ false, & \text{otherwise} . \end{cases} \quad (3)$$

Those variables can be calculated by a dynamic programming in $O(m^2n^2 \cdot \min\{m^2, n^2\})$ time. □

Corollary 1. *Given two vectors $H \in \mathbb{N}^m$ and $V \in \mathbb{N}^n$ it is possible to reconstruct a discrete set $F \in \mathcal{HV}^c$ (if exists) in polynomial time such that $\mathcal{H}(F) = H$ and $\mathcal{V}(F) = V$ and F has the maximal (minimal) number of components.*

Proof. Since the number of components can be at most $\min\{m, n\}$ one simply has to iterate Theorem 2 for $k = 1, \ldots, \min\{m, n\}$. □

We close with two further observations concerning the number of possible solutions. In [2] it was shown that the number of discrete sets in the class $\mathcal{HV}_8^c$ having the same horizontal and vertical projections can be at most $\min\{m,n\}$, i.e. the ambiguity of the reconstruction is always bounded by a polynomial. This no longer holds in the class $\mathcal{HV}^c$.

Theorem 3. *For some vectors H and V there can be exponentially many discrete sets in the class $\mathcal{HV}^c$ having H and V as horizontal and vertical projections, respectively.*

Proof. It follows from the fact that the number of hv-convex polyominoes having the same horizontal and vertical projections can be exponentially large [7] and any such polyomino can serve as a component of a canonical hv-convex discrete set. □

The following result complements Theorem 3 assuming that there is no ambiguity at component-level. Let $\mathcal{HV}^{c+}$ denote the class of canonical hv-convex discrete sets whose components are uniquely determined and consider the following problem

#CONSISTENCY($\mathcal{HV}^{c+}, H, V$)
Instance Two non-negative vectors $H \in \mathbb{N}^m$ and $V \in \mathbb{N}^n$.
Question How many discrete sets of $\mathcal{HV}^{c+}$ exists with the horizontal and vertical projections H and V, respectively?

Theorem 4. #CONSISTENCY($\mathcal{HV}^{c+}, H, V$) $\in \mathbf{P}$.

Proof. Algorithm REC-HVC can be modified in a straightforward way to solve this problem. Instead of storing one of the *true* conjunction terms (if exists) for each $B_{i,j}$, in the array L we store the number of such terms. □

4   Experimental Results

The worst case time complexity of Algorithm REC-HVC is quite big. In order to investigate the average running time of this algorithm, we generated hv-convex canonical discrete sets (having more than one components) from uniform random distributions by using an altered version of the method presented in [1]. In this way we obtained 100-100 canonical hv-convex discrete sets of size 10×10, 20×20, 30×30, 40×40, and 50×50. The algorithm were implemented in C++ and the test run on a PC with Intel(R) Core(TM)2 Duo CPU of 2x2.4 GHz and 2 GB RAM under SuSE Linux. We must also note that for technical reasons the code was not optimized. The running times for each data set are given in Table 1. From the entries of this table we can deduce that, fortunately, the average running time of the algorithm does not grow as rapidly as one can expect from the formula of the worst case time complexity.

Table 1. The average running time in seconds (rounded to two digits) of Algorithm REC-HVC as it depends on the size of the discrete set to be reconstructed

Size	Avg. time (s)
10×10	1.35
20×20	5.92
30×30	21.75
40×40	39.61
50×50	55.06

5   Conclusion

We introduced an intermediate subclass of hv-convex discrete sets that lies in be-tween the class of hv-convex 8-connected sets and the class of general hv-convex discrete sets. While in the former class the reconstruction from two projections can be solved in polynomial time, in the latter one this problem is known to be NP-hard. We showed that hv-convex canonical discrete sets can also be re-constructed from two projections in polynomial time, and they form one of the broadest known classes of discrete sets having this property. Thus, our work narrows the gap between problems of the complexity classes **P** and **NP**, from the viewpoint of discrete tomography. Beside giving a theoretical result for the worst case time complexity of the presented problem, we also conducted experi-ments to investigate the average running time, too. The results can be adapted to hv-convex anticanonical discrete sets in a straightforward way but further generalization seems to be difficult. However, we think that with a more effec-tive implementation of the algorithm and a deeper analysis of its behaviour we can gain some further important theoretical and experimental observations.

References

1. Balázs, P.: A framework for generating some discrete sets with disjoint components by using uniform distributions. Theor. Comput. Sci. 406, 15–23 (2008)
2. Balázs, P., Balogh, E., Kuba, A.: Reconstruction of 8-connected but not 4-connected hv-convex discrete sets. Discr. Appl. Math. 147, 149–168 (2005)
3. Balogh, E., Kuba, A., Dévényi, C., Del Lungo, A.: Comparison of algorithms for reconstructing hv-convex discrete sets. Lin. Alg. Appl. 339, 23–35 (2001)
4. Barcucci, E., Del Lungo, A., Nivat, M., Pinzani, R.: Reconstructing convex poly-ominoes from horizontal and vertical projections. Theor. Comput. Sci. 155, 321–347 (1996)
5. Brunetti, S., Del Lungo, A., Del Ristoro, F., Kuba, A., Nivat, M.: Reconstruction of 4- and 8-connected convex discrete sets from row and column projections. Lin. Algebra Appl. 339, 37–57 (2001)
6. Chrobak, M., Dürr, C.: Reconstructing hv-convex polyominoes from orthogonal projections. Inform. Process. Lett. 69, 283–289 (1999)

7. Del Lungo, A.: Polyominoes defined by two vectors. Theoret. Comput. Sci. 127, 187–198 (1994)
8. Herman, G.T., Kuba, A. (eds.): Discrete Tomography: Foundations, Algorithms and Applications. Birkhäuser, Boston (1999)
9. Herman, G.T., Kuba, A. (eds.): Advances in Discrete Tomography and Its Applications. Birkhäuser, Boston (2007)
10. Kak, A.C., Slaney, M.: Principles of Computerized Tomographic Imaging. IEEE Press, New York (1988)
11. Kuba, A.: The reconstruction of two-directionally connected binary patterns from their two orthogonal projections. Comp. Vision, Graphics, and Image Proc. 27, 249–265 (1984)
12. Kuba, A.: Reconstruction in different classes of 2D discrete sets. In: Bertrand, G., Couprie, M., Perroton, L. (eds.) DGCI 1999. LNCS, vol. 1568, pp. 153–163. Springer, Heidelberg (1999)
13. Woeginger, G.W.: The reconstruction of polyominoes from their orthogonal projections. Inform. Process. Lett. 77, 225–229 (2001)

About the Complexity of Timetables and 3-Dimensional Discrete Tomography: A Short Proof of NP-Hardness

Yan Gerard

Univ. Clermont 1, LAIC, Campus des Cézeaux, 63172 Aubière, France
gerard@laic.u-clermont1.fr

Abstract. We consider the problem of 3-dimensional Discrete Tomography according to three linearly independent directions. Consistency of this problem has been proved to be NP-compete by M. Irving and R.W. Jerrum in 1993 [9] but there exists since 1976 a very close result of NP-hardness in the framework of Timetables which is due to S. Even, A. Itai, and A. Shamir [2]. The purpose of this paper is to provide a new result of NP-hardness for a very restricted class of 3D Discrete Tomography which is common with Timetables. Hence NP-hardness of 3D Discrete Tomography and of Timetables both follow from this new stronger result that we obtain with a short proof based on a generic principle.

Keywords: Discrete Tomography, NP-complete, Timetable, Flow problems.

1 Introduction

1.1 Why Now and Not 8 Years Ago?

The starting point of this paper is a talk that I gave in December 2000 during a workshop organized by R.J. Gardner and P. Gritzmann at the Mathematisches Forschungsinstitut Oberwolfach. The subject of my presentation was the NP-hardness of a restricted class of 3-dimensional Discrete Tomography. Every member of the conference knew that the whole class was NP-hard -a result proved by M. Irving, R.W. Jerrum in 1993 [9]. The restricted class that I introduced was using a 3-dimensional lattice with a height reduced to 3 and one of the matrices of projections -or X-rays- with constant coefficients all equal to 1. I proved the NP-hardness of this subclass of problems during the session and at the end of my talk, as it can happen, a member of the workshop came to explain to me that this result was already known. The reference that he gave was the one of a well-known and important paper about Timetables [2] that result is cited in [7].

Back home, one of my duties was of course to read this paper and to determine whether there was one ounce of originality in my result: my answer was "no". The first conclusion was that my work was useless and the second one that the good reference for the NP-hardness of 3D Discrete Tomography was not [9] as

P. Wiederhold and R.P. Barneva (Eds.): IWCIA 2009, LNCS 5852, pp. 289–301, 2009.

everybody thought but the older reference about Timetables [2]. 8 years later, I had a discussion on the subject with Sara Brunetti. I explained to her my point of view but she was not satisfied by my explanations and she came to the conclusion that I was wrong. I must confess that she was right: In 2000, I red the paper about Timetables [2] with the idea that the problem was a kind of problem of Discrete Tomography and it is true that they are very close. They are very close BUT different! Hence the old result that I have rejected as already known is in fact stronger than the older results and my ambition in this paper is obviously to prove that it is true but also that it remains very interesting for several reasons that I will explain!

1.2   Relation between 3D Discrete Tomography and Polyatomic DT

Let us start by saying some words about Discrete Tomography: the investigation of this field has started in the fifties with the works of D. Gale [4] and H.J. Ryser [10]. They have proved that the reconstruction of a binary matrix having prescribed numbers of 1s in each row and column can be solved in polynomial time. This problem is the starting point of many questions ranging from the computation of the exact number of solutions to many multi-dimensional generalizations related with practical problems (see for instance [8] for an introduction to the field). The 3-dimensional generalization consists in constructing a three dimensional lattice set having prescribed numbers of points in the lines parallel with the axis (Fig. 1). Consistency of this problem is denoted 3DT.

Problem 1 (3DT). **Given:** three positive integers m, n, l and three positive integer matrices $(r_{i,k})_{1 \leq i \leq m, 1 \leq k \leq l}$, $(c_{j,k})_{1 \leq j \leq n, 1 \leq k \leq l}$ and $(v_{i,j})_{1 \leq i \leq m, 1 \leq j \leq n}$.

Question: existence of a lattice set of $[1..m] \times [1..n] \times [1..l]$ having for any $(i, j, k) \in [1..m] \times [1..n] \times [1..l]$ exactly $r_{i,k}$ points verifying $x = i$ and $z = k$, $c_{j,k}$ points verifying $y = j$ and $z = k$ and $v_{i,j}$ points verifying $x = i$ and $y = j$?

M. Irving, R.W. Jerrum have proved in 1993 that this class of problems is NP-hard [9]:

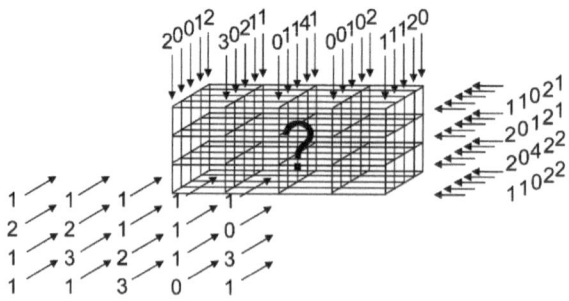

Fig. 1. An instance of 3DT

Theorem 1 (Irving, Jerrum). *3DT is NP-complete.*

The proof of this result is not direct in [9]. It requires a rather long work to obtain the NP-hardness of constrained latin square reconstruction as intermediary result. Some other results from which 3DT NP-hardness can be derived have been obtained more recently in the framework of Discrete Tomography. It is in particular the case of the polyatomic problem in 2D:

Problem 2 (a-atomic 2DT). **Given:** two positive integers m, n and two positive integer matrices $(r_{i,k})_{1 \leq i \leq m, 1 \leq k \leq a}$, $(c_{j,k})_{1 \leq j \leq n, 1 \leq k \leq a}$ (k is the index of the kind of atoms considered namely $k \in [1..a]$).

Question: existence of a disjoint lattice sets S_k of $[1..m] \times [1..n]$ with $k \in [1..a]$ such that for any $(i, j, k) \in [1..m] \times [1..n] \times [1..a]$, S_k has exactly $r_{i,k}$ points verifying $x = i$ and $c_{j,k}$ points verifying $y = j$?

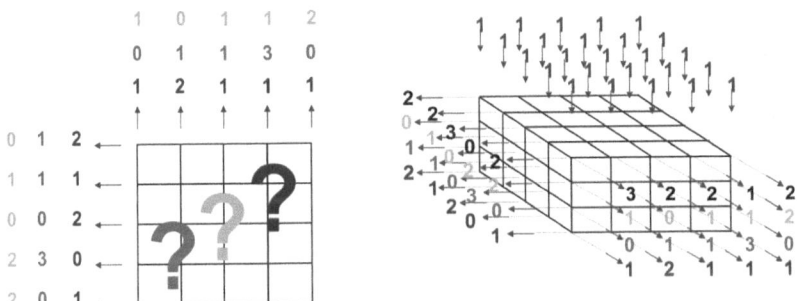

Fig. 2. An instance of a-atomic 2DT with $m = 5$, $n = 5$, $a = 3$ atoms (on the left) and its translation in an equivalent instance of 3DT with $m = 5$, $n = 5$, $l = a + 1 = 4$. Note that the two matrices $(r_{i,k})_{1 \leq i \leq m, 1 \leq k \leq a}$, $(c_{j,k})_{1 \leq j \leq n, 1 \leq k \leq a}$ of the 3-atomic 2DT instance have been completed by a line $r_{i,a+1} = n - \sum_{k=1}^{a} r_{i,k}$ and $c_{j,a+1} = m - \sum_{k=1}^{a} c_{j,k}$ while the disjunction condition on the sets S_k is given by the matrix $(v_{i,j})_{1 \leq i \leq m, 1 \leq j \leq n}$ that coefficients are all fixed to 1.

R.J. Gardner, Peter Gritzmann and D. Prangenberg have proved in [5] that a-atomic 2DT was NP-hard for $a \geq 6$. This result has been improved by M. Chrobak and C. Dürr in [1].

Theorem 2 (Chrobak, Dürr). *a-atomic 2DT is NP-complete for $a \geq 3$.*

As we know since a long time that the problem is polynomial with $a = 1$ (only one kind of atoms) [4,10], it remains only one unknown case : $a = 2$. The complexity of 2-atomic 2DT is one of the main open problems of complexity in DT.

About the consequences of Theorem 2 in the framework of 3DT, it is rather easy to see that any instance of a-atomic 2DT can be translated in an equivalent instance of 3DT with a 3D lattice of length m, width n and height $l = a + 1$ (Fig. 2). It follows that 3DT remains NP-hard even by fixing the height l with

$l \geq 4$. For $l = 1$, we are in the 2D case which means that the problem is polynomial [4,10]. Hence the question is: what is the complexity of 3DT if the height l is fixed equal to 2 or 3? The case $l = 2$ is considered in the next section: we prove that it can be solved in polynomial time. The case $l = 3$ is solved in Sect. 3, we reduce it to 3-SAT and prove that it is NP-complete. It is an original result in DT but also in the framework of Timetables.

1.3   Relation between 3D Discrete Tomography and Timetables

From [2], Theoretical Timetable problems can be formulated as follows:

Problem 3 (TT). **Given:** three positive integers m, n, l, two binary matrices $(r_{i,k})_{1\leq i\leq m,1\leq k\leq l}$, $(c_{j,k})_{1\leq j\leq n,1\leq k\leq l}$ and an integer matrix $(v_{i,j})_{1\leq i\leq m,1\leq j\leq n}$.

Question: existence of a lattice set S of $[1..m] \times [1..n] \times [1..l]$ such that for any $(i,k) \in [1..m] \times [1..l]$ its number of points verifying $x = i$ and $z = k$ is less than $r_{i,k}$ (namely $card((\{i\} \times [1..n] \times \{k\}) \cap S) \leq r_{i,k})$, for any $(j,k) \in [1..n] \times [1..l]$ its number of points verifying $y = j$ and $z = k$ is less than $c_{j,k}$ (namely $card(([1..m] \times \{j\} \times \{k\}) \cap S) \leq c_{j,k})$ and for any $(i,j) \in [1..m] \times [1..n]$ its number of points verifying $x = i$ and $y = j$ is exactly $v_{i,j}$ (namely $card((\{i\} \times \{j\} \times [1..l]) \cap S) = v_{i,j})$?

Problem TT has exactly the same input than 3DT (with a binary condition on two matrices but it is useless for our purpose). It differs mainly from 3DT by the constraints $card((\{i\}\times[1..n]\times\{k\})\cap S) \leq r_{i,k}$ and $card(([1..m]\times\{j\}\times\{k\})\cap S) \leq c_{j,k}$ which are inequalities while they are equalities in 3DT. We can however notice that the interesting instances of 3DT all verify a necessary condition of consistency (i):

- $\forall i \in [1..m]$, $\sum_{k=1}^{l} r_{i,k} = \sum_{j=1}^{n} v_{i,j}$,
- $\forall j \in [1..n]$, $\sum_{k=1}^{l} c_{j,k} = \sum_{i=l}^{m} v_{i,j}$,
- $\forall k \in [1..l]$, $\sum_{i=1}^{m} r_{i,k} = \sum_{j=1}^{n} c_{j,k}$.

If we consider now an instance of TT verifying this condition, the equalities $card((\{i\} \times \{j\} \times [1..l]) \cap S) = v_{i,j}$ implies that all inequalities $card((\{i\} \times [1..n] \times \{k\}) \cap S) \leq r_{i,k}$ and $card(([1..m] \times \{j\} \times \{k\}) \cap S) \leq c_{j,k}$ are in fact equalities. It means that the solutions of such an instance of TT are exactly the solutions of the equivalent instance of 3DT. Hence the two problems 3DT and TT differ only on instances which are obviously not consistent as 3DT instances. It leads to conclude that 3DT is in fact the subclass of problem TT with condition (i) (if we do not take into account the useless binary condition on matrices $r_{i,k}$ and $c_{j,k}$).

Hence it is not really surprising that the class of problems TT has been proved to be NP-hard more than one decade before 3DT. This result has been published in 1976 but the original paper [2] of this result contains more. It proves the NP-hardness of a very restricted subclass of TT with $l = 3$ which is very close to the problem of complexity that we plan to prove in Sect. 3. This subclass of TT is called Restricted-TT:

Problem 4 (Restricted-TT). Restricted-TT is the subclass of instances of TT where

- $l = 3$,
- $\forall (i, k) \in [1..m] \times [1..3]$, $r_{i,k} = 1$,
- Matrix $v_{i,j}$ is binary (all the coefficients of the input matrices are 0 or 1).
- $\forall j \in [1..n]$, $\sum_{k=1}^{l} c_{j,k} = \sum_{i=1}^{m} v_{i,j}$ and $\sum_{k=1}^{l} c_{j,k} \in \{2; 3\}$,

The strong result of complexity of [2] is:

Theorem 3 (Even, Itai, Shamir). *Restricted-TT is NP-complete.*

Notice that condition (i) is only partially considered in Restricted-TT which means that it is not a subclass of 3DT but we can complete it and hence obtain a subclass of instances of Restricted-TT which is also a subclass of 3DT instances (Fig. 3). We call it Restricted-3DT:

Problem 5 (Restricted-3DT). Restricted-3DT (Fig. 4) is the subclass of problems of 3DT where

- $l = 3$,
- the three input matrices are binary,
- $\forall (i, k) \in [1..m] \times [1..3]$, $r_{i,k} = 1$,
- $\forall i \in [1..m]$, $\sum_{k=1}^{l} r_{i,k} = \sum_{j=1}^{n} v_{i,j} = 3$,
- $\forall j \in [1..n]$, $\sum_{k=1}^{l} c_{j,k} = \sum_{i=l}^{m} v_{i,j} \in \{2; 3\}$,
- $\forall k \in [1..l]$, $\sum_{i=1}^{m} r_{i,k} = \sum_{j=1}^{n} c_{j,k}$.

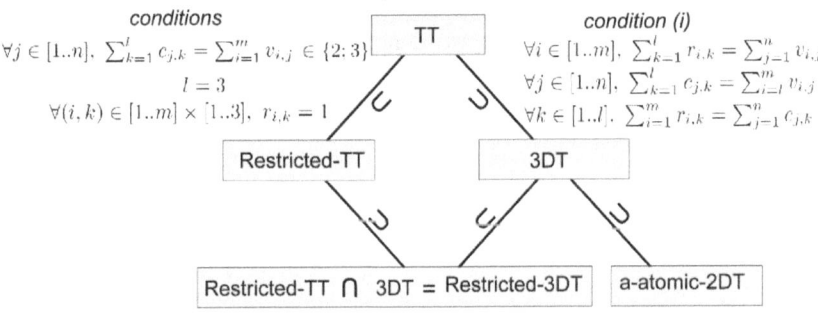

Fig. 3. Diagramm of inclusion of the class TT and its subclasses 3DT, Restricted-TT, Restricted-3DT of *a*-atomic 3DT (for convenience, we do not take into account the binary condition on matrices $r_{i,k}$ and $c_{j,k}$ in TT)

Hence Restricted-3DT is both a subclass of 3DT and of Restricted-TT (and therefore also of TT). By proving that it is NP-hard as we plan to do in Sect. 3, we obtain an original result and a proof of NP-hardness for all these problems. In order to avoid a last possible confusion between the 3D translation of a 2-atomic 2DT instance and Restricted-3DT, notice that the constant matrix of coefficients 1 is not the same in both problems (Fig. 4).

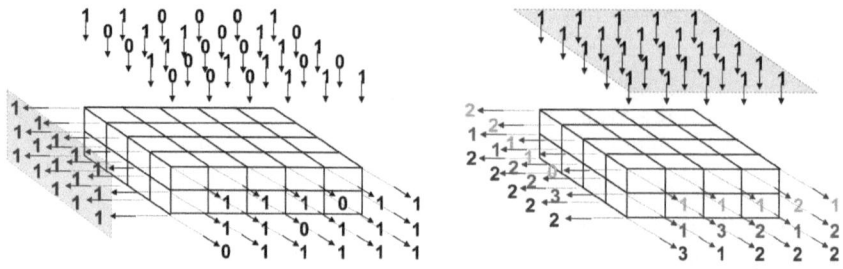

Fig. 4. An instance of Restricted-3DT (on the left) and a 3D translation of a 2-atomic 2DT instance (on the right). The constant matrix of 1 is placed differently for the two problems. It shows how Restricted-3DT and 2-atomic 2DT are different.

2   3DT with $l = 2$ Is in P

We consider in this section the problem 3DT with a height $l = 2$. It means that we have only two *levels*, one with $k = 1$, the other with $k = 2$. In this framework, we can assume that the values of coefficients $v_{i,j}$ belong to the set $\{0; 1; 2\}$ (v is the vertical projection on the 3D figures of the section) and that the condition (i) is satisfied because otherwise, the instance has no solution.

If a coefficient $v_{i,j}$ is equal to zero, there can not be any point in the positions $(i; j; 1)$ and $(i; j; 2)$ of the lattice while if $v_{i,j}$ is equal to 2, there are necessarily two points in these two positions (Fig. 5). These points which can be directly fixed have some projections that we denote $r_{i,k}^{fixed}$, $c_{j,k}^{fixed}$ and $v_{i,j}^{fixed}$ for any (i, j, k). We point out the fact that there is no ambiguity about these positions and that we can focus our attention on the undetermined positions.

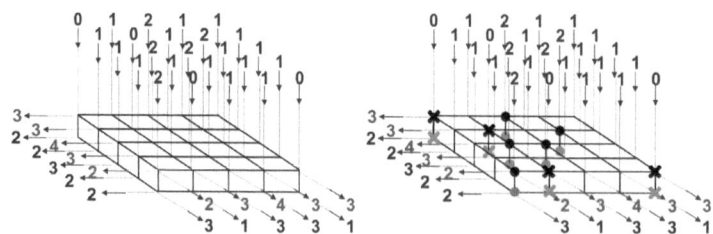

Fig. 5. An instance of 3DT with $l = 2$ (on the left). On the right, the crosses indicate the positions where there can be no point (due to the vertical projection equal to 0) while some other points are already fixed (due to the vertical projections equal to 2). The projections of the points already fixed are $(r_{i,1}^{fixed})_{1 \le i \le 5} = (r_{i,2}^{fixed})_{1 \le i \le 5} = (1; 1; 2; 1; 0)$ while $(c_{j,1}^{fixed})_{1 \le i \le 5} = (c_{j,2}^{fixed})_{1 \le i \le 5} = (1; 1; 2; 0; 1)$. If we consider now the uncertain positions, the projections of the set that it remains to determine to complete the solution should be equal to $r_{i,k} - r_{i,k}^{fixed}$ and $c_{j,k} - c_{j,k}^{fixed}$ (Fig. 6).

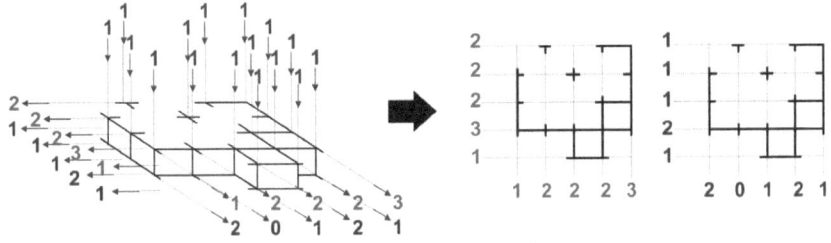

Fig. 6. The projections of the instance of Fig. 5 have been decreased and the determined positions removed (on the left). It remains the incomplete lattice $L \times \{1; 2\}$. We obtain two related problems of DT in 2D on the sub-lattice L, one in level $k = 1$ and another one in level $k = 2$.

We denote $L \times \{1; 2\}$ the incomplete lattice of the undetermined positions. In other words, L is the subset of $[1..m] \times [1..n]$ of the points (i, j) with $v_{i,j} = 1$. Now that we consider only the sublattice L with its two levels $k = 1$ and $k = 2$, we should decrease the two matrices $r_{i,k}$ and $c_{j,k}$ from the projections $r_{i,k}^{fixed}$, $c_{j,k}^{fixed}$ of the points which are already determined (Fig. 6). Instead of considering the problem of reconstruction of a subset $S \subset L \times \{1; 2\}$ in 3D, we can decompose it on its two levels of height $k = 1$ and $k = 2$ (Fig. 6). We obtain two problems in 2D which should not be solved independently. A solution of the problem at level $k = 1$ is denoted S_1 while a solution at level $k = 2$ is denoted S_2. According to the 3D problem, these two solutions should be related by the vertical projection equal to 1 on the whole sub-lattice L. It means exactly $S_1 \cap S_2 = \emptyset$ and $S_1 \cup S_2 = L$ namely $S_2 = L/S_1$. At last, we can notice that if S_1 is solution of the problem of reconstruction at level $k = 1$, it follows from condition (i) -decreased from the positions already determined- that its complementary $S_2 = L/S_1$ has exactly the right projections to be a solution the problem at level $k = 2$.

As conclusion, for solving the problem, we just have to determine if there exists a subset S_1 of the 2 dimensional sublattice L with the prescribed row and column sums given by the projections at level $k = 1$ (the set S_2 is just its

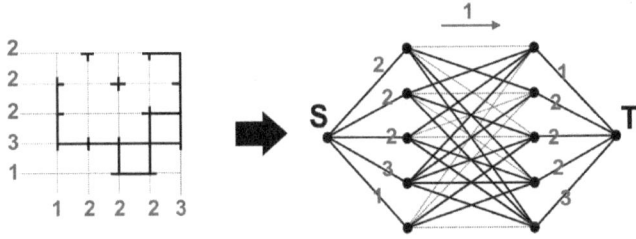

Fig. 7. The problem of reconstruction in the 2 dimensional sublattice L and its translation in a problem of flow in a bipartite graph

complementary in the sublattice: $S_2 = L/S_1$). This problem of computing S_1 can be considered as a problem of max-flow in a bipartite graph and solved in polynomial time by one of the well-known algorithms of this field (Fig. 7) [3]. A solution of the initial 3DT problem exists if and only if there exists a flow of value equal to the sum of the horizontal or vertical projections. As this question can be solved in a polynomial time, it provides the following proposition:

Proposition 1. *3DT with $l = 2$ is in P.*

3   Restricted-3DT **Is NP-Hard**

The main purpose of this section is to reduce Restricted-3DT to 3-SAT. As it is obvious that Restricted-3DT is in NP, since it is a subclass of problems already known in NP, it would provide the NP-completeness of Restricted-3DT. We decompose the reduction in two steps. In the first subsection, we present the combinatorial construction which represents the boolean variables of an instance of 3-SAT. The second step is the encoding of the clauses and we end the section by proving the equivalence between the instance of Restricted-3DT that we build and the initial instance of 3-SAT.

3.1   Encoding the Boolean Variables

The principle is to encode the boolean variables of a 3-SAT formula in combinatorial structures called "switches".

Definition 1. *A Δ-switch is a pair of disjoint lattice sets S_0 and S_1 having the same number of points on straight lines belonging to a set of lines Δ. We add usually that the pair S_0 and S_1 should be minimal under inclusion.*

For convenience, we avoid the letter Δ before the word "switch". As defined, switches are a generalization of "switching components" which correspond to the case where Δ contains all the lines in a given subset of directions. In our framework, the set Δ contains the whole set of lines parallel to the three axis of the lattice except some lines in direction j. These missing lines are called *clause lines* (Fig. 8) since they are used in the next Subsection to encode the clauses of the 3-SAT formula. With the partial data of 3DT drawn in Fig. 8, there are only two possible cases: a solution contains S_0 or it contains S_1. This binary choice encodes one boolean variable. This construction can be repeated as much as necessary so that in our final instance of Restricted-3DT, we would have exactly one switch per boolean variable of the initial 3-SAT formula.

Notice now that the lower part of the switch is made from pairs of points (one belonging to S_0, the other to S_1) in direction k that we call *swings*. The switch drawn in Fig. 8 can be enlarged an arbitrary number of times, so that it can hold an arbitrary large number of swings (the number of swings remains a

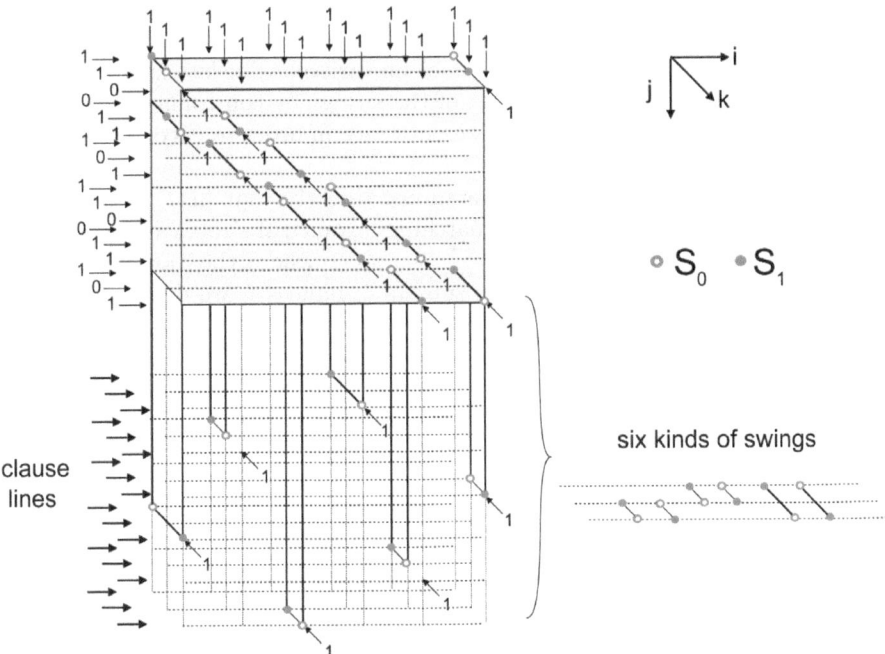

Fig. 8. A 3-dimensional switch S_0, S_1. The arrows indicate the lines with prescribed numbers of points equal to 1 and the clause lines. All the other lines parallel in direction k have prescribed numbers of points equal to zero. This switch is "holding in suspension" some pairs of points (one in both feasible set) that we call swings. The height of each swing can be easily adjusted by modifying the height of the 1 given in the instance.

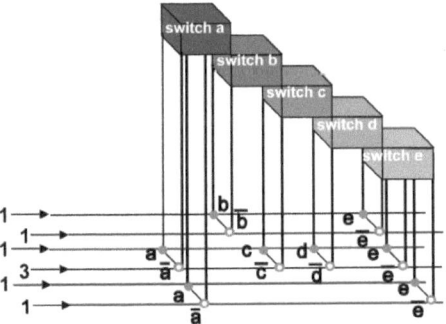

Fig. 9. Encoding of three constraints $b + e = 1$, $a + c + d + e = 1$, $a + e = 1$ on five binary variables a, b, c, d, and e by using the clause lines of five switches. As we have six kind of swings (not represented here), an elementary solution is to encode six times each clause which means that the construction of the figure is repeated six times.

multiple of 6 because the construction creates six different kinds of swings and the same number of swings of each kind). It allows to build large switches with 6 swings for each clause containing the boolean variable. The last important point is the possibility to adjust the height of the swings by controlling the height of the transverse 1 in the lower part of the instance in order to encode the clauses.

This construction allows to prove almost directly that 3DT with $l = 3$ is NP-hard: We just recall that a 3D-matching instance of size h can be expressed under the form of $3h^2$ constraints $\sum_{i \in I} a_i = 1$ where the variables a_i with $i \in [1..h^3]$ are binary and where the subsets of indices I are depending on the instance. Such constraints can be easily encoded in such an instance of 3DT (Fig. 9). The attentive reader could be convinced that the reduction of 3DT with $l = 3$ to 3D-matching is easy to obtain but as a stronger reduction is provided in next section, we do not give more details here about this construction.

3.2   Encoding the Clauses of a 3-SAT Formula

Let a be a boolean variable of our initial 3-SAT formula. We associate to a a switch with 6α swings, where α is the number of 3-clauses containing a or its complementary. Any 3-clause can be encoded by using three different swings for each variable as shown Fig. 11. For using all the swings of the switches, each constraint should be encoded two times, the second encoding being "symmetric" to the first one.

3.3   Equivalence between a 3-SAT Formula and the Corresponding Restricted-3DT Instance

We can check easily that the 3DT instance built according to the principle explained previously (Fig. 10) verifies the conditions of Restricted-3DT: we have $l = 3$, the three input matrices are binary and by construction we have $\forall (i, k) \in [1..m] \times [1..3]$, $r_{i,k} = 1$, $\forall i \in [1..m]$, $\sum_{k=1}^{l} r_{i,k} = \sum_{j=1}^{n} v_{i,j} = 3$, $\forall j \in [1..n]$, $\sum_{k=1}^{l} c_{j,k} = \sum_{i=1}^{m} v_{i,j} \in \{2; 3\}$ and at last $\forall k \in [1..l]$, $\sum_{i=1}^{m} r_{i,k} = \sum_{j=1}^{n} c_{j,k}$.

If we assume now that for this instance of Restricted-3DT there exists a solution S, then for each switch i, we have either $S_0^i \subset S$ or $S_1^i \subset S$. If we consider the three points of S belonging to the vertical plane P of a clause (see the detail of Fig. 11 drawn Fig. 12), at least one of these three points is in the vertical plane $k = 1$. It follows that we can not have all the swings of the boolean variables a, b and c in positions $\bar{a}$, $\bar{b}$ and $\bar{c}$. At least one of them is in a position of truth which means that the corresponding clause ($a \vee b \vee c$ on the figure) is satisfied.

If we assume now that we have a solution of the 3-SAT instance, we can see that the set S obtained by putting all the switches in the positions of their corresponding variables is a solution of our Restricted-3DT instance: we just have to check on Fig. 11 that points of S can be chosen in the vertical planes

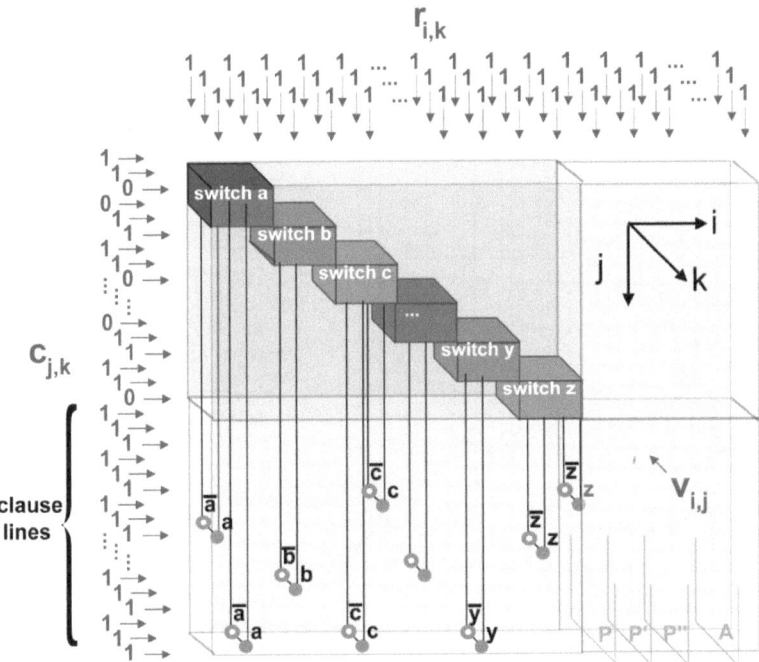

Fig. 10. A built instance of Restricted-3DT: each boolean variable is encoded by a switch. Each switch holds 6α swings (only 1 or 2 of each switch are represented on the figure). 6 swings of each variable a are used to encode each 3-clauses containing a as drawn Fig. 11, the height of the swings being controlled by the projections $v_{i,j}$.

P, P', P'', Q, Q', Q'', A, ..., F so that all constraints of projections can be satisfied. Hence the question is: the swings of a, b and c being not all in false positions, can we complete the right drawing of Fig. 11. The answer is yes, because we can choose easily the points in plane P since one of the three swings is in position true. Solutions in P', P'' and Q, Q', Q'' are obtained by rotations and symmetry, so that we can easily complete the last vertical planes A, ..., F by rotation of the point of each highest non-empty line. It proves that there exists a solution of the initial 3-SAT instance if and only if the corresponding Restricted-3DT instance has a solution. As the construction of the Restricted-3DT can be done in a polynomial time -linear, in fact- it proves the following proposition:

Proposition 2. *Restricted-3DT is NP-hard.*

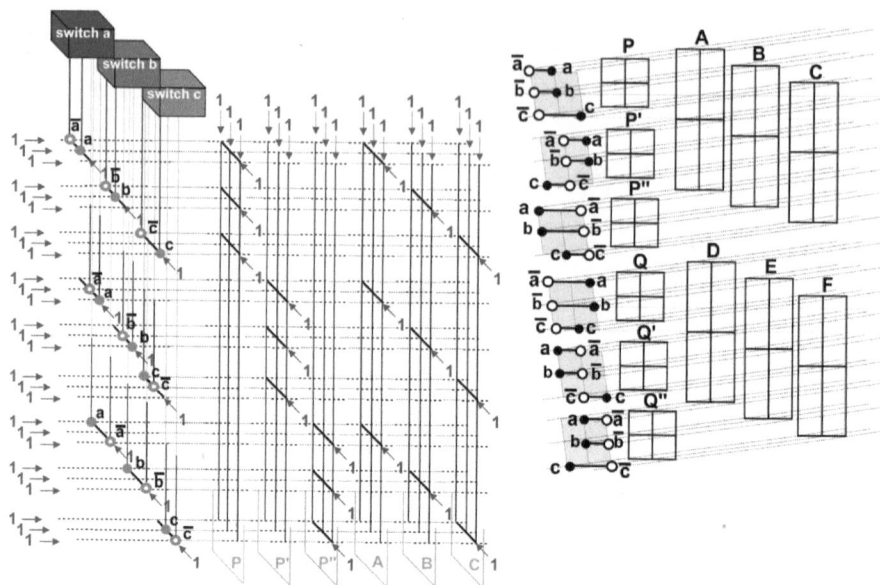

Fig. 11. Encoding of the 3−clause $a \vee b \vee c$ (on the left) by using projections equal to 1 for all clause lines and the swings of three switches associated with the boolean variables a, b and c (all the lines which are not drawn have projections equal to 0 in the instance). Six extra planes P, P', P'', A, B, C are introduced to obtain the equivalence between the clause $a \vee b \vee c$ and this Restricted-3DT data. This construction is however incomplete because it uses only one half of the swings. It is shown on the right how, by symmetry, we use exactly 6 different swings of each switch for encoding exactly the clause $a \vee b \vee c$ (notice that there should be exactly one point per line).

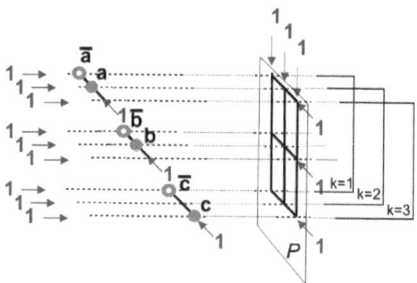

Fig. 12. A detail of Fig. 11. It appears clearly that if there exists a solution of the Restricted-3DT instance, the solution has three points in the vertical plane P and necessarily, the three swings of S can not be all in false positions $\bar{a}$, $\bar{b}$ and $\bar{c}$. At least one of the swings is in a position of truth.

4 Conclusion

If we consider the state of the art of Discrete Tomography in three dimension namely 3DT with a height l fixed a-priori, the complexity of the problem was unknown for $l = 2$ and $l = 3$, knowing that if $l = 1$ the problem is in P while if $l \geq 3$ it is NP-complete. We have proved in the paper that for $l = 2$, the problem remains polynomial and that for $l = 3$, it is NP-hard. We have even improved a result known since a longtime in the framework of Timetables (Restricted-TT is NP-hard) by reducing again the class of problems considered in this framework. The consequence is that this single proof provides in fact a cascade of results -some of them having previously complex proofs. About the perspectives, they are of two kinds: we could consider other problems of Discrete Tomography where one dimension of the problem is fixed a-priori with the aim to obtain their complexity. We could also use the powerful principle of encoding boolean variables in switches in order to try to obtain new results of complexity...

References

1. Chrobak, M., Dürr, C.: Reconstructing polyatomic structures from discrete X-rays: NP-completeness proof for three atoms. TCS 259(1-2), 81–98 (2001)
2. Even, S., Itai, A., Shamir, A.: On the complexity of timetable and multicommodity flow problems. SIAM J. Computing 5, 691–703 (1976)
3. Ford, L.R., Fulkerson, D.R.: Maximal flow through a network. Canadian Journal of Mathematics 8, 399–404 (1956)
4. Gale, D.: A theorem on flows in networks. Pacific J. Math. 7, 1073–1082 (1957)
5. Gardner, R.J., Gritzmann, P., Prangenberg, D.: On the computational complexity of determining polyatomic structures by X-rays. TCS 233(1-2), 91–106 (2000)
6. Gardner, R.J., Gritzmann, P., Prangenberg, D.: On the computational complexity of reconstructing lattice sets from their X-rays. Discrete Math. 202, 45–71 (1999)
7. Garey, M.R., Johnson, D.S.: Computers and Intractability: A Guide to the Theory of NP-Completeness. W. H. Freeman, New York (1979)
8. Herman, G.T., Kuba, A.: Discrete Tomography: Foundations, Algorithms and Applications. Birkhäuser, Basel (1999)
9. Irving, M., Jerrum, R.W.: Three-dimensional statistical data security problems. SIAM Journal on Computing 23 (1993)
10. Ryser, H.J.: Combinatorial properties of zeros and ones. Canada J. Math. 9, 371–377 (1957)

Theoretical Issues of Cluster Pattern Interfaces*

Reneta P. Barneva[1], Valentin E. Brimkov[2], and Kamen K. Kanev[3]

[1] Department of Computer Science, SUNY Fredonia,
NY 14063, USA
`barneva@cs.fredonia.edu`
[2] Mathematics Department, SUNY Buffalo State College,
Buffalo, NY 14222, USA
`brimkove@buffalostate.edu`
[3] Research Institute of Electronics, Shizuoka University,
Hamamatsu 432-8011, Japan
`kanev@rie.shizuoka.ac.jp`

Abstract. Since the 80s, when the computer mouse was first invented, point-and-click functionality became widely adopted, in particular, for work with images and GUIs. It would be useful to provide such functionality for printed materials, as well. A direct point-and-click functionality should not require tablet-like devices, but should rather use an embedded marking which defines a coordinate system within the image. An original approach and related technology for direct interface, known as Cluster Pattern Interface (CLUSPI), was proposed [12]. It is based on an unobtrusive layer of the image defining a coordinate system, a camera-based pointing device, and decoding software. CLUSPI technology was invented and patented by one of the authors of this paper. Here we present the theoretical (mathematical) foundations of the methods developed. We also describe some theoretical challenges and propose solutions to them.

Keywords: carpet encoding, cluster pattern interface CLUSPI, parallel algorithm, CRCW-PRAM model.

1 Introduction

Since the 80s, when the computer mouse was first invented, point-and-click functionality became widely adopted, in particular, for work with images and GUIs. The main points are that the user employs a pointing device with clickable buttons invoking actions, and has feedback (cursor). It would also be useful to provide such functionality for printed materials, such as books, handouts, and labels. The idea is to have direct point-and-click functionality that does not require feedback on the computer screen or tablet-like devices, but rather uses an embedded marking which defines a coordinate system within the image. The main requirements to such a marking are:

---

* The work was partly supported by the Cooperative Research Project of the Research Institute of Electronics, RIE, Shizuoka University, Japan.

P. Wiederhold and R.P. Barneva (Eds.): IWCIA 2009, LNCS 5852, pp. 302–315, 2009.

- To be as fine as possible, preferably to alter only a few pixels/dots;
- Not to obscure part of the image and to be almost invisible;
- To make even very small parts of the object identifiable;
- To be easily and efficiently recognized and decoded almost instantaneously.

The widely used 1D and 2D codes on printed materials and surfaces of physical objects – such as bar-codes and QR-codes [18] – are usually based on dots, bars, and other graphical primitives. They obscure part of the printed material and small defects can make them non-decodable.

The above issues were to a great extent resolved by an original approach, known as CLUSPI, that was invented and patented by one of the authors of the present paper, Kanev, and Kimura [12]. The CLUSPI methodology addresses the problems discussed above by employing clusters of graphical primitives that are adjustable to different contents and surfaces. It overcomes the drawbacks of the known 2D codes through an unobtrusive layer of the image, a pointing device, and decoding software.

The basic principles of CLUSPI technology, together with an application for language acquisition, were reported at IWCIA'08 Application Track [2]. For a detailed account of the considered practical problems, the proposed approaches for their solution, and the obtained results presented from an applied point of view, the reader is referred to [1].

In the present work we describe the theoretical (mathematical) foundations of the developed methods related to CLUSPI which have not been published in the scientific literature so far due to the patenting process. We also consider some theoretical challenges important from a practical perspective and propose solutions to some of them.

The paper is organized as follows. In the next section we present the scheme for object encoding and decoding used in CLUSPI. In Section 3, we consider some theoretical questions related to searching a pattern in a noisy environment. In Section 4, we analyze the possibility of speeding-up the search by parallelization. In Section 5 we outline some of the succesful implementations of CLUSPI in various applications. We conclude in Section 6 with some final remarks about further theoretical challenges and applications.

In the presentation, we conform in part to the traditional terminology of combinatorial pattern matching. In particular, a sequence (string, pattern) of length m will be called an *m-sequence/string/pattern*, for short.

2 Encoding and Decoding Scheme

In this section we present the theoretical background for CLUSPI technology.

2.1 General Idea of the Code

In many cases, it is useful to directly point on the surface of a printed material or object and have the absolute coordinates of the pointed location sent to the

									0	100
0,3	1,3	2,3	3,3		0	$b_1\, b_2\, ...\, b_m$			1	001
0,2	1,2	2,2	3,2		1	$b_2\, b_3\, ...\, b_{m+1}$			2	011
0,1	1,1	2,1	3,1		...	...			3	110
0,0	1,0	2,0	3,0		$n\text{-}m$	$b_{n\text{-}m+1}\, b_{n\text{-}m+2}\, ...\, b_n$			4	101

Fig. 1. *Left:* Cartesian grid. *Middle:* Encoding of the numbers $0, 1, \dots, n - m$ through m-subsequences of the Σ-sequence $b_1, b_2, \dots, b_n$. *Right:* Encoding of the numbers $0, 1, \dots, 4$ by the sequence 1001101 for $m = 3$.

computer. The coordinates may be used for identifying parts of the picture/text and invoking a hyperlink. This cannot be done with a mouse, because it returns only relative coordinates and it is not a very precise pointing device. Devices such as tablets restrict the size of the document and require placing it over the tablet. Some shortcomings of bar-codes and carpet encodings have already been mentioned in the Introduction. To overcome these difficulties, an absolute coordinate system on the surface of the document must be established. One should also have a specialized reading device (camera) that recognizes this coordinate system.

The simplest coordinate system is the Cartesian coordinate system. With it, any rectangular array of the surface can be covered with a square grid, as every unit of the grid is enumerated with two coordinates – x and y (see Fig. 1, left).

The requirements to the grid are as follows:

1. The squares should be as small as possible, so that even very small details of the picture can be pointed out;
2. The code should be easily recognizable, even if the camera is rotated over the surface;
3. The code should blend with the page content on the surface in a way to avoid deteriorating the image appearance.

In order to fulfill the above requirements, it seems the best way to encode the coordinate system would be via a binary code through pixels/printer dots, because they are small enough and can blend well with the picture. However, having a fine grid, e.g., on A4 paper format (210×297 mm) with a unit of one square millimeter, means to deal with x and y coordinates in the range of $[0, 296]$. Then, we would need 9 bits to encode each coordinate, that is, in one square millimeter we should use 18 dots for the code: 9 for the x-coordinate and other 9 for the y-coordinate. Clearly, such a code would cover a significant part of the page content and affect its quality. Besides, such an encoding will not be error proof, since even minor noise of one dot may cause a serious error in the code recognition.

On the other hand, the span of the reading camera is usually significantly larger than a square millimeter. Thus, one can easily read simultaneously the code in the adjacent grid cells.

1,1	0,1	0,1	1,1	1,1	0,1	1,1
1,0	0,0	0,0	1,0	1,0	0,0	1,0
1,1	0,1	0,1	1,1	1,1	0,1	1,1
1,1	0,1	0,1	1,1	1,1	0,1	1,1
1,0	0,0	0,0	1,0	1,0	0,0	1,0
▨	0,0	0,0	1,0	1,0	0,0	1,0
1,1	0,1	0,1	1,1	1,1	0,1	1,1

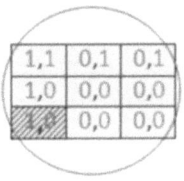

Fig. 2. *Left:* Encoding of the light grey region by the sequence 1001101. Note that the white region is also necessary for decoding the topmost/rightmost row/column. *Right:* Area read by the camera.

CLUSPI approach to encoding is based on two main heuristics:

(A) The code of the coordinates of one cell depends on the coordinates of the adjacent cells;

(B) The cells that define the code are not in a linear sequence, but rather, they form a two-dimensional array.

Let us first elaborate on heuristic (A). Assume that we have a sequence of bits $b_1, b_2, b_3, \ldots, b_n$, such that no two subsequences of m consecutive bits are identical.[1] We will call this type of sequences Σ-sequences. For example, the sequence 1001101 has this property for $m = 3$, as the possible subsequences of three consecutive elements are $100, 001, 011, 110$, and 101, and they are all different. This leads to the encoding illustrated in Fig. 1, middle. For the example above, the generated code is shown in Fig. 1, right.

A cell code is based on the codes in the adjacent cells, as shown in Fig. 2. In order to find the position of the hatched cell in Fig. 2, left, we read the 3×3 matrix originating at that cell. The x-coordinate is encoded by the binary string 100, that corresponds to 0, while the y-coordinate is encoded by the string 001, that corresponds to 1. Thus, the encoding reads $(0, 1)$. (Use the table in Fig.1, right.) In the above example, we used a 3×3 matrix, because the code length m was equal to 3. Note however that, for larger m (e.g., for $m \geq 9$) used for A4 size paper, the matrix to be read becomes too large and may not fit into the camera's span. On the other hand, as we have already observed, only part of the code (consisting of the cells above and to the right of the pointed cell) was used to find out the cell code. The rest of the cells in the camera's span were not used at all. Thus, we come to heuristic (B) which allows us to use all the cells in the camera's span.

The idea here is to "fold" the code in such a way that it fills in a $p \times q$ matrix with $p \cdot q \geq m$ for a code of length m. This is achieved by shifting each consecutive row in x-direction with p positions, as shown in Fig. 3, left. For the sequence 1001101 from the example considered above, and for a 2×2 matrix, the sequence will be "folded" as shown in Fig. 3, middle. The array in grey is

---

[1] We will see how such a sequence can be constructed in the next section.

b_{3u+1}	b_{3u+2}	b_{3u+3}	b_{3u+4}	b_{3u+5}	...
b_{2u+1}	b_{2u+2}	b_{2u+3}	b_{2u+4}	b_{2u+5}	...
b_{u+1}	b_{u+2}	b_{u+3}	b_{u+4}	b_{u+5}	...
b_1	b_2	b_3	b_4	b_5	...

1	0	1					
0	1	1	0	1			
1	0	0	1	1	0	1	

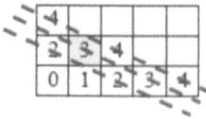

Fig. 3. *Left:* "Folding" the sequence. *Middle:* Folding of 1001101 for $p = 2$, $q = 2$. *Right:* Decimal values corresponding to the values in the middle. Dashed lines connecting the cells with the same value are parallel to x-axis.

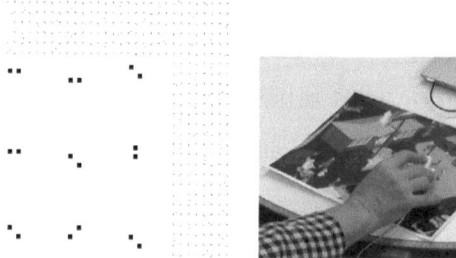

Fig. 4. *Left:* Encoding of a page. The left bottom part is zoomed to better show the structure of the code. *Right:* Work with CLUSPI. (Picture is courtesy of R. Mitzel, mitzel.us).

read as 110. (Note: the actual sequence is 1101, however, we chose $m = 3$ and, thus, we omit the last digit.) In fact, as it is easy to see, the binary table of Fig. 3, middle, corresponds to the decimal table of Fig. 3, right. We connect the cells with the same value and obtain lines parallel to one of the coordinate axes. Note that the axes are now oblique. In the example of Fig. 3, right, one of them is parallel to the dashed lines. The other axis is obtained in a similar fashion. In such a way, every cell is encoded with two binary digits for x and y coordinates. These digits are represented by clusters of dots. For example, in our practical implementation, the configuration $\bullet\ \bullet$ stands for $(0,0)$, $\begin{smallmatrix}\bullet\\\bullet\end{smallmatrix}$ - for $(1,1)$, $\bullet\ {}^{\bullet}$ - for $(0,1)$, and ${}^{\bullet}\bullet$ - for $(1,0)$.

As a result, the image is covered by a very fine carpet of dots, which blends with the printed content and is practically invisible. See Fig. 4 for illustration.

With this preparation, we are now in a position to summarize the encoding and decoding phases, as follows.

Encoding

First, a binary sequence $S \in \Sigma$ is created. Then a rectangle $p \times q$ is chosen, whose dimensions satisfy the condition $p \cdot q \geq m$. The code is written as shown in Fig. 3, left, once for the x-coordinate and once for the y-coordinate. The (x, y) pairs are encoded with clusters of dots as explained above. Finally, the obtained carpet is blended with the original page content and printed. The carpet color is chosen in such a way that it is not met elsewhere in the picture.

Decoding

As a first stage, all dots in the reading area with the code color are recognized and kept in a list. Then clusters of dots are formed. The clusters are recognized as one of the four coding configurations for $(0,0), (1,1), (0,1)$ and $(1,0)$. Thus, two tables, such as in Fig. 3, left, are restored (one for x and one for y-coordinate). Next, for each of them, an m-digit binary sequence P is constructed. The occurrence of P is sought in S. Thus, the decimal value of the coordinate is obtained and, as a result, the cell is located.

3   Searching Noisy Patterns

In this section we first explain how a Σ-sequence is built. Then we deduce some properties about searching noisy patterns.

3.1   Generating a Shift-Register Σ-Sequence

Σ-sequences are a special class of shift-register sequences [9]. Let $P(x) = a_m x^m + a_{m-1}x^{m-1} + \cdots + a_1 x + a_0$ be an irreducible polynomial of degree m, i.e., such that cannot be factored into a product of a number of polynomials with integer coefficients and of degree greater than or equal to one. We will suppose that the coefficients of $P(x)$ are 0's or 1's. Since the degree of $P(x)$ is m, we have $a_m = 1$. Since $P(x)$ is irreducible, we also have $a_0 = 1$. We will use the polynomial $P(x)$ to build a binary sequence S of length $|S| = 2^m + m - 2$ with the property that every subsequence of S of length m is unique within S.

Let $x^\alpha, x^\beta, \ldots, x^\gamma$ be the nonzero monomials of $P(x)$, i.e., those with coefficients $a_\alpha(= a_m) = a_\beta = \cdots = a_\gamma = 1$. (So, $a_i = 0$ for $i \neq \alpha, \beta, \ldots, \gamma$, $1 \leq i \leq m$.) Since a_0 is always equal to 1, for the sake of simplicity it is excluded from consideration in the construction that follows. Thus, we have $\alpha = m > \beta > \cdots > \gamma \geq 1$.

For a given binary sequence $y_1, y_2, \ldots, y_m$, we will refer to its indexes $\alpha, \beta, \ldots, \gamma$ as to *degree positions*.

For $y_1, y_2, \ldots, y_m$, define an *Exclusive OR (XOR)* operator *relative to indexes* $\alpha, \beta, \ldots, \gamma$, as follows:

$$XOR_{\alpha,\beta,\ldots,\gamma}(y_1, y_2, \ldots, y_m) = XOR(y_\alpha, y_\beta, \ldots, y_\gamma).$$

Recall that the XOR-operator applied to a number of binary values returns 1, if the number of 1's among these values is odd, and 0 otherwise.

With the above settings, the construction of a shift-register Σ-sequence S starts from an arbitrary binary sequence $a_1, a_2, \ldots, a_m$, $a_i = 0$ or 1 for $1 \leq i \leq m$, such that not all a_i's are 0's. This m-tuple constitutes a prefix of S. Then the next element of S is found as

$$a_{m+1} = XOR_{\alpha,\beta,\ldots,\gamma}(a_1, a_2, \ldots, a_m).$$

More in general, assume that $a_1, a_2, \ldots, a_m, \ldots, a_k$ are the first k elements of S, where $m \leq k < 2^m + m - 2$. Then

$$a_{k+1} = XOR_{\alpha,\beta,\ldots,\gamma}(a_{k-m+1}, a_{k-m+2}, \ldots, a_m). \tag{1}$$

The generation of S ends when $2^m - 2$ iterations are performed. It is proved [9] that in the so-constructed sequence every m-string is met only once, and if another, $2^m + m - 1$'th element a_{2^m+m-1} is concatenated to S, then the m-prefix of the obtained sequence $S|a_{2^m+m-1}$ will be the same as the m-suffix of S.

Example 1. Consider the irreducible polynomial $P(x) = x^3 + x + 1$. In terms of the denotations used above, we have $\alpha = m = 3$ and $\beta = \gamma = 1$. Starting, e.g., from the binary string 001 of length $m = 3$, after six iterations of the described generation process, we obtain the shift-register Σ-sequence 001110100.

3.2   Maximal Length of a Unique Pattern with a Gap

As already discussed in Section 2, when a click is made with the CLUSPI device, the camera reads a $p \times q$-rectangular pattern. This pattern is a folded representation of an $m = p \times q$ pattern sequence P. Then the pattern is searched in the Σ-sequence S. Note however that, for alignment and orientation independency, the camera reads a number of neighboring positions to the pattern P. (See Fig. 2, right.)

   After this observation, we next count the possibility of having some position (or positions) of the pattern damaged and therefore non-readable. Consider an m-pattern P that contains a non-readable element (a "gap"), which we will denote by $*$. Then, clearly, this pattern will have two occurrences in S rather than a single occurrence. For example, in the Σ-sequence of Example 1, the pattern $0 * 1$ has two occurrences: one for $* = 0$ and another for $* = 1$. On the other hand, it may be possible that a pattern longer than m has a unique occurrence in S. Thus, if we are able to identify a pattern size that guarantees its uniqueness, we could extend the pattern with certain extra elements read by the camera and thus assure successful identification of the searched position. In the rest of the present section we investigate this issue. More precisely, we have the following proposition.

Proposition 1. *Let P be an m-pattern that has a gap $(*)$ in its j^*'th position. Let j' be the closest degree position to the left of j^*. Then all patterns that have P as their prefix and are of length greater than $m + j^* - j' - 1$, are unique.*

Proof. Consider the two occurrences $P_1[k_1, k_1 + m - 1]$ and $P_2[k_2, k_2 + m - 1]$ of pattern P in S, where k_1 and k_2 are the positions within S where P_1 and P_2 start. Let us perform a series of operations $O_1, O_2, \ldots$ of type (1) to both patterns, thus extending them to the right. One can observe that the corresponding elements that are added to P_1 and to P_2, respectively, after performing an operation O_i, will be the same as long as no degree position interferes with the values at positions $k_1 + j^* - 1$ and $k_2 + j^* - 1$. This is because the newly computed values are obtained by performing the same operation O_i of type (1) on identical inputs.

After performing a number of operations causing a number of shifts, $k_1 + j^* - 1$ and $k_2 + j^* - 1$ will simultaneously become degree positions. Since $P[k_1 + j^* - 1] \neq P[k_2 + j^* - 1]$, the next values in the sequence S will certainly be different. This completes the proof.   □

Proposition 1 implies a simple way of identifying the shortest unique pattern that contains P as a prefix. For this, after detecting the position of a gap in P, one has to locate the closest degree position to the left of the gap. Obviously, this can be achieved with $O(m)$ arithmetic operations. Then, the required dimension is implied by Proposition 1.

4   Parallelization

In the practical implementation of CLUSPI, the binary sequence S has been generated by an irreducible polynomial of degree $m = 12$. That is, the sequence length equals $n = |S| = 2^{12} + 10 = 4106$. With these comparatively small dimensions, the search of an m-pattern within the n-sequence is immediate, even if a naive search strategy is applied. The values $m = 12$ and $n = 4106$ have been chosen with practical arguments in mind. Specifically, these values serve well for encoding and decoding of information on printed materials of A4 up to A0+ paper size. However, for printed materials of larger size (such as large posters, for example), these values would be insufficiently small. To assure reasonable functionality of the CLUSPI technology applied to materials of any possible size met in practice, one can choose the value of m to be as large as $m = 20$ or even $m = 30$. Then, patterns should be searched in considerably longer sequences of length in the order of $2^{20} > 10^6$ to $2^{30} > 10^9$. In that case, the search may not be immediate even if advanced sequential pattern matching algorithms are used.

The need of speeding-up the computational process suggests to resort to efficient parallelization of the computation. In this section we analyze the possibility for such a parallelization. The algorithms considered work within the concurrent-read concurrent-write parallel machine model (CRCW-PRAM). In order to make the paper self-contained, below we briefly recall some basic points related to parallel models of computation. For a detailed accounting of the matter the reader is referred to [6,7].

4.1   CRCW-PRAM Model

PRAM is a synchronous model of parallel computation. It is a natural generalization of the random access memory model (RAM), which is the model commonly used in sequential computation. PRAM consists of m shared memory cells

$M_1, \ldots, M_m$ and p processors $P_1, \ldots, P_p$. Each processor can be considered as a RAM computational model with a private local memory. Processors communicate via the shared memory. Thus, restricting the amount of the shared memory corresponds to restricting the amount of information which can be communicated between processors in one step. A step of computation consists of read, compute, and write phases. In the *read phase* every processor may read from a shared memory cell. In the *compute phase* every processor may perform a local operation. In the *write phase* every processor may write to one shared memory cell. Any number of processors can simultaneously read from or attempt to simultaneously write to the same shared memory cell.

An important issue in CRCW-PRAM is how to resolve the write conflict, which occurs when a number of processors are concurrently attempting to write to the same shared memory cell. This leads to several variants of CRCW-PRAM (see [6] for a survey). The algorithms discussed below work in the framework of the so-called COMMON model [16], in which more than one processor are allowed to simultaneously write to the same memory cell only if they are all writing a common value. This is the weakest version of CRCW-PRAM. Thus, they also work without any change under any other version of CRCW-PRAM (such as the PRIORITY model [8] and the ARBITRARY model [21]). Furthermore, the complexity bounds which hold for the COMMON model apply to the other basic models as well. For further details and results about the relationship between different PRAM models, the reader is referred to [6].

Work of a PRAM-algorithm is defined as the product of its running time and the number of processors used.

A parallel algorithm is said to be *optimal* if its work equals the running time of the fastest possible sequential algorithm for the considered problem.

4.2   Cole et al. Parallel Pattern Matching Algorithm

In the next section we use as a subroutine the parallel pattern matching algorithm of Cole et al. [5], denoted further Cole-et-al.-algorithm. Remember that the string matching problem consists of finding all the occurrences of a pattern string $P[0 \ldots m-1]$ in a text string $T[0 \ldots n-1]$. Similarly to all known pattern matching algorithms, the Cole-et-al.-algorithm works in two main phases.

- *Preprocessing phase.* The pattern P is analyzed and as a result useful information is extracted and saved in appropriate data structures to be used in the following phase.
- *Search phase.* The text T is searched using the data structures created in the first phase.

In order to provide the reader with an idea how such algorithms are designed, we briefly recall some basic notions of combinatorial pattern matching theory and the most important ingredients of the Cole-et-al.-algorithm.

A possible starting position of an occurrence of the pattern in the text is called a *candidate*. Candidates are *eliminated* by proving that an occurrence of the pattern cannot start from them.

A string u is called a *period* of a string w if w is a prefix of u^k for some positive integer k. The shortest period of w is called *the period* of w. P is *periodic* if its period is shorter than half of P's length.

Let p be the period length of a pattern string P. For every i, $0 < i < p$, there exists at least one position h in P with $h > i$, such that $P[h] \neq P[h-i]$. Such a position is called a *witness* against i. In the preprocessing phase an *array of witnesses* is generated. For this, one finds $r = \min(m/2, p)$ and then for every $i, 0 < i < r$, a position $witness[i] = h$ is computed, such that $P[h] \neq P[h-i]$. Using the witnesses' array, Vishkin [22] proposed a method called a *duel*, aimed at eliminating at least one of two candidates. Specifically, suppose that i and j are two candidates with $0 < j - i < p$, so that $h = witness[j - i]$ is defined. Since $P[h] \neq P[h+i-j]$, at most one of the candidates is equal to $T[i+h]$ and, consequently, at least one of them can be eliminated. The witnesses' array of the pattern P is used in the search phase to eliminate many candidates concurrently, through duels. It can be generated in $O(\log \log m)$ CRCW-PRAM time and $O(m)$ work.

Another useful concept introduced by Vishkin [20] is that of *deterministic sample* (DS for short). For a given nonperiodic pattern string P of length m, a DS for k shifts, $k \leq m/2$, is an ordered set A of positions of the pattern and a number f, such that if the positions are verified to match a candidate i in the text, then all other candidates in the interval $[i - f + 1, i + k - f]$ can be eliminated. Such an elimination is possible because a mismatch with P is guaranteed to exist with at least one of the A's positions. The *size* of a DS is the size of the ordered set A.

It has been shown by Breslauer and Galil [4] that if one has the witnesses precomputed, it suffices to find DS of appropriate size for a nonperiodic substring of P of size $r = m^{1/3}$, if $m = r^3$ processors and space are available. Cole et al. [5] have shown how to construct a $\log k$-size DS of a string x of length r, for k shifts, $k \leq r/2$, in constant time using r^3 processors and r^2 space. This computation uses the well-known constant time deterministic polynomial approximate compaction (d-PAC) of Ragde [19] or its improvement by Hagerup [10].

The results outlined above have constituted the basic elements of the string matching Cole-et-al.-algorithm. The preprocessing phase of the latter consists of generating an array of witnesses (as mentioned, this can be done in $O(\log \log m)$ time and $O(m)$ work) and a DS (which can be found in constant time and $O(m)$ work). Then the array of witnesses and the DS are used in the search phase to find all the occurrences of the pattern in the text in constant time and $O(n)$ work.

The above bounds on the time and work of the Cole-et-al.-algorithm hold within the COMMON CRCW-PRAM model.

4.3   Application to CLUSPI Pattern Matching Problems

In the evaluation of the time and work performance of the proposed algorithm we will use the following classical result.

Lemma 1. *(Brent's Lemma [3]) Any parallel algorithm of time t that consists of a total of x elementary operations can be implemented on p processors in $\lceil \frac{x}{p} \rceil + t$ time.*

It is well-known (see, e.g., [5]) that an m-pattern can be searched in an n-string in *optimal* sequential time $T = O(n)$. Let M be the satisfactory time limit for a computation. As already discussed, in the considered practical application, T may considerably exceed M.

Assume that a CRCW-PRAM with p processors is available. We have the following proposition.

Proposition 2. *The search of any m-pattern in the corresponding shift-register n-sequence ($n = 2^m + m - 2$) can be performed with p processors in $O(n/p + \log\log\log n)$ parallel time and $O(n)$ work.*

Proof. Let us apply the Cole-et-al. pattern matching algorithm for searching an m-pattern in the n-seqience. The preprocessing phase takes $O(\log\log m)$ time and $O(m)$ work, while the search phase is done in constant time c and $O(n)$ work.

Using Brent's Lemma, one can slow-down the computation of the search phase by implementing it on p processors in $\lceil T/p \rceil + c = O(n/p + c)$ time. Then the overall time of both phases will be

$$O(n/p + c) + O(\log\log m) = O(n/p + \log\log\log n),$$

while the total work will amount to $O(m) + O(n) = O(n)$.   □

Corollary 1. *Under the terms of Proposition 2, the computation can be performed in $O(\log\log\log n)$ time with $O(n/\log\log\log n)$ processors.*

The above theoretical result suggests that a reasonable trade-off between number of processors and parallel computation time should be sought. While the search will be immediate if implemented as in Corollary 1, it might be problematic to assure the necessary number of parallel processors, e.g., for $m = 30$. However, if we chose $p = 2^{13} = 8,192$, then the parallel computation time would be of the order of $2^{17} = 131,072$ arithmetic operations. This seems both satisfactory and feasible, as relevant hardware is readily available on the market[2].

5   Applications

CLUSPI technology has already been employed for different purposes in a number of projects. Below we outline some of them wich together with the system

---

[2] Such as parallel graphical chips, like the hardware managed NVIDIA GeForce-8 Series that is a massively parallel computing platform with 12,288 concurrent threads.

for language aquisition reported at IWCIA'08 [2] (see also its extended version [1]) illustrate the variety of applications.

Based on CLUSPI a parking simulation and guidance system employing 1/10 replica car models has been built [15]. In this system, positions and movements of remote controlled car models are tracked by analyzing CLUSPI codes embedded in the model parking environment. The system employs wireless cameras for imaging of the embedded codes and transmission to a centralized server supporting a virtual model of the parking environment and the cars. Obtained tracking data, when mapped to the geometry of the model parking facility, allows for early detection of potential collisions and thus for appropriate steering guidance. By employing virtual models of real cars and parking facilities realistic visualization and simulations are provided and enhancements of learning and comprehension are achieved.

A Flexible Scanning Tablet (FST) based on CLUSPI is also considered as a basis for constructing portable nondestructive testing devices. The FST consists of a thin film with a digitally encoded coordinate system, applicable to flat and curved surfaces, that allows probe positions to be tracked by a specialized optical reader. In [17] the FST idea is applied to handheld Eddy Current Testing (ECT) and Electro-Magnetic Acoustic Transducer (EMAT) probes for 2D and 3D image reconstruction.

In a work devoted to tangible interfaces for steering and control of interactive multimedia presentations [11], various methods for digital encoding of physical objects are considered and their applicability in surface encoding for tangible interface components is discussed. Experiments with presentation controls, based on direct interaction with digitally encoded printed handouts are reported and an approach for transferring presentation controls from printed handouts to surfaces of real physical objects is introduced. USB and wireless cameras are employed as CLUSPI readers for implementing surface based interactions and a portable communication device with an embedded camera is considered as a possible truly mobile solution.

In [14] the general concept of newsputers as digitally enhanced printouts, allowing real-time access to computer-based multimedia information through paper-based interfaces is introduced. Newsputers could serve as conventional newspapers, books, magazines, etc. and in addition as clickable layouts and templates for invoking special functions and for accessing complementary information in different pervasive environments. A software system for providing newsputer-enabling support to publishers and readers is proposed, its design and development is discussed. A CLUSPI based implementation and experiments are reported.

CLUSPI technology has also been employed in a project dedicated to Computer-Supported Collaborative Learning (CSCL) with dynamic groups, where at different stages students work independently, interact with each other in pairs, and conduct joint work in larger groups with varying number of participants. A Dynamic Group Environment for Collaborative Learning (DGE/CL) [13] supports students in making informed and intelligent choices about how, when, and with whom to collaborate. This is a face-to-face collaborative environment, where all students are in the same room, can move freely around and interact with each other while using digitally enhanced printed materials with direct point-and-click functionality. Flexible and efficient support for dynamic group management is ensured through the adopted CLUSPI technology, which, while preserving the original touch-and-feel of printed educational materials, supports additional affordances and allows employment of new, non-traditional paper-based interactions.

6   Concluding Remarks

In this paper we presented the mathematical foundations of the Cluster Pattern Interface technology and proposed solutions to some open theoretical questions. Work in progress is devoted to experiments designed to test the obtained theoretical results. Further research will be aimed at generalizing the theory in several regards. One possible problem of interest is related to searching noisy patterns with more than one gap. Extending the considerations to higher dimensions is seen as another important task. We are also planning to apply CLUSPI in various other fields.

Acknowledgements

We thank Boris Brimkov for proofreading the text and the four anonymous referees for the useful remarks.

References

1. Barneva, R.P., Brimkov, V.E., Kanev, K.: Combining Ubiquitous Direction-Sensitive Digitizing with a Multimedia Electronic Dictionary for Enhanced Understanding. Int. J. on Imaging Systems and Technology 19, 39–49 (2009)
2. Barneva, R.P., Brimkov, V.E., Kanev, K.: Electronic Multimedia Dictionary with Direct-Access Printed Interface. In: Barneva, R.P., Brimkov, V.E. (eds.) Image Analysis: From Theory to Applications, pp. 39–46. RPS, Singapore (2008)
3. Brent, R.P.: Evaluation of General Arithmetic Expressions. J. ACM 21, 201–206 (1974)
4. Breslauer, D., Galil, Z.: An Optimal $O(\log \log n)$ Time Parallel String Matching Algorithm. SIAM J. Comput. 19, 1051–1058 (1990)
5. Cole, R., et al.: Optimally Fast Parallel Algorithms for Preprocessing and Pattern Matching in One and Two Dimensions. In: 34th IEEE Symp. Found. Computer Science, Palo Alto, CA, pp. 248–258 (1993)
6. Fich, F.: The Complexity of Computation on the Parallel Random Access Machine. In: Reif, J. (ed.) Synthesis of Parallel Algorithms, ch. 21. Morgan-Kaufmann, San Francisco (1993)
7. Fich, F., Ragde, P., Wigderson, A.: Relations Between CR-models of Parallel Computations. SIAM J. Comput. 17(3), 606–627 (1988)
8. Goldschlager, L.: A Unified Approach to Models of Synchronous Parallel Machines. J. ACM 29, 1073–1086 (1982)

9. Golomb, S.W.: Shift Register Sequences. Holden-Day, Inc., San Francisco (1967)
10. Hagerup, T.: On a Compaction Theorem of Ragde. Inf. Proc. Lett. 43, 335–340 (1992)
11. Kanev, K.: Tangible Interfaces for Interactive Multimedia Presentations. Int. Journal of Mobile Information Systems 4(3), 183–193 (2008)
12. Kanev, K., Kimura, S.: Digital Information Carrier. Patent Registration No 3635374, Japan Patent Office (2005)
13. Kanev, K., Kimura, S., Orr, T.: A Framework for Collaborative Learning in Dynamic Group Environments. Int. Journal of Distance Education Technologies 7(1), 58–77 (2009)
14. Kanev, K., Mirenkov, N., Hasegawa, A.: Newsputers: Digitally Enhanced Printouts Supporting Pervasive Multimodal Interactions. In: Proc. First IEEE International Conference on Ubi-media Computing (U-Media 2008), Lanzhou, China, pp. 1–7 (2008)
15. Kanev, K., Mirenkov, N., Urata, A.: Position Sensing for Parking Simulation and Guidance. The Journal of Three Dimensional Images 21(1), 66–69 (2007)
16. Kucera, L.: Parallel Computation and Conflicts in Memory Access. Inf. Proc. Lett. 14(2), 93–96 (1982)
17. Nishimura, Y., Kanev, K., Sasamoto, A., Suzuki, T., Inokawa, H.: Toward a Portable Electromagnetic Testing Scanner with Advanced Image Reconstruction Capabilities. In: Proc. Eleventh Int. Conf. on Humans and Computers HC 2008, Nagaoka, Japan, pp. 83–89 (2008)
18. QR-code Standartization, http://www.denso-wave.com/qrcode/qrstandard-e.html (accessed April 17, 2009)
19. Ragde, P.: The Parallel Simplicity of Compaction and Chaining. J. Algorithms 14, 371–380 (1993)
20. Vishkin, U.: Deterministic Sampling – a New Technique for Fast Pattern Matching. SIAM J. Comput. 20, 22–40 (1991)
21. Vishkin, U.: Implementation of Simultaneous Memory Access in Models that Forbid it. J. Algorithms 4, 45–50 (1983)
22. Vishkin, U.: Optimal Parallel Pattern Matching in Strings. Inform. and Control 67, 91–113 (1985)

Ω-Arithmetization: A Discrete Multi-resolution Representation of Real Functions*

Agathe Chollet[1], Guy Wallet[1], Laurent Fuchs[2], Eric Andres[2],
and Gaëlle Largeteau-Skapin[2]

[1] Laboratoire MIA, Université de La Rochelle,
Avenue Michel Crépeau 17042 La Rochelle cedex, France
{Guy.Wallet,achollet01}@univ-lr.fr
[2] Laboratoire XLIM SIC, UMR 6172, Université de Poitiers,
BP 30179 86962 Futuroscope Chasseneuil cédex, France
{fuchs,glargeteau,andres}@sic.univ-poitiers.fr

Abstract. Multi-resolution analysis and numerical precision problems are very important subjects in fields like image analysis or geometrical modeling. In the continuation of previous works of the authors, we expose in this article a new method called the Ω-arithmetization. It is a process to obtain a multi-scale discretization of a continuous function that is a solution of a differential equation. The constructive properties of the underlying theory leads to algorithms which can be exactly translated into functional computer programs without uncontrolled numerical errors. An important part of this work is devoted to the definition and the study of the theoretical framework of the method. Some significant examples of applications are described with details.

Keywords: discrete geometry, nonstandard analysis, multi-resolution analysis, constructive mathematics.

1   Introduction

In some previous works [6,3], the authors have systematically studied a method of discretization called the arithmetization method. The principle of this method has led Reveillès to the definition of the discrete analytical line [17,18,19]. This arithmetization process is a way to discretize a continuous curve solution of a differential equation. The informal point of view [7,8] that *the real line $\mathbb{R}$ is the same thing as the discrete line $\mathbb{Z}$ seen from far away* is the intuitive basis of this method. This could seem quite surprising since $\mathbb{Z}$ is denumerable while $\mathbb{R}$ is not but the theoretical framework we are working in is nonstandard analysis. Hence, denumerable sets are not exactly as in the "usual" framework The transformation from $\mathbb{Z}$ to $\mathbb{R}$ corresponds to a rescaling which induces a strong deformation of space. Moreover, in computer science, practical real numbers are only constructive ones that cannot be other than denumerable. So, from a practical point of view, our approach leads to nothing else than what we can expect

---

* Partially supported by the PPF GIC.

P. Wiederhold and R.P. Barneva (Eds.): IWCIA 2009, LNCS 5852, pp. 316–329, 2009.

in computer science. The arithmetization is obtained by transforming the usual integration Euler scheme used to compute the curves, solution to differential equations, into an equivalent integer scheme.

A rigorous implementation of this approach requires a model of the set $\mathbb{Z}$ of integer numbers together with a notion of infinitely large number (i.e. a scale on $\mathbb{Z}$). In the papers [6,3], such a model was introduced with the help of an axiomatic version of nonstandard analysis. The major drawback of this approach is that the infinitely large integers which arise in the corresponding method have only an axiomatic status. Consequently, in applications with *concrete* computations, it is impossible to give an exact numerical representation of these numbers; in such a situation, we are forced to choose sufficiently large values in an arbitrary manner[1]. Hence, this choice is only a metaphoric representation of the theoretical framework.

In the present paper, we propose to rebuild the arithmetization method with the notion of Ω-numbers introduced by Laugwitz and Schmieden [12,11,10]. Roughly speaking, an Ω-number (natural, integer or rational) is a sequence of numbers of the same nature together with an adapted equality relation. The sets of Ω-numbers are extending the corresponding sets of usual numbers with the added advantage of providing a natural concept of infinitely large integer numbers: for instance, an Ω-integer α represented by a sequence (α_n) of integers is such that $\alpha \simeq +\infty$ if $\lim_{n\to+\infty} \alpha_n = +\infty$ in the usual meaning. Clearly, these infinite numerical entities are effectively constructive.

After having chosen an Ω-integer ω such that $\omega \simeq +\infty$, we can define the Harthong-Reeb line $\mathcal{HR}_\omega$ [4] which is a numerical system consisting of Ω-integers with the additional property of being roughly equivalent to the real line system. Not only the elements of $\mathcal{HR}_\omega$ have a constructive flavor, but we can show that the structure of this system partially fits with the constructive axiomatic developed by Bridges [1].

With this, it is possible to develop the Ω-arithmetization as an arithmetization method based on this new framework. The principle of this method is unchanged and the resulting algorithm is formally the same. The new and crucial facts are the following:

- The algorithm operates on Ω-numbers in a completely constructive way and consequently, in the applications, we can represent adequately all the entities present in the theory.
- The result of the algorithm appears to be an exact discrete multi-resolution representation of the continuous function on which the method is applied. See figure 1.

From the first point, we deduce that the implementation of the method does not lead to uncontrolled approximation errors. Even for the authors, the second point was a (good) surprise.

---

[1] For instance, for the figure of [3] page 2024, we took $\beta = 50$ as an infinitely large number.

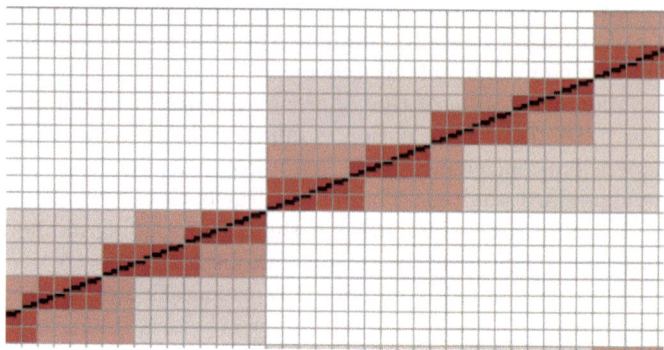

Fig. 1. Plot of the multi-resolution aspects of the Ω-arithmetization of the real function $X(T) = 2T/5$. (Full explanation in section 4).

Actually, this multi-resolution aspect is a normal consequence of the Ω-arithmetization: this is in relation with the nature of the scaling parameter β used in the method (see section 4). This parameter is an infinitely large Ω-integer that encodes an infinity of increasing scales. The arithmetization process gives simultaneously a discretization of the initial continuous function at each of these scales.

Since nowadays many developments in image analysis, geometric modeling, etc. use multi-resolution approaches and must deal with numerical precision problems, the Ω-arithmetization is a new tool which has the interesting property of taking into account these two aspects.

Let us note that our goal is not to define a discretization method that produce "better" images and "faster" algorithms. On the bases of a significant theoretical analysis, our originality is to propose a constructive and exact discrete representation of continuous functions[2]. Moreover, this framework naturally leads to a multi-resolution representation.

The paper is organized as follow: in part 2, we introduce the Ω-numbers and study their general mathematical and logical properties, in part 3, we use the Ω-numbers to define an Harthong-Reeb line $\mathcal{HR}_\omega$ and finally, in part 4 we present the Ω-arithmetization.

2   The Ω-Numbers of Laugwitz and Schmieden

In this section we will present the notion of Ω-numbers introduced by Laugwitz and Schmieden [12,11,10]. For the most part, we follow the presentations of these authors, but on some points, we have introduced new developments and, from our point of view, important distinctions. The Ω-numbers are nonstandard numbers

---

[2] For instance, we argue that the figure 4 (a) is a graphical illustration of an exact representation of the exponential function.

but the encompassing theory has two complementary characteristics: it seems theoretically weaker than the usual versions of nonstandard analysis [21,16,5] but it has an undeniable flavour of constructivity suggesting the possibility of explicit and exact computations. The principal goal of Laugwitz and Schmieden was to build a new approach to real analysis based only on the introduction of a set of Ω-rational numbers which is an extension of the usual set $\mathbb{Q}$. In our case and in view of the arithmetization process, we are mainly interested in the Ω-integers but we will occasionally use the Ω-rationals.

The first step is to extend a given formal theory T (unspecified but including an elementary theory of integer and rational numbers) by introducing a new number constant Ω and a new rule (BD) described thereafter. This leads to a new theory $T\langle\Omega\rangle$ in which we can form new statements depending on Ω for which the truth is given by the following *Basic Definition* (BD):

> *Let $S(n)$ be a statement of T depending on $n \in \mathbb{N}$. If $S(n)$ is true for almost $n \in \mathbb{N}$, then $S(\Omega)$ is true.*

We specify that here and in all this article, the expression "*almost $n \in \mathbb{N}$*" means "for all $n \in \mathbb{N}$ from some level", i.e. "$(\exists N \in \mathbb{N})$ such that $(\forall n \in \mathbb{N})$ with $n > N$". Since Ω can be substituted by any natural number, it denotes an Ω-number which is the first example of Ω-integer. Immediately, we can verify that Ω is infinitely large, i.e. greater than every element of $\mathbb{N}$. Indeed, for $p \in \mathbb{N}$, we apply (BD) to the statement $p < n$ which is true for almost $n \in \mathbb{N}$; thus $p < \Omega$ for each $p \in \mathbb{N}$.

The second step is to describe a world of mathematical objects which is a realization of the extended theory $T\langle\Omega\rangle$. For this purpose, we consider the set of sequences of integers or rational numbers. On this set, we introduce the equivalence relation $\mathcal{R}$ such that, for $a = (a_n)$ and $b = (b_n)^3$, we have $a\mathcal{R}b$ if and only if $a_n = b_n$ for *almost $n \in \mathbb{N}$*. Then, we introduce the following definition:

> *Each equivalence class for the relation $\mathcal{R}$ is called an Ω-number.*

In the general case, an Ω-number is also called an Ω-rational number. We agree to identify each sequence of numbers $a = (a_n)$ with the Ω-number equal to the equivalence class of a. Given a sequence $a = (a_n)$ such that $a_n \in \mathbb{Z}$ for all $n \in \mathbb{N}$, we can say that $a = (a_n)$ is an Ω-integer. Finally, we decide that the symbol Ω is the name of the particular Ω-number $(n)_{n\in\mathbb{N}}$. The following development will show that these choices are coherent.

Let $\mathbb{Z}_\Omega$ be the set of Ω-integers, $\mathbb{N}_\Omega$ be the set of Ω-integers $c = (c_n)$ such that $c_n \geqslant 0$ for almost $n \in \mathbb{N}$ and $\mathbb{Q}_\Omega$ be the set of Ω-rational numbers. We consider the embedding $i : \mathbb{Z} \to \mathbb{Z}_\Omega$ which associates to each $p \in \mathbb{Z}$ the constant sequence of value p. An Ω-integer $a = (a_n)$ is said to be *standard* if a belongs to the image of the preceding embedding, i.e. if there exists $p \in \mathbb{N}$ such that $a_n = p$ for almost $n \in \mathbb{N}$. Any sequence of integers $f = (f(n))$ is a map $f : \mathbb{N} \to \mathbb{Z}$ which has a

---

[3] Although this is not always indicated, in our sequences, the index n takes all the values $0, 1, \ldots$ in $\mathbb{N}$.

natural extension $f : \mathbb{N}_\Omega \to \mathbb{Z}_\Omega$ defined by $f(a) =_{def} (f(a_n))_{n \in \mathbb{N}}$ for $a = (a_n)$. For each Ω-integer $b = (b_n)$, we can extend the underlying sequence to $\mathbb{N}_\Omega$ and we obtain in particular $b_\Omega = (b_n) = b$. Applying this property to $(n)_{n \in \mathbb{N}}$, we find again $\Omega = (n)_{n \in \mathbb{N}}$, which partly shows the consistency of our previous choice. We do the same for the Ω-rational numbers.

Any operation or relation defined on $\mathbb{Z}$ (or $\mathbb{Q}$) naturally extends to $\mathbb{Z}_\Omega$ (or $\mathbb{Q}_\Omega$). For instance, for every Ω-numbers $a = (a_n)$ and $b = (b_n)$ let us set:

- $a + b =_{def} (a_n + b_n)$ and $-a =_{def} (-a_n)$ and $a \times b =_{def} (a_n \times b_n)$;
- $a > b =_{def} [(\exists N \forall n > N)\ a_n > b_n]$ and $a \geqslant b =_{def} [(\exists N \forall n > N)\ a_n \geqslant b_n]$;
- $|a| =_{def} (|a_n|)$.

It is easy to check that $(\mathbb{Z}_\Omega, +, \times)$ is a commutative ring with the constant sequence of value 0 as zero and the constant sequence of value 1 as unit. The previous map $i : \mathbb{Z} \to \mathbb{Z}_\Omega$ is an injective ring homomorphism which allows to identify $\mathbb{Z}$ with the subring of standard elements of $\mathbb{Z}_\Omega$. From now on, we identify any integer $p \in \mathbb{Z}$ with the Ω-integer $i(p)$ equal to the sequence of constant value p.

For the implementation of an arithmetization process based on Ω-integers, we need an extension of the Euclidean division to the Ω-integers and of the floor and the fractional part functions to the Ω-rational numbers.

- Given two Ω-integers $a = (a_n)$ and $b = (b_n)$ verifying $b > 0$, there is an unique $(q, r) \in \mathbb{Z}_\Omega^2$ such that $a = bq + r$ and $0 \leqslant r < b$. Indeed, since $b_n > 0$ from some level $N \in \mathbb{N}$, we can set $q = (q_n)$ and $r = (r_n)$ where, for $n \geqslant N$, q_n is the quotient of a_n by b_n and r_n is the remainder of this Euclidean division, and for $n < N$ the values of q_n and r_n are arbitrary (for instance 0). We will use the usual notations $a \div b$ for the quotient q and $a \bmod b$ for the remainder r.
- Given an Ω-rational number $r = (r_n)$, there is a unique $\lfloor r \rfloor \in \mathbb{Z}_\Omega$ and a unique $\{r\} \in \mathbb{Q}_\Omega$ such that $(0 \leqslant \{r\} < 1) \wedge (r = \lfloor r \rfloor + \{r\})$. Indeed, we can choose $\lfloor r \rfloor = (\lfloor r_n \rfloor)$ and similarly $\{r\} = (\{r_n\})$.

Regarding the order relation, the usual properties true on $\mathbb{Z}$ are not always verified on $\mathbb{Z}_\Omega$. For instance

$$(\forall a, b \in \mathbb{Z}_\Omega) \quad (a \geqslant b) \vee (b \geqslant a) \tag{1}$$

is not valid as we can see for the particular Ω-integers $a = ((-1)^n)_{n \in \mathbb{N}}$ and $b = ((-1)^{n+1})_{n \in \mathbb{N}}$. Nevertheless, given two arbitrary Ω-integers $a = (a_n)$ and $b = (b_n)$, we have

$$(\forall n \in \mathbb{N}) \quad (a_n \geqslant b_n) \vee (b_n \geqslant a_n). \tag{2}$$

Using (BD), we obtain $(a_\Omega \geqslant b_\Omega) \vee (b_\Omega \geqslant a_\Omega)$ and thus (1) since $a_\Omega = a$ and $b_\Omega = b$. Hence, there is a contradiction. To avoid it, we might admit that the application of (BD) leads to a notion of truth weaker than the usual notion. Hence, we introduce an important logical distinction:

> A statement is said to be weakly true in case it derives from (BD).

In contrast to the weak truth, we may use the terms of strong truth for the usual truth. For instance, (1) is weakly true but not strongly true, and the weak truth of (1) means exactly that (2) is (strongly) true. In the sequel, we will use the following properties.

Proposition 1. *The following statements are weakly true on $\mathbb{Z}_\Omega$:*
(1) $\forall(x, y) \in \mathbb{Z}_\Omega^2 \quad (x < y) \vee (x \geq y);$
(2) $\forall(x, y, z) \in \mathbb{Z}_\Omega^3 \quad (x + y > z) \Rightarrow (2x > z) \vee (2y > z).$

Proof. Let $x = (x_n)$, $y = (y_n)$ and $z = (z_n)$. For each $n \in \mathbb{N}$, we have

$$(x_n < y_n) \vee (x_n \geq y_n) \text{ and } (x_n + y_n > z_n) \Rightarrow (2x_n > z_n) \vee (2y_n > z_n)$$

Thus, we can apply (BD) and we get the two statements.

Let us remark that the first statement says that the order relation on $\mathbb{Z}_\Omega$ is (weakly) decidable.

Returning to the Ω-rational numbers, we can check that $(\mathbb{Q}_\Omega, +, \times, \geqslant)$ is a commutative ordered field for the weak truth. Given two Ω-integers $a = (a_n)$ and $b = (b_n)$, if $b \neq 0$ in the weak meaning, then b has an inverse b^{-1} in $\mathbb{Q}_\Omega$ and $a/b =_{def} a \times b^{-1}$ is an Ω-rational number. Conversely, if $r \in \mathbb{Q}_\Omega$ is weakly different from 0, then there is a unique pair $(a, b) \in \mathbb{Q}_\Omega^2$ with $b > 0$ such that $r = a/b$; then, it is easy to check that we have the usual relations $\lfloor r \rfloor = a \div b$ and $\{r\} = (a \bmod b)/b$.

An Ω-rational number $a = (a_n)$ is said to be *limited* in case there is a standard $p \in \mathbb{N}$ such that $|a| \leqslant p$ where $|a| = (|a_n|)$; this means that $|a_n| \leqslant p$ for almost $n \in \mathbb{N}$. Let $\mathbb{Q}_\Omega^{lim}$ be the set of limited Ω-rational numbers. In the same way, we say that a is *infinitely small* and we write $a \simeq 0$ in case $p|a| \leqslant 1$ for every $p \in \mathbb{N}$. For $a, b \in \mathbb{Q}_\Omega$, we write $a \simeq b$ when $a - b \simeq 0$ and $a \lesssim b$ when $p(a - b) \leq 1$ for every $p \in \mathbb{N}$. ; it is easy to check that $\simeq$ is an equivalence relation and that $\lesssim$ is an order relation on $\mathbb{Q}_\Omega$. This leads to the numerical system $(\mathbb{Q}_\Omega^{lim}, \simeq, \lesssim, +, \times)$ which is, for Laugwitz and Schmieden [11], an equivalent of the classical system of the real numbers $(\mathbb{R}, =, \leqslant, +, \times)$.

3   A Harthong-Reeb Line Based on Ω-Integers

The Harthong-Reeb line is a numerical line which is, in some meaning, both discrete and continuous. For obtaining such a paradoxical space, the basic idea is to make a strong contraction on the set $\mathbb{Z}$ such that the prescribed infinitely large $\omega \in \mathbb{N}$ becomes the new unit; the result of this scaling is a line which looks like the real one (See figure 2).

Historically, this system is at the origin of the definition of the analytic discrete line proposed by J.P. Reveillès [17,18] in discrete geometry. For a rigorous implementation of this idea, we must have a mathematical concept of infinitely large numbers. In previous works [4,6,3] on this subject, this was done with the help of an axiomatic version of nonstandard analysis in the spirit of the Internal

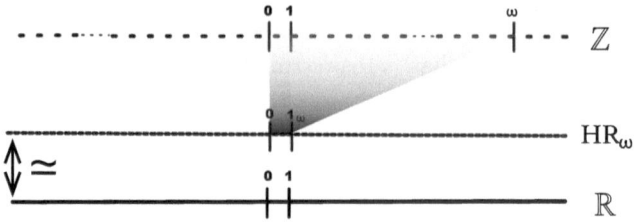

Fig. 2. An intuitive representation of the Harthong-Reeb line

Set Theory [16]. Our purpose in the present section is to define a Harthong-Reeb line based on the notion of Ω-integers introduced in the previous section. Our main motivation is to obtain a more constructive version of the Harthong-Reeb line allowing an exact translation of the arithmetization process into computer programs.

Although the definition has already been stated in the previous section, we recall that an Ω-number a is infinitely large if, for all $p \in \mathbb{N}$, we have $p \leq |a|$. If a is infinitely large and $a > 0$ we note $a \simeq +\infty$. We already know that $\Omega = (n)_{n \in \mathbb{N}} \simeq +\infty$. More generally, for $a = (a_n)$, it is easy to check that $a \simeq +\infty$ if and only if $\lim_{n \to +\infty} a_n = +\infty$.

> In the present and the next section, the symbol ω denotes a fixed Ω-integer such that $\omega \simeq +\infty$.

Let us remark that ω may be different from Ω. We only know that there is a sequence (ω_n) of natural numbers such that $\omega = (\omega_n)$ and $\lim_{n \to +\infty} \omega_n = +\infty$. Now, we are going to give the definition of the Harthong-Reeb line which results in the scaling on $\mathbb{Z}_\Omega$ such that ω becomes the new unit.

Definition 1. *We consider the following set*

$$\mathcal{HR}_\omega = \{x \in \mathbb{Z}_\Omega, \ \exists p \in \mathbb{N}, \ |x| \leq p\omega\}$$

and the relations, operations and constants on $\mathcal{HR}_\omega$ described by the following definitional equalities: for all $(x, y) \in \mathcal{HR}_\omega^2$, we set

- $(x =_\omega y) =_{def} (\forall p \in \mathbb{N}) \, (p | x - y| \leq \omega)$;
- $(x >_\omega y) =_{def} (\exists p \in \mathbb{N}) \, (p(x - y) \geq \omega)$;
- $(x \neq_\omega y) =_{def} (x >_\omega y) \vee (x <_\omega y)$;
- $(x \leq_\omega y) =_{def} (\forall z \in \mathcal{HR}_\omega) \, (z <_\omega x \Rightarrow z <_\omega y)$;
- $(x +_\omega y) =_{def} (x + y)$ *and* $0_\omega =_{def} 0$ *and* $-_\omega x =_{def} -x$;
- $(x \times_\omega y) =_{def} ((x \times y) \div \omega)$ *and* $1_\omega =_{def} \omega$ *and* $x^{(-1)_\omega} =_{def} (\omega^2 \div x)$ *for* $x \neq_\omega 0$.

Then, the Harthong-Reeb line is the numerical system $(\mathcal{HR}_\omega, =_\omega, \leq_\omega, +_\omega, \times_\omega)$.

We can say that $\mathcal{HR}_\omega$ is the set of Ω-integers which are limited at the scale ω. Note that the way of introducing separately the two order relations and the non-equality relation is quite traditional from a constructive point of view.

Proposition 2. *For every $x = (x_n)$ and $y = (y_n)$ in $\mathcal{HR}_\omega$, we have the following equivalences:*

(1) $x =_\omega y \iff \forall p \in \mathbb{N}\ \exists M_p \in \mathbb{N}\ \forall n \geq M_p\quad p|x_n - y_n| \leq \omega_n$
(2) $x >_\omega y \iff \exists p \in \mathbb{N}\ \exists M_p \in \mathbb{N}\ \forall n \geq M_p\quad p(x_n - y_n) \geq \omega_n$
(3) $x \leq_\omega y \iff \forall p \in \mathbb{N},\ p(x - y) \leq \omega$

Proof. The points (1) and (2) result of the definition of the order relation $\leq$ on $\mathbb{Z}_\Omega$. We will only give the outline of a proof of (3).
Let us suppose that $x \leq_\omega y$. For every $p \in \mathbb{N} \setminus \{0\}$, we consider $z_p =_{def} x - \lfloor \omega/p \rfloor$. Since $z_p <_\omega x$, we obtain $z_p <_\omega y$. Thus, there is $k \in \mathbb{N}$ such that $k(y - x + \lfloor \omega/p \rfloor) \geq \omega$. Hence, for every $p \in \mathbb{N}$

$$p(x - y) \leq p\lfloor \omega/p \rfloor - p\omega/k = p(\omega/p - \{\omega/p\}) - p\omega/k \leq \omega$$

Let us suppose now that $p(x - y) \leq \omega$ for each $p \in \mathbb{N}$. We consider an arbitrary $z \in \mathcal{HR}_\omega$ such that $z <_\omega x$. Thus, there is $k \in \mathbb{N}$ such that $k(x - z) \geq \omega$. We obtain $k(y - z) \geq k(y - x) + \omega$ and since $2k(y - x) \geq -\omega$ we get $2k(y - z) \geq \omega$ and thus $z <_\omega y$.

Now, we want to show that the Harthong-Reeb line is equivalent to the system of real numbers. In this context, the appropriate model for the real line is the system $(\mathbb{Q}_\Omega^{lim}, \simeq, \lesssim, +, \times)$ of limited Ω-rational numbers of Laugwitz and Schmieden described in the previous section. To this end, we introduce the two following maps:

$$\left\{ \begin{array}{c} \varphi_\omega : \mathcal{HR}_\omega \to \mathbb{Q}_\Omega^{lim} \\ x \mapsto x/\omega \end{array} \right\} \text{ and } \left\{ \begin{array}{c} \psi_\omega : \mathbb{Q}_\Omega^{lim} \to \mathcal{HR}_\omega \\ u \mapsto (\lfloor \omega u \rfloor) \end{array} \right\}$$

The proof of the following properties is straightforward.

Proposition 3. *For every $x, y \in \mathcal{HR}_\beta$ and $u \in \mathbb{Q}_\Omega^{lim}$, we have :*

- $x \leq_\omega y \quad \Rightarrow \quad \varphi_\omega(x) \lesssim \varphi_\omega(y)$;
- $\varphi_\omega(x +_\omega y) \simeq \varphi_\omega(x) + \varphi_\omega(y)$;
- $\varphi_\omega(x \times_\omega y) \simeq \varphi_\omega(x) \times \varphi_\omega(y)$;
- $\varphi_\omega(0_\omega) \simeq 0$ and $\varphi_\omega(1_\omega) \simeq 1$;
- $x =_\omega y \Leftrightarrow \varphi_\omega(x) \simeq \varphi_\omega(y)$;
- $\forall u \in \mathbb{Q}_\Omega^{lim}\ \exists x \in \mathcal{HR}_\omega \quad \varphi_\omega(x) \simeq u$;
- $\psi_\omega \circ \varphi_\omega(x) =_\omega x$ and $\varphi_\omega \circ \psi_\omega(u) \simeq u$.

We can summarize these properties by saying that φ_ω *is an isomorphism from* $(\mathcal{HR}_\omega, =_\omega, \leq_\omega, +_\omega, \times_\omega)$ *to* $(\mathbb{Q}_\Omega^{lim}, \simeq, \lesssim, +, \times)$ *and that ψ_ω is the inverse isomorphism.*

Since the Harthong-Reeb line $\mathcal{HR}_\omega$ is a kind of model of the real line, it is natural to wonder about the constructive content of this new numerical system. With regards to the constructivism, we only recall that these mathematics are characterized by the BHK-interpretation of the logical constants[4] and, for more precisions we refer to the excellent description given in [1]. Although we do not develop this point in this article, we have shown that the Harthong-Reeb line satisfied the axiomatic presentation of the constructive real line proposed by Bridges [1,2]. Of course, for $\mathcal{HR}_\omega$ some of the axioms of Bridges are only weakly true. The proof is long and technical and will appear in a future paper.

4   Arithmetization with Ω-Integers

In the previous sections, we have seen that, given an infinitely large Ω-integer ω, the corresponding Harthong-Reeb line $\mathcal{HR}_\omega$ is a relatively constructive numerical system equivalent to the real number system. This equivalence gives us a way to represent continuous entities (real numbers, real function, etc.) in the discrete model $\mathcal{HR}_\omega$. Moreover, this representation comes with a strong computational content that allows us to derive concrete algorithms.

To illustrate this, we give the arithmetization of the well-known Euler integration scheme. Then, we use it to produce arithmetized solutions of differential equations. These solutions are respectively linear functions, exponential functions and arcs of parabola. These examples demonstrate important facts about our approach. Firstly, we can produce arithmetization of continuous functions (or objects) using only integer computations. Secondly, the arithmetized objects are exact. This means that arithmetized objects contain as much computable informations as the continuous ones.

For our purpose, we recall the developments presented in [3]. The new point is that we are now working with the rich structure of the Ω-integers. To do so, consider a real function $X : T \mapsto X(T)$ which is solution of the Cauchy problem $X' = F(T, X)$ with the initial condition $X(A) = B$. Approximations of the function X are obtained using the Euler scheme with integration step h and real variables T_k and X_k:

$$\begin{cases} T_0 = A \; ; X_0 = B \\ T_{k+1} = T_k + h \\ X_{k+1} = X_k + F(T_k, X_k) \times h. \end{cases} \tag{3}$$

This defines a sequence of points $(T_k, X_k)_{k \in \mathbb{N}}$ that approximate the solution curve X. It is well know that we obtain more accurate sequences when the integration step h becomes smaller.

The arithmetization method transfers the Euler approximation scheme to the discrete world $\mathcal{HR}_\omega$. To this end, we choose ω such that there is $\beta \simeq +\infty$ in $\mathbb{Z}_\Omega$ with $\omega = \beta^2$. Here, the product we use to define the multiplication β^2 is the one from $\mathbb{Z}_\Omega$. Now, we consider that $h = 1/\beta$.

---

[4] The interpretation of Brouwer, Heyting and Kolmogorov which defines the intuitionistic logic.

Since the map $\psi_\omega : U \mapsto \lfloor U\omega \rfloor$ is an isomorphism between the real line and the Harthong-Reeb line $\mathcal{HR}_\omega$, it is natural to introduce in (3) the change of variables $t_k =_{def} \lfloor \omega T_k \rfloor$ and $x_k =_{def} \lfloor \omega X_k \rfloor$. Then, neglecting the terms τ such that $\tau =_\omega 0$, we get the following scheme which is an arithmetic analogue of (3) with Ω-integer variables t_k, x_k:

$$\begin{cases} t_0 = a \; ; \; x_0 = b \\ t_{k+1} = t_k + \beta \\ x_{k+1} = x_k + f(t_k, x_k) \div \beta \end{cases} \tag{4}$$

where $f(t_k, x_k) = \lfloor \omega F(t_k/\omega, x_k/\omega) \rfloor$, $a = \lfloor \omega A \rfloor$ and $b = \lfloor \omega B \rfloor$.

Because the integration step is an infinitesimal ($h = 1/\beta$ with $\beta \simeq +\infty$), the discrete function whose graph is the set of the points (t_k, x_k) is an exact[5] representation of the initial real continuous function $X : T \mapsto X(T)$.

This arithmetized solution suffers however from a major imperfection: its domain is not connected at all, since $t_{k+1} - t_k = \beta$ that cannot be neglected at the scale ω. In order to correct this defect, we perform the following scaling:

$$\begin{array}{ll} \psi_\beta \circ \varphi_\omega : \mathcal{HR}_\omega \longrightarrow \mathcal{HR}_\beta \\ \quad x \longmapsto \lfloor x\omega/\beta \rfloor = x \div \beta. \end{array}$$

whose meaning is that we observe now the arithmetized solution at the intermediate scale β.

In order to compute the effect of this scaling, it is convenient to introduce the following notation: for every $x \in \mathcal{HR}_\omega$, we write $x = \widetilde{x}\beta + \widehat{x}$, where $\widetilde{x} =_{def} x \div \beta$ and $\widehat{x} =_{def} x \mod \beta$. The operations $\div$ and mod are the quotient and the remainder in $\mathbb{Z}_\Omega$. Using these notations, we see that $\widetilde{x} \in \mathcal{HR}_\beta$ is the result of the scaling on $x \in \mathcal{HR}_\omega$. Hence, from (4) we obtain the following Ω-arithmetization of the Euler Scheme at the intermediary scale β

$$\begin{cases} \widetilde{t}_0 = a \div \beta, \; \widetilde{x}_0 = b \div \beta \text{ and } \widehat{x}_0 = b \mod \beta \\ \widetilde{t}_{k+1} = \widetilde{t}_k + 1 \\ \widetilde{x}_{k+1} = \widetilde{x}_k + (\widehat{x}_k + \widetilde{f}_k) \div \beta \\ \widehat{x}_{k+1} = (\widehat{x}_k + \widetilde{f}_k) \mod \beta \end{cases} \tag{5}$$

where $\widetilde{f}_k = f(\widetilde{t}_k \beta + a \mod \beta, \widetilde{x}_k \beta + \widehat{x}_k) \div \beta$ and $f(t, x) = \lfloor \omega F(t/\omega, x/\omega) \rfloor$.

Now, the relevant variables are $\widetilde{t}_k$ and $\widetilde{x}_k$ while the $\widehat{x}_k$ are auxiliary variables that manage the remainder coming from the Euclidean division. The important outcome of this scaling is that the discrete function whose graph is the set of points $(\widetilde{t}_k, \widetilde{x}_k)$ is now defined over a connected domain, i.e. $\widetilde{t}_{k+1} - \widetilde{t}_k = 1$ that can be neglected at the scale β.

This discrete function $(\widetilde{t}_k, \widetilde{x}_k)$ is the arithmetization of the initial real function $X : T \mapsto X(T)$ at the intermediate scale β. It is a discrete and exact representation of X.

From a practical point of view the Ω-integers used in the algorithm associated to (5) are nothing else than sequences of integers as defined in section 2. So in the

---

[5] This arithmetized solution contains the same information as the original solution.

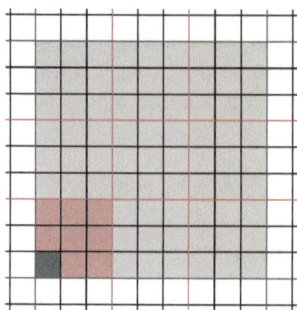

Fig. 3. Representations of an Ω-integer point at levels 1, 3 and 9 with $\beta : n \mapsto n$. From one level to another, each square is divided into 9 squares

implementation, Ω-integers are easily represented by functions from the positive integer type to the integers type[6]. Operations over Ω-integers are defined exactly how they are defined in section 2. As an example, for a fixed level n, the sum of two Ω-integers $a = (a_n)$ and $b = (b_n)$ is the function that computes $a_n + b_n$.

In order to plot the results of computations, we have to give a correct interpretation of the different levels n of Ω-integers. As we are working in $\mathcal{HR}_\beta$ all the Ω-integers must be interpreted at the scale β. This means that, for an Ω-integer $a = (a_n)$, the level n must be interpreted at the scale β_n.

As an example, let us define Ω-integer points in the plane by two Ω-integers $a = (a_n)$ and $b = (b_n)$. At level 1, the (Ω-)point can be represented by a square of side of size l. Hence, at level n the point is represented by a square of side of size l/β_n. Thus, if we choose $\beta : n \mapsto n$, at level 1 the point (a, b) can be represented by a square of side of size l, at level 3 the point (a, b) is represented by a square of side of size $l/3$, at level 9 it is represented by a square of side of size $l/9$ and so on[7]. The graphical representation we obtain is a sequence of overlapping squares the we have drawn on figure 3. Note that at level $3i$, 9 squares fit into the level $3(i-1)$ for $i \in \mathbb{N}$.

Other choices can be made for β. For example, if we choose $\beta : n \mapsto 2^n$, the usual quad-tree representation is obtained, see figure 4(a). We can also choose logarithmic scales ($\beta : n \mapsto \log n$) like we do on figure 4(b).

Now, using the Ω-integer implementation by integer functions, translation of the Euler scheme (5) into a computer programming language provides the algorithm that computes the Ω-arithmetization of the function X solution to the Cauchy's problem $X' = F(T, X)$.

The program we obtain computes the discrete function $(\widetilde{t}_k, \widetilde{x}_k)$ and we plot this function at different levels. On the figure 5(a), we chose $\beta : n \mapsto n$ and we give some plots of the arithmetization of the exponential function $t \mapsto e^t/3$.

---

[6] As we do in our implementation using the functional programming language O'Caml [9].

[7] Level 0 can be seen as an initialization step.

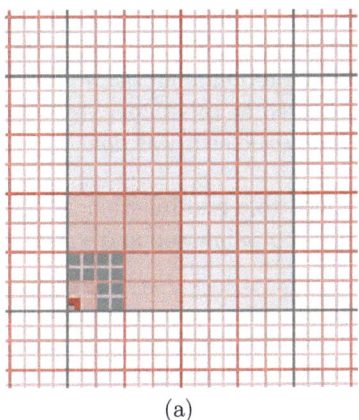
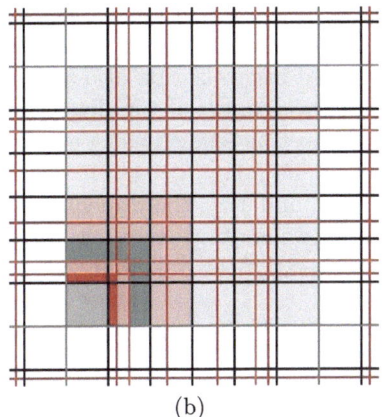

(a)   (b)

Fig. 4. Different choices for the β. On 4(a) $\beta : n \mapsto 2^n$, the point is represented from levels 1 to 6, this is the usual quad-tree representation. On 4(b) $\beta : n \mapsto \log n$, the point is represented at levels 10, 10^2, 10^3, 10^4, 10^5 and 10^6. The square of level 10^n divides the square of level 10 into n^2 squares.

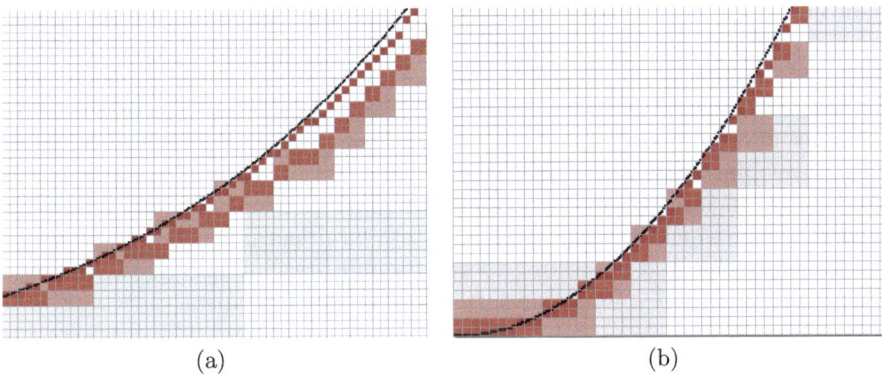

(a)   (b)

Fig. 5. Plots of the Ω-arithmetization of an exponential function (a) and a parabolic function (b) with $\beta : n \mapsto n$

In that case, we have $\widetilde{f}_k := \widetilde{x}_k$ in the Euler scheme (5). Figure 5(b) gives the corresponding plots for the parabolic function $t \mapsto t^2$ (here we have $\widetilde{f}_k := \widetilde{t}_k \div 2$).

The different colors encode the level n in the Ω-integers: the most light-colored for the first considered level (here $n = 5$) and the darkest color for the last considered level (here $n = 200$, intermediary levels are $n = 10$, $n = 20$ and $n = 40$).

The different discrete curves that are plotted for each level are different approximations of the function. It works like if we had computed the Euler scheme for $h = 1/\beta_n$ for $n = 5, 10, 20, 40$ and 200. We observe that accuracy grows when h becomes smaller like it is expected.

We must keep in mind that, when we compute with Ω-integers in $\mathcal{HR}_\beta$, the computations are made at the same time for all scales β_n. Consequently, the obtained discretization also appears to be a multi-resolution analysis.

Connectedness of the discretization is not addressed in this paper, but this issue must be studied using appropriate intermediate scale as it was done by the authors in [20].

To summarize, the principle of the method applied to a real function X is to compute simultaneously, for every $n \in \mathbb{N}$, a discrete approximation and a scaling of ratio β_n of this function.

5   Conclusion

In the present paper, we have introduced the Ω-arithmetization as a method which gives a discrete and multi-scale representation of a continuous function solution of a differential equation. Due to the structure of the Ω-integers, we obtain completely constructive algorithms which can be exactly translated into functional computer programs. As a consequence, these programs do not generate any numerical error. Moreover, the result appears to be a new tool for a multi-resolution analysis of discrete functions arising from continuous ones.

In future works on this subject, we plan to study systematically this form of multi-resolution analysis and its applications to discrete geometry. In addition, we intend to change our general theoretical framework; we want to move to the formalism of constructive type theory of P. Martin-Löf [13,14]. The first reason is that this stark approach of mathematics and computer science is well suited for both developing constructive mathematics and writing programs. Furthermore, Martin-Löf has already developed a nonstandard extension of constructive type theory [15] in which we dispose of infinitely large natural numbers. Hence, it would be possible and interesting to build a multi-resolution analysis and more generally a theory of scaling transformations in this formalism.

Acknowledgement

The authors would like to thank Christophe Saint-Jean for his precious programming help.

References

1. Bridges, D.S.: Constructive mathematics: a foundation for computable analysis. Theoretical Computer Science 219, 95–109 (1999)
2. Bridges, D., Reeves, S.: Constructive mathematics, in theory and programming practice. Technical Report CDMTCS-068, Centre for Discrete Mathematics and Theorical Computer Science (1997)
3. Chollet, A., Wallet, G., Fuchs, L., Largeteau-Skapin, G., Andres, E.: Insight in discrete geometry and computational content of a discrete model of the continuum. Pattern Recognition 42, 2220–2228 (2009)

4. Diener, M.: Application du calcul de Harthong-Reeb aux routines graphiques. In: Salanskis, J.M., Sinaceurs, H. (eds.) Le Labyrinthe du Continu, pp. 424–435. Springer, Heidelberg (1992)

5. Diener, F., Reeb, G.: Analyse Non Standard. Hermann, Paris (1989)

6. Fuchs, L., Largeteau-Skapin, G., Wallet, G., Andres, E., Chollet, A.: A first look into a formal and constructive approach for discrete geometry using nonstandard analysis. In: Coeurjolly, D., Sivignon, I., Tougne, L., Dupont, F. (eds.) DGCI 2008. LNCS, vol. 4992, pp. 21–32. Springer, Heidelberg (2008)

7. Harthong, J.: Éléments pour une théorie du continu. Astérisque 109/110, 235–244 (1983)

8. Harthong, J.: Une théorie du continu. In: Barreau, H., Harthong, J. (eds.) La mathématiques non standard, Éditions du CNRS, pp. 307–329 (1989)

9. INRIA: The caml language, http://www.ocaml.org

10. Laugwitz, D.: Leibniz' principle and omega calculus. In: Salanskis, J., Sinacoeur, H. (eds.) Le Labyrinthe du Continu, pp. 144–155. Springer, France (1992)

11. Laugwitz, D.: Ω-calculus as a generalization of field extension. In: Hurd, A. (ed.) Nonstandard Analysis - Recent developments. Lecture Notes in Mathematics, pp. 144–155. Springer, Heidelberg (1983)

12. Laugwitz, D., Schmieden, C.: Eine Erweiterung der Infinitesimalrechnung. Mathematische Zeitschrift 89, 1–39 (1958)

13. Martin-Löf, P.: Constructive mathematics and computer programming. In: Logic, Methodology and Philosophy of Science VI, pp. 153–175 (1980)

14. Martin-Löf, P.: Intuitionnistic Type Theory. Bibliopolis, Napoli (1984)

15. Martin-Löf, P.: Mathematics of infinity. In: Martin-Löf, P., Mints, G. (eds.) COLOG 1988. LNCS, vol. 417, pp. 146–197. Springer, Heidelberg (1990)

16. Nelson, E.: Internal set theory: A new approach to nonstandard analysis. Bulletin of the American Mathematical Society 83, 1165–1198 (1977)

17. Reveillès, J.P.: Géométrie discrète, Calcul en nombres entiers et algorithmique. PhD thesis, Université Louis Pasteur, Strasbourg, France (1991)

18. Reveillès, J.P.: Mathématiques discrètes et analyse non standard. In: Salanskis, J.M., Sinaceurs, H. (eds.) Le Labyrinthe du Continu, pp. 382–390. Springer, Heidelberg (1992)

19. Reveillès, J.P., Richard, D.: Back and forth between continuous and discrete for the working computer scientist. Annals of Mathematics and Artificial Intelligence, Mathematics and Informatic 16, 89–152 (1996)

20. Richard, A., Wallet, G., Fuchs, L., Andres, E., Largeteau-Skapin, G.: Arithmetization of a circular arc. In: DGCI 2009, Montreal, Canada (2009) (to be published)

21. Robinson, A.: Non-standard analysis, 2nd edn. American Elsevier, New York (1974)

Tree-Based Encoding for Cancellations on Morse Complexes

Lidija Čomić[1] and Leila De Floriani[2]

[1] Faculty of Engineering, University of Novi Sad, Serbia
comic@uns.ac.rs
[2] Department of Computer Science, University of Genova, Italy
deflo@disi.unige.it

Abstract. A scalar function f, defined on a manifold M, can be simplified by applying a sequence of removal and contraction operators, which eliminate its critical points in pairs, and simplify the topological representation of M, provided by Morse complexes of f. The inverse refinement operators, together with a dependency relation between them, enable a construction of a multi-resolution representation of such complexes. Here, we encode a sequence of simplification operators in a data structure called an *augmented cancellation forest*, which will enable procedural encoding of the inverse refinement operators, and reduce the dependency relation between modifications of the Morse complexes. In this way, this representation will induce a high flexibility of the hierarchical representation of the Morse complexes, producing a large number of Morse complexes at different resolutions that can be obtained from the hierarchy.

Keywords: Morse theory, Morse complexes, simplification operators, graph-based representation, augmented cancellation forest.

1 Introduction

Morse and Morse-Smale complexes are a widely-used topological representation which describes the subdivision of a manifold M into meaningful parts, characterized by uniform flow of the gradient of a scalar function f defined on M. In scientific visualization, large data sets are usually interpolated by a continuous function, and then topological features are extracted. Morse and Morse-Smale complexes can be applied for segmenting the graph of a scalar field for terrain modeling in 2D, and recently some algorithms have been developed for segmenting three-dimensional scalar fields through Morse-Smale complexes [10,4,8].

One of the major issues when computing a representation of a scalar field as a Morse, or a Morse-Smale complex is the over-segmentation, due to the presence of noise in the data sets. *Simplification algorithms* have been developed in order to eliminate less significant features from the Morse or the Morse-Smale complex [14,7,2,13,9]. Simplification is achieved by applying an operator called *cancellation* of critical points. A cancellation transforms one Morse complex into

P. Wiederhold and R.P. Barneva (Eds.): IWCIA 2009, LNCS 5852, pp. 330–343, 2009.

another, with fewer number of cells, and enables the creation of a hierarchical representation. In 2D, a cancellation consists of collapsing a maximum-saddle pair into a maximum, or a minimum-saddle pair into a minimum, while maintaining the consistency of the underlying complex. In higher dimensions, a cancellation which does not involve an extremum may increase the number of pairs of cells which become incident to each other, although the number of cells in the Morse complexes decreases by two. In [3], we defined two operators which we called *removal* and *contraction*, as a special case of a general cancellation. These two operators can be viewed as merging of cells in the Morse complexes.

Here, we represent a sequence of such simplification operators in a data structure which we call an *augmented cancellation forest*. This representation is dimension independent, and generalizes the one in [1], introduced for a 2D case. Our aim is to define a data structure which will enable an easy implementation of simplification operators, and of inverse refinement operators, thus providing the means of defining a variable resolution adaptive representation of the morphology of the Morse complexes. A general hierarchical model, based on cell complexes, is presented in [5]. This model, applied directly on the sequence of Morse complexes simplified by removal and contraction operators, implies that the inverse refinement operator is an undo of the corresponding simplification. This induces many dependencies between modifications of the Morse complexes, and thus reduces the number of different representations of the Morse complexes which can be obtained from the model. Here, we propose a different approach. We introduce an augmented cancellation forest, which will enable procedural encoding of refinements, and will allow us to reduce the dependency relation between modifications. In this way, a more flexible adaptive multiscale model of the Morse complexes will be obtained.

The remainder of the paper is organized as follows. In Section 2, we review some basic notions on Morse theory. In Section 3, we recall the definition of a removal and contraction, and how a Morse complex is affected by those operators. In Section 4, we recall the notion of a cancellation forest in 2D. In Section 5, we introduce a new notion of an augmented cancellation forest as a data structure to encode a sequence of simplification operators on Morse complexes, and we give an algorithm for the construction of an augmented cancellation forest starting from a sequence of removals and contractions on a Morse complex. Conclusion and future work are given in Section 6.

2   Morse Theory

Morse theory studies the relationship between the topology of a manifold, and the critical points of a scalar (real-valued) function defined on the manifold (for more details on Morse theory, see [11,12]).

Let f be a C^2 real-valued function defined over a closed compact n-manifold M. A point p is a *critical point* of f if and only if the gradient $\nabla f = (\frac{\partial f}{\partial x_1}, ..., \frac{\partial f}{\partial x_n})$ (in some local coordinate system around p) of f vanishes at p. Function f is a *Morse function* if all its critical points are non-degenerate (the Hessian matrix

$Hess_p f$ of the second derivatives of f at p is non-singular). The number of negative eigenvalues of $Hess_p f$ is called the *index* of critical point p, and p is called an *i-saddle*. A 0-saddle, or an n-saddle, is also called a *minimum*, or a *maximum*, respectively. An *integral line* of f is a maximal path which is everywhere tangent to the gradient of f. Each integral line connects two critical points of f, called its *origin* and *destination*. Each 1-saddle ($(n-1)$-saddle) is connected by an integral line to exactly two, not necessarily distinct, minima (maxima).

Integral lines that converge to (originate at) a critical point p of index i form an *i-cell* ($(n-i)$-cell) p called a *descending (ascending) manifold*, or cell, of p. The descending and ascending manifolds decompose M into *descending* and *ascending Morse complexes*, denoted by Γ_d and Γ_a, respectively, illustrated in 2D in Figure 1. If two cells p and q in the descending complex Γ_d are such that p is on the boundary of q, then the critical point q has a higher f value than the critical point p. A Morse function f is called a *Morse-Smale function* if the descending and the ascending manifolds intersect transversally. If f is a Morse-Smale function, then the complexes Γ_a and Γ_d are dual to each other.

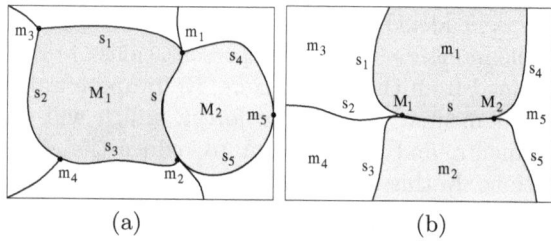

(a)   (b)

Fig. 1. (a) A 2D descending Morse complex. The descending 2-cells of maxima M_1 and M_2 are shaded. (b) The dual ascending Morse complex. The ascending 2-cells of minima m_1 and m_2 are shaded.

Simplification of a Morse function f can be achieved by an operator called *cancellation* of (a pair of) critical points. Two critical points p and q can be canceled if and only if

1. p and q are of index i and $i + 1$, respectively, $0 \leq i \leq n - 1$, and
2. there is a unique integral line connecting p and q.

In the descending Morse complex, after a cancellation of an i-saddle p and an $(i+1)$-saddle q, each cell r which was on the boundary of $(i+1)$-cell q in Γ_d becomes incident to each cell t which was in the co-boundary of i-cell p in Γ_d. Thus, if the cancellation does not involve an extremum, the number of pairs of incident cells in the Morse complexes increases, although the number of cells in the Morse complexes decreases by two.

3   Simplification of Morse Complexes

In [3], we have defined two dual simplification operators in arbitrary dimensions, which we call *removal* and *contraction*. We have obtained them by imposing additional constraints on a general cancellation, which ensure that the two operators do not enlarge the incidence relation on the Morse complexes, and that they can be viewed as merging of cells in the Morse complexes.

A *removal* (a *contraction*) (of index i) of an i-saddle s and an $(i+1)$-saddle $((i-1)$-saddle) p is defined if s is connected by a unique integral line to

1. $(i+1)$-saddle $((i-1)$-saddle) p, and exactly one other $(i+1)$-saddle $((i-1)$-saddle) q different from p, or
2. exactly one $(i+1)$-saddle $((i-1)$-saddle) p.

In the first case, a removal (contraction) of s and p is denoted by $rem(p, s, q)$ $(con(p, s, q))$, and in the second case by $rem(p, s, \emptyset)$ $(con(p, s, \emptyset))$.

The effect of a removal $rem(p, s, q)$ on the descending Morse complex Γ_d is that i-cell s is deleted and $(i+1)$-cell p is merged into $(i+1)$-cell q $((i+1)$-cell q is, after removal, the union of itself with $(i+1)$-cell p before removal). Thus, the boundary of p is merged into the boundary of q in Γ_d. An example of the effect of a removal $rem(M_1, s, M_2)$ on a 2D descending Morse complex is illustrated in Figure 2 (a).

A contraction $con(p, s, q)$ deletes i-cell s and merges $(i-1)$-cell p into $(i-1)$-cell q in Γ_d (i-cell s is deleted, and each i-cell in the co-boundary of p is extended to include a copy of i-cell s, i.e., each i-cell in the co-boundary of p is, after contraction, the union of itself with i-cell s before contraction). Thus, the co-boundary of p is merged into the co-boundary of q in Γ_d. An example of a contraction $con(e_1, f, e_2)$ on a 3D descending Morse complex is illustrated in Figure 2 (b).

A removal $rem(p, s, q)$ (contraction $con(p, s, q)$) of the second kind, deletes i-cell s and $(i+1)$-cell $((i-1)$-cell) p from Γ_d.

When $i = 0$ (and dually when $i = n$), a special case of a removal $rem(p, s, q)$ of index 0 is obtained, which is defined when a 0-cell s is incident once to exactly two different 1-cells p and q in Γ_d, and which has the same geometric effect as

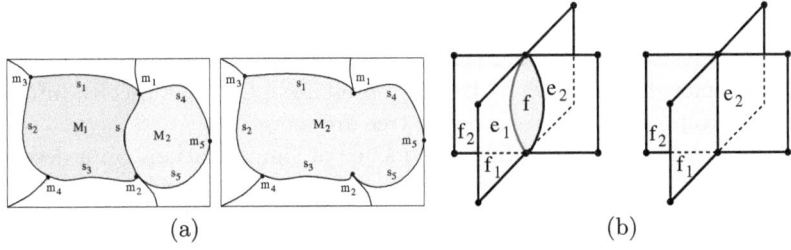

(a)   (b)

Fig. 2.   (a) Part of a 2D descending Morse complex before and after a removal $rem(M_1, s, M_2)$. (b) Part of a 3D descending Morse complex before and after a contraction $con(e_1, f, e_2)$.

a contraction $con(s, p, s')$, where s' is the other 0-cell on the boundary of 1-cell p. The effect of both operators is that 0-cell s' on the boundary of p, different from 0-cell s, becomes incident to cells in the co-boundary of 0-cell s, different from p. Depending on the application, such cases may be forbidden or allowed.

For the two types of operators, consistency of the simplified complexes from the point of view of function values can be guaranteed. For a feasible removal of index i, from the two possible removals with middle cell s, $rem(p, s, q)$, and $rem(q, s, p)$, the first one is performed if $f(p) < f(q)$, and the second one is performed if $f(p) > f(q)$. In the first case, all i-cells t in Γ_d which were incident to $(i + 1)$-cell p before a removal, and which become incident to $(i + 1)$-cell q after the removal, have a lower function value than q, i.e., $f(t) < f(q)$, and symmetrically for the second case. Conditions for a contraction are dual. For a removal $rem(p, s, \emptyset)$ or a contraction $con(p, s, \emptyset)$ of the second kind, consistency from the point of view of function values is trivially satisfied.

4   Cancellation Forest in 2D

In 2D, a general cancellation involves either a saddle s and a maximum M, or a saddle s and a minimum m, and each saddle s is connected to exactly two maxima, and exactly two minima. Thus, in 2D, a general cancellation of a saddle s and a maximum M reduces to a removal $rem(M, s, M')$, where M' is a unique other maximum (different from M) connected to s. In the descending Morse complex Γ_d, a removal $rem(M, s, M')$ merges the descending 2-cell M into the descending 2-cell M'. A general cancellation of a saddle s and a minimum m reduces to a contraction $con(m, s, m')$, where m' is a unique other minimum (different from m) connected to s. In Γ_d, a contraction $con(m, s, m')$ merges the descending 0-cell m into the descending 0-cell m'. A sequence of removals and contractions on an initial full-resolution Morse complex can be encoded in a data structure called cancellation forest, introduced in [1], in which each tree encodes either a set of removals, or a set of contractions.

A removal (contraction) tree is a rooted tree. Nodes of a removal (contraction) tree correspond to maxima (minima) in the initial highest-resolution descending Morse complex. Each arc in a removal (contraction) tree represents a saddle, i.e., a 1-cell s, such that a removal $rem(M, s, M')$ (contraction $(con(m, s, m'))$ belongs to the set of simplifications on the Morse complex, and both nodes M and M' (nodes m and m'), and arc s, belong to the same tree in the forest. The root of a removal (contraction) tree T represents a 2-cell (a 0-cell), into which all cells corresponding to nodes of the tree are merged.

A cancellation forest encoding a sequence of simplifications on a descending Morse complex is built starting from a graph (forest), in which each removal (contraction) tree has only one node, corresponding to a maximum (a minimum). Each simplification merges two trees into one tree, by introducing an arc in the forest, and the inverse refinement operator removes an arc from the forest [1].

Figure 3 shows a sequence of removals on a 2D descending Morse complex, and the construction of the corresponding removal tree encoding that sequence. This example is taken from [1], but is expressed in terms of a descending Morse complex instead of a Morse-Smale complex. First, removals $rem(M_1, s_1, M_2)$ and $rem(M_2, s_2, M_3)$ merge 2-cells M_1 and M_2 into M_3, and removals $rem(M_4, s_3, M_5)$ and $rem(M_6, s_4, M_5)$ merge 2-cells M_4 and M_6 into M_5, as shown in Figure 3 (a). Then, removals $rem(M_7, s_6, M_5)$ and $rem(M_3, s_5, M_5)$ merge 2-cells M_7 and M_3 into M_5. Node M_3 is not connected to node M_5 in the cancellation forest, because descending 2-cells M_3 and M_5 are not adjacent in the initial Morse complex. Instead, the tree rooted at node M_3 is connected to the tree rooted at node M_5 through an arc connecting its node M_2 to node M_5 (since descending 2-cells M_2 and M_5 are adjacent in the initial Morse complex), as shown in Figure 3 (b). The coarsest-resolution complex, and the corresponding cancellation forest (which in this example is a tree), are shown in Figure 3 (c).

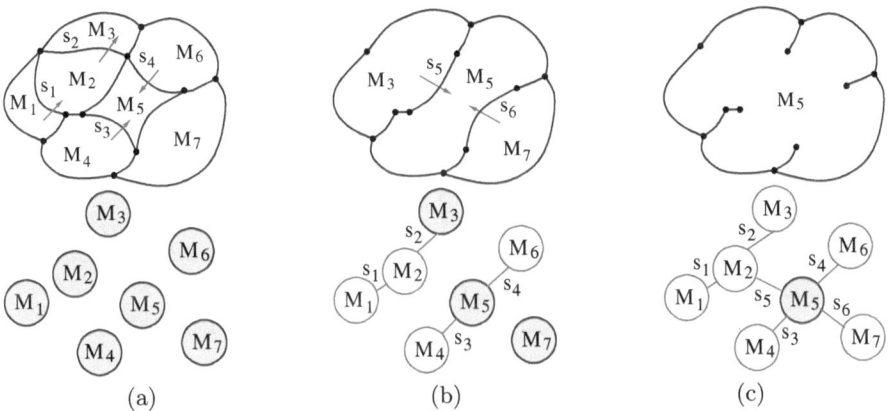

Fig. 3. A sequence of removals on a 2D descending Morse complex, and the process of building the cancellation forest, consisting of removal trees. The root of each tree created in the process is shaded.

5   Augmented Cancellation Forest

In this Section, we extend the notion of the cancellation forest to arbitrary dimensions. In Section 5.1, we discuss some issues related to the extension to higher dimensions. In Section 5.2, we introduce the notion of an augmented cancellation forest. In Section 5.3, we give an algorithm for constructing an augmented cancellation forest, starting from a sequence of simplifications on an n-dimensional Morse complex.

5.1   Augmented Cancellation Forest - Motivation

As we have seen, in 2D a cancellation reduces either to a removal or to a contraction, depending on index of the canceled points, i.e., on the dimension of the corresponding descending cells. A cancellation of a 2-cell and a 1-cell reduces to a removal, and a cancellation of a 0-cell and a 1-cell reduces to a contraction. In other words, the role of a critical point p in the cancellation forest is determined by the index of p: a maximum (minimum) may only correspond to a node in a removal (contraction) tree, and a saddle may only correspond to an arc in a removal or a contraction tree.

The extension of the tree-based encoding of a sequence of simplification operators to arbitrary dimensions is non-trivial due to the following reasons.

– In higher dimensions, a simplification is not uniquely determined by the index of the eliminated critical points, as a critical point p of index i and a critical point q of index $i+1$ may be involved either in a removal $rem(q, p, r)$, or in a contraction $con(p, q, r)$. Thus, a critical point of index i may correspond either to a node or to an arc in either a removal or a contraction tree. For example, a simplification eliminating 1-cell p and 2-cell q may be a removal $rem(q, p, r)$, shown in Figure 4 (a), or a contraction $con(p, q, t)$, shown in Figure 4 (b).

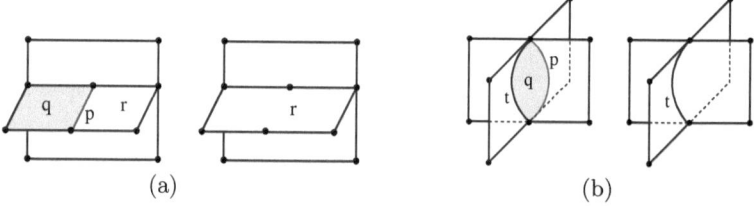

(a)   (b)

Fig. 4. The type of a simplification is not determined by the index of eliminated critical points. (a) Removal $rem(q, p, r)$ of 1-cell p is feasible, while contraction of 2-cell q is not feasible. (b) Contraction $con(p, q, t)$ of 2-cell q is feasible, while removal of 1-cell p is not feasible.

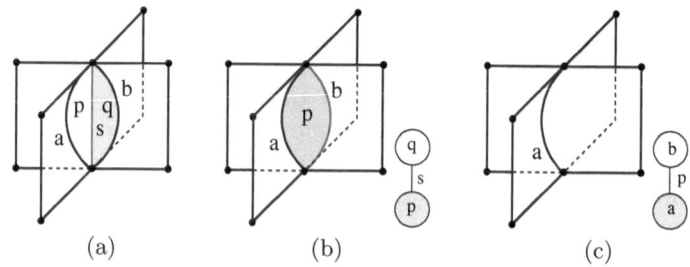

(a)   (b)   (c)

Fig. 5. (a) Initial 3D descending complex. (b) After a removal $rem(q, s, p)$. (c) After a contraction $con(b, p, a)$. The root of each tree is shaded.

- The role of a critical point may change after a cancellation. An example is illustrated in Figure 5. Initially, 2-cells p and q, and 1-cells a, b, and s, are the roots of trees, each having only one node. After a removal $rem(q, s, p)$, p and q are nodes, and s is an arc in a removal tree rooted at p, as shown in Figure 5 (b). After a contraction $con(b, p, a)$, b and a are nodes, and p is an arc in a contraction tree, rooted at a, as shown in Figure 5 (c).
- In higher dimensions, removals and contractions of the second kind are not ignored, as is the case in 2D. Thus, we will have two kinds of removal (contraction) trees, depending on the type of a simplification they encode.
- We want a tree-based data structure which will be able to handle the inverse refinements efficiently, namely as deletion of arcs in the forest, like in the 2D case.

5.2   Augmented Cancellation Forest - Definition

In this Section, we introduce the notions of an augmented tree and an augmented cancellation forest.

Informally, in an augmented tree each node and each arc may encode another augmented tree. An *augmented tree* is a tree which may have regular nodes and hyper-nodes, and regular arcs and hyper-arcs. A hyper-node (a hyper-arc) is a node (an arc) which encodes an augmented tree. Regular nodes (arcs) of augmented trees encoded by hyper-nodes (hyper-arcs) of T_p are called *hidden* nodes (arcs) of T_p. We will consider augmented trees with finite number of (regular and hidden) nodes.

An augmented cancellation forest is a forest of augmented trees such that all nodes and all arcs of each augmented tree are labeled consistently, as explained below. We assign a label $Type(T)$ to each tree T in an augmented cancellation forest, depending on the type of a sequence of simplifications it encodes (removals or contractions). We call such trees removal and contraction trees. All nodes in a tree correspond to critical points of the same index i, and all arcs correspond to critical points of index $i-1$ (index $i+1$), if the tree is a removal (contraction) tree. Thus, we partition the set N of labels of nodes and arcs in the forest into $n+1$ subsets N_i, $0 \leq i \leq n$, such that labels in N_i correspond to critical points of index i. We introduce a hyper-node in the forest when a root of a removal tree labels a node in a contraction tree, and when a root of a contraction tree labels a node in a removal tree. We introduce a hyper-arc in the forest when a root of a removal or a contraction tree labels an arc of another tree in the cancellation forest. We also introduce a special tree T_{empty}, rooted at node $empty$, which encodes all removals and contractions of the second kind (of the form $rem(p, s, \emptyset)$ and $con(p, s, \emptyset)$). Tree T_{empty} does not have any label indicating its type.

More formally, an *augmented cancellation forest ACF* is a forest of rooted augmented trees, such that

1. all nodes and arcs in ACF have a label l in the set $N = \{empty\} \cup N_0 \cup N_1 \cup .. \cup N_n$,
2. there is a bijection between the set N and the set of regular nodes and regular arcs (hidded or not) of ACF,

3. label *empty* labels the root of an augmented tree T_{empty}, which is not represented by any hyper-node or hyper-arc,

4. each augmented tree T_p has an associated label $Type(T_p)$, which is set to *void* (if T_p has only one node p, or if $T_p = T_{empty}$), or to r or c (if T_p has more than one node),

5. the labels l of nodes in each tree $T_p \neq T_{empty}$ in ACF belong to the same set N_i, $0 \leq i \leq n$,

6. the labels of arcs in each tree T_p rooted at node p with $l(p) \in N_i$, $0 \leq i \leq n$, all belong either to N_{i-1}, in which case $Type(T_p) = r$ (and T_p is called a removal tree), or to N_{i+1}, in which case $Type(T_p) = c$ (and T_p is called a contraction tree),

7. if q_{new} is a hyper-node in an augmented tree T_p with $Type(T_p) = r$, representing an augmented tree T_q, then $Type(T_q) = c$, and $l(q_{new}) = l(q)$,

8. if q_{new} is a hyper-node in an augmented tree T_p with $Type(T_p) = c$, representing an augmented tree T_q, then $Type(T_q) = r$, and $l(q_{new}) = l(q)$,

9. if the root *empty* is deleted from T_{empty}, then the tree T_{empty} decomposes into subtrees T_q that all satisfy conditions 4. to 8.

5.3   Augmented Cancellation Forest - Construction

An augmented cancellation forest ACF, representing a sequence of simplification operators on Morse complexes, is built bottom up, starting from the full-resolution Morse complex, over which a sequence of simplifications is applied. An ACF by itself is not sufficient for the recovery of all the information encoded in the sequence of simplified Morse complexes. Connectivity between the cells needs also to be stored. Specifically, for each i-cell p in the initial Morse complex, we need to store all $(i-1)$-cells on its direct boundary, and all $(i+1)$-cells in its direct co-boundary. This can be done using an *incidence graph* [6]. A regular node with label in N_i in an augmented tree represents an i-cell in an initial Morse complex. The root p of each removal or contraction tree T_p represents all the cells, corresponding to nodes of the tree, merged into it through cells, corresponding to arcs of T_p. Recall that a removal $rem(p, s, q)$ merges the boundary of p into the boundary of q, and does not change the co-boundary of q, and dually a contraction $con(p, s, q)$ merges the co-boundary of p into the co-boundary of q, and does not change the boundary of q. This knowledge enables us to keep track of the direct boundaries and co-boundaries of each intermediate cell p created in the simplification process, and represented by a root p of some tree T_p.

The construction process of an ACF actually describes the effect of a cancellation applied on an ACF corresponding to the current Morse complex which is being simplified. Initially, when the two Morse complexes are at full resolution, each cell p (corresponding to a critical point p) defines a tree T_p with only one regular node, labeled p, which is a root of T_p. Thus, labels in N_i correspond to i-saddles, i.e., to i-cells of the initial descending Morse complex. Each cell p is initially a representative only of itself. This is illustrated in 2D in Figure 6 (a), and in 3D in Figure 7 (a). For each tree T_p, the label $Active(T_p)$ is set to 1, meaning that cell p corresponding to the root p of tree T_p is present in the Morse

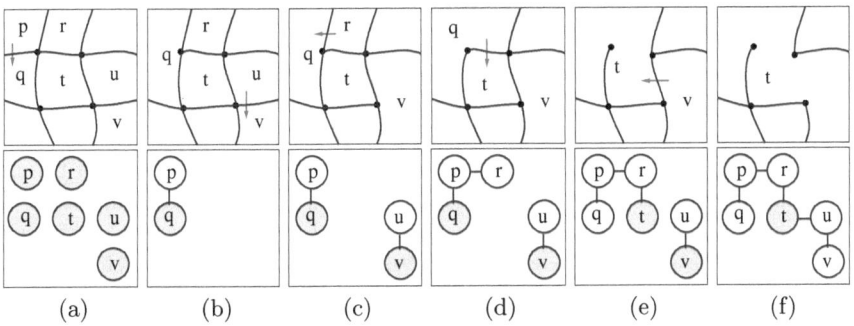

Fig. 6. A sequence of removals on a 2D descending Morse complex, and the construction of the cancellation forest. The root of each tree is shaded.

complex, and the label $Type(T_p)$ is set to void, meaning that no simplification eliminating cell p has yet been applied. A tree T_{empty}, rooted at node *empty*, without an *Active* or a *Type* label is initialized.

Let $sim(p, s, q)$ be a feasible simplification (removal or contraction) on a current Morse complex. Then p, q, and s are roots of trees T_p, T_q, and T_s, respectively. The label $Type(sim(p, s, q))$ is set to r or c if $sim(p, s, q)$ is a removal or a contraction, respectively. If $Type(sim(p, s, q)) = r$ ($Type(sim(p, s, q)) = c$), the set of all cells in the initial Morse complex which constitute the boundary (co-boundary) of a current cell s, represented by the root s of tree T_s, will be denoted by S. The trees T_p and T_s rooted at nodes p and s, respectively, are deactivated from the forest, by changing the labels $Active(T_p)$ and $Active(T_s)$ from 1 to 0, indicating that cells p and s are no longer present in the Morse complex.

Let $sim(p, s, q)$ be a removal (a contraction). Then, a regular or a hidden node p' in tree T_p, and a regular or a hidden node q' in T_q are found, such that cells p' and q' are in the co-boundary (on the boundary) of cell s, i.e., such that $p', q' \in S$. Nodes p' and q' (if they are not hidden in T_p and T_q, respectively), or the hyper-nodes containing them, are connected through an arc labeled s, if s is a cell in the initial Morse complex, or through a hyper-arc encoding augmented tree T_s otherwise. The $Type$ label of tree T_q depends on the $Type$ labels of trees T_p and T_q, and of simplification $sim(p, s, q)$. We describe in more detail how the nodes p' and q', and the label $Type(T_q)$, are determined.

- If $Type(T_p) = Type(T_q) = void$, as it happens at the beginning of the construction of the augmented cancellation forest, then nodes p and q are the only (regular) nodes of trees T_p and T_q, respectively. The two trees are merged into a new tree T_q through an arc labeled s (or a hyper-arc encoding T_s), connecting nodes p and q, as illustrated in Figure 6 (b) and (c), and in Figure 7 (b). The root of the new tree T_q with two regular nodes p and q is node q, and the label $Type(T_q)$ is set to be equal to the label $Type(sim(p, s, q))$.

- If $Type(T_p) = void$, and $Type(T_q) = Type(sim(p, s, q))$, then a node q' is found in tree T_q, such that cell q' is in S. If q' is a regular node, then trees T_p and T_q are merged into a new tree T_q through an arc labeled s or T_s, connecting nodes p and q', as illustrated in Figure 6 (d). If node q' is hidden, then trees T_p and T_q are merged into a new tree T_q through an arc connecting node p to a hyper-node containing node q'. The root of the new tree T_q remains node q, and the label $Type(T_q)$ remains unchanged.

- If $Type(T_p) = Type(sim(p, s, q))$, and $Type(T_q) = void$, we proceed analogously to the previous case. A node p' is found in tree T_p, such that cell p' is in S. If p' is a regular node, then trees T_p and T_q are merged into a new tree T_q through an arc labeled s or T_s, connecting nodes p' and q, as illustrated in Figure 6 (e). If node p' is hidden, then trees T_p and T_q are merged into a new tree T_q through an arc connecting node q to a hyper-node containing node p'. The root of the new tree T_q remains node q, and the label $Type(T_q)$ is set to be equal to the label $Type(sim(p, s, q))$.

- If $Type(T_p) = Type(T_q) = Type(sim(p, s, q))$, then nodes p' and q' are found in trees T_p and T_q, respectively, such that cells p' and q' are in S. If p' and q' are regular nodes, then trees T_p and T_q are merged into a new tree T_q through an arc labeled s or T_s, connecting nodes p' and q', as illustrated in Figure 6 (f). If p' or q' are hidden, then the corresponding hyper-nodes are connected, as explained above. The root of the new tree T_q remains node q, and the label $Type(T_q)$ remains unchanged.

- If $Type(T_p) \neq Type(sim(p, s, q))$, $Type(T_p) \neq void$, and $Type(T_q) = Type(sim(p, s, q))$, then a new hyper-node p_{new}, associated to tree T_p is constructed. A node q' is found in tree T_q such that cell q' is in S. If node q' is regular, then trees $T_{p_{new}}$ (consisting of a single hyper-node p_{new}) and T_q are merged into a new tree T_q through an arc labeled s or T_s, connecting nodes p_{new} and q', as illustrated in Figure 7 (c). Otherwise, a hyper-node containing q' is connected to hyper-node p_{new}. The root of the new tree T_q is node q, and the label $Type(T_q)$ remains equal to the label $Type(sim(p, s, q))$.

- If $Type(T_p) = Type(sim(p, s, q))$, $Type(T_q) \neq Type(sim(p, s, q))$ and $Type(T_q) \neq void$, then tree T_q is deactivated from the forest, by setting its $Active$ label to 0, and a new hyper-node q_{new}, associated to tree T_q is constructed. A node p' is found in tree T_p such that cell p' is in S. If node p' is regular, then trees T_p and $T_{q_{new}}$ (consisting of a single hyper-node q_{new}) are merged into a new tree $T_{q_{new}}$ through an arc labeled s or T_s, connecting nodes p' and q_{new}. Otherwise, a hyper-node containing p' is connected to hyper-node q_{new}, as illustrated in Figure 7 (d). The root of the new tree $T_{q_{new}}$ is node q_{new}, the label $Active(T_{q_{new}})$ is set to 1, and the label $Type(T_{q_{new}})$ is set to be equal to the label $Type(sim(p, s, q))$.

- If $Type(T_p) \neq void$, $Type(T_p) \neq Type(sim(p, s, q))$ $Type(T_q) \neq void$, and $Type(T_q) \neq Type(sim(p, s, q))$, then tree T_q is deactivated from the forest by setting the label $Active(T_q)$ to 0. Two new hyper-nodes p_{new} and q_{new}, associated to trees T_p and T_q, respectively, are constructed. Trees $T_{p_{new}}$ and $T_{q_{new}}$ (consisting of a single hyper-node p_{new} and q_{new}, respectively) are merged into a new tree $T_{q_{new}}$ through an arc labeled s or T_s, connecting hyper-nodes p_{new} and q_{new}. The root of the new tree $T_{q_{new}}$ is node q_{new}, the label $Active(T_{q_{new}})$ is set to 1, and the label $Type(T_{q_{new}})$ is set to be equal to the label $Type(sim(p, s, q))$.

In Figure 7, a sequence of cancellations, consisting of a removal cancellation merging 1-cell p into 1-cell q, a removal cancellation merging 1-cell r into 1-cell t, a contraction cancellation merging 1-cell t into 1-cell u, and a contraction cancellation merging 1-cell q into 1-cell u, is illustrated. After the first two cancellations, two removal trees, rooted at q and t are constructed. A third cancellation is a contraction, and T_t is a removal tree. Thus, a new hyper-node t_{new}, associated to tree T_t is created, and it is connected through an arc to node u. The fourth cancellation is a contaction, and node q is a root of a removal tree. So again, a new hyper-node q_{new} is created, and it is connected through an arc to hyper-node t_{new} of tree T_u.

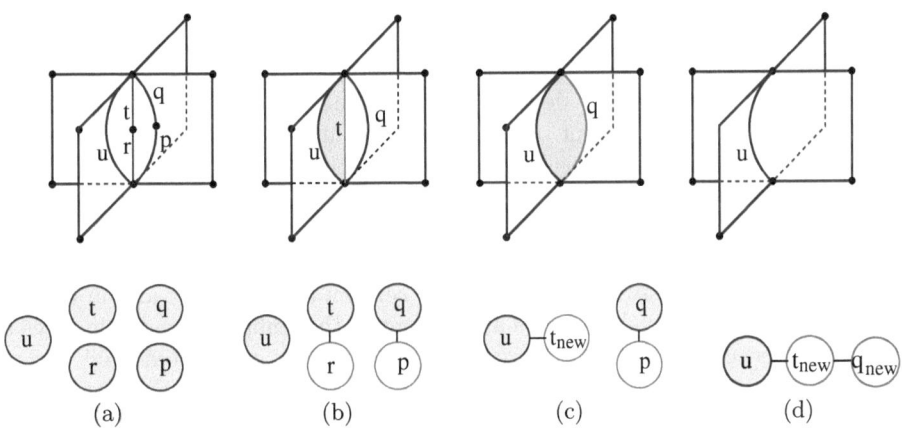

Fig. 7. The effect on the initial Morse complex and on the corresponding ACF (a) of a sequence of simplifications. It consists of two removals, which merge 1-cells p and r into 1-cells q and t, respectively (b), followed by two contractions. The first merges 1-cell t into 1-cell u (c), and the second merges 1-cell q into 1-cell u (d). The root of each tree is shaded.

If a simplification is a removal $rem(p, s, \emptyset)$ or a contraction $con(p, s, \emptyset)$ of the second kind, then trees T_p and T_s are deactivated by setting their $Active$ labels to 0. An arc labeled s or T_s, connecting node p to node $empty$ is added to the set of arcs of the forest.

If we take care of the consistency of the simplified complexes with regard to function values when we construct an augmented cancellation forest, then we will assign an f value to each hyper-node p_{new} representing a tree T_p to be equal to the f value of the root p of T_p, i.e., $f(p_{new}) = f(p)$. Then the root of each removal tree T_p will be the regular node or the hyper-node of T_p with highest f value among all its nodes, and dually for contraction trees. This is important for two reasons. First, in this way, each feasible simplification is determined only by its type (removal or contraction) and its middle-cell s (cell which is removed or contracted). Second, the inverse refinement operators are also determined only by the middle cell s, and can be implemented as deletion of arcs in the augmented cancellation forest. After the deletion, the root of each of the two new removal (contraction) trees will be its node or hyper-node with highest (lowest) f value.

6   Concluding Remarks

We have introduced a new graph-based data structure for encoding a set of removal and contraction simplification operators on Morse complexes, called an augmented cancellation forest, and we have presented an algorithm for its construction starting from a full-resolution Morse complex, on which a sequence of simplification operators is applied.

We plan to build a multi-resolution representation of the Morse complexes based on simplification operators, and their inverse refinement operators, and the augmented cancellation forest. To this aim, a dependency relation between the refinements needs to be defined. One approach, proposed in [5], defines a modification as a replacement of one set of cells in a complex with another set of cells, and two modifications are said to be dependent if the geometric intersection of the cells involved in the two modifications is non-empty. Applying this approach directly on removal and contraction operators, and their inverse refinement operators, implies that a refinement is an undo of a simplification (a refinement can be performed only if all the cells on the boundary and in the co-boundary of the reintroduced cells are present in the complex), and it induces a large number of dependencies between refinements. The augmented cancellation forest constitutes a first and important step towards reducing the dependency relation, and enabling a construction of a more flexible multi-resolution representation of the Morse complexes.

Acknowledgements

This work has been partially supported by the National Science Foundation through grant CCF-0541032, by the MIUR-FIRB project SHALOM under contract number RBIN04HWR8, and by the Ministry of Science of the Republic of Serbia through Project 23036.

References

1. Bremer, P.-T., Pascucci, V., Hamann, B.: Maximizing Adaptivity in Hierarchical Topological Models. In: Belyaev, A.G., Pasko, A.A., Spagnuolo, M. (eds.) Proc. of the Int. Conf. on Shape Modeling and Applications 2005 (SMI 2005), pp. 298–307. IEEE Computer Society Press, Los Alamitos (2005)

2. Bremer, P.-T., Edelsbrunner, H., Hamann, B., Pascucci, V.: A Topological Hierarchy for Functions on Triangulated Surfaces. IEEE Trans. on Visualization and Computer Graphics 10(4), 385–396 (2004)
3. Čomić, L., De Floriani, L.: Modeling and Simplifying Morse Complexes in Arbitrary Dimensions. In: Int. Workshop on Topological Methods in Data Analysis and Visualization (TopoInVis), Snowbird, Utah, February 23-24 (2009)
4. Danovaro, E., De Floriani, L., Mesmoudi, M.M.: Topological Analysis and Characterization of Discrete Scalar Fields. In: Asano, T., Klette, R., Ronse, C. (eds.) Geometry, Morphology, and Computational Imaging. LNCS, vol. 2616, pp. 386–402. Springer, Heidelberg (2003)
5. De Floriani, L., Magillo, P., Puppo, E.: Multiresolution Representation of Shapes Based on Cell Complexes. In: Bertrand, G., Couprie, M., Perroton, L. (eds.) DGCI 1999. LNCS, vol. 1568, pp. 3–18. Springer, Heidelberg (1999)
6. Edelsbrunner, H.: Algorithms in Combinatorial Geometry. Springer, Berlin (1987)
7. Edelsbrunner, H., Harer, J., Zomorodian, A.: Hierarchical Morse Complexes for Piecewise Linear 2-Manifolds. In: Proceedings 17th ACM Symposium on Computational Geometry, pp. 70–79 (2001)
8. Gyulassy, A., Bremer, P.-T., Hamann, B., Pascucci, V.: A Practical Approach to Morse-Smale Complex Computation: Scalability and Generality. IEEE Trans. on Visualization and Computer Graphics 14(6), 1619–1626 (2008)
9. Gyulassy, A., Natarajan, V., Pascucci, V., Bremer, P.-T., Hamann, B.: Topology-Based Simplification for Feature Extraction from 3D Scalar Fields. In: Proceedings IEEE Visualization 2005, pp. 275–280. ACM Press, New York (2005)
10. Gyulassy, A., Natarajan, V., Pascucci, V., Hamann, B.: Efficient Computation of Morse-Smale Complexes for Three-dimensional Scalar Functions. IEEE Trans. on Visualization and Computer Graphics 13(6), 1440–1447 (2007)
11. Matsumoto, Y.: An Introduction to Morse Theory, American Mathematical Society, Translations of Mathematical Monographs 208 (2002)
12. Milnor, J.: Morse Theory. Princeton University Press, New Jersey (1963)
13. Takahashi, S.: Algorithms for Extracting Surface Topology from Digital Elevation Models. In: Rana, S. (ed.) Topological Data Structures for Surfaces, pp. 31–51. John Wiley & Sons Ltd., Chichester (2004)
14. Wolf, G.W.: Topographic Surfaces and Surface Networks. In: Rana, S. (ed.) Topological Data Structures for Surfaces, pp. 15–29. John Wiley & Sons Ltd., Chichester (2004)

Parallel Contextual Hexagonal Array Grammars and Languages

D.G. Thomas, M.H. Begam, and N.G. David

Department of Mathematics, Madras Christian College,
Tambaram, Chennai - 600 059, India
dgthomasmcc@yahoo.com

Abstract. Hexagonal patterns are known to occur in the literature on picture processing and image analysis. Siromoney et al. constructed hexagonal array grammars for generating hexagonal arrays and hexagonal patterns. On the other hand, Marcus introduced a class of grammars called contextual grammars in contrast to Chomskian grammars that generate words by starting with an initial word and adding iteratively pairs of words called contexts associated to a set of words called selector to the words already obtained.

In this paper, we develop a new method of generating hexagonal arrays based on an extension of contextual grammars called parallel contextual hexagonal array grammars. We present necessary conditions for classification of parallel contextual hexagonal array languages, compare with other hexagonal array generating models and study closure properties under union.

Keywords: Hexagonal arrays, hexagonal array grammars, parallel contextual array grammars, Z-direction parallel contextual hexagonal array grammars.

1 Introduction

Formal language theory is one of the fundamental areas of theoretical computer science. This study has its origin in Chomskian grammars. Theoretical development in the area of formal languages has given rise to many deep and interesting results [17]. Contextual grammars which are different from Chomskian grammars, have been intensively investigated [3,4,9,14] by formal language theorists, as they offer novel insight into a number of issues central to formal language theory. In a total contextual grammar, a context is adjoined depending on the whole current string. Two special cases of total contextual grammars, called internal and external are very natural and have been extensively investigated. An external contextual grammar generates a language starting from a finite set of strings (the base) and iteratively adjoining to its contexts. In other families of contextual grammars, such as internal contextual grammars [3], the contexts are adjoined inside the current string.

There has been a continued interest in adapting the techniques of formal string language theory for developing methods to study the problem of picture

P. Wiederhold and R.P. Barneva (Eds.): IWCIA 2009, LNCS 5852, pp. 344–357, 2009.

generation and description, where pictures are considered as connected, digitized finite arrays in the two-dimensional plane [16]. Recently, extensions of string contextual grammars to array structures and hyper graphs have been made [1,5,7,8,15,20]. In [7], a new and simple method of description of pictures of digitized rectangular arrays is introduced based on contextual grammars.

Hexagonal patterns are known to occur in the literature on picture processing and scene analysis. Siromoney et al. [18] constructed hexagonal ko-lam array grammars for generating hexagonal arrays and hexagonal patterns on triangular grids which can be treated as two-dimensional representation of three-dimensional blocks. For the study of two-dimensional representations of three-dimensional blocks, we refer to [12,13]. Recently, the hexagons and the hexagonal tiling have been addressed by a symmetric coordinate frame in [6,11] and possible link to applications in [10]. In this paper, we develop a new method of generating hexagonal arrays based on an extension of contextual grammars called parallel contextual hexagonal array grammars. These systems yield languages of hexagons using parallel rewriting relations. In fact, we make use of 'window movement' on arrow heads to decide whether they are generated by array contexts of choice mappings by the applications of array contextual operations parallelly. We present necessary conditions for classification of parallel contextual hexagonal array languages, compare with other hexagonal array generating models and study closure properties under language theoretic operation like union.

2   Hexagonal Arrays

In this section, we review some basic definitions introduced in [2]. A rectangular array of size $m \times n$ is of the form
$$
\begin{array}{cccc}
a_{11} & a_{12} & \cdots & a_{1n} \\
a_{21} & a_{22} & \cdots & a_{2n} \\
\cdots & & \cdots & \\
a_{m1} & a_{m2} & \cdots & a_{mn}
\end{array}.
$$

We consider hexagons as shown in fig 1

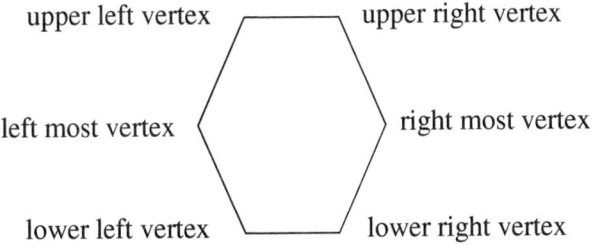

Fig. 1.

Let Σ be a finite alphabet of symbols. A hexagonal picture p over Σ is a hexagonal array of symbols of Σ. For example, a hexagonal picture over the alphabet $\{a, b\}$ is :

$$\begin{array}{ccc} & a \ \ a & \\ a & b & b \\ & b \ \ a & \end{array}$$

The set of all hexagonal arrays over the alphabet Σ is denoted by Σ^{**H}. A hexagonal picture language L over Σ is a subset of Σ^{**H}.

Given a picture $p \in \Sigma^{**H}$, let $l_1(p)$ denote the number of elements in the border of p from upper left vertex to left most vertex in the direction $\nearrow$ called x direction, $l_2(p)$ denote the number of elements in the border of p from upper right vertex to right most vertex in the direction $\searrow$ called y direction and $l_3(p)$ denote the number of elements in the border of p from upper left vertex to upper right vertex in the direction $\rightarrow$ called z direction.

The directions are fixed with origin of reference as the upper left vertex, having coordinates $(1, 1, 1)$. The triple $(l_1(p), l_2(p), l_3(p))$ is called the size of the picture p.

Definition 1. *For $k \geq 2$, $a_j, b_j \in \Sigma$, $(j \geq 1)$, let*

1. *$xyTr$ be a trapezium array of type* $\begin{array}{ccccc} a_1 & a_2 & \cdots & a_k & \\ b_1 & b_2 & \cdots & b_k & b_{k+1} \end{array}$

2. *$yxTr$ be a trapezium array of type* $\begin{array}{ccccc} a_1 & a_2 & \cdots & a_k & a_{k+1} \\ b_1 & b_2 & \cdots & b_k & \end{array}$

3. *$xxPA$ be a parallelogram array of type* $\begin{array}{cccc} a_1 & a_2 & \cdots & a_k \\ b_1 & b_2 & \cdots & b_k \end{array}$ *and*

4. *$yyPA$ be a parallelogram array of type* $\begin{array}{cccc} a_1 & a_2 & \cdots & a_k \\ b_1 & b_2 & \cdots & b_k \end{array}$

Remark 1

1. A $xxPA$ parallelogram array $\begin{array}{cccc} a_1 & a_2 & \cdots & a_k \\ b_1 & b_2 & \cdots & b_k \end{array}$ is denoted by $\begin{bmatrix} u_1 \\ u_2 \end{bmatrix}$ where $u_1 = a_1 a_2 \ldots a_k$ and $u_2 = b_1 b_2 \ldots b_k$ are rectangular arrays of size $1 \times k, k \geq 2$.

2. A $yyPA$ parallelogram array $\begin{array}{cccc} a_1 & a_2 & \cdots & a_k \\ b_1 & b_2 & \cdots & b_k \end{array}$ is denoted by $\begin{bmatrix} u_1 \\ u_2 \end{bmatrix}$ where $u_1 = a_1 a_2 \ldots a_k$ and $u_2 = b_1 b_2 \ldots b_k$ are rectangular arrays of size $1 \times k, k \geq 2$.

3. Let PA_x be the set of all $xxPA$ parallelogram arrays including the type of arrays $\begin{array}{c} a \\ b \end{array}$, denoted by $\begin{bmatrix} a \\ b \end{bmatrix}$ where $a, b \in \Sigma$.

4. Let PA_y be the set of all $yyPA$ parallelogram arrays including the type of arrays $\begin{array}{c} a \\ b \end{array}$, denoted by $\begin{bmatrix} a \\ b \end{bmatrix}$ where $a, b \in \Sigma$.

5. Let $xyTR$ be the set of all $xyTr$ trapezium arrays and $yxTR$ be the set of all $yxTr$ trapezium arrays.

Definition 2. *For,* $xyTr_1 = \begin{array}{ccccc} a_1 & a_2 & \cdots & a_k & \\ b_1 & b_2 & \cdots & b_k & b_{k+1} \end{array}$ *and*

$xyTr_2 = \begin{array}{ccccc} b_1 & b_2 & \cdots & b_{k+1} & \\ c_1 & c_2 & \cdots & c_k & c_{k+2} \end{array}$, *the operation* $\ominus$ *is defined as*

$xyTr_1 \ominus xyTr_2 = \begin{array}{ccccc} a_1 & a_2 & \cdots & a_k & \\ b_1 & b_2 & \cdots & b_k & b_{k+1} \\ c_1 & c_2 & \cdots & \cdots & c_{k+1} & c_{k+2} \end{array}$, *where* $k \geq 2$.

Remark 2. Similarly we can define $yxTr_1 \ominus yxTr_2$, $xxPA_1 \ominus xxPA_2$, $yyPA_1 \ominus yyPA_2$, $xyTr \ominus yxTr$, $xyTr \ominus xxPA$, $xyTr \ominus yyPA$, $xxPA \ominus yxTr$, $yyPA \ominus yxTr$.
(Here the compatibility of the operation $\ominus$ is to be taken care of).

Lemma 1. *Every Hexagonal array H of size (ℓ, m, n) can be represented as*

1. $H = xyTr_1 \ominus xyTr_2 \ominus \cdots \ominus xyTr_{\ell-1} \ominus yxTr_1 \ominus yxTr_2 \ominus \cdots \ominus yxTr_{m-1}$ *if* $\ell = m$.
2. $H = xyTr_1 \ominus xyTr_2 \ominus \cdots \ominus xyTr_{\ell-1} \ominus yyPA_1 \ominus yyPA_2 \ominus \cdots \ominus yyPA_{m-l} \ominus yxTr_1 \ominus \cdots \ominus yxTr_{\ell-1}$ *if* $\ell < m$.
3. $H = xyTr_1 \ominus xyTr_2 \ominus \cdots \ominus xyTr_{m-1} \ominus xxPA_1 \ominus xxPA_2 \ominus \cdots xxPA_{\ell-m} \ominus yxTr_1 \ominus yxTr_2 \ominus \cdots \ominus yxTr_{m-1}$ *if* $\ell > m$.

Example 1. Let $H_1 = \begin{array}{ccccc} & a & a & a & a & \\ & b & b & b & b & b \\ c & c & c & c & c & c. \\ & d & d & d & d & d \\ & & e & e & e & e \end{array}$

Then $H_1 = xyTr_1 \ominus xyTr_2 \ominus yxTr_1 \ominus yxTr_2$ where

$xyTr_1 = \begin{array}{ccccc} a & a & a & a & \\ b & b & b & b & b \end{array}$, $\quad xyTr_2 = \begin{array}{ccccc} b & b & b & b & b \\ c & c & c & c & c & c \end{array}$

$yxTr_1 = \begin{array}{cccccc} c & c & c & c & c & c \\ d & d & d & d & d & \end{array}$, $\quad yxTr_2 = \begin{array}{ccccc} d & d & d & d & d \\ e & e & e & e & \end{array}$

Definition 3

1. *A xy arrow-head is a picture obtained from the representation*
   $xxPA_1 \ominus xxPA_2 \ominus \cdots \ominus xxPA_r \ominus yyPA_1 \ominus yyPA_2 \ominus \cdots \ominus yyPA_s$
   where $r, s \geq 1$.
2. *A yx arrow-head is a picture obtained from the representation*
   $yyPA_1 \ominus yyPA_2 \ominus \cdots \ominus yyPA_m \ominus xxPA_1 \ominus xxPA_2 \ominus \cdots \ominus xxPA_n$
   where $m, n \geq 1$.

3. *A xy arrow is a picture of the form* $a_n = b_1$

$$
\begin{matrix}
 & & a_1 \\
 & a_2 & \\
 & \cdots & \\
b_1 & & \\
 & b_2 & \\
 & \cdots & \\
 & & b_m
\end{matrix}
$$

, $n, m \geq 2.$

4. *A yx arrow is a picture of the form* $c_k = d_1$, $k, s \geq 2.$

$$
\begin{matrix}
c_1 & & \\
 & c_2 & \\
 & \cdots & \\
 & & c_k \\
 & d_2 & \\
 & \cdots & \\
d_s & &
\end{matrix}
$$

Similarly we can define yz, zy, zx, xz arrow-heads and arrows.

Lemma 2. *Let H be a hexagonal array of size (ℓ, m, n). Then, H can be written as $H = X_1 \otimes X_2 \oslash X_3$ where X_1 is either a xy arrow-head or xy arrow, X_3 is either a yx arrow-head or yx arrow, X_2 is a picture, which may or may not be a hexagonal array, $\otimes$ is either xy arrow-head catenation or xy arrow catenation and $\oslash$ is either yx arrow-head catenation or yx arrow catenation.*

3   Z-Direction Parallel Contextual Hexagonal Array Grammars

Definition 4. *Let u_1, u_2 be rectangular arrays of size $1 \times k, k \geq 1$ and v_1, v_2 be rectangular arrays of size $1 \times r, r \geq 1$ and $\$_{xy}, \$_{yx}, \$_{xx}$ and $\$_{yy}$ be special symbols not in Σ.*

1. *A xy array context over Σ is of the form*

$$
xy = \begin{bmatrix} & u_1 \\ u_2 & \end{bmatrix} \$_{xy} \begin{bmatrix} v_1 & \\ & v_2 \end{bmatrix} \in PA_x \$_{xy} PA_y
$$

2. *A yx array context over Σ is of the form*

$$
yx = \begin{bmatrix} u_1 & \\ & u_2 \end{bmatrix} \$_{yx} \begin{bmatrix} & v_1 \\ v_2 & \end{bmatrix} \in PA_y \$_{yx} PA_x
$$

3. *A xx array context over Σ is of the form*

$$
xx = \begin{bmatrix} & u_1 \\ u_2 & \end{bmatrix} \$_{xx} \begin{bmatrix} & v_1 \\ v_2 & \end{bmatrix} \in PA_x \$_{xx} PA_x \quad and
$$

4. *A yy array context over Σ is of the form*

$$
yy = \begin{bmatrix} u_1 & \\ & u_2 \end{bmatrix} \$_{yy} \begin{bmatrix} v_1 & \\ & v_2 \end{bmatrix} \in PA_y \$_{yy} PA_y
$$

We now define parallel internal xy, xx, yx, yy array contextual operations.

Definition 5. *Consider a hexagonal array $H = [a_{ijk}]$ of size (ℓ, m, n). Suppose H can be written as $H = X_1 \oslash X_2 \oslash X_3$ where X_1 is a xy arrow-head or a xy arrow, X_3 is a yx arrow-head or a yx arrow, $X_2 = [a_{i'j'k'}]$ is a hexagonal array of size (ℓ, m, n') where $n' < n$.*
We write $H \Rightarrow_{z-in} H'$ if there exists

1. *a finite number of xy array contexts $x_i \in X (1 \le i \le m - 1$ if $\ell > m$ or $1 \le i \le \ell - 1$ if $m \ge \ell)$ not all need be distinct, where X is a finite subset of $PA_x \$_{xy} PA_y$,*

2. *a finite number of yx array contexts $y_i \in Y (1 \le i \le m - 1$ if $\ell > m$ or $1 \le i \le \ell - 1$ if $m \ge \ell)$ not all need be distinct, where Y is a finite subset of $PA_y \$_{yx} PA_x$,*

3. *a finite number of xx array contexts $z_i \in Z (1 \le i \le \ell - m$ if $\ell > m)$ not all need be distinct or no xx array context in Z if $m \ge \ell$ where Z is a finite subset (may be empty also) of $PA_x \$_{xx} PA_x$,*

4. *a finite number of yy array contexts $t_i \in T (1 \le i \le m - l$ if $m > \ell)$ not all need be distinct or no yy array context in T if $\ell \ge m$ where T is a finite subset (may be empty also) of $PA_y \$_{yy} PA_y$,*

5. *a choice mapping $\varphi_{xy} : xyTR \to 2^X$ such that*
$$x_i = \begin{bmatrix} u_i \\ u_{i+1} \end{bmatrix} \$_{xy} \begin{bmatrix} v_i \\ v_{i+1} \end{bmatrix} \in \varphi_{xy}(xyTr_1) \quad (1 \le i \le m - 1 \text{ if } \ell > m \text{ or } 1 \le i \le \ell - 1 \text{ if } m \ge \ell),$$

6. *a choice mapping $\varphi_{yx} : yxTR \to 2^Y$ such that*
$$y_i = \begin{bmatrix} u'_i \\ u'_{i+1} \end{bmatrix} \$_{yx} \begin{bmatrix} v'_i \\ v'_{i+1} \end{bmatrix} \in \varphi_{yx}(yxTr_1) \quad (1 \le i \le m - 1 \text{ if } \ell > m \text{ or } 1 \le i \le \ell - 1 \text{ if } m \ge \ell),$$

7. *a choice mapping $\varphi_{xx} : PA_x \to 2^Z$ such that*
$$z_i = \begin{bmatrix} u_i \\ u_{i+1} \end{bmatrix} \$_{xx} \begin{bmatrix} v_i \\ v_{i+1} \end{bmatrix} \in \varphi_{xx}(xxPA_i) \quad i - 1, 2, \ldots, \ell - m \text{ if } \ell > m,$$

8. *a choice mapping $\varphi_{yy} : PA_y \to 2^T$ such that*
$$t_i = \begin{bmatrix} u_i \\ u_{i+1} \end{bmatrix} \$_{yy} \begin{bmatrix} v_i \\ v_{i+1} \end{bmatrix} \in \varphi_{yy}(yyPA_i) \quad i = 1, 2, \ldots, m - \ell \text{ if } m > \ell$$

with $H' = X_1 \oslash L \oslash X_2 \oslash R \oslash X_3$ where L is a xy arrow-head or xy arrow :

$$
\begin{array}{l}
\quad\quad u_1 \\
\quad u_2 \\
\quad\ \cdots \\
u_\ell = u'_1 \\
\quad u'_2 \\
\quad\ \cdots \\
\quad\quad u'_m
\end{array}
\qquad (u_i, u'_j \text{ are rectangular arrays of same size } 1 \times k \ (k \ge 1)) \text{ and}
$$

R is a yx arrow-head or yx arrow :

v_1

$\quad v_2$

$\quad\quad \cdots$

$\quad\quad\quad v_m = v'_1$ *(v_i, v'_j are rectangular arrays of same size $1 \times r$ ($r \geq 1$)).*

$\quad\quad v'_2$

$\quad\quad \cdots$

v'_ℓ

Now we define the central notion of parallel internal contextual hexagonal array grammars.

Definition 6. *A Z-direction parallel internal contextual hexagonal array grammar with choice is an ordered system $G_{ZI} = (\Sigma, B, X, Y, Z, T, \varphi_{xy}, \varphi_{yx}, \varphi_{xx}, \varphi_{yy})$ where*

(i) Σ is a finite alphabet,
(ii) B is a finite subset of Σ^{**H} called the base of G_{ZI},
(iii) X is a finite subset of $PA_x \$_{xy} PA_y$ called set of xy array contexts,
(iv) Y is a finite subset of $PA_y \$_{yx} PA_x$ called set of yx array contexts,
(v) Z is a finite subset of $PA_x \$_{xx} PA_x$ called set of xx array contexts,
(vi) T is a finite subset of $PA_y \$_{yy} PA_y$ called set of yy array contexts,
(vii) $\varphi_{xy} : xyTR \to 2^X$, $\varphi_{yx} : yxTR \to 2^Y$, $\varphi_{xx} : PA_x \to 2^Z$, $\varphi_{yy} : PA_y \to 2^T$ are the choice mappings which perform the parallel internal xy, yx, xx, yy contextual operations respectively. When $\varphi_{xy}, \varphi_{yx}, \varphi_{xx}, \varphi_{yy}$ are omitted we call G_{ZI} as a Z-direction parallel internal hexagonal array grammar without choice.

The direct derivation with respect to G_{ZI} is a binary relation $\Rightarrow_{z-in}$ on Σ^{**H}. It is defined as $H \Rightarrow_{z-in} H'$ where $H, H' \in \Sigma^{**H}$ with $H = X_1 \oslash X_2 \oslash X_3$, $H' = X_1 \oslash L \oslash X_2 \oslash R \oslash X_3$ for some $X_2 \in \Sigma^{**H}$, X_1 is a xy arrow-head or xy arrow, X_3 is a yx arrow-head or yx arrow and L, R are contexts, which are respectively xy arrow-head or xy arrow and yx arrow-head or yx arrow, performed by the parallel internal xy, yx, xx, yy contextual operations according to the choice mappings. $\Rightarrow^*_{z-in}$ is a reflexive and transitive closure of the relation $\Rightarrow_{z-in}$.

Let $G_{ZI} = (\Sigma, B, X, Y, Z, T, \varphi_{xy}, \varphi_{yx}, \varphi_{xx}, \varphi_{yy})$ be a Z-direction parallel internal contextual hexagonal array grammar. The language generated by G_{ZI}, denoted by $L(G_{ZI})$ is defined as $L(G_{ZI}) = \{H' \in \Sigma^{**H} /$ there exists $H \in B$ such that $H \Rightarrow^*_{z-in} H'\}$. The family of all Z-direction parallel internal contextual hexagonal array languages is denoted by $ZPIHAL$.

Example 2. Let $G_{ZI} = (\Sigma, B, X, Y, Z, T, \varphi_{xy}, \varphi_{yx}, \varphi_{xx}, \varphi_{yy})$ be a Z-direction parallel internal contextual hexagonal array grammar where $\Sigma = \{a\}, B = \{H\}$

$$\text{with } H = \begin{matrix} & a & a & a & a & \\ a & a & a & a & a \\ a & a & a & a & a \\ & a & a & a & a \end{matrix}, \quad X = \left\{ \begin{bmatrix} a \\ a \end{bmatrix} \$_{xy} \begin{bmatrix} a \\ a \end{bmatrix} \right\}, \quad Y = \left\{ \begin{bmatrix} a \\ a \end{bmatrix} \$_{yx} \begin{bmatrix} a \\ a \end{bmatrix} \right\},$$

$$T = \left\{ \begin{bmatrix} a \\ a \end{bmatrix} \$_{yy} \begin{bmatrix} a \\ a \end{bmatrix} \right\}, \quad Z = \emptyset, \quad \varphi_{xy} \left(\begin{matrix} a & a \\ a & a & a \end{matrix} \right) = \begin{bmatrix} a \\ a \end{bmatrix} \$_{xy} \begin{bmatrix} a \\ a \end{bmatrix},$$

$$\varphi_{yy}\left(\begin{bmatrix} a & a & a \\ a & a & a \end{bmatrix}\right) = \begin{bmatrix} a \\ a \end{bmatrix} \$_{yy} \begin{bmatrix} a \\ a \end{bmatrix}, \quad \varphi_{yx}\left(\begin{bmatrix} a & a & a \\ a & a \end{bmatrix}\right) = \begin{bmatrix} a \\ a \end{bmatrix} \$_{yx} \begin{bmatrix} a \\ a \end{bmatrix}$$

$B = X_1 \otimes X_2 \otimes X_3$ where

$$X_1 = \begin{matrix} & a & \\ a & & \\ & a & \\ & & a \end{matrix}, \quad X_2 = \begin{matrix} a & a \\ a & a & a \\ a & a & a \\ a & a \end{matrix}, \quad X_3 = \begin{matrix} & a & \\ a & \\ & a & \\ & a \end{matrix}, \quad L = \begin{matrix} & a & \\ a & \\ & a & \\ & a \end{matrix}, \quad R = \begin{matrix} & a & \\ & a \\ & a \\ & a \end{matrix}$$

$$H \Rightarrow_{z-in} H' = \begin{matrix} & a & a & a & a & a & a & \\ a & a & a & a & a & a & a \\ a & a & a & a & a & a & a \\ & a & a & a & a & a & \end{matrix}$$

$L(G_{ZI})$ is the set of all hexagonal arrays over $\{a\}$ of sizes $(2, 3, 2n), n \geq 2$.

Definition 7. *A Z-direction parallel external contextual hexagonal array grammar is a system*

$$G_{ZE} = (\Sigma, B, X, Y, Z, T, \varphi_{xy}, \varphi_{yx}, \varphi_{xx}, \varphi_{yy})$$

where the components are same as in Definition 6.
The direct derivation with respect to G_{ZE} is a binary relation $\Rightarrow_{z-ex}$ on Σ^{**H}.
It is defined as $H \Rightarrow_{z-ex} H'$ where $H, H' \in \Sigma^{**H}$ if and only if
$H' = L \otimes H \otimes R$ and L, R are contexts as explained in definition 6.
We denote by $\Rightarrow_{z-ex}^{*}$ the reflexive and transitive closure of the relation $\Rightarrow_{z-ex}$.
The language generated by G_{ZE}, denoted by $L(G_{ZE})$, is defined as

$$L(G_{ZE}) = \{H' \in \Sigma^{**H} / \text{ there exists } H \in B \text{ such that } H \Rightarrow_{z-ex}^{*} H'\}.$$

The family of all Z-direction parallel external contextual hexagonal array languages is denoted by $ZPEHAL$.

Example 3. Let $G = (\Sigma, B, X, Y, Z, T, \varphi_{xy}, \varphi_{yx}, \varphi_{xx}, \varphi_{yy})$ be a Z-direction parallel external contextual hexagonal array grammar where $\Sigma = \{a\}, \quad B = \{H\}$

with $H = \begin{matrix} & a & a & \\ a & a & a \\ a & a & a \\ & a & a \end{matrix}, \quad X = \left\{ \begin{bmatrix} a \\ a \end{bmatrix} \$_{xy} \begin{bmatrix} a \\ a \end{bmatrix} \right\}, \quad Y = \left\{ \begin{bmatrix} a \\ a \end{bmatrix} \$_{yx} \begin{bmatrix} a \\ a \end{bmatrix} \right\},$

$T = \left\{ \begin{bmatrix} a \\ a \end{bmatrix} \$_{yy} \begin{bmatrix} a \\ a \end{bmatrix} \right\}, \quad Z = \emptyset, \quad \varphi_{xy}\left(\begin{bmatrix} a^n \\ a^{n+1} \end{bmatrix}_{n \geq 2}\right) = \begin{bmatrix} a \\ a \end{bmatrix} \$_{xy} \begin{bmatrix} a \\ a \end{bmatrix},$

$\varphi_{yy}\left(\begin{bmatrix} a^n \\ a^n \end{bmatrix}_{n \geq 2}\right) = \begin{bmatrix} a \\ a \end{bmatrix} \$_{yy} \begin{bmatrix} a \\ a \end{bmatrix}, \quad \varphi_{yx}\left(\begin{bmatrix} a^{n+1} \\ a^n \end{bmatrix}_{n \geq 2}\right) = \begin{bmatrix} a \\ a \end{bmatrix} \$_{yx} \begin{bmatrix} a \\ a \end{bmatrix}$

where

$$\begin{bmatrix} a^n \\ a^{n+1} \end{bmatrix} = \begin{bmatrix} a & \cdots & a \\ a & a & \cdots & a \end{bmatrix}, \quad \begin{bmatrix} a^n \\ a^n \end{bmatrix} = \begin{bmatrix} a & \cdots & a \\ a & \cdots & a \end{bmatrix}$$

$$\begin{bmatrix} a^n \\ & a^n \end{bmatrix} = \begin{bmatrix} a & \cdots & a \\ & a & \cdots & a \end{bmatrix}, \quad \begin{bmatrix} a^{n+1} \\ & a^n \end{bmatrix} = \begin{bmatrix} a & a & \cdots & a \\ & a & a & \cdots & a \end{bmatrix}$$

$L(G_{ZE})$ is the set of hexagonal arrays over $\{a\}$ of sizes $(2, 3, 2n)$, $n \geq 2$.

By suitably changing the definitions of G_{ZI} and G_{ZE}, we can have X-direction parallel internal contextual hexagonal array grammar (G_{XI}), Y-direction parallel internal contextual hexagonal array grammar (G_{YI}), X-direction parallel external contextual hexagonal array grammar (G_{XE}) and Y-direction parallel external contextual hexagonal array grammar (G_{YE}). We can define a parallel internal contextual hexagonal array grammar as $G_{IN} = (\Sigma, B, G_{XI}, G_{YI}, G_{ZI})$ where $\Sigma = \Sigma_1 \cup \Sigma_2 \cup \Sigma_3$, $B = B_1 \cup B_2 \cup B_3$ with alphabets Σ_1, Σ_2 and Σ_3 and bases B_1, B_2 and B_3 of G_{XI}, G_{YI}, G_{ZI} respectively and consider the applications of G_{XI}, G_{YI}, G_{ZI} sequentially in any order on elements of B yielding the growth of hexagonal arrays of B in all directions. Similarly we can define a parallel external contextual hexagonal array grammar as $G_{EX} = (\Sigma, B, G_{XE}, G_{YE}, G_{ZE})$. We obtain correspondingly the class of parallel internal contextual hexagonal array languages (PIHAL) and the class of parallel external contextual hexagonal array languages (PEHAL).

4   Properties

In this section, we study some properties of languages belonging to the classes ZPIHAL and ZPEHAL. This study can be extended to the classes PIHAL and PEHAL.

We need the following notations:

If H is a hexagonal array of size (ℓ, m, n), then let $|H|_Z = n$. If X is a xy arrow-head, then $|X|_Z$ denotes the number of elements in the border of X in Z direction. If X is a xy arrow, then $|X|_Z = 1$.

We can similarly define $|Y|_Z$ if Y is either a yx arrow-head or a yx arrow.

Definition 8. *A language $M \subseteq \Sigma^{**H}$ has the Z-external bounded step property (ZEBS) if there is a constant p such that for each $H \in M$, $|H|_Z > p$, there is a $H' \in M$ such that*
$H = L \ominus H' \ominus R$ *and* $|L|_Z + |R|_Z \leq p$.

Definition 9. *A language $M \subseteq \Sigma^{**H}$ has the Z-internal bounded step property (ZIBS) if there is a constant p such that for each $H \in M$, $|H|_Z > p$, there is a $H' \in M$ such that*
$H = X_1 \ominus L \ominus X_2 \ominus R \ominus X_3$, $H' = X_1 \ominus X_2 \ominus X_3$ *and* $0 < |L|_Z + |R|_Z \leq p$.

Theorem 1. *Every ZPIHAL language has the ZIBS property.*

Proof. Let $M \in ZPIHAL$ and
$G = (\Sigma, B, X, Y, Z, T, \varphi_{xy}, \varphi_{yx}, \varphi_{xx}, \varphi_{yy})$ be a Z-direction parallel internal contextual hexagonal array grammar generating M.
Let $p_1 = max\{|H|_Z / H \in B\}$

$p_2 = max\{|L|_Z + |R|_Z\}$ where L is either a xy arrow-head or a xy arrow and R is either a yx arrow-head or a yx arrow obtained from some $H, H' \in M$ with $H = X_1 \oslash L \oslash X_2 \oslash R \oslash X_3$, $H' = X_1 \oslash X_2 \oslash X_3$ and $H' \Rightarrow_{z-in} H$.
Let $p = max\{p_1, p_2\}$.
If $H = \langle a_{ijk} \rangle \in M$ is such that $|H|_Z > p$ then $H \notin B$.

Hence $H = X_1 \oslash L \oslash X_2 \oslash R \oslash X_3$ for some L and R and $H' = X_1 \oslash X_2 \oslash X_3 \in M$. Hence there is a constant p such that for each $H \in M$ with $|H|_Z > p$, there exists $H' = X_1 \oslash X_2 \oslash X_3$ such that $H = X_1 \oslash L \oslash X_2 \oslash R \oslash X_3$ and $0 < |L|_Z + |R|_Z \le p$. Hence M has $ZIBS$ property.   □

Theorem 2. *Every ZPEHAL language has the ZEBS property.*

Proof is similar to Theorem 1.

Definition 10. *A language $M \subseteq \Sigma^{**H}$ has the Z-bounded hexagonal array length increasing property (ZBHALI) if there is a constant p such that for each $H \in M$ there is a $H' \in M$ with $-p \le |H|_Z - |H'|_Z \le p$.*

Theorem 3. *Every ZPIHAL as well as ZPEHAL language has the ZBHALI property.*

Proof. Let $M \in ZPIHAL$ and
$G = (\Sigma, B, X, Y, Z, T, \varphi_{xy}, \varphi_{yx}, \varphi_{xx}, \varphi_{yy})$ be a Z-direction parallel internal contextual hexagonal array grammar generating M.
Let $p_1 = max\{|H|_Z/H \in B\}$
$p_2 = max\{|L|_Z + |R|_Z\}$ where L is either a xy arrow-head or a xy arrow and R is either a yx arrow-head or a yx arrow obtained from some $H, H' \in M$ with $H = X_1 \oslash L \oslash X_2 \oslash R \oslash X_3$, $H' = X_1 \oslash X_2 \oslash X_3$ and $H' \Rightarrow_{z-in} H$.
Let $p = max\{p_1, p_2\}$.
Let $H = \langle a_{ijk} \rangle \in M$

1. If $H \in B$ then there exists $H' \in M$ such that
   $H' = X_1 \oslash L \oslash X_2 \oslash R \oslash X_3$ and $H = X_1 \oslash X_2 \oslash X_3$.
   $0 < |H'|_Z - |H|_Z = |L|_Z + |R|_Z = p_2 \le p$.
   $-p \le |H|_Z - |H'|_Z < 0 \le p$
   i.e., $-p \le |H|_Z - |H'|_Z \le p$.
2. If $H \notin B$ then there exists $H' \in M$ such that
   $H = X_1 \oslash L \oslash X_2 \oslash R \oslash X_3$ for some L and R and
   $H' = X_1 \oslash X_2 \oslash X_3$.
   $0 < |H|_Z - |H'|_Z = |L|_Z + |R|_Z \le p$.
   $-p < |H|_Z - |H'|_Z \le p$

From (1) and (2) we see that for each $H \in M$ there is a $H' \in M$ such that $-p \le |H|_Z - |H'|_Z \le p$.
Hence M has the $ZBHALI$ property.   □
The proof is similar for $ZPEHALI$.

Theorem 4. *If $M \in ZPIHAL$, then there is a constant p such that every $H \in M$, $|H|_Z > p$ can be written in the following form $H = X_1 \otimes L \otimes X_2 \otimes R \otimes X_3$ with $X_2 \in \Sigma^{**H}$, X_1 and L are either xy arrow-heads or xy arrows, X_2 and R are either yx arrow-heads or yx arrows and $X_1 \otimes L^i \otimes X_2 \otimes R^i \otimes X_3 \in M$ for all $i \geq 0$.*

Proof. Let $G = (\Sigma, B, X, Y, Z, T, \varphi_{xy}, \varphi_{yx}, \varphi_{xx}, \varphi_{yy})$ be a Z-direction parallel internal contextual hexagonal array grammar with $L(G) = M$.
We define $p = max\{|H|_Z / H \in B\}$
If $H \in M = L(G)$ is such that $|H|_Z > p$ then $H \notin B$.
There is $H' \in M$ such that $H' \Rightarrow_{z-in} H$ with $H' = X_1 \otimes X_2 \otimes X_3$ and
$H = X_1 \otimes L \otimes X_2 \otimes R \otimes X_3$
Hence $X_1 \otimes L^i \otimes X_2 \otimes R^i \otimes X_3 \Rightarrow_{z-in} X_1 \otimes L^{i+1} \otimes X_2 \otimes R^{i+1} \otimes X_3$ for $i \geq 0$.
All these hexagonal arrays are in M.   □

Theorem 5. *If $M \in \Sigma^{**H}$ and $M \in ZPEHAL$, then there is a constant p such that every $H \in M$, $|H|_Z > p$ can be written in the following form $H = L \otimes H' \otimes R$ where $H' \in \Sigma^{**H}$, L is either a xy arrow-head or a xy arrow, R is either a yx arrow-head or a yx arrow and $L^i \otimes H' \otimes R^i \in M$ for all $i \geq 0$.*

Proof is similar to Theorem 4.

Theorem 6. *$ZPEHAL$ is closed under union.*

Proof. Let $L_1, L_2 \in ZPEHAL$ be generated by the two Z-direction parallel external contextual hexagonal array grammars
$G'_{ZE} = (\Sigma_1, B_1, X_1, Y_1, Z_1, T_1, \varphi'_{xy}, \varphi'_{yx}, \varphi'_{xx}, \varphi'_{yy})$
$G''_{ZE} = (\Sigma_2, B_2, X_2, Y_2, Z_2, T_2, \varphi''_{xy}, \varphi''_{yx}, \varphi''_{xx}, \varphi''_{yy})$
where the components are same as in Definition 6.
Assume $\Sigma_1 \cap \Sigma_2 = \emptyset$.
Construct $G_{ZE} = (\Sigma, B, X, Y, Z, T, \varphi_{xy}, \varphi_{yx}, \varphi_{xx}, \varphi_{yy})$ where
$\Sigma = \Sigma_1 \cup \Sigma_2$, $B = B_1 \cup B_2$, $X = X_1 \cup X_2$
$Y = Y_1 \cup Y_2$, $Z = Z_1 \cup Z_2$, $T = T_1 \cup T_2$
Let $U = xyTr_1 \ominus \cdots \ominus xyTr_n \ominus xxPA_1 \ominus \cdots \ominus xxPA_s \ominus yxTr_1 \ominus \ldots \ominus yxTr_m$ or
$U = xyTr_1 \ominus \cdots \ominus xyTr_n \ominus yyPA_1 \ominus \cdots \ominus yyPA_t \ominus yxTr_1 \ominus \ldots \ominus yxTr_m$
Suppose $U \in B_1$.
Let $\varphi_{xy}(xyTr_i) = \varphi'_{xy}(xyTr_i)$ $i = 1, ..., n$
$\varphi_{xx}(xxPA_i) = \varphi'_{xx}(xxPA_i)$ $i = 1, ..., s$
$\varphi_{yx}(yxTr_i) = \varphi'_{yx}(yxTr_i)$ $i = 1, ..., m$
$\varphi_{yy}(yyPA_i) = \varphi'_{yy}(yyPA_i)$ $i = 1, ..., t$
Similarly, if $U \in B_2$ then we can define $\varphi_{xy}, \varphi_{yx}, \varphi_{xx}, \varphi_{yy}$ as follows:
$\varphi_{xy}(xyTr_i) = \varphi''_{xy}(xyTr_i)$ $i = 1, ..., n$
$\varphi_{xx}(xxPA_i) = \varphi''_{xx}(xxPA_i)$ $i = 1, ..., s$
$\varphi_{yx}(yxTr_i) = \varphi''_{yx}(yxTr_i)$ $i = 1, ..., m$
$\varphi_{yy}(yyPA_i) = \varphi''_{yy}(yyPA_i)$ $i = 1, ..., t$

If $L(G_{ZE}) = L$, then we can prove that $L = L_1 \cup L_2$.

First we will show that $L \subseteq L_1 \cup L_2$. Let $U \in L$

If $U \in B$, then $U \in B_1$ or $U \in B_2$ and hence $U \in L_1 \cup L_2$.

If $U \notin B$, then there exists $V \in B$ such that $V \Rightarrow^*_{z-ex} U$. Since $V \in B, V \in B_1$ or $V \in B_2$.

If $V \in B_1$ then $U \in L_1$, since $V_1 \cup V_2 = \emptyset$ and hence U is in $L_1 \cup L_2$.

Similarly, $V \in B_2$ implies $U \in L_2$ and hence, $U \in L_1 \cup L_2$. Thus, $L \subseteq L_1 \cup L_2$.

To show that $L_1 \cup L_2 \subseteq L$,

Let $U \in L_1 \cup L_2$. Then $U \in L_1$ or $U \in L_2$. Suppose $U \in L_1$. If $U \in B_1$ then $U \in B$ and hence $U \in L$.

If $U \notin B_1$ then there exists some $V \in B_1$ such that $V \Rightarrow^*_{z-ex} U$.

i.e., $U = L^i \oslash V \oslash R^i$ where L and R are arrow-heads or arrows which are obtained through the contexts from X_1, Y_1, Z_1, T_1 and hence from the context X, Y, Z, T.

Thus $U \in L$. Similarly, the case $U \in L_2$ can be dealt.

So, we obtain $L = L_1 \cup L_2$. □

5   Comparison Results

In this section, we compare ZPEHAL with some known classes of hexagonal array languages.

Siromoney et al. [18] introduced two classes of hexagonal array languages, viz., hexagonal kolam array grammar languages ($HKAGL$) and controlled table 0L hexagonal array languages ($CT0HAL$). Subramanian [19] studied another class of hexagonal array languages ((X, Y) HAL) generating more interesting hexagonal arrays on triangular grids where $X, Y \in \{R, CF, CS\}$ with R the class of regular languages, CF the class of context-free languages and CS the class of context-sensitive languages. For definitions and studies of these classes, we refer to [18,19].

Theorem 7

1. *The families $HKAGL$ [18] and ZPEHAL have nonempty intersection.*
2. *The families $CT0HAL$ [18]and ZPEHAL have nonempty intersection.*
3. *The families (X, Y) HAL [19] and ZPEHAL have nonempty intersection where $X, Y \in \{R, CF, CS\}$.*

Proof is omitted.

Recently Dersanambika et al. [2] have introduced two interesting classes of hexagonal picture languages, viz., local hexagonal picture languages (HLOC) and recognizable hexagonal picture languages (HREC) and studied their properties. For definitions and details we refer to [2].

Theorem 8. *The classes $HLOC$ and $ZPEHAL$ are incomparable but not disjoint.*

Proof. Let $G = (\Sigma, B, X, Y, Z, T, \varphi_{xy}, \varphi_{yx}, \varphi_{xx}, \varphi_{yy})$ be a Z-direction parallel external contextual hexagonal array grammar where $\Sigma = \{1,2,3\}$, $B = \{H\}$ with

$$H = \begin{matrix} & 1 & 1 \\ 2 & 2 & 2 \\ & 3 & 3 \end{matrix}$$

$$X = \left\{ \begin{bmatrix} 1 \\ 2 \end{bmatrix} \$_{xy} \begin{bmatrix} \epsilon \\ \epsilon \end{bmatrix} \right\}, \qquad Y = \left\{ \begin{bmatrix} 2 \\ 3 \end{bmatrix} \$_{yx} \begin{bmatrix} \epsilon \\ \epsilon \end{bmatrix} \right\}, \quad Z = T = \emptyset$$

$$\varphi_{xy}\left(\begin{bmatrix} 1^n \\ 2^{n+1} \end{bmatrix}_{n\geq 2} \right) = \begin{bmatrix} 1 \\ 2 \end{bmatrix} \$_{xy} \begin{bmatrix} \epsilon \\ \epsilon \end{bmatrix}, \quad \varphi_{yx}\left(\begin{bmatrix} 2^{n+1} \\ 3^n \end{bmatrix}_{n\geq 2} \right) = \begin{bmatrix} 2 \\ 3 \end{bmatrix} \$_{yx} \begin{bmatrix} \epsilon \\ \epsilon \end{bmatrix}$$

Then $L = \begin{matrix} & 1 \\ 2 & \\ & 3 \end{matrix}$, $\quad R = \begin{matrix} & \epsilon \\ \epsilon & \end{matrix}$ and $H \Rightarrow_{z-ex} H' = \begin{matrix} & 1 & 1 & 1 \\ 2 & 2 & 2 & 2 \\ & 3 & 3 & 3 \end{matrix}$

$L(G)$ is the set of all hexagonal arrays over $\{1,2,3\}$ of sizes $(2,2,k)$, $k \geq 2$. This language is in $HLOC$.

Consider the local hexagonal array language L' given in [2].

$L' \in HLOC$, but $L' \notin PEHAL$ since it is clear from the definition of Z-direction parallel external contextual hexagonal array language, only hexagonal arrays of size $\ell = m = n$ cannot be generated by Z-direction parallel external contextual hexagonal array grammar.

Let $G_{ZE} = (\Sigma, B, X, Y, Z, T, \varphi_{xy}, \varphi_{yx}, \varphi_{xx}, \varphi_{yy})$ where $\Sigma = \{a\}$, $B = \begin{matrix} & a & a \\ a & a & a \\ & a & a \end{matrix}$,

$$X = \left\{ \begin{bmatrix} a \\ a \end{bmatrix} \$_{xy} \begin{bmatrix} \epsilon \\ \epsilon \end{bmatrix} \right\} \qquad Y = \left\{ \begin{bmatrix} a \\ a \end{bmatrix} \$_{yx} \begin{bmatrix} \epsilon \\ \epsilon \end{bmatrix} \right\}, \quad Z = T = \emptyset,$$

$$\varphi_{xy}\left(\begin{bmatrix} a^n \\ a^{n+1} \end{bmatrix}_{n\geq 2} \right) = X \qquad \varphi_{yx}\left(\begin{bmatrix} a^{n+1} \\ a^n \end{bmatrix}_{n\geq 2} \right) = Y$$

Then $L'' = L(G_{ZE})$ is the set of all hexagonal arrays over $\{a\}$ of sizes $(2, 2, k)$ $k \geq 2$ which is not in $HLOC$.   □

6   Conclusion

A new method of generating hexagonal arrays based on an extension of contextual grammars called parallel contextual hexagonal array grammars is presented in this paper. In the grammar introduced here, contexts along arrow heads are allowed and the contextual rules are finite. Necessary conditions for classification of parallel contextual hexagonal array languages, comparison with other hexagonal array generating models and closure properties are studied. It is worth examining further properties of these parallel contextual hexagonal array grammars, closure properties under intersection, complement, homomorphism and geometric operations like reflection about the base lines and rotation of hexagonal arrays by 60°, 120°, etc.

References

1. Dediu, A.-H., Klempien-Hinrichs, R., Kreowski, H.-J., Nagy, B.: Contextual hypergraph grammars - a new approach to the generation of hypergraph languages. In: Ibarra, O.H., Dang, Z. (eds.) DLT 2006. LNCS, vol. 4036, pp. 327–338. Springer, Heidelberg (2006)
2. Dersanambika, K.S., Krithivasan, K., Martin-Vide, C., Subramanian, K.G.: Hexagonal pattern languages. In: Klette, R., Žunić, J. (eds.) IWCIA 2004. LNCS, vol. 3322, pp. 52–64. Springer, Heidelberg (2004)
3. Ehrenfeucht, A., Păun, G., Rozenberg, G.: Contextual grammars and formal languages. In: Rozenberg, G., Salomaa, A. (eds.) Handbook of Formal Languages, vol. 2, pp. 237–293 (1997)
4. Fernau, H., Freund, R., Holzer, M.: Representations of recursively enumerable array languages by contextual array grammars. Fundamenta Informaticae 64, 159–170 (2005)
5. Helen Chandra, P., Martin-Vide, C., Subramanian, K.G., Van, D.L., Wang, P.S.P.: Parallel contextual array grammars and trajectories. In: Chen, C.H., Wang, P.S.P. (eds.) Handbook of Pattern Recognition and Computer Vision, 3rd edn., pp. 55–70. World Scientific, Singapore (2004)
6. Her, I.: Geometric transformations on the hexagonal grid. IEEE Transactions on Image Processing 4, 1213–1222 (2005)
7. Helen Chandra, P., Subramanian, K.G., Thomas, D.G.: Parallel contextual array grammars and languages. Electronic Notes in Discrete Mathematics 14, 537–550 (2000)
8. Krithivasan, K., Balan, M.S., Rama, R.: Array contextual grammars. In: Martin-Vide, C., Păun, G. (eds.) Recent Topics in Mathematical and Computational Linguistics, pp. 154–168 (2000)
9. Marcus, S.: Contextual grammars. Rev. Roum. Math. Pures Appl. 14, 1525–1534 (1969)
10. Middleton, L., Sivaswamy, J.: Hexagonal Image Processing: A Practical Approach. Springer, Heidelberg (2005)
11. Nagy, B.: Shortest path in triangular grids with neighbourhood sequences. Journal of Computing and Information Technology 11, 111–122 (2003)
12. Nagy, B.: Generalized triangular grids in digital geometry. Acta Mathematica Academiae Paedagogicae Nyiregyhaziensis 20, 63–78 (2004)
13. Nagy, B., Strand, R.: A connection between $\mathbb{Z}^n$ and generalized triangular grids. In: Bebis, G., Boyle, R., Parvin, B., Koracin, D., Remagnino, P., Porikli, F., Peters, J., Klosowski, J., Arns, L., Chun, Y.K., Rhyne, T.-M., Monroe, L. (eds.) ISVC 2008, Part II. LNCS, vol. 5359, pp. 1157–1166. Springer, Heidelberg (2008)
14. Paun, G.: Marcus Contextual Grammars. Kluwer, Dordrecht (1997)
15. Rama, R., Smitha, T.A.: Some results on array contextual grammars. International Journal Pattern Recognition and Artificial Intelligence 14, 537–550 (2000)
16. Rosenfeld, G., Siromoney, R.: Picture languages - A survey. Languages of Design 1, 229–245 (1993)
17. Rozenberg, G., Salomaa, A.: Handbook of Formal Languages, vol. 1, 2, 3. Springer, Heidelberg (1997)
18. Siromoney, G., Siromoney, R.: Hexagonal arrays and rectangular blocks. Computer Graphics and Image Processing 5, 353–381 (1976)
19. Subramanian, K.G.: Hexagonal array grammars. Computer Graphics and Image Processing 10, 388–394 (1979)
20. Subramanian, K.G., Van, D.L., Helen Chandra, P., Quyen, N.D.: Array Grammars with contextual operations. Fundamenta Informaticae 83, 411–428 (2008)

Connected Viscous Filters

Ana M. Herrera-Navarro[1], Israel Santillán[1,2], Jorge D. Mendiola-Santibáñez[2], and Iván R. Terol-Villalobos[3]

[1] Facultad en Ingeniería, Universidad Autónoma de Querétaro, 76000, México
[2] Universidad Politécnica de Querétaro, Carretera Estatal 420 S/N, el Rosario el Marqués, 76240 Querétaro, México
[3] CIDETEQ, S.C., Parque Tecnológico Querétaro S/N, SanFandila-Pedro Escobedo, 76700, Querétaro Mexico
famter@ciateq.net.mx

Abstract. This paper deals with the notion of connectivity in viscous lattices. In particular, a new family of morphological connected filters, called connected viscous filters is proposed. Connected viscous filters are completely determined by two criteria: size parameter and connectivity. The connection of these filters is defined on viscous lattices in such a way that they verify several properties of the traditionally known filters by reconstruction. Moreover, reconstruction algorithms used to implement filters by reconstruction can also be used to implement these new filters. The interest of these new connected filters is illustrated with different examples.

Keywords: Viscous lattices, connected viscous filters, filters by reconstruction, geodesic reconstruction, transformations with reconstruction criteria.

1 Introduction

Connectivity notion plays an important role in morphological image processing, particularly, its wide application in image segmentation and filtering. In image processing and analysis, connectivity is defined by means of either topological or graph-theoretic frameworks. Nevertheless, although these concepts have been extensively applied in image processing, they are incompatible. In addition to incompatibilities between these approaches, there are conceptual limitations that restrict the type of objects to which connectivity can be applied. These main problems of classical connectivity approaches inspired Matheron and Serra [13] to propose a purely algebraic axiomatic approach to connectivity, known as connectivity classes. Intensive work has been done on the study of connectivity classes ([1,4,8,9,14,16,20,23], just to mention a few). Image filtering, on the other hand, is one of the most interesting topics in image processing and analysis. The main goal in image filtering consists in removing undesirable regions from the image. In this sense, filtering is closely related to image segmentation. Particularly, the class of connected filters called by reconstruction [2,11,12,22], has been successfully applied in image segmentation. One of the most interesting characteristics

P. Wiederhold and R.P. Barneva (Eds.): IWCIA 2009, LNCS 5852, pp. 358–369, 2009.

of the filters by reconstruction is that they enable the complete extraction of the marked objects by preserving the edges. However, the main difficulty with these filters, more precisely with its associated connectivity (arcwise connectivity), is that it connects too much. Thus, this connectivity frequently hinders the identification of the different regions in an image. To attenuate this inconvenience, several works in the last ten years have been addressing this problem, [5,6,10,15,16,18,19,20], among others. Meyer [5,6] takes into account the interesting characteristics of filters by reconstruction to define monotone planings and flattenings in order to introduce levelings; these definitions enable to have an algebraic framework to propose the notion of opening (also closing) with reconstruction criteria (Terol-Villalobos and Vargas-Vázquez, [18]). Recently, Meyer [7] studied levelings as a tool for simplifying and segmenting images. In addition, Serra extended these concepts by means of a marker approach and the activity mappings notion [15]. However, among these works, the most interesting is that proposed by Serra [16], in which the author characterizes the concept of viscous propagation proposed by Vachier and Meyer [21], by means of geodesic reconstruction. To carry out this characterization, the usual working space $\mathcal{P}(E)$, the set of all subsets, is replaced by a more convenient framework given by the dilated sets (viscous lattices). Thus, a connection on viscous lattices which does not connect too much is defined, allowing the separation of arc-wise connected components into a set of connected components in the viscous lattice sense. These results give rise to the working space for the transformations with reconstruction criteria, which is given by viscous lattices (Terol-Villalobos et al. [19]). Based on the works developed by Serra [16] on viscous lattices, the current paper aims to introduce the connected viscous filters notion.

This paper is organized as follows. One first gives a review of some morphological filters and the connectivity class notion in Section 2. In Section 3, a study of connectivity on viscous lattices is carried out in order to introduce connected viscous filters. Next, in Section 4, an analysis of connected viscous filters is performed, showing that these filters have similar properties than filters by reconstruction.

2   Some Basic Concepts of Morphological Filtering

2.1   Basic Notions of Morphological Filtering

Mathematical morphology is mainly based on the increasing and idempotent transformations [3,13,17]. The use of both properties plays a fundamental role in the theory of morphological filtering. In fact, one calls morphological filter all increasing and idempotent transformation. The basic morphological filters are the morphological opening $\gamma_{\mu B}$ and the morphological closing $\varphi_{\mu B}$ with a given structuring element. In general, in the present work, a square structuring element is employed. Then, B represents the elementary structuring element (*3x3* pixels, for example) containing its origin; $\check{B}$ is its transposed set ($\check{B} = \{-x : x \in B\}$) and μ is an homothetic parameter. Thus, the morphological opening and closing are given, respectively, by equation (1):

$$\gamma_{\mu B}(f) = \delta_{\mu \check{B}}(\varepsilon_{\mu B}(f)) \quad \varphi_{\mu B}(f) = \varepsilon_{\mu \check{B}}(\delta_{\mu B}(f)) \tag{1}$$

where the morphological erosion $\varepsilon_{\mu B}$ and dilation $\delta_{\mu B}$ are expressed by $\varepsilon_{\mu B}(f)(x) = \bigwedge\{f(y) : y \in \mu \check{B}_x\}$ and $\delta_{\mu B}(f)(x) = \bigvee\{f(y) : y \in \mu \check{B}_x\}$. $\bigwedge$ is the inf operator and $\bigvee$ is the sup operator.

Henceforth, the set B is suppressed, i.e., the expressions γ_μ, $\gamma_{\mu B}$ are equivalent ($\gamma_\mu = \gamma_{\mu B}$). When the parameter μ is equal to one, all parameters are suppressed ($\delta_B = \delta$). Another interesting class morphological filters is derived from the notion of reconstruction. Reconstruction transformations enable the elimination of some undesirable regions without considerably affecting the remaining structures. Geodesic transformations are used to build the reconstruction transformations [22]. In the binary case, the geodesic dilation of size 1 of a set Y (called the marker) inside the set X is defined as $\delta_X^1(Y) = \delta(Y) \cap X$. To build a geodesic dilation of size m, the geodesic dilation of size 1 is iterated m times. Similarly, a geodesic erosion $\varepsilon_X^m(Y)$ is computed by iterating m times the geodesic erosion of size 1, $\varepsilon_X^1(Y) = \varepsilon(Y) \cup X$. When filters by reconstruction are built, the geodesic transformations are iterated until idempotence is reached. The reconstruction transformation in the gray-level case is a direct extension of the binary one. In this case, the geodesic dilation $\delta_f^1(g)$ (resp. the geodesic erosion $\varepsilon_f^1(g)$) with $g \leq f$ (resp. $g \geq f$) of size 1 given by $\delta_f^1(g) = f \wedge \delta_B(g)$ (resp. $\varepsilon_f^1(g) = f \vee \varepsilon_B(g)$) is iterated until idempotence. Consider two functions f and g, with $f \geq g$ ($f \leq g$). Reconstruction transformations of the marker function g in f, using geodesic dilations and erosions, expressed by $R(f,g)$ and $R^*(f,g)$, respectively, are defined by:

$$R(f, g) = \delta_f^n(g) = \underbrace{\delta_f^1 \cdots \delta_f^1 \delta_f^1}_{Until\ stability} (g) \quad R^*(f, g) = \varepsilon_f^n(g) = \underbrace{\varepsilon_f^1 \cdots \varepsilon_f^1 \varepsilon_f^1}_{Until\ stability} (g) \tag{2}$$

where n means until stability. When the marker function g is equal to the erosion or the dilation of the original function in equation (2), the opening and the closing by reconstruction are obtained:

$$\tilde{\gamma}_\mu(f) = \delta_f^n(\varepsilon_\mu(f)) \qquad \tilde{\varphi}_\mu(f) = \varepsilon_f^n(\delta_\mu(f)) \tag{3}$$

2.2   Connected Class

One of the most interesting concepts proposed in mathematical morphology (MM) is the notion of connectivity class. Let E be a non-empty space and $\mathcal{P}(E)$ the set of all subsets of E. In MM a connection, or connected class [13] on E is a set family $\mathcal{C} \subseteq \mathcal{P}(E)$ that satisfies the three following axioms

$$i/\ \ \emptyset \in \mathcal{C}, \qquad ii/\ \ x \in E \Rightarrow \{x\} \in \mathcal{C},$$

$$iii/\ \ \{X_i, i \in I\} \subseteq \mathcal{C} \quad and \quad \bigcap X_i \neq \emptyset \Rightarrow \bigcup X_i \in \mathcal{C}$$

An equivalent definition to the connected class is the point connected opening expressed by the following theorem, which provides an operating way to act on connections:

Theorem 1. *(Point connected opening) The datum of a connected class C on $\mathcal{P}(E)$ is equivalent to the family $\{\gamma_x, x \in E\}$ [13] of the so called "point connected opening" such that*

iv/ for all $x \in E$, we have $\gamma_x(x) = \{x\}$,

v/ for all $A \subseteq E$, x, y in E, $\gamma_x(A)$ and $\gamma_y(A)$ are equal or disjoint,

vi/ for all $A \subseteq E$ and for all $x \in E$, we have $x \notin A \Rightarrow \gamma_x(A) = \emptyset$

3   Connectivity and Filtering on Viscous Lattices

3.1   Connectivity and Connected Viscous Filters

Before introducing a new class of connected filters, let us remember an interesting concept that has been recently proposed by Serra [16]. In order to characterize the notion of viscous propagations [21], Serra replaces the usual space $P(E)$ by a viscous lattice structure. By establishing that the family $\mathcal{L} = \{\delta_\lambda(X), \quad \lambda > 0, \quad X \in \mathcal{P}(E)\}$ is both the image $\mathcal{P}(E)$ under an extensive dilation δ_λ and under the opening $\delta_\lambda \varepsilon_\lambda$, Serra proposes the viscous lattice as follows:

Proposition 1 (Viscous Lattice [16]). *The set $\mathcal{L}$ is a complete lattice regarding the inclusion ordering. In this lattice, the supremum coincides the set union, whereas the infimum $\wedge$ is the opening according to $\gamma_\lambda = \delta_\lambda \varepsilon_\lambda$ of the intersection.*

$$\wedge\{X_i, \quad i \in I\} = \gamma_\lambda(\cap\{X_i, i \in I\}) \quad \{X_i, \quad i \in I\} \in \mathcal{L}$$

The extreme elements of $\mathcal{L}$ are E and the empty set $\emptyset$. $\mathcal{L}$ is said to be the viscous lattice of dilation δ_λ.

$\mathcal{L}_\lambda$ denotes the image of $\mathcal{P}(E)$ under the opening (or the dilation) size λ, and $\mathcal{L}$ the image of $\mathcal{P}(E)$ under the opening or (dilation) for all $\lambda > 0$. Now, since dilation δ_λ is extensive by definition, it preserves the whole class C and the adjoint erosion ε treats the connected component independently of each other, i.e.

$$X = \bigcup\{X_i, X_i \in C\} \Rightarrow \varepsilon_\lambda(X) = \bigcup \varepsilon_\lambda(X_i)$$

Given that Serra proposes to define connections on $\mathcal{P}(E)$ before the dilation $\delta_{\lambda B}$, these connections can be established as follows.

Theorem 2 (Serra [16]). *Let C be a connection on $\mathcal{P}(E)$ and $\delta_\lambda \colon \mathcal{P}(E) \to \mathcal{P}(E)$ be an extensive dilation, of adjoint erosion ε_λ, that generates the lattice $\mathcal{L} = \delta_\lambda(\mathcal{P})$. If the closing $\epsilon_\lambda \delta_\lambda$ preserves the connection C, i.e. $\epsilon_\lambda \delta_\lambda(C) \subseteq C$, then the image $C' = \delta_\lambda(C)$ of the connected sets under δ_λ is a connection on the lattice $\mathcal{L}$.*

Thus, one expresses the $\mathcal{C}'$ components of a given set A as a function of its $\mathcal{C}$ components (arc-wise connected components), as follows:

Proposition 2 (Serra [16]). *Let γ_x (resp. $\gamma_{\delta(x)}$) be the point connected opening at point $x \in E$ (resp. at $\delta_x \in \mathcal{L}$) for the connection $\mathcal{C}$ (resp. $\mathcal{C}'$) of the above theorem. Then, the two openings γ_x and $\gamma_{\delta(x)}$ are linked by the relationship*
$$\gamma_{\delta(x)} = \delta_\lambda \gamma_x \varepsilon_\lambda$$

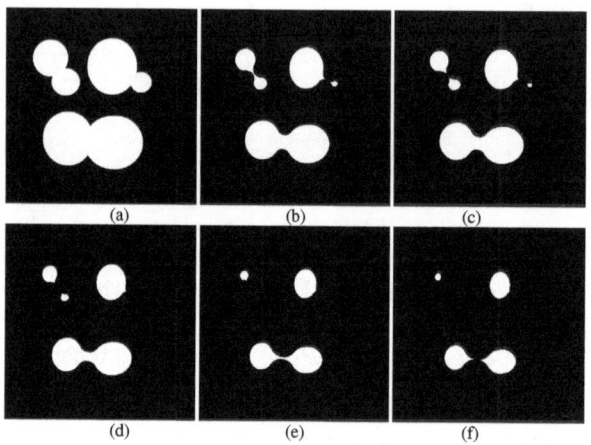

Fig. 1. (a) Original image X, (b),(c),(d), (e) and (f) Connected components at viscosity $\lambda = 20$, $\lambda = 22$, $\lambda = 27$, $\lambda = 36$, and $\lambda = 39$, respectively

According to this proposition, the number of connected components depends on the viscosity parameter λ. In Fig 1(a) one illustrates the original image composed of three arc-wise connected components or three components at viscosity $\lambda = 0$. Figures 1(b)-(f) show the eroded images by discs of sizes 20, 22, 27, 36 and 39, respectively. Then, at viscosity $\lambda = 20$, one has four connected components, whereas at viscosity $\lambda = 22$, the image is composed of five connected components. This is a most interesting connection since it exploits the goal in binary image segmentation which is to split the connected components into a set of elementary shapes. Now, let us introduce new connected openings and closings into the viscous lattices. Let γ_o and φ_o be the trivial opening and closing, respectively [2]:

$$\gamma_o(A) = \begin{cases} A, \text{ if A satisfies an} \\ \quad \text{increasing criterion} \\ \emptyset, \text{ otherwise} \end{cases} \qquad \varphi_o(A) = \begin{cases} E, \text{ if A satisfies an} \\ \quad \text{increasing criterion} \\ A, \text{ otherwise} \end{cases}$$

Now, the following connected opening and closing on the viscous lattice can be introduced.

Definition 1. *An opening* $\tilde{\gamma}_{\lambda,o}(f)$ *(closing* $\tilde{\varphi}_{\lambda,o}$*) is a connected viscous opening (a closing) at viscosity* λ *if and only if it is connected in the viscous lattice sense and*

$$\tilde{\gamma}_{\lambda,o}(X) = \bigvee_x \delta_\lambda \gamma_o \gamma_x \varepsilon_\lambda(X) = \delta_\lambda \bigvee_x \gamma_o \gamma_x \varepsilon_\lambda(X) = \delta_\lambda \tilde{\gamma} \varepsilon_\lambda(X)$$

$$\tilde{\varphi}_{\lambda,o}(X) = \bigwedge_x \varepsilon_\lambda \varphi_o \varphi_x \delta_\lambda(X) = \varepsilon_\lambda \bigwedge_x \varphi_o \varphi_x \delta_\lambda(X) = \varepsilon_\lambda \tilde{\varphi} \delta_\lambda(X) \quad (4)$$

These new filters are established in a general form, thus $\tilde{\gamma}$ and $\tilde{\varphi}$ are openings and closings by reconstruction in a general sense-area openings and closings, volume openings and closings, etc. However, the present study is focused on the use of increasing criteria given by the morphological erosion ε_μ and dilation δ_μ.

3.2   Geodesic Reconstruction and Filtering

Connected filters are frequently implemented by means of reconstruction algorithms. Similarly, connected viscous filters on viscous lattices can also be implemented using these algorithms. In this section, it is assumed that the increasing criteria of the trivial opening and closing are established by the morphological erosion and dilation, respectively. Thus, the connected viscous opening and closing at viscosity λ are given by:

$$\tilde{\gamma}_{\lambda,\mu}(f) = \delta_\lambda R(\varepsilon_\lambda(f), \varepsilon_{\mu-\lambda}\varepsilon_\lambda(f)) = \delta_\lambda R(\varepsilon_\lambda(f), \varepsilon_\mu(f))$$

$$\tilde{\varphi}_{\lambda,\mu}(f) = \tilde{\varepsilon}_\lambda R^*(\delta_\lambda(f), \delta_{\mu-\lambda}\delta_\lambda(f)) = \varepsilon_\lambda R^*(\delta_\lambda(f), \delta_\mu(f)) \quad (5)$$

Both transformations are now expressed in terms of reconstruction transformations. The parameter λ determines the viscosity of the transformation and μ the size criterion. Then, to compute $\tilde{\gamma}_{\lambda,\mu}$, one selects from the erosion $\varepsilon_\lambda(f)$, the regional maxima of size greater than or equal to $\mu - \lambda$, which is equivalent to apply an opening by reconstruction of this size, whereas for computing $\varphi_{\lambda,\mu}$, the minima structures of size greater than or equal to $\mu - \lambda$ are selected from $\delta_\lambda(f)$ using a closing by reconstruction. Thus,

$$\tilde{\gamma}_{\lambda,\mu}(f) = \delta_\lambda \tilde{\gamma}_{\mu-\lambda} \varepsilon_\lambda(f) \qquad \tilde{\varphi}_{\lambda,\mu}(f) = \varepsilon_\lambda \tilde{\varphi}_{\mu-\lambda} \delta_\lambda(f) \quad (6)$$

Now, from equations (5) and (6), it is easy to note that ordering exists. For $\lambda = \mu$; $\tilde{\gamma}_{\mu,\mu}(f) = \gamma_\mu(f)$ and $\tilde{\varphi}_{\mu,\mu}(f) = \varphi_\mu(f)$, whereas for $\lambda = 0$; $\tilde{\gamma}_{0,\mu}(f) = \tilde{\gamma}_\mu(f)$ and $\tilde{\varphi}_{0,\mu}(f) = \tilde{\varphi}_\mu(f)$. Then, the following inclusion relationships can be established:

$$\gamma_\mu(f) = \tilde{\gamma}_{\mu,\mu}(f) \leq \tilde{\gamma}_{\mu-1,\mu}(f) \cdots \leq \tilde{\gamma}_{1,\mu}(f) \leq \tilde{\gamma}_{0,\mu}(f) = \tilde{\gamma}_\mu(f)$$

$$\varphi_\mu(f) = \tilde{\varphi}_{\mu,\mu}(f) \geq \tilde{\varphi}_{\mu-1,\mu}(f) \cdots \geq \tilde{\varphi}_{1,\mu}(f) \geq \tilde{\varphi}_{0,\mu}(f) = \tilde{\varphi}_\mu(f) \quad (7)$$

The interest of these transformations is illustrated in Fig. 2. To remove the skull from the image in Fig. 2(a), the morphological opening, the opening by reconstruction and the connected viscous opening were applied. Figures 2 (b) and (c) show the output images computed by the morphological opening of size 5 and 15,

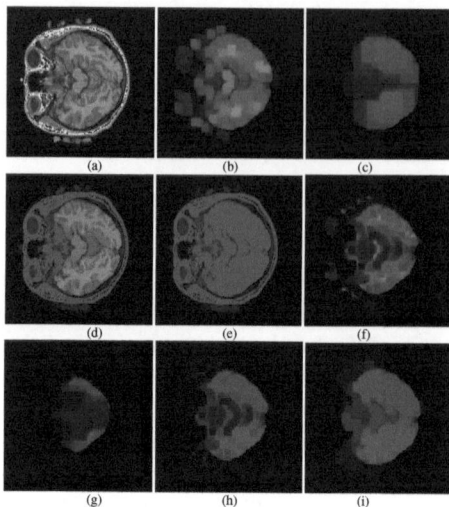

Fig. 2. (a) Original image, (b) and (c) Morphological opening γ_μ with μ, *5* and *15*, respectively, (d) and (e) Opening by reconstruction $\tilde{\gamma}_\mu$ with μ, *5* and *15*, respectively, (f) and (g) Eroded images ε_μ of sizes 5 and 15, respectively, (h) Geodesic reconstruction of reference image (f) from marker image (g), (i) Dilated image computed from (h) size

respectively, whereas Figs. 2 (d) and (e) illustrate the output images computed by the opening by reconstruction of sizes 5 and 15. The morphological opening removes the skull but causes a considerable modification in the remaining structures, while the opening by reconstruction does not remove the skull. Both parameter values, 5 and 15, were used to build the connected viscous opening. Figures 2(f) and (g) show the eroded images of sizes 5 and 15, which were used as the reference and marker images, respectively, to obtain the image in Fig. 2(h) $(R(\varepsilon_5(f), \varepsilon_{15}(f)))$. Finally, this last image was dilated with $\mu = 5$ in order to compute de image in Fig. 2(i). Observe that the skull has been removed by preserving the remaining structures.

4   Some Properties of Connected Viscous Filters

Under this new definition (equation 6), several interesting properties fulfilled by filters by reconstruction are also satisfied by connected viscous filters. As already expressed in Section 3, the connectivity of an image depends on the viscosity λ, hence let us classify the connected components in the viscous lattice. One will say that the connected components of $\mathcal{L}_\lambda$ are called λ-connected components. It is clear that a connected component of $\mathcal{L}_{\lambda_2}$ is a connected component of $\mathcal{L}_{\lambda_1}$ with $\lambda_1 \leq \lambda_2$; for all connected components $A \in \mathcal{L}_{\lambda_2}$, i.e., $\delta_{\lambda_2}\varepsilon_{\lambda_2}(A) = A \Rightarrow \delta_{\lambda_1}\varepsilon_{\lambda_1}(A) = A$. Let us study the effect of these filters on connected components of the input set. Particularly, the behavior of these filters when they form granulometries. Then, for two elements of a family of connected viscous openings, one establishes the following property.

Property 1. *Let μ_1 and μ_2 be two parameters such that $\lambda \leq \mu_1 \leq \mu_2$. Then, if* $\gamma_{\delta(x)}\widetilde{\gamma}_{\lambda,\mu_1} \neq \emptyset$ *and* $\gamma_{\delta(x)}\widetilde{\gamma}_{\lambda,\mu_2}(X) \neq \emptyset \Rightarrow \gamma_{\delta(x)}\widetilde{\gamma}_{\lambda,\mu_1}(X) = \gamma_{\delta(x)}\widetilde{\gamma}_{\lambda,\mu_2}(X)$

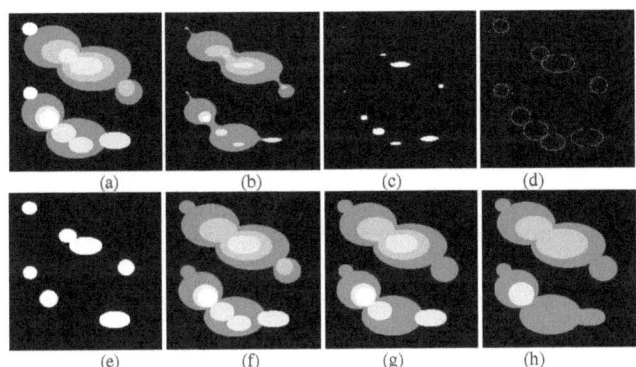

Fig. 3. (a) Input image, (b) Eroded image size 20, (c) Maxima of image in (b), (d) Contours of λ-maxima, (e) Regional maxima of image in input image, (f), (g) and (h) Output images computed by $\widetilde{\gamma}_{\lambda,\mu}$ with $\lambda = 20$ and μ values 25, 28, and 32, respectively

Thus, for all k, λ-connected components in $\widetilde{\gamma}_{\lambda,\mu_k}(X)$ are identical and they belong to the lattice $\mathcal{L}_\lambda$ (belong to the input set). This property can be extended to the function case by working with maxima (also minima) notion and the flat zone concept. In order to be congruent with the λ-connected components, one calls these features λ-flat zones and λ-maxima. In Fig. 3 the difference between λ-maxima and regional maxima is illustrated. Figures 3(b) and (c) show the eroded image size *20*, computed from 3(a) and its maxima, respectively. The eroded image contains nine maxima, hence the original image in Fig. 3(a) has nine λ-maxima (at viscosity $\lambda = 20$), as illustrated in Fig. 3(d). Only their contours are shown to better illustrate λ-maxima. Compare these λ-maxima, with the regional maxima shown in Fig. 3(e). Finally, Figs. 3(f), (g) and (h) illustrate the output images computed by $\widetilde{\gamma}_{\lambda,\mu}$ with $\lambda = 20$ and μ values 25, 28 and 32, respectively. Thus, for all k, the λ-connected components (λ-flat zones) in $\widetilde{\gamma}_{\lambda,\mu_k}(f)$ are passed to coarser levels or eliminated.

Property 2. λ-*Flat zones of the output* $\widetilde{\gamma}_{\lambda,\mu_k}(f)$ *grow as the parameter* μ_k *increases.*

Since flat zones of $\widetilde{\gamma}_{\lambda,\mu_k}(f)$ increase with the parameter μ_k, they are merged. This property explains the effect of simplification while preserving the contour information and it remembers us the pyramid notion of connected operators [11,12]. In other words, if Ψ_μ denotes a pyramid of connected operators acting on functions, then, for any function f, the flat zones of Ψ_μ increase with μ. Thus,

Property 3. *The family $\{\widetilde{\gamma}_{\lambda,\mu_k}(f)\}$ form a pyramid of connected operators in the viscous lattice sense of parameter μ_k.*

Now, before expressing an interesting property of the connected viscous opening, let us define the image difference in the viscous lattice. This is necessary in order to remain inside the viscous space. Thus, to obtain an element of the viscous lattice let us propose the following image difference.

Definition 2 (Viscous residue). *Let $\widetilde{\gamma}_{\lambda,\mu_1}$ and $\widetilde{\gamma}_{\lambda,\mu_2}$ be two connected viscous openings with $\mu_1 \leq \mu_2$. The residue between $\widetilde{\gamma}_{\lambda,\mu_1}(f)$ and $\widetilde{\gamma}_{\lambda,\mu_2}(f)$, is given by:*

$$\widetilde{\gamma}_{\lambda,\mu_1}(f) \div \widetilde{\gamma}_{\lambda,\mu_2}(f) = \delta_\lambda[\epsilon_\lambda(\widetilde{\gamma}_{\lambda,\mu_1}(f)) - \epsilon_\lambda(\widetilde{\gamma}_{\lambda,\mu_2}(f))]$$

The symbol $-$ denotes the arithmetical difference and $\div$ the viscous difference.

For example, Figs. 4(a) and (b) illustrate the output images computed from the original image in Fig. 1(a), by the openings $\widetilde{\gamma}_{\lambda,\mu_1}$ and $\widetilde{\gamma}_{\lambda,\mu_2}$, with viscosity parameter $\lambda = 22$, and size values, $\mu_1 = 27$, $\mu_2 = 39$, respectively. Figures 4(c) and (d) illustrate the eroded images size 22 computed from images in Figs. 4(a) and (b), respectively. Figure 4(e) shows the arithmetical difference between images 4(a) and (b), whereas 4(f) shows the viscous residue. The output image in Fig. 4(e) is not an element of the viscous lattice $\mathcal{L}_\lambda$. In Fig. 4(g), the λ-connected components of the original set in Fig. 1(a) are shown (only their contours are illustrated). Then, connected viscous openings satisfy the following property.

Property 4. *Let $\widetilde{\gamma}_{\lambda,\mu_1}$ and $\widetilde{\gamma}_{\lambda,\mu_2}$ be two connected viscous openings with $\lambda \leq \mu_1 \leq \mu_2$. Then, for all image f the difference $\widetilde{\gamma}_{\lambda,\mu_1}(f) \div \widetilde{\gamma}_{\lambda,\mu_2}(f)$ is an invariant of $\widetilde{\gamma}_{\lambda,\mu_1}$, i.e., $\widetilde{\gamma}_{\lambda,\mu_1}[\widetilde{\gamma}_{\lambda,\mu_1}(f) \div \widetilde{\gamma}_{\lambda,\mu_2}(f)] = \widetilde{\gamma}_{\lambda,\mu_1}(f) \div \widetilde{\gamma}_{\lambda,\mu_2}(f)$.*

These properties express that connected viscous filters do not introduce new structures.

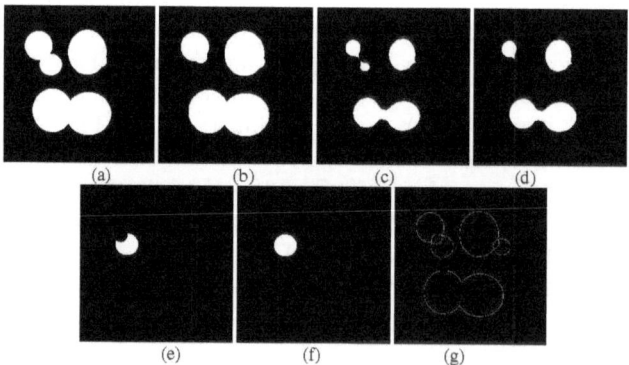

Fig. 4. (a) and (b) Output images computed by $\widetilde{\gamma}_{\lambda,\mu}$ with $\lambda = 22$ and μ 27 and 39, respectively, (c) and (d) Eroded images size 22 computed from (a) and (b), respectively, (e) Arithmetical difference between images (a) an (b), (f) Viscous difference between images (a) and (b), (g) λ-connected components of the original set in 1 (a)

4.1   Alternating Sequential Filters

Even though both families, $\{\widetilde{\gamma}_{\lambda,\mu_i}\}$ and $\{\widetilde{\varphi}_{\lambda,\mu_i}\}$, have similar properties than $\{\widetilde{\gamma}_{\mu_i}\}$ and $\{\widetilde{\varphi}_{\mu_i}\}$, little can be said about the properties of the families $\{\widetilde{\gamma}_{\lambda,\mu_i}\widetilde{\varphi}_{\lambda,\mu_i}\}$ and $\{\widetilde{\varphi}_{\lambda,\mu_i}\widetilde{\gamma}_{\lambda,\mu_i}\}$ when compared with filters by reconstruction. For example, one has that $\widetilde{\gamma}_{\lambda,\mu_i}\widetilde{\varphi}_{\lambda,\mu_i}\widetilde{\gamma}_{\lambda,\mu_i} \neq \widetilde{\varphi}_{\lambda,\mu_i}\widetilde{\gamma}_{\lambda,\mu_i}$. However, from a practical point of view, these filters have interesting characteristics. It is well-known that when structures (dark and bright) that are to be removed from the image have a wide range of scales, the use of a sequence of an opening (closing) followed by a closing (opening) does not lead to acceptable results. A solution to this problem is the use of alternating sequential filters (ASF) (see [3,13,17]). Serra defines and characterizes four operators $m_\mu(f) = \gamma_\mu\varphi_\mu(f)$, $n_\mu(f) = \varphi_\mu\gamma_\mu(f)$ $r_\mu(f) = \varphi_\mu\gamma_\mu\varphi_\mu(f)$, $s_\mu(f) = \gamma_\mu\varphi_\mu\gamma_\mu(f)$, where the size of the structuring element μ is indexed over a size distribution with $1 \leq \mu \leq \nu$. Let us take two of these operators defined as follows: $m_{\lambda,\mu}(f) = \widetilde{\gamma}_{\lambda,\mu}\widetilde{\varphi}_{\lambda,\mu}(f)$ and $n_{\lambda,\mu}(f) = \widetilde{\varphi}_{\lambda,\mu}\widetilde{\gamma}_{\lambda,\mu}(f)$. Since the connected viscous opening and closing take two parameters, let us use the μ parameter as the size of the structures of the image to be preserved and the λ parameter as the size of those structures linked to the structures to be preserved. While the size parameter μ detects the main structures, the λ parameter permits the smoothing of these structures. Then, for a family $\{\lambda_i\}$ with $\lambda_j < \lambda_k$ if $j < k$, the following alternating sequential filters are defined:

$$M_{\lambda_n,\mu}(f) = m_{\lambda_n,\mu}\dots m_{\lambda_2,\mu}m_{\lambda_1,\mu}(f) \quad N_{\lambda_n,\mu}(f) = n_{\lambda_n,\mu}\dots n_{\lambda_2,\mu}n_{\lambda_1,\mu}(f)$$

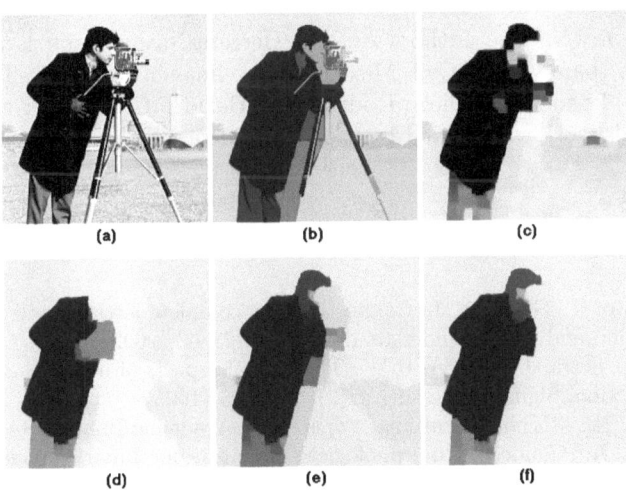

Fig. 5.   (a) Original image, (b) Alternating filter $\widetilde{\gamma}_\mu\widetilde{\varphi}_\mu(f)$ with $\mu = 25$, (c) ASF $\gamma_4\varphi_4\gamma_3\varphi_3\gamma_2\varphi_2\gamma_1\varphi_1(f)$, (d) Alternating filter $\widetilde{\gamma}_{\lambda,\mu}\widetilde{\varphi}_{\lambda,\mu}(f)$ with $\mu = 25$ and $\lambda = 4$, (e) ASF $\widetilde{\gamma}_{4,\mu}\widetilde{\varphi}_{4,\mu}\widetilde{\gamma}_{3,\mu}\widetilde{\varphi}_{3,\mu}\widetilde{\gamma}_{2,\mu}\widetilde{\varphi}_{2,\mu}\widetilde{\gamma}_{1,\mu}\widetilde{\varphi}_{1,\mu}(f)$ with $\mu = 25$, (f) ASF $\widetilde{\gamma}_{4,\mu}\widetilde{\varphi}_{4,\mu}\widetilde{\gamma}_{2,\mu}\widetilde{\varphi}_{2,\mu}(f)$

with the condition $\lambda_n \leq \mu$. Figures 5 illustrates the differences between two kinds of ASF. The image in Fig. 5(b) shows the output image computed from the original image in Fig. 5(a) by the alternate filter using an opening and closing by reconstruction $\widetilde{\gamma}_\mu \widetilde{\varphi}_\mu(f)$ with $\mu = 25$, while the image in Fig. 5(c) was computed by an ASF $\gamma_4 \varphi_4 \gamma_3 \varphi_3 \gamma_2 \varphi_2 \gamma_1 \varphi_1(f)$. Observe that the alternating filter, using reconstruction filters, enables us to extract the main structure of the image by removing all regions that are not connected to this structure; while the ASF, using morphological openings and closings, allows the removal of some structures linked to the main region (the cameraman). The idea of using connected viscous openings and closings consists in taking into account both behaviors. Then, let us apply a sequence of connected viscous openings and closings. The images in Figs. 5(d) to (f) illustrate the use of ASF using connected viscous openings and closings. The image in Fig. 5(d) was computed by using the filter $\widetilde{\gamma}_{\lambda,\mu} \widetilde{\varphi}_{\lambda,\mu}(f)$ with $\mu = 25$ and $\lambda = 4$, whereas in Fig. 5(e) the result of using the ASF $\widetilde{\gamma}_{4,\mu} \widetilde{\varphi}_{4,\mu} \widetilde{\gamma}_{3,\mu} \widetilde{\varphi}_{3,\mu} \widetilde{\gamma}_{2,\mu} \widetilde{\varphi}_{2,\mu} \widetilde{\gamma}_{1,\mu} \widetilde{\varphi}_{1,\mu}(f)$ with $\mu = 25$ is shown. Finally, the image in Fig. 5(f) was computed with the ASF $\widetilde{\gamma}_{4,\mu} \widetilde{\varphi}_{4,\mu} \widetilde{\gamma}_{2,\mu} \widetilde{\varphi}_{2,\mu}(f)$ and $\mu = 25$.

5   Conclusion and Future Work

In the present work, connected viscous filters were introduced. The connection of these filters is defined on viscous lattices in such a way that they verify several properties of the traditionally called connected filters by reconstruction. Moreover, reconstruction algorithms used to implement filters by reconstruction are used to implement these new filters. The interest of these new connected filters was illustrated with different examples.

Acknowledgements. The authors Ana M. Herrera-Navarro and Israel Santillan would like to thank CONACyT México for the financial support. The author I. Terol would like to thank Diego Rodrigo and Darío T.G. for their great encouragement. This work was funded by the government agency CONACyT (58367), México.

References

1. Braga-Neto, U., Goutsias, J.: Connectivity on complete lattices: new results. Computer Vision and Image Understanding 85(1), 23–53 (2002)
2. Crespo, J., Serra, J., Schafer, R.W.: Theoretical aspects of morphological filters by reconstruction. Signal processing 47(2), 201–225 (1995)
3. Heijmans, H.: Morphological image operators. Academic Press, USA (1994)
4. Heijmans, H.: Connected morphological operators for binary images. Computer Vision and Image Understanding 73(1), 99–120 (1999)
5. Meyer, F.: From connected operators to levelings. In: Mathemathical Morphology and Its Applications to Image and Signal Processing, pp. 191–198. Kluwer, Dordrecht (1998)
6. Meyer, F.: The levelings. In: Mathemathical Morphology and Its Applications to Image and Signal Processing, pp. 199–206. Kluwer, Dordrecht (1998)

7. Meyer, F.: Levelings, image simplification filters for segmentation. J. of Math., Imaging and Vis. 20, 59–72 (2004)

8. Ronse, R.: Set-theoretical algebraic approaches to connectivity in continuous or digital spaces. J. of Math. Imaging and Vis. 8, 41–58 (1998)

9. Ronse, C., Serra, J.: Geodesy and connectivity in lattices. Fundadenta Informaticae 46, 349–395 (2001)

10. Salembier, P., Oliveras, A.: Practical extensions of connected operators. In: Maragos, P., Schafer, R.W., Butt, M.A. (eds.) Mathematical Morphology and Its Applications to Image and Signal Processing, pp. 97–110. Kluwer Academic Publishers, The Netherlands (1996)

11. Salembier, P., Serra, J.: Flat zones filtering, connected operators, and fiters by reconstruction. IEEE Trans. on Image Processing 4(8), 1153–1160 (1995)

12. Serra, J., Salembier, P.: Connected operators and pyramids. In: SPIE Image Algrebra and Mathematical Morphology, San Diego California, USA, vol. 2030, pp. 65–76 (1993)

13. Serra, J.: Image Analysis and Mathematical Morphology. In: Theoretical advances, vol. 2. Academic Press, New York (1988)

14. Serra, J.: Connectivity on complete lattices. J. of Math. Imaging and Vis. 9(3), 231–251 (1998)

15. Serra, J.: Connection for sets and functions. Fundamenta Informaticae 41, 147–186 (2000)

16. Serra, J.: Viscous lattices. J. of Math. Imaging and Vis. 22(2-3), 269–282 (2005)

17. Soille, P.: Morphological Image Analysis: Principles and Applications. Springer, Berlin (1999)

18. Terol-Villalobos, I.R., Várgas-Vázquez, D.: Openings and closings by reconstruction with reconstruction criteria: a study of a class of lower and upper levelins. J. of Electronic Imaging 14(1), 013006 (2005)

19. Terol-Villalobos, I.R., Mendiola-Santibañez, J.D., Canchola-Magdaleno, S.L.: Image segmentation and filtering based on transformations with reconstruction criteria. J. of Vis. Commun. and Image Represent. 17, 107–130 (2006)

20. Tzafestas, C.S., Maragos, P.: Shape connectivity: multiscale analysis and application to generalized granulometries. J. of Math. Imaging and Vis. 17, 109–129 (2002)

21. Vachier, C., Meyer, F.: The viscous watershed transform. J. of Math. Imaging and Vis. 22(2-3), 251–267 (2005)

22. Vincent, L.: Morphological grayscale reconstruction in image analysis: applications and efficient algorithms. IEEE Trans. on Image Processing 2(2), 176–201 (1993)

23. Wilkinson, M.: Connected filtering by reconstruction: basis and new advances. In: IEEE International Conference on Image Processing, pp. 2180–2183 (2008)

Signatures of Combinatorial Maps

Stéphane Gosselin, Guillaume Damiand, and Christine Solnon*

Université de Lyon, CNRS
Université Lyon 1, LIRIS, UMR5205, F-69622, France
{stephane.gosselin,guillaume.damiand,christine.solnon}@liris.cnrs.fr

Abstract. In this paper, we address the problem of computing a canonical representation of an n-dimensional combinatorial map. To do so, we define two combinatorial map signatures: the first one has a quadratic space complexity and may be used to decide of isomorphism with a new map in linear time whereas the second one has a linear space complexity and may be used to decide of isomorphism in quadratic time. Experimental results show that these signatures can be used to recognize images very efficiently.

Keywords: Combinatorial map, canonical representation, signature, linear isomorphism.

1 Motivations

Combinatorial maps are good data structures for modelling space subdivisions. First defined in 2D [7,16,9,3], they have been extended to nD [2,11,12] and model the subdivision cells and their adjacency relations in any dimension.

There are many different data structures for modelling the partition in regions of an image. All these structures are more or less derived from Region Adjacency Graphs (RAG) [15]. It has been shown that using combinatorial maps allows a precise description of the topology of the image partition (for example in 2D [1] or in 3D [4]) and there are efficient image processing algorithms using this topological information.

Our general goal is to define new algorithms for combinatorial maps allowing new and efficient image processing algorithms. Among these operations, we are interested in image classification. In particular, we propose to characterize image classes by extracting patterns which occur frequently in these classes. When modelling images with combinatorial maps, this involves finding frequent submaps. Finding frequent patterns in large databases is a classical data mining problem, the tractability of which highly depends on the existence of efficient algorithms for deciding if two patterns are actually different or if they are two occurrences of a same object. Hence, if finding frequent subgraphs is intractable in the general case, it may be solved in incremental polynomial time when considering classes of graphs for which subgraph isomorphism may be solved in polynomial time, such as trees or outerplanar graphs [8].

* The authors acknowledge an ANR grant BLANC 07-1-184534: this work was done in the context of project SATTIC.

P. Wiederhold and R.P. Barneva (Eds.): IWCIA 2009, LNCS 5852, pp. 370–382, 2009.

In this paper, we address the problem of computing a canonical representation of combinatorial maps which may be used to efficiently search a map in a database. This work may be related to [10], which proposes a polynomial algorithm for deciding of the isomorphism of ordered graphs, based on a vertex labelling. Recently, this work has been extended to deal with combinatorial maps by proposing a polynomial algorithm for map and submap isomorphism based on a traversal of the map [6]. We use these principles of labelling and traversal to define canonical representations of combinatorial maps.

More precisely, we define two map signatures. Both signatures may be computed in quadratic time with respect to the number of darts of the map. The first signature is a lexicographic tree and has a quadratic space complexity. It allows one to decide in linear time if a new map is isomorphic to a map described by this signature. The second signature is a word and has a linear space complexity. As a counterpart, isomorphism with a new map becomes quadratic.

Notions of combinatorial maps are introduced in section 2. Section 3 presents an algorithm for labelling maps. Two signatures for connected maps are defined in section 4 and section 5. These signatures are extended to non connected maps in section 6. Finally, we experimentally evaluate our work in section 7.

2 Recalls on Combinatorial Maps

Definition 1 (Combinatorial map [12]). *An nD combinatorial map, (or n-map) is defined by a tuple $M = (D, \beta_1, \ldots, \beta_n)$ where*

- *D is a finite set of darts;*
- *β_1 is a permutation on D, i.e., a one-to-one mapping from D to D;*
- *$\forall 2 \leq i \leq n$, β_i is an involution on D, i.e., a one-to-one mapping from D to D such that $\beta_i = \beta_i^{-1}$;*
- *$\forall 1 \leq i \leq n-2$, $\forall i+2 \leq j \leq n$, $\beta_i \circ \beta_j$ is an involution on D.*

We note β_0 for β_1^{-1}. Two darts i and j such that $i = \beta_k(j)$ are said to be k-sewn. Fig. 1 gives an example of a $2D$ combinatorial map.

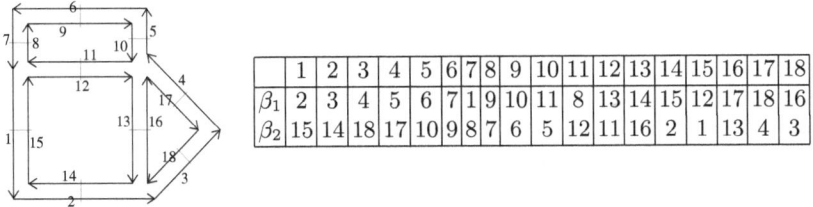

	1	2	3	4	5	6	7	8	9	10	11	12	13	14	15	16	17	18
β_1	2	3	4	5	6	7	1	9	10	11	8	13	14	15	12	17	18	16
β_2	15	14	18	17	10	9	8	7	6	5	12	11	16	2	1	13	4	3

Fig. 1. Combinatorial map example. Darts are represented by numbered black segments. Two darts 1-sewn are drawn consecutively, and two darts 2-sewn are concurrently drawn and in reverse orientation, with little grey segment between the two darts.

In some cases, it may be useful to allow some β_i to be partially defined, thus leading to open combinatorial maps. The basic idea is to add a new element ϵ to the set of darts, and to allow darts to be i-sewn with ϵ. By definition, $\forall 0 \leq i \leq n$, $\beta_i(\epsilon) = \epsilon$. Fig. 2 gives an example of open map (see [14] for precise definitions).

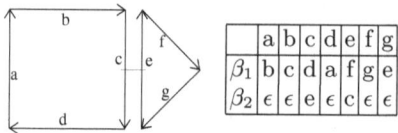

	a	b	c	d	e	f	g
β_1	b	c	d	a	f	g	e
β_2	ϵ	ϵ	e	ϵ	c	ϵ	ϵ

Fig. 2. Open combinatorial map example. Darts a, b, d, f and g are not 2-sewn.

Lienhardt has defined isomorphism between two combinatorial maps as follows.

Definition 2 (Map isomorphism [13]). *Two maps* $M = (D, \beta_1, \ldots, \beta_n)$ *and* $M' = (D', \beta'_1, \ldots, \beta'_n)$ *are isomorphic if there exists a one-to-one mapping* $f : D \to D'$, *called isomorphism function, such that* $\forall d \in D, \forall i, 1 \leq i \leq n$ $f(\beta_i(d)) = \beta'_i(f(d))$.

This definition has been extended to open maps in [6] by adding that $f(\epsilon) = \epsilon$, thus enforcing that, when a dart is i-sewn with ϵ, then the dart matched to it by f is i-sewn with ϵ.

Finally, Def. 3 states that a map is connected if there is a path of sewn darts between every pair of darts.

Definition 3 (Connected map). *A combinatorial map* $M = (D, \beta_1, \ldots, \beta_n)$ *is connected if* $\forall d \in D, \forall d' \in D$, *there exists a path* $(d_1, \ldots, d_k)$ *such that* $d_1 = d$, $d_k = d'$ *and* $\forall 1 \leq i < k, \exists j_i \in \{0, \ldots, n\}, d_{i+1} = \beta_{j_i}(d_i)$.

3   Map Labelling

Our signatures are based on map labellings, which associate a different label with every different dart. By definition, the label associated with ϵ is 0.

Definition 4 (Labelling). *Given a map* $M = (D, \beta_1, \ldots, \beta_n)$ *a labelling of* M *is a bijective function* $l : D \cup \{\epsilon\} \to \{0, \ldots, |D|\}$ *such that* $l(\epsilon) = 0$.

Example 1. $l = \{\epsilon : 0, a : 3, b : 1, c : 5, d : 7, e : 2, f : 6, g : 4\}$ is a labelling of the map displayed in Fig. 2.

One may compute a labelling of a map by performing a map traversal and labelling darts with respect to the order of their discovery. Different labellings may be computed, depending on (i) the initial dart from which the traversal is started, (ii) the strategy used for memorizing the darts that have been discovered

---

Algorithm 1. $BFL(M, d)$

Input: an open connected map $M = (D, \beta_1, \ldots, \beta_n)$, and a dart $d \in D$
Output: a labelling $l : D \cup \{\epsilon\} \to \{0, \ldots, |D|\}$

1 **for** *each* $d' \in D$ **do** $l(d') \leftarrow -1$
2 $l(\epsilon) \leftarrow 0$
3 let Q be an empty queue
4 add d at the end of Q
5 $l(d) \leftarrow 1$
6 *nextLabel* $\leftarrow 2$
7 **while** Q *is not empty* **do**
8 | remove d' from the head of Q
9 | **for** i *in* $0 \ldots n$ **do**
10 | | **if** $l(\beta_i(d')) = -1$ **then**
11 | | | $l(\beta_i(d')) \leftarrow nextLabel$
12 | | | $nextLabel \leftarrow nextLabel + 1$
13 | | | add $\beta_i(d')$ at the end of Q

14 **return** l

---

but that have not yet been used to discover new darts (e.g., FIFO or LIFO), and (iii) the order in which the β_i functions are used to discover new darts.

We define below the labelling corresponding to a breadth first traversal of a map where β_i functions are used in increasing order.

Definition 5 (Breadth first labelling (BFL)). *Given a connected map* $M = (D, \beta_1, \ldots, \beta_n)$ *and a dart* $d \in D$ *the breadth first labelling associated with* (M, d) *is the labelling returned by the function* $BFL(M, d)$ *described in algorithm 1.*

Example 2. The breadth first labellings associated with the map of Fig. 2 for darts a and e respectively are

$$BFL(M, a) = \{\epsilon : 0, \; a : 1, \; b : 3, \; c : 4, \; d : 2, \; e : 5, \; f : 7, \; g : 6\}$$
$$BFL(M, e) = \{\epsilon : 0, \; a : 7, \; b : 5, \; c : 4, \; d : 6, \; e : 1, \; f : 3, \; g : 2\}$$

Proposition 1. *Algorithm 1 returns a labelling.*

Proof

- $l(\epsilon)$ is set to 0 in line 2.
- $\forall d, d' \in D, d \neq d' \Rightarrow l(d) \neq l(d')$. Indeed, each time a label is assigned to a dart (line 11), *nextLabel* is incremented (line 12).
- $\forall d \in D, 1 \leq l(d) \leq |D|$. Indeed, each dart enters exactly once in the queue because (i) the map is connected and (ii) a dart enters the queue only if it has not yet been labelled, and it is labelled just before entering it.   □

Proposition 2. *The time complexity of algorithm 1 is* $\mathcal{O}(n \cdot |D|)$.

Proof. The while loop (line 7-13) is iterated $|D|$ times as (i) exactly one dart d is removed from the queue at each iteration; and (ii) each dart $d \in D$ enters the queue exactly once. The for loop (lines 9-13) is iterated $n + 1$ times.   □

Given a map M and a labelling l, one may describe M (i.e., its functions β_1 to β_n) by a sequence of labels of l. The idea is to first list the n labels of the n darts which are i-sewn with the dart labelled by 1 (*i.e.*, $l(\beta_1(1)), \ldots, l(\beta_n(1)))$, and then by 2 (*i.e.*, $l(\beta_1(2)), \ldots, l(\beta_n(2)))$, etc. More formally, we define the word associated with a map and a labelling as follows.

Definition 6 (Word). *Given a connected map* $M = (D, \beta_1, \ldots, \beta_n)$ *and a labelling* $l : D \cup \{\epsilon\} \rightarrow \{0, \ldots, |D|\}$ *the word associated with* (M, l) *is the sequence*

$$W(M, l) = < w_1, \ldots, w_{n \cdot |D|} >$$

such that $\forall i \in \{1, \ldots, n\}$, $\forall k \in \{1, \ldots, |D|\}$, $w_{i \cdot k} = l(\beta_i(d_k))$ *where* d_k *is the dart labelled with* k, *i.e.*, $d_k = l^{-1}(k)$.

Algorithm. Given a map $M = (D, \beta_1, \ldots, \beta_n)$ and a labelling l, the word $W(M, l)$ is computed by considering every dart of D in increasing label order and enumerating the labels of its n i-sewn darts. This is done in $\mathcal{O}(n \cdot |D|)$.

Notation. The word associated with the breadth first labelling of a map M, starting from a dart d, is denoted by $W_{BFL}(M, d)$, i.e.,

$$W_{BFL}(M, d) = W(M, BFL(M, d))$$

Example 3. The words associated with the map of Fig. 2 for the two labellings of example 2 respectively are

$$W_{BFL}(M, a) = < 3, 0, 1, 0, 4, 0, 2, 5, 7, 4, 5, 0, 6, 0 >$$
$$W_{BFL}(M, e) = < 3, 4, 1, 0, 2, 0, 6, 1, 4, 0, 7, 0, 5, 0 >$$

The key point which allows us to use words for building signatures is that two maps are isomorphic if and only if they share a word for a breadth first labelling, as stated in theorem 1.

Theorem 1. *Two connected maps* $M = (D, \beta_1, \ldots, \beta_n)$ *and* $M' = (D', \beta'_1, \ldots, \beta'_n)$ *are isomorphic iff there exist* $d \in D$ *and* $d' \in D'$ *such that* $W_{BFL}(M, d) = W_{BFL}(M', d')$

Proof. ⇒ Let us first consider two isomorphic maps $M = (D, \beta_1, \ldots, \beta_n)$ and $M' = (D', \beta'_1, \ldots, \beta'_n)$, and let us show that there exist two darts d and d' such that $W_{BFL}(M, d) = W_{BFL}(M', d')$. If M and M' are isomorphic then there exists $f : D \rightarrow D'$ such that $\forall d \in D, \forall i \in \{1, \ldots, n\}, f(\beta_i(d)) = \beta'_i(f(d))$. Let d_1 be a dart of D, and let us note l (resp. l') the labellings returned by $BFL(M, d_1)$ (resp. $BFL(M', f(d_1))$). <u>Claim 1</u>: l and l' are such that $\forall d_i \in$

$D, l(d_i) = l'(f(d_i))$. This is true for the initial dart d_1 as both d_1 and $f(d_1)$ are labelled with 1 at the beginning of each traversal. This is true for every other dart $d_i \in D$ as the traversals of M and M' performed by BFL are completely determined by the fact that (i) they consider the same FIFO strategy to select the next labelled dart which will be used to discover new darts and (ii) they use the β_i functions in the same order to discover new darts from a selected labelled dart. <u>Claim 2</u>: $\forall k \in \{1, \ldots, |D|\}, f(l^{-1}(k)) = l'^{-1}(k)$. This is a direct consequence of Claim 1. <u>Conclusion</u>: $\forall i \in \{1, \ldots, n\}, \forall k \in \{1, \ldots, |D|\}$, the $i.k^{th}$ element of $W_{BFL}(M, d_1)$ is equal to the $i.k^{th}$ element of $W'_{BFL}(M', f(d_1))$, i.e., $l(\beta_i(l^{-1}(k))) = l'(\beta'_i(l'-1(k)))$. Indeed,

$$\begin{aligned}
l(\beta_i(l^{-1}(k))) &= l'(f(\beta_i(l^{-1}(k)))) && \text{(because of Claim 1)} \\
&= l'(\beta'_i(f(l^{-1}(k)))) && \text{(because } f \text{ is an isomorphism function)} \\
&= l'(\beta'_i(l'^{-1}(k))) && \text{(because of Claim 2)}
\end{aligned}$$

$\Leftarrow$ Let us now consider two maps $M = (D, \beta_1, \ldots, \beta_n)$ and $M' = (D', \beta'_1, \ldots, \beta'_n)$ and two darts d and d' such that $W_{BFL}(M, d) = W_{BFL}(M', d')$, and let us show that M and M' are isomorphic. Let us note l (resp. l') the labellings returned by $BFL(M, d)$ (resp. $BFL(M', d')$), and let us define the function $f : D \to D'$ which matches darts with same labels, i.e., $\forall d_j \in D, f(d_j) = l'^{-1}(l(d_k))$. Note that this implies as well that $l(d_j) = l'(f(d_j))$. <u>Claim 3</u>: $\forall i \in \{1, \ldots, n\}, \forall k \in \{1, \ldots, |D|\}, l(\beta_i(l^{-1}(k))) = l'(\beta'_i(l'-1(k)))$. This comes from the fact that $W_{BFL}(M, d) = W_{BFL}(M', d')$ so that the $i.k^{th}$ element of $W_{BFL}(M, d_1)$ is equal to the $i.k^{th}$ element of $W'_{BFL}(M', f(d_1))$. <u>Conclusion</u>: $\forall i \in \{1, \ldots, n\}, \forall d_j \in D$,

$$\begin{aligned}
f(\beta_i(d_j)) &= l'^{-1}(l(\beta_i(d_j))) && \text{(by definition of } f) \\
&= l'^{-1}(l'(\beta'_i(l'^{-1}(l(d_j))))) && \text{(because of Claim 3)} \\
&= \beta'_i(l'^{-1}(l(d_j))) && \text{(by simplification)} \\
&= \beta'_i(l'^{-1}(l'(f(d_j)))) && \text{(by definition of } f) \\
&= \beta'_i(f(d_j)) && \text{(by simplification)}
\end{aligned}$$

Hence, f is an isomorphism function and M and M' are isomorphic. $\square$

4   Set Signature of a Connected Map

A map is characterized by the set of words associated with all possible breadth first labellings. This set defines a signature.

Definition 7 (Set Signature). *Given a map* $M = (D, \beta_1, \ldots, \beta_n)$ *the Set Signature associated with* M *is* $SS(M) = \{W_{BFL}(M, d) | d \in D\}$

Algorithm. $SS(M)$ is built by performing $W_{BFL}(M, d)$ for each $d \in D$ and collecting all different returned words in $SS(M)$. Hence, the overall time complexity is $\mathcal{O}(n \cdot |D|^2)$.

Theorem 2. $SS(M)$ *is a signature, i.e., two connected maps* M *and* M' *are isomorphic if and only if* $SS(M) = SS(M')$.

Proof. ⇒ Let us consider two isomorphic maps $M = (D, \beta_1, \ldots, \beta_n)$ and $M' = (D', \beta'_1, \ldots, \beta'_n)$, and let us show that $SS(M) = SS(M')$. This is a direct consequence of theorem 1, where we have shown that given an isomorphism function f between M and M' we have, for every dart $d \in D$, $W_{BFL}(M, d) = W_{BFL}(M', f(d))$. Hence, every word of $SS(M)$, computed from any dart of D, necessary belongs to $SS(M')$ (and conversely).

⇐ Let us consider two maps $M = (D, \beta_1, \ldots, \beta_n)$ and $M' = (D', \beta'_1, \ldots, \beta'_n)$ such that $SS(M) = SS(M')$, and let us show that M and M' are isomorphic. Indeed, there exist two words $W \in SS(M)$ and $W' \in SS(M')$ such that $W = W'$, thus M and M' are isomorphic due to theorem 1.   □

Note that a direct consequence of theorem 1 and theorem 2 is that for two non isomorphic maps M and M', $SS(M) \cap SS(M') = \emptyset$.

Property 1. The space complexity of the set signature of a map is $\mathcal{O}(n \cdot |D|^2)$.

Proof. The set signature contains at most $|D|$ words (it may contains less than $|D|$ words in case of automorphisms). Each word contains exactly $n \cdot |D|$ labels.

Lexicographical tree. The set signature of a map may be represented by a lexicographical tree which groups common prefixes of words. For example, the lexicographical tree of the set signature of the map displayed in Fig. 2 is displayed in Fig. 3.

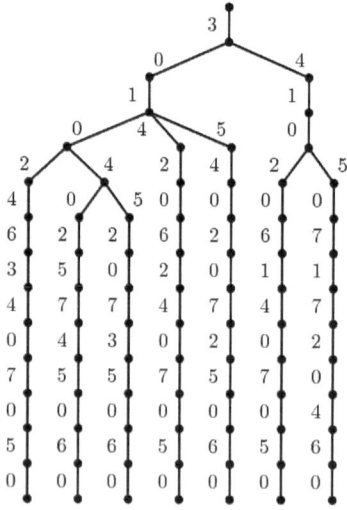

Fig. 3. Lexicographical tree of the set signature of the map of Fig. 2

Property 2. Given a map $M = (D, \beta_1, \ldots, \beta_n)$ and the lexicographical tree of the set signature $SS(M')$ of another map M', one determine the isomorphism between M and M' in $\mathcal{O}(n \cdot |D|)$.

Proof. To decide of the isomorphism, one has to build a breadth first labelling, starting from any dart of $d \in D$, and check if $W_{BFL}(M, d)$ is a branch of the lexicographical tree. Note that a node of the tree may have more than one son but, as the number of branches of the tree is bounded by the number of darts, deciding if a breadth first labelling corresponds to a branch of the tree may be done in linear time. □

5   Word Signature of a Connected Map

The lexicographical order is a strict total order on a set signature, and we have shown that if two set signatures share one word, then they are equal. Hence, one may define a map signature by considering the smallest word of the set signature.

Definition 8 (Word Signature). *Given a map M, the Word Signature of M is, $WS(M) = w \in SS(M)$ such that $\forall w' \in SS(M), w \leq w'$.*

Example 4. The word signature of the map displayed in Fig. 2 is

$$WS(M) = \langle 3, 0, 1, 0, 2, 4, 6, 3, 4, 0, 7, 0, 5, 0 \rangle$$

Property 3. The space complexity of a word signature is $\mathcal{O}(n \cdot |D|)$.

Algorithm. The word signature of a map M is built by calling $W_{BFL}(M, d)$ for each dart $d \in D$, and keeping the smallest returned word with respect to the lexicographical order. The time complexity for computing the word signature is $\mathcal{O}(n \cdot |D|^2)$. Note that this process may be improved (without changing the worst case complexity) by incrementally comparing the word in construction with the current smallest word and stopping the construction as soon as it becomes greater.

Property 4. Given a map $M = (D, \beta_1, \ldots, \beta_n)$ and the word signature $WS(M')$ of another map M', one determine the isomorphism between M and M' in $\mathcal{O}(n \cdot |D|^2)$.

Proof. To decide of the isomorphism, one has to build breadth first labellings, starting from every different dart of $d \in D$, and check if $W_{BFL}(M, d) = WS(M')$. In the worst case, one has to build $|D|$ labellings so that the overall time complexity is in $\mathcal{O}(n \cdot |D|^2)$. □

6   Signatures of Non Connected Maps

When a map is not connected, it may be decomposed in a set of disjoint connected maps in linear time with respect to the number of darts by performing successive map traversals until all darts have been discovered. The signature of a non connected map is built from the signatures of its different connected components.

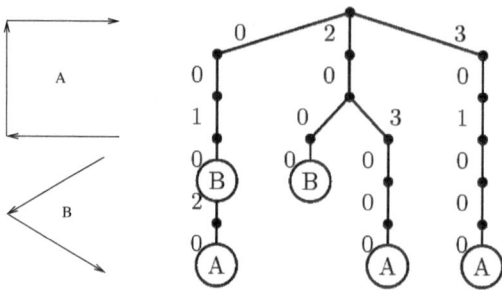

Fig. 4. Example of non connected map and its Set Signature

Word signature. To build a signature from the word signatures of the connected components, we sort the signatures by lexicographical order to obtain a list of words.

Set signature. To build a signature from the set signatures of the connected components, we merge the lexicographical trees as displayed in Fig. 4. Each node in the merged tree corresponding to a leaf in the tree of the ith connected component is labelled by i. Note that a node in the merged tree may have several labels as some connected components may be isomorphic. Note also that labelled nodes in the merged tree may be internal nodes as a branch of the tree of a connected component may be a prefix of the branch of another tree. To decide isomorphism between a non connected map and a Set Signature, we have to compute a breadth first labelling of each connected component starting from any dart, and check that each corresponding word ends in a node of the merged tree which is labelled by a different connected component.

7   Experiments

Using map signatures to search images. Maps may be extracted from segmented images by using the linear algorithm described in [5]. We obtain the same map whatever we submit the image to a rotation or a scale-up. Hence, map signatures may be used to identify images even if they have been rotated or scaled-up. Table 1 shows 5 images and gives the number of darts and faces of maps extracted from these images, and then gives the CPU time needed to compute set and word signatures.

Scale-up properties of signature constructions. To compare scale-up properties of set and word signatures, we have performed experiments on randomly generated maps with exponentially growing sizes (with 1000, 2000, 4000 and 8000 darts). Table 2 first compares time complexities for constructing set and word signatures. To build the set signature, one has to perform a complete breadth

Table 1. From images to signatures: the first line displays images, the next two lines give the number of darts and faces in the corresponding maps; the last two lines give the CPU time in seconds for computing the set and word signatures of these maps

Image					
Darts	3410	6060	1728	4224	1590
Faces	590	1044	295	716	275
$SS(M)$	0.83	2.21	0.26	1.14	0.26
$WS(M)$	0.26	0.53	0.15	0.32	0.16

first traversal for each dart so that the total number of visited darts is always equal to $|D|^2$ and the time complexity does not depend on the initial dart chosen to start the traversal. To build a word signature, one also has to perform a breadth first traversal for each dart but each traversal may be stopped as soon as the corresponding word is greater than the smallest word computed so far. Hence, if the worst case complexity is quadratic, Table 2 shows that the CPU time needed to compute a word signature is sub quadratic in practice. Indeed, the average number of darts visited for each traversal varies from 19.48 for the map with 1000 darts to 26.91 for the map with 8000 darts. Note that, if the number of visited darts actually depends on the order in which initial darts are chosen, standard deviations are rather low.

Scale-up properties of isomorphism. We now compare set and word signatures for deciding if a new map M' is isomorphic to a map M described by its signature.

Table 2. Comparison of time complexities for computing set and word signatures of a map. Each line successively gives the number of darts $|D|$ of the map and, for each signature, the CPU time (in seconds) and the ratio between the number of visited darts and $|D|$. We give average results (and standard deviations) obtained with different initial darts.

	Set Signature			Word Signature		
	Time	$\frac{Visited\ darts}{|D|}$		Time	$\frac{Visited\ darts}{|D|}$	
$|D|$	avg	avg	(sdv)	avg	avg	(sdv)
1000	0.054	1000	(0)	0.047	19.48	(3.24)
2000	0.228	2000	(0)	0.084	19.27	(3.71)
4000	1.056	4000	(0)	0.262	23.78	(5.31)
8000	4.088	8000	(0)	0.352	26.91	(4.88)

When using the set signature $SS(M)$, the worst case complexity is $\mathcal{O}(n \cdot |D|)$. Table 3 shows that, when M' and M are isomorphic (when the percentage of different darts is 0%), the algorithm visits each dart exactly once. However, when M and M' are not isomorphic, the breadth first traversal of M' may be stopped as soon as no branch of the lexicographical tree matches the word under construction. Table 3 shows that the more different M and M', the smaller the number of visited darts.

Table 3. Comparison of scale-up properties of set and word signatures for deciding if a new map M' is isomorphic to a map M given the signature of M. M and M' have the same number of darts, but M' is obtained from M by removing and then adding a given percentage of darts. When this percentage is 0%, M and M' are isomorphic. Each line successively gives: the number of darts of M, the percentage of different darts between M and M', and, for each signature, the time and the ratio between the number of visited darts and the number of darts of M. We give average results (and standard deviations) obtained when changing the initial dart of M'.

$|D|$		Set Signature			Word Signature		
		Time	$\frac{Visited\ darts}{|D|}$		Time	$\frac{Visited\ darts}{|D|}$	
		avg	avg	(sdv)	avg	avg	(sdv)
1000	0%	0.000099	1.000	(0.000)	0.035	2.13	(0.64)
	1%	0.000091	0.298	(0.214)	0.060	3.71	(1.48)
	10%	0.000086	0.026	(0.021)	0.059	3.41	(1.34)
	50%	0.000072	0.015	(0.006)	0.056	1.88	(1.19)
	99%	0.000068	0.011	(0.004)	0.050	1.59	(0.90)
2000	0%	0.000215	1.000	(0.000)	0.084	2.59	(1.47)
	1%	0.000161	0.069	(0.081)	0.095	3.08	(1.79)
	10%	0.000130	0.019	(0.032)	0.076	2.92	(1.76)
	50%	0.000098	0.006	(0.005)	0.073	1.77	(1.40)
	99%	0.000097	0.006	(0.003)	0.069	1.38	(0.83)
4000	0%	0.000341	1.000	(0.000)	0.262	2.46	(1.30)
	1%	0.000292	0.015	(0.037)	0.434	3.09	(1.89)
	10%	0.000222	0.005	(0.005)	0.329	2.57	(1.81)
	50%	0.000178	0.005	(0.006)	0.286	2.03	(1.41)
	99%	0.000164	0.005	(0.003)	0.273	1.43	(0.85)
8000	0%	0.000697	1.000	(0.000)	0.352	2.23	(1.04)
	1%	0.000556	0.032	(0.178)	1.451	3.11	(1.86)
	10%	0.000439	0.003	(0.009)	1.343	3.05	(1.81)
	50%	0.000296	0.002	(0.003)	1.042	2.44	(1.25)
	99%	0.000353	0.003	(0.003)	0.993	1.53	(1.02)

When using the word signature $WS(M)$, the worst case complexity is $\mathcal{O}(n \cdot |D|^2)$ as one has to perform a breadth first traversal starting from every dart of M'. However, one may stop each breadth first traversal as soon as the word under construction is different from the signature. Hence, Table 3 shows that the more different M and M', the smaller the number of visited darts. In practice, each dart is visited between 2 to 4 times. Interestingly, this ratio does not significantly vary when increasing the size of the map.

8   Conclusion

In this paper, we have defined two signatures of combinatorial maps, corresponding to canonical representations of n-dimensional combinatorial map. The memory complexity of the first signature is quadratic, but allows one to decide isomorphism in linear time in worst case and faster in average time. The memory complexity of the second signature is linear, but the complexity of the algorithm to decide isomorphism is quadratic in worst case and linear in average case.

The results of our experiments show that the signatures can be used to characterize images. This method is resistant to rotation and scale up. Moreover, both signatures can be used to find a map in a database. The Set Signature is faster but the size of the maps can be limited due to the memory size required to store the lexicographical trees. The Word Signature solves the memory problem, but the query's runtime is longer. We need to make several experiments in order to find the good compromise depending on the needs of the applications.

Now, we plan to use these signatures to compute a similarity measure between two combinatorial maps. In order to do that, we have to modify the signature so that it becomes error-tolerant. In further works, we will use those signatures to search frequent submaps in a database of maps. Our objective is to apply such method to chemical compound or to images to make a classification.

References

1. Braquelaire, J.-P., Brun, L.: Image segmentation with topological maps and inter-pixel representation 9(1), 62–79 (March 1998)
2. Brisson, E.: Representing geometric structures in d dimensions: Topology and order. In: Proc. 5th Annual ACM Symposium on Computational Geometry, Saarbrücken, Germany, pp. 218–227 (1989)
3. Cori, R.: Un code pour les graphes planaires et ses applications. In: Astérisque, Paris, France. Soc. Math. de France, vol. 27 (1975)
4. Damiand, G.: Topological model for 3d image representation: Definition and incremental extraction algorithm. Computer Vision and Image Understanding 109(3), 260–289 (2008)
5. Damiand, G., Bertrand, Y., Fiorio, C.: Topological model for two-dimensional image representation: definition and optimal extraction algorithm. Computer Vision and Image Understanding 93(2), 111–154 (2004)
6. Damiand, G., De La Higuera, C., Janodet, J.-C., Samuel, E., Solnon, C.: Polynomial Algorithm for Submap Isomorphism: Application to searching patterns in images. In: Torsello, A., Escolano, F., Brun, L. (eds.) GbRPR 2009. LNCS, vol. 5534, pp. 102–112. Springer, Heidelberg (2009)
7. Edmonds, J.: A combinatorial representation for polyhedral surfaces. Notices of the American Mathematical Society 7 (1960)

8. Horvath, T., Ramon, J., Wrobel, S.: Frequent subgraph mining in outerplanar graphs. In: KDD 2006, pp. 197–206 (2006)
9. Jacques, A.: Constellations et graphes topologiques. Combinatorial Theory and Applications 2, 657–673 (1970)
10. Jiang, X., Bunke, H.: Optimal quadratic-time isomorphism of ordered graphs. Pattern Recognition 32(7), 1273–1283 (1999)
11. Lienhardt, P.: Subdivision of n-dimensional spaces and n-dimensional generalized maps. In: Proc. 5th Annual ACM Symposium on Computational Geometry, Saarbrücken, Germany, pp. 228–236 (1989)
12. Lienhardt, P.: Topological models for boundary representation: a comparison with n-dimensional generalized maps. Computer-Aided Design 23(1), 59–82 (1991)
13. Lienhardt, P.: N-dimensional generalized combinatorial maps and cellular quasi-manifolds. Int. Journal of Computational Geometry and Applications 4(3), 275–324 (1994)
14. Poudret, M., Arnould, A., Bertrand, Y., Lienhardt, P.: Cartes combinatoires ouvertes. Research Notes 2007-1, Laboratoire SIC E.A. 4103, F-86962 Futuroscope Cedex - France (October 2007)
15. Rosenfeld, A.: Adjacency in digital pictures. Information and Control 26(1), 24–33 (1974)
16. Tutte, W.T.: A census of planar maps. Canad. J. Math. 15, 249–271 (1963)

Using Membrane Computing for Obtaining Homology Groups of Binary 2D Digital Images

Hepzibah A. Christinal[1,2], Daniel Díaz-Pernil[1], and Pedro Real Jurado[1]

[1] Research Group on Computational Topology and Applied Mathematics
University of Sevilla, Avda. Reina Mercedes s/n, 41012, Sevilla, Spain
[2] Karunya University, Coimbatore, Tamilnadu, India
{hepzi,sbdani,real}@us.es

Abstract. Membrane Computing is a new paradigm inspired from cellular communication. Until now, P systems have been used in research areas like modeling chemical process, several ecosystems, etc. In this paper, we apply P systems to Computational Topology within the context of the Digital Image. We work with a variant of P systems called *tissue-like P systems* to calculate in a general maximally parallel manner the homology groups of 2D images. In fact, homology computation for binary pixel-based 2D digital images can be reduced to connected component labeling of white and black regions. Finally, we use a software called *Tissue Simulator* to show with some examples how these systems work.

Keywords: computational topology, homology groups, membrane computing, P systems.

1   Introduction

Natural Computing studies new computational paradigms inspired from Nature. It abstracts the way in which Nature "computes", conceiving new computing models. There are several fields in Natural Computing that are now well established. To mention a few of these, Genetic Algorithms introduced by J. Holland[22] which is inspired by natural evolution and selection in order to find an optimal solution in a large set of feasible candidate solutions; Neural Networks introduced by W.S. McCulloch and W. Pitts[24] which is based on the interconnections of neurons in the brain; or DNA-based molecular computing, that was initiated when L. Adleman[1] published a solution to an instance of the Hamiltonian path problem by manipulating DNA strands in a lab.

Membrane Computing[1] is a theoretical model of computation inspired by the structure and functioning of cells like living organisms able to process and generate information. The computational devices in Membrane Computing are called *P systems*. Roughly speaking, a P system consists of a membrane structure,

---

[1] A layman-oriented introduction can be found in [32], a comprehensive presentation can be found in [30] and further updated bibliography in [40]. A presentation of applications can be found in [5].

P. Wiederhold and R.P. Barneva (Eds.): IWCIA 2009, LNCS 5852, pp. 383–396, 2009.

in the compartments of which one places multisets of objects which evolve according to given rules. In the most extended model, the rules are applied in a synchronous non-deterministic maximally parallel manner, but some other semantics are being explored. Ever since the seminal paper [29] was introduced, different models of P systems have been studied. According to their architecture, these models can be split into two sets: cell-like P systems and tissue-like P systems[33,7]. In cell-like P systems, membranes are hierarchically arranged in a tree-like structure. The inspiration for such architecture is the set of vesicles inside the cell. All of them perform their biological processes in parallel and life is the consequence of the harmonious conjunction of such processes.

This paper is devoted to the second approach: tissue-like P systems. According to the architecture, the main difference with cell-like P systems is that the structure of membranes is defined by a general graph instead of a tree-like graph. These models were first presented by Martín–Vide *et al.* in [25] and it has two biological inspirations (see [26]): intercellular communication and cooperation between neurons. The common mathematical model of these two mechanisms is a network of processors dealing with symbols and communicating these symbols along channels specified in advance. The communication among cells is based on symport/antiport rules. This way of communication for P systems was introduced in [31] on the basis of the communication between cells. Symport rules move objects across a membrane together in one direction, whereas antiport rules move objects across a membrane in opposite directions.

On the other hand, homology groups related to the "different" n-dimensional holes (connected component, tunnels, cavities,...) are invariants from Algebraic Topology which are frequently used in Digital Image Analysis and Structural Pattern Recognition. In some sort, they reflect the topological nature of the object in terms of the number and characteristics of its holes. In a binary 2D image, the computation of homology groups can be reduced to a process of black and white connected components labeling. The different black connected components are the generators of the 0-dimensional homology group which is the "black" part of the image. On the other hand, the closed "black" curves surrounding the different white connected components of the image are the generators of its 1-dimensional homology group.

J. Chao and J. Nakayama connected Natural Computing and Algebraic Topology using Neural Networks[4] (extended Kohonen mapping). Moreover, the idea to relate P systems and image processing already appeared in [3,6]. Here, we use for the first time, the power and efficiency of a variant of P systems called tissue-like P systems[8,9] to calculate the homology groups to binary pixel-based 2D images. The parallelism is massive in this model (see [20,23]), so the time used to obtain the homology groups does not depend on the number of black and white connected components, but only on the thickness of them.

The paper is structured as follows: in the next section we present the definition of basic P systems with input. In Section 3, we design two systems for calculating H_0 and H_1 for any binary pixel-based 2D digital image (having $n \times n$ pixels) and we show how both systems calculate the homology groups to two specific

8×8 images in the following section. In final part of the paper, we present some conclusions and future work.

2   Description of a Model of Membranes

In the first definition of tissue P systems in [25,26] the membrane structure did not change along the computation. Based on the cell-like model of P systems with active membranes, Gh. Păun et al. presented in [33] a new model of tissue-like P systems *with cell division*. The biological inspiration is clear: alive tissues are not *static* network of cells, since cells are duplicated via mitosis in a natural way. Díaz-Pernil presented in [7] a formalization of *Tissue-like P systems* (without cellular division), and these are the systems that we use in this paper.

The main features of this model, from the computational point of view, are that cells do not have polarizations (the contrary holds in the cell-like model of P systems, see [30]) and the membrane structure is a general graph, not a tree (i.e., not a cell-like model).

Formally, a *tissue-like P system* of degree $q \geq 1$ with input is a tuple of the form

$$\Pi = (\Gamma, \Sigma, \mathcal{E}, w_1, \ldots, w_q, \mathcal{R}, i_\Pi, o_\Pi),$$

where

1. Γ is a finite *alphabet*, whose symbols will be called *objects*,
2. $\Sigma(\subset \Gamma)$ is the input alphabet,
3. $\mathcal{E} \subseteq \Gamma$ (the objects in the environment),
4. $w_1, \ldots, w_q$ are strings over Γ representing the multisets of objects associated with the cells at the initial configuration,
5. $\mathcal{R}$ is a finite set of communication rules of the following form: $(i, u/v, j)$, for $i, j \in \{0, 1, 2, \ldots, q\}, i \neq j, u, v \in \Gamma^*$,
6. $i_\Pi \in \{0, 1, 2, \ldots, q\}$,
7. $o_\Pi \in \{0, 1, 2, \ldots, q\}$.

A tissue-like P system of degree $q \geq 1$ can be seen as a set of q cells (each one consisting of an elementary membrane) labeled by $1, 2, \ldots, q$. We will use 0 to refer to the label of the environment, i_Π and o_Π denote the input region and the output region (which can be the region inside a cell or the environment) respectively.

The strings $w_1, \ldots, w_q$ describe the multisets of objects placed in the q cells of the system. We interpret that $\mathcal{E} \subseteq \Gamma$ is the set of objects placed in the environment, each one of them available in an arbitrary large amount of copies.

The communication rule $(i, u/v, j)$ can be applied over two cells labeled by i and j such that u is contained in cell i and v is contained in cell j. The application of this rule means that the objects of the multisets represented by u and v are interchanged between the two cells. Note that if either $i = 0$ or $j = 0$ then the objects are interchanged between a cell and the environment.

Rules are used as usual in the framework of membrane computing, that is, in a maximally parallel way (a universal clock is considered). In one step, each object in a membrane can only be used for one rule (non-deterministically chosen when there are several possibilities), but any object which can participate in a rule of any form must do it, i.e, in each step we apply a maximal set of rules.

Now, to understand how we can obtain a computation of one of these P systems we present an example of them:

Consider us the following tissue-like P system

$$\Pi' = (\Gamma, \Sigma, \mathcal{E}, w_1, w_2, \mathcal{R}, i_\Pi, o_\Pi)$$

where

1. $\Gamma = \{a, b, c, d, e\}$,
2. $\Sigma = \emptyset$,
3. $\mathcal{E} = \{a, b, e\}$,
4. $w_1 = a^3 e$, $w_2 = b^2 c d$,
5. $\mathcal{R}$ is the following set of communication rules
   (a) $(1, a/b, 2)$,
   (b) $(2, c/b^2, 0)$,
   (c) $(2, d/e^2, 0)$,
   (d) $(1, e/\lambda, 0)$,
6. $i_\Pi = 1$,
7. $o_\Pi = 0$

We can observe the initial configuration of this system in the Figure 1 (a). We have four rules to apply. First rule is $(1, a/b, 2)$. The rule can be applied whenever an object 'a' is founded in cell 1 and one copy of 'b' appear in cell 2. This rule sends 'a' to cell 2 and 'b' from cell 2 to cell 1. Rule 2 is $(2, c/b^2, 0)$ and implies that when symbol 'c' present in cell 2 then this rule takes two copies of 'b' from environment and sends 'c' to the environment (i.e. cell 0). Rule 3 is similar to rule 2. Rule 4, $(1, e/\lambda, 0)$, sends the object 'e' to the environment. So, as we have 3 copies of 'a' and 1 copy of 'e' in cell 1 and 2 copies of 'b', one copy of 'c' and two copies of 'd' appear in cell 2. Then, all the rules can be applied in a parallel manner. Figure 1(b) show the next configuration of the system after applying the rules. If reader observes the initial elements in the environment of a tissue-like P systems (in this

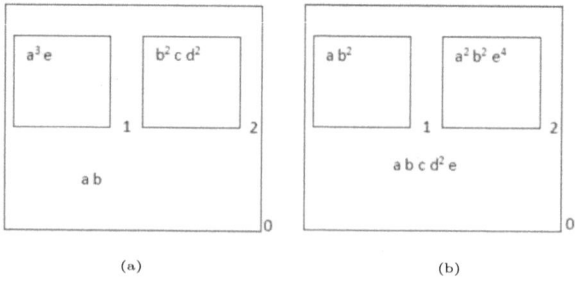

(a)   (b)

Fig. 1. (a) Initial Configuration of system Π' (b) Following Configuration of Π'

case a, b), one can observe the number of the copies of these elements always appear as one, because we have an arbitrary large amount of copies of them. The only objects changing its number of copies in the environment during a computation are the elements were not appear there initially. In this example, d has two copies because it is not an initial element of the environment.

3   Calculating Homology Groups

In the following, we will try to calculate homology groups, H_0 and H_1, to a digital image of two dimensions. The image is given by multiple pixels forming a $\mathbb{N}^2$ mesh. We suppose each pixel has associated with one of the two possible colors, black and white. Then, the black or white pixel in the position (i, j) is codified by the object B_{ij} or W_{ij}.

H_0 is given by the number of connected components formed by the black pixels and H_1 is given by the number of the holes created by the black pixels; i.e., the number of connected components of white pixels surrounded by black pixels. So, we consider 4-adjacency to see which are the neighboring pixels (if we consider 8-adjacency the systems will be very similar to the systems appear in this paper).

3.1   A Family of Tissue-Like P Systems to Obtain H_0

At this point, we want to know the number of connected components formed by the black pixels. We define a family of tissue-like P systems and for all digital images with n^2 pixels ($n \in \mathbb{N}$) we take a tissue-like P system whose input is given by two types of elements: B_{ij} codifying a black pixel, W_{ij} codifying a white pixel of the input image. The output is given by the number of objects C that appear in the output cell when the system stops (the number of connected components).

Below, we describe the rules of the family of systems in a schematic manner. For each type of rules we show a representative rule. For example, we describe the rules of type 1 as follows:

$$
\begin{array}{ccc}
W\ K & & W\ K \\
(1,\ W\ B\ B & / & W\ W\ B\ ,0) \\
W\ B & & W\ B
\end{array}
$$

where K could be B or W.

We describe the rules of each type depending on the position of the black pixels (up, down, left and right) respect to white pixels. For example, with the above schema we represent 8 subtypes of rules of type 1: 4 for each possible position and we must consider the possible values of K.

So, we can define a family of tissue-like P systems to calculate H_0 to any 2D image. For each $n \in \mathbb{N}$ we will consider the tissue-like P system of the family with input of degree 2:

$$
\Pi_0(n) = (\Gamma, \Sigma, \mathcal{E}, w_1, w_2, \mathcal{R}, i_\Pi, o_\Pi),
$$

defined as follows

a) $\Gamma = \Sigma \cup \{G_{ij} : 1 \le i, j \le n\} \cup \{C\}$,
b) $\Sigma = \{B_{ij}, W_{ij} : 1 \le i, j \le n\}$,
c) $\mathcal{E} = \Sigma \cup \{C\}$,
d) $w_1 = \{W_{ij} : (i = 0 \wedge 1 \le j \le n) \vee$
$\qquad\qquad (i = n+1 \wedge 1 \le j \le n) \vee$
$\qquad\qquad (j = 0 \wedge 1 \le i \le n) \vee$
$\qquad\qquad (j = n+1 \wedge 1 \le i \le n)\}$,

$w_2 = \emptyset$,

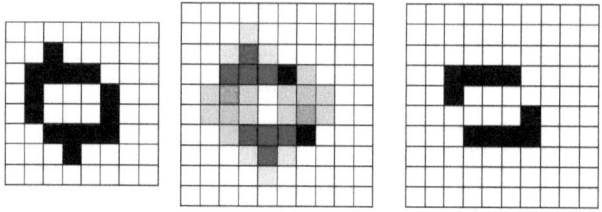

Fig. 2. Cutting branches of two black connected components

e) R is the following set of communication rules:

$$
1. \; (1, \begin{matrix} W & K \\ W & B & B \\ & W & B \end{matrix} \; / \; \begin{matrix} W & K \\ W & W & B \\ & W & B \end{matrix} ,0), \text{ where } K = B \text{ or } K = W.
$$

$$
2. \; (1, \begin{matrix} W & W \\ W & B & B \\ & W & W \end{matrix} \; / \; \begin{matrix} W & W \\ W & W & B \\ & W & W \end{matrix} ,0)
$$

The above two types of rules are used to eliminate single points, i.e. branches of black connected components, as seen in Figure 2 where the necessary pixels to apply two rules of type 1 are colored in red, and then colored in green when pixels are used to apply rules of type 2.

$$
3. \; (1, \begin{matrix} W & W \\ W & B & B \\ & W & B & B \end{matrix} \; / \; \begin{matrix} W & W \\ W & W & B \\ & W & B & B \end{matrix} ,0)
$$

$$
4. \; (1, \begin{matrix} & W \\ W & B & B & B \\ & B & W & W \\ & & W \end{matrix} \; / \; \begin{matrix} & W \\ W & W & B & B \\ & B & B & W \\ & & W \end{matrix} ,0)
$$

$$
5. \; (1, \begin{matrix} W & W & W \\ B & B & B & W \\ & B & W & B \\ & & W \end{matrix} \; / \; \begin{matrix} W & W & W \\ B & W & W & W \\ & B & B & B \\ & & W \end{matrix} ,0)
$$

$$
6. \; (1, \begin{matrix} B & B & B \\ B & W & B \\ B & B & B \end{matrix} \; / \; \begin{matrix} B & B & B \\ B & B & B \\ B & B & B \end{matrix} ,0)
$$

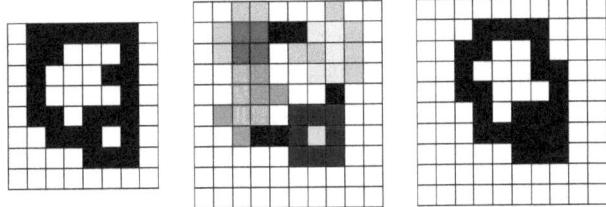

Fig. 3. Reducing black connected components with rules of type 2 to 6

These of rules are used to reduce the dimensions of the black connected components with white pixels inside them, as seen in Figure 3 where we have colored pixels used by rules of types 3 to 6 with different colors: red, blue, yellow and green, respectively.

$$7.\ (1, \begin{matrix} & W & \\ W & B & W \\ & W & \end{matrix} \ /\ \begin{matrix} & W & \\ W & G & W \\ & W & \end{matrix} ,0)$$

The 7-th type of rule is used when the system reduces the black connected components to only one pixel. So, these rules change the color of the pixel to green (codify with the object G_{ij}).

8. $(1, G_{ij}/W_{ij}\, C, 0)$, for $1 \leq i, j \leq n$
The 8-th type of rules brings an object C and W_{ij} to membrane 1 and sends G_{ij} to the environment.

9. $(1, C/\lambda, 2)$
The 9-th type of rule sends one copy of the object C to the output cell. Then, so much copies of C as connected components of black pixel arrive to cell 2.

f) $i_\Pi = 1$,
g) $o_\Pi = 2$.

Overview of a Computation: Given an image as input data whose size is $n \times n$, there exists a system of this family working in a parallel manner: First, it eliminates the branches of the black connected components that appear in the image in 4 steps. For this, the system uses rules of type 1 and 2. Secondly, the system reduces the size of the black connected components from four directions—up, down, left and right. The system takes the rules of types 3 to 6 to realize this task, and needs a logarithmic number of steps proportionate to the size of the biggest black connected component and reduce each component to only one black pixel. In this manner, we have obtain the complexity of the problem to obtain homology group H_0 of binary 2D digital image using tissue-like P systems.

Complexity and Necessary Resources: Taking account the size of the input data is $O(n^2)$, the amount of necessary resources for defining the systems of our family and the complexity of our problem can be observed in the following table:

H_0 Problem	
Complexity	
Number of steps of a computation	$O(n)$
Necessary Resources	
Size of the alphabet	$3n^2 + 1$
Initial number of cells	2
Initial number of objects	$4n - 3$
Number of rules	$O(n^2)$
Upper bound for the length of the rules	22

3.2   A Family of Tissue-Like P System to Obtain H_1

Now, we want to know the number of white connected components surrounded by one or more black connected components. So, we define a family of tissue-like P systems and for each digital image with n^2 pixels ($n \in \mathbb{N}$), we take a specific tissue-like P system of the family for all the images with size $n \times n$.

For each $n \in \mathbb{N}$ we will consider the tissue-like P system

$$\Pi_1(n) = (\Gamma, \Sigma, \mathcal{E}, w_1, w_2, \mathcal{R}, i_\Pi, o_\Pi),$$

defined as follows

a) $\Gamma = \Sigma \cup \{b_{ij}, g_{ij}, w_{ij} : 1 \le i, j \le n\} \cup \{C\}$,

b) $\Sigma = \{B_{ij}, W_{ij} : 1 \le i, j \le n\} \cup \{P_{ij} : (i = 0 \wedge 1 \le j \le n) \vee (i = n+1 \wedge 1 \le j \le n) \vee (j = 0 \wedge 1 \le i \le n) \vee (j = n+1 \wedge 1 \le i \le n)\}$,

c) $\mathcal{E} = \Gamma$,

d) $w_1 = \{a_1\} \cup \{P_{ij} : (i = 0 \wedge 1 \le j \le n) \vee (i = n+1 \wedge 1 \le j \le n) \vee (j = 0 \wedge 1 \le i \le n) \vee (j = n+1 \wedge 1 \le i \le n)\}$,

   $w_2 = \emptyset$,

e) R is the following set of communication rules:

   1. $(1, a_i/a_{i+1}^2, 0)$, for $i = 1 \dots n/2$
   It is a counter used to decide when the objects codifying pixels are sent to cell 2.

   2. $(1, P\,W\,/\,P\,P, 0)$
   The system eliminates all the white pixels (pass to be colored in pink) that are not inside black connected component.

3. $(1, a_{2\lceil \lg n\rceil} K_{ij}/k_{ij}, 0)$, for $1 \le i, j \le n$ and $K = B \vee W$.

When the objects $a_{2\lceil \lg n\rceil}$ appear in the cell 1 system sends all the objects codifying the black or white pixels to the cell 2.

The rest of the rules are the same of the system Π_0, but exchanging the white pixels by black pixels and in the other way:

$$4. \quad \left(1, \begin{matrix} & w\ k \\ w & b\ b \\ & w\ b \end{matrix} \right) \Big/ \begin{matrix} & w\ k \\ w\ w & b\ ,0 \\ & w\ b \end{matrix}, \text{ where } k = b \text{ or } k = w.$$

$$5. \quad \left(1, \begin{matrix} & w\ w \\ w & b\ b \\ & w\ w \end{matrix} \right) \Big/ \begin{matrix} & w\ w \\ w\ w & b\ ,0 \\ & w\ w \end{matrix}$$

$$6. \quad \left(1, \begin{matrix} & w\ w \\ w & b\ b \\ & w\ b\ b \end{matrix} \right) \Big/ \begin{matrix} & w\ w \\ w\ w & b\ ,0 \\ & w\ b\ b \end{matrix}$$

$$7. \quad \left(1, \begin{matrix} & w \\ w\ b\ b\ b \\ b\ w\ w \\ & w \end{matrix} \right) \Big/ \begin{matrix} & w \\ w\ w\ b\ b\ ,0 \\ b\ b\ w \\ & w \end{matrix}$$

$$8. \quad \left(1, \begin{matrix} & w\ w\ w \\ b\ b\ b\ w \\ b\ w\ b \\ & w \end{matrix} \right) \Big/ \begin{matrix} & w\ w\ w \\ b\ w\ w\ w\ ,0 \\ b\ b\ b \\ & w \end{matrix}$$

$$9. \quad \left(1, \begin{matrix} b\ b\ b \\ b\ w\ b \\ b\ b\ b \end{matrix} \right) \Big/ \begin{matrix} b\ b\ b \\ b\ b\ b\ ,0 \\ b\ b\ b \end{matrix}$$

$$10. \quad \left(1, \begin{matrix} w \\ w\ b\ w \\ w \end{matrix} \right) \Big/ \begin{matrix} w \\ w\ g\ w\ ,0 \\ w \end{matrix}$$

11. $(1, g_{ij}/w_{ij} C, 0)$, for $1 \le i, j \le n$

12. $(1, C/\lambda, 2)$

f) $i_\Pi = 1$,

g) $o_\Pi = 2$.

Overview of a Computation: Using a tissue-like P system, to compute H_1 of a digital image is similar to compute H_0. There exists a system of this family working in a parallel manner: First, it takes the white pixels not contained in black connected components and transforms these pixels in pink (type of rules 2). Using the counter a_i, white and black pixels are transformed in other objects (small letters) in $n/2+1$ steps (types of rules 1 and 3). In this form, we can apply the rest of rules (those similar to Π_0). So system eliminates the branches of the white connected components that appear in the image in 4 steps. For this, the system uses types of rules 4 and 5. Then, the system reduces the size of the white

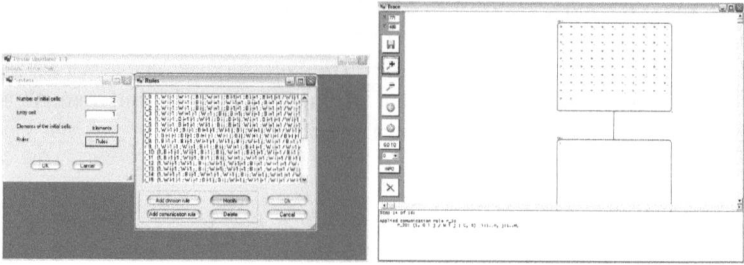

Fig. 4. Two images about Tissue Simulator

connected components from four directions: up, down, left and right. The system takes the rules of types 6 to 9 to realize this task, and needs a logarithmic number of steps proportionate to the size of the biggest white connected component and reduce each component to only one white pixel (less than $O(n)$). In this manner, we have obtained the complexity of the problem to obtain homology group H_1 of binary 2D digital image using tissue-like P systems.

Complexity and Necessary Resources: Taking account the size of the input data is $O(n^2)$, the amount of necessary resources to construct the tissue-like P systems of our family and the cellular complexity respect to time of our problem can be observed in the following table:

H_1 **Problem**	
Complexity	
Number of steps of a computation	$O(n)$
Necessary Resources	
Size of the alphabet	$5n^2 + 4n - 2$
Initial number of cells	2
Initial number of objects	$4n - 2$
Number of rules	$O(n^2)$
Upper bound for the length of the rules	22

4 Some Examples

In this section, we check the tissue-like P systems in section 3 above with some images that appear in Figure using a specific sequential software, called *Tissue Simulator* (see [39]) and developed by R. Borrego-Ropero et al. in [2]. This software was developed to help researchers to understand how these systems obtain a possible computation. Although, this program was developed in Java, it was not meant to be used in Digital Image. So, we do not work with images

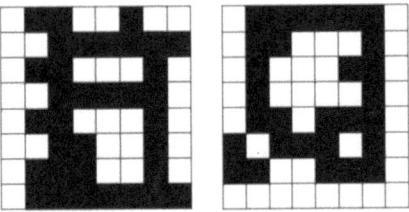

Fig. 5. Two images to check

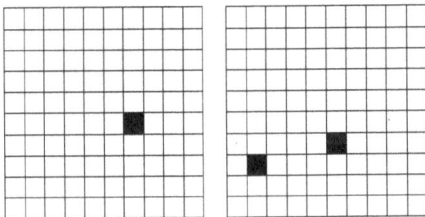

Fig. 6. Number of black connected components in the previous images

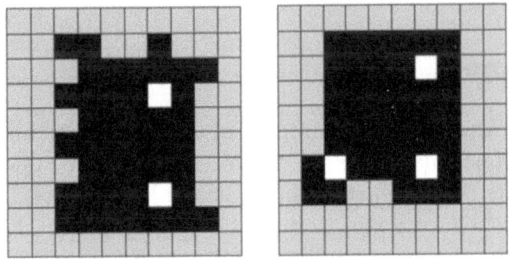

Fig. 7. Number of white holes in the previous images

directly, but we work with elements (of an alphabet) codifying the pixels of an image, and the output is given by these elements (see Figure 4).

First, we are going to obtain the different connected components for the images given by Figure 5.

After a logarithmic number of steps with respect to the input data, the Tissue Simulator stops and gives the output data that appears in the output cell (cell 2) of the system Π_1 (created to calculate the number of black connected components). This output is given using elements codifying the images which are shown in the Figure 6.

On the other hand, using a logarithmic number of steps again with respect to the input data, Tissue Simulator calculates the number of white connected components inside black connected components. It is shown in Figure 7, the output codified by the elements that appear in the output cell for each one of images of Figure 5.

5   Conclusions and Future Work

We have shown in this paper that the homology for 2D digital objects (using 4-connectivity or 8-connectivity for neighbor pixels) can be efficiently obtained using P systems. The most important issue we want to deal with the near future, is to use P systems for getting homological and cohomological information (Reeb graphs[17], AT-models ([13,14,15,16,18]), homology gradient vector field [28,27,37,36], representative (co)cycles of (co)homology generators [19,11,12], cohomology algebra [21,11], cohomology operations [38,10], torsion numbers [21], homotopy groups [34,35] for $3D$ and $4D$ geometric objects. The complexity in time for most of the algorithms previously cited ranges from linear (for connected component labeling), passing through cubical (for homology gradient vector fields or homology groups), and $O(n^5)$ for cohomology algebra and cohomology operations, up to exponential and more in the case of homotopy groups. The predictable drastic improvements in complexity that P-systems could mean in Computational Algebraic Topology methods. This would allow in the future to handle with optimism the computation processes of these complex topological invariants.

Acknowledgement

The first author acknowledge the support of the project "Computational Topology and Applied Mathematics" PAICYT research project FQM-296. The second author acknowledge the support of the project TIN2006-13425 of the Ministerio de Educación y Ciencia of Spain, cofinanced by FEDER funds, and the support of the Project of Excellence with *Investigador de Reconocida Valía* of the Junta de Andalucía, grant P08-TIC-04200. The third author acknowledge the support of the project MTM2006-03722 of the Ministerio español de Educación y Ciencia and the project PO6-TIC-02268 of Excellence of Junta de Andalucía.

References

1. Adleman, L.M.: Molecular computations of solutions to combinatorial problems. Science 226, 1021–1024 (1994)
2. Borrego–Ropero, R., Díaz–Pernil, D., Pérez–Jiménez, M.J.: Tissue Simulator: A Graphical Tool for Tissue P Systems. In: Proc. International Workshop, Automata for Cellular and Molecular Computing, MTA SZTAKI, Budapest, pp. 23–34 (2007)
3. Ceterchi, R., Mutyam, M., Păun, G., Subramanian, K.G.: Array-rewriting P systems. Natural Computing 2, 229–249 (2003)
4. Chao, J., Nakayama, J.: Cubical Singular Simples Model for 3D Objects and Fast Computation of Homology Groups. In: Proc. ICPR 1996, pp. 190–194. IEEE, Los Alamitos (1996)

5. Ciobanu, G., Păun, G., Pérez–Jiménez, M.J. (eds.): Applications of Membrane Computing. Springer, Heidelberg (2005)
6. Dersanambika, K.S., Krithivasan, K., Subramanian, K.G.: P Systems Generating Hexagonal Picture Languages. In: Martín-Vide, C., Mauri, G., Păun, G., Rozenberg, G., Salomaa, A. (eds.) WMC 2003. LNCS, vol. 2933, pp. 168–180. Springer, Heidelberg (2004)
7. Díaz–Pernil, D.: Sistemas P de Tejido: Formalización y Eficiencia Computacional. PhD Thesis, University of Seville (2008)
8. Díaz–Pernil, D., Gutiérrez, M.A., Pérez–Jiménez, M.J., Riscos–Núñez, A.: A uniform family of tissue P systems with cell division solving 3-COL in a linear time. Theoretical Computer Science 404(1–2), 76–87 (2008)
9. Díaz–Pernil, D., Pérez–Jiménez, M.J., Romero, A.: Computational efficiency of cellular division in tissue-like P systems. Romanian Journal of Information Science and Technology 11(3), 229–241 (2008)
10. González Díaz, R., Real, P.: Computation of Cohomology Operations of Finite Simplicial Complexes. Homology Homotopy and Applications 2, 83–93 (2003)
11. Gonzalez-Diaz, R., Real, P.: Towards Digital Cohomology. In: Nyström, I., Sanniti di Baja, G., Svensson, S. (eds.) DGCI 2003. LNCS, vol. 2886, pp. 92–101. Springer, Heidelberg (2003)
12. Gonzalez-Diaz, R., Real, P.: On the cohomology of 3D digital images. Discrete Applied Mathematics 147, 245–263 (2005)
13. Gonzalez-Diaz, R., Medrano, B., Real, P., Sanchez-Pelaez, J.: Algebraic Topological Analysis of Time-sequence of Digital Images. In: Ganzha, V.G., Mayr, E.W., Vorozhtsov, E.V. (eds.) CASC 2005. LNCS, vol. 3718, pp. 208–219. Springer, Heidelberg (2005)
14. Gonzalez-Diaz, R., Medrano, B., Real, P., Sánchez, J.: Reusing Integer Homology Information of Binary Digital Images. In: Kuba, A., Nyúl, L.G., Palágyi, K. (eds.) DGCI 2006. LNCS, vol. 4245, pp. 199–210. Springer, Heidelberg (2006)
15. Gonzalez-Diaz, R., Medrano, B., Real, P., Sánchez-Peláez, J.: Simplicial Perturbation Technique and Effective Homology. In: Ganzha, V.G., Mayr, E.W., Vorozhtsov, E.V. (eds.) CASC 2006. LNCS, vol. 4194, pp. 166–177. Springer, Heidelberg (2006)
16. Gonzalez-Diaz, R., Jiménez, M.J., Medrano, B., Real, P.: Extending AT-Models for Integer Homology Computation. In: Escolano, F., Vento, M. (eds.) GbRPR 2007. LNCS, vol. 4538, pp. 330–339. Springer, Heidelberg (2007)
17. Gonzalez-Diaz, R., Jiménez, M.J., Medrano, B., Real, P.: A Graph-with-Loop Structure for a Topological Representation of 3D Objects. In: Kropatsch, W.G., Kampel, M., Hanbury, A. (eds.) CAIP 2007. LNCS, vol. 4673, pp. 506–513. Springer, Heidelberg (2007)
18. Gonzalez-Diaz, R., Jimenez, M.J., Medrano, B., Molina-Abril, H., Real, R.: Integral Operators for Computing Homology Generators at Any Dimension. In: Ruiz-Shulcloper, J., Kropatsch, W.G. (eds.) CIARP 2008. LNCS, vol. 5197, pp. 356–363. Springer, Heidelberg (2008)
19. Gonzalez-Diaz, R., Jimenez, M.J., Medrano, B., Real, P.: Chain homotopies for object topological representations. Discrete Applied Mathematics 157, 490–499 (2009)
20. Gutiérrez-Naranjo, M.A., Pérez-Jiménez, M.J.: Riscos-Núñez: On the degree of parallelism in membrane systems. Theoretical Computer Science 372, 183–195 (2007)
21. Hatcher, A.: Algebraic Topology. Cambridge Univ. Press, Cambridge (2001)
22. Holland, J.H.: Adaptation in Natural and Artificial Systems. University of Michigan Press, Ann Arbor (1975)
23. Loos, R., Nagy, B.: Parallelism in DNA and Membrane Computing. In: CiE, Local Proceedings, pp. 283–287 (2007)

24. McCulloch, W.S., Pitts, W.: A logical calculus of the ideas immanent in nervous activity. Bulletin of Mathematical Biophysics 5, 115–133 (1943)
25. Martín–Vide, C., Pazos, J., Păun, G., Rodríguez Patón, A.: A New Class of Symbolic Abstract Neural Nets: Tissue P Systems. In: Ibarra, O.H., Zhang, L. (eds.) COCOON 2002. LNCS, vol. 2387, pp. 290–299. Springer, Heidelberg (2002)
26. Martín–Vide, C., Pazos, J., Păun, G., Rodríguez Patón, A.: Tissue P systems. Theoretical Computer Science 296, 295–326 (2003)
27. Molina-Abril, H., Real, P.: Cell AT-models for digital volumes. In: Torsello, A., Escolano, F., Brun, L. (eds.) GbRPR 2009. LNCS, vol. 5534, pp. 314–323. Springer, Heidelberg (2009)
28. Molina-Abril, H., Real, P.: Advanced Homological information on 3D Digital volumes. In: da Vitoria Lobo, N., Kasparis, T., Roli, F., Kwok, J.T., Georgiopoulos, M., Anagnostopoulos, G.C., Loog, M. (eds.) S+SSPR 2008. LNCS, vol. 5342, pp. 361–371. Springer, Heidelberg (2008)
29. Păun, G.: Computing with membranes. Journal of Computer and System Sciences 61(1), 108–143 (2000)
30. Păun, G.: Membrane Computing. An Introduction. Springer, Berlin (2002)
31. Păun, A., Păun, G.: The power of communication: P systems with symport/antiport. New Generation Computing 20(3), 295–305 (2002)
32. Păun, G., Pérez–Jiménez, M.J.: Recent computing models inspired from biology: DNA and membrane computing. Theoria 18(46), 72–84 (2003)
33. Păun, G., Pérez–Jiménez, M.J., Riscos–Núñez, A.: Tissue P System with cell division. In: Second Brainstorming Week on Membrane Computing, Sevilla, Report RGNC 01/2004, pp. 380–386 (2004)
34. Real, P.: An Algorithm Computing Homotopy Groups. Mathematics and Computers in Simulation 42(4-6), 461–465 (1996)
35. Real, P.: Homological Perturbation Theory and Associativity. Homology Homotopy and Applications, pp. 51–88 (2000)
36. Real, P.: Connectivity forests for homological analysis of digital volumes. In: Cabestany, J., et al. (eds.) IWANN 2009, Part I. LNCS, vol. 5517, pp. 415–423. Springer, Heidelberg (2009)
37. Real, P., Molina-Abril, H., Kropatsch, W.: Homological tree-based strategies for image analysis. In: Computer Analysis and Image Patterns, CAIP (2009)
38. Sergeraert, F.: The computability problem in algebraic topology. Advances in Mathematics 104, 1–29 (1994)
39. The Tissue Simulator Website: http://www.tissuesimulator.es.kz
40. The P Systems Website: http://ppage.psystems.eu/

Collapses and Watersheds in Pseudomanifolds

Jean Cousty, Gilles Bertrand, Michel Couprie, and Laurent Najman

Université Paris-Est, Laboratoire d'Informatique Gaspard-Monge, A3SI, ESIEE
{j.cousty,g.bertrand,m.couprie,l.najman}@esiee.fr

Abstract. This work is settled in the framework of abstract simplicial complexes. We propose a definition of a watershed and of a collapse for maps defined on pseudomanifolds of arbitrary dimension. Through an equivalence theorem, we establish a deep link between these two notions: any watershed can be obtained by collapse iterated until idempotence, and conversely any collapse iterated until idempotence induces a watershed. We also state an equivalence result which links the notions of a watershed and of a collapse with the one of a minimum spanning forest.

Keywords: watershed, collapse, topology, segmentation.

1   Introduction

For topographic purposes, the watershed has been extensively studied during the 19th century [19]. One hundred years later, the watershed transform [16,23,3,14] was introduced by Digabel and Lantuéjoul for image segmentation and is now used as a fundamental step in many powerful segmentation procedures [6]. Intuitively, a watershed of a map, seen as a topographic surface, may be thought of as a separating subset of the domain, from which a drop of water can flow down towards several distinct minima.

Topology-preserving transformations, such as homotopic skeletonization [18,9], are used in many applications of image analysis to transform an object while leaving unchanged its topological characteristics. Applications in 2D and 3D are already widely spread. In particular, skeletons are often used as a simplification of the original data, which facilitates shape recognition, registration, or animation. A notion of a homotopic skeleton of a map has also been investigated and was proved to be useful for image processing [5,10,21]. Intuitively, a map H is said to be homotopic to a map F, if every level set of H is homotopic (in the sense of sets) to the corresponding level set of F. An (homotopic) ultimate skeleton H of F is a map homotopic to F such that lowering any value of H leads to a map which is no more homotopic to F.

Our main result in this paper is an equivalence theorem (Theorem 12) which establishes a deep link between watersheds and homotopy. Intuitively, it states that a set X is a watershed of a map F if and only if there exists an ultimate skeleton H of F such that X is exactly the set of points adjacent to several distinct minima of H. This result holds true in a large family of n-dimensional discrete spaces, namely the pseudomanifolds. This study is developed in the

P. Wiederhold and R.P. Barneva (Eds.): IWCIA 2009, LNCS 5852, pp. 397–410, 2009.

framework of simplicial complexes (triangulated objects) of arbitrary dimension. The notion of watershed that we use here is based on the drop of water principle [14,15] and the one of homotopy relies on the collapse operation [24], a topology-preserving transformation known in algebraic topology. In this context, we furthermore establish (Theorem 13) that any nonempty watershed in a pseudomanifold of dimension n is a simplicial complex of dimension $n-1$. At last, we present a strong link between watersheds, collapses and minimum spanning forests (Theorem 15).

The proofs of the properties presented in this article will be given in an extended version [13]. Notice that all notions and properties can be easily transposed (see [4]) to the framework of cubical complexes, which allows for handling digital images.

2   Simplicial Complexes and Pseudomanifolds

We call *(abstract) simplex* any finite nonempty set. The *dimension* of a simplex x is the number of its elements minus one. In the following, a simplex of dimension d will also be called a *d-simplex*. If x is a simplex, we set $\hat{x} = \{y \mid y \subseteq x, y \neq \emptyset\}$. A finite set X of simplices is a *cell* if there exists $x \in X$ such that $X = \hat{x}$.

Fig. 1a (resp. b,c and d) graphically represents a simplex x of dimension 0 (resp. 1, 2 and 3). Fig. 1e shows a cell composed of one 2-simplex, three 1-simplices and three 0-simplices.

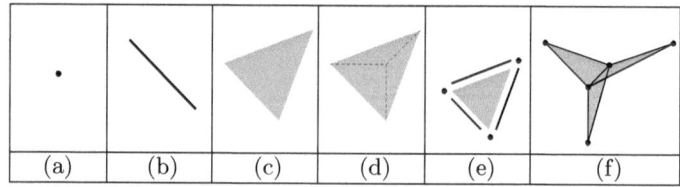

Fig. 1. (a, b, c, d) Simplices of dimension 0, 1, 2 and 3. (e) A 2-cell. (f) A complex.

If X is a finite set of simplices, we write $X^- = \bigcup_{x \in X} \hat{x}$, the set X^- is called the *(simplicial) closure of* X. A finite set X of simplices is a *(simplicial) complex* if $X = X^-$. Let X be a complex. Any element in X is a *face of* X and we call *d-face of* X any face of X whose dimension is d. Any d-face of X that is not contained in any $(d+1)$-face of X is called a *(d-) facet of* X. The *dimension* of X is the largest dimension of its faces. If d is the dimension of X, we say that X is *pure* whenever the dimension of all its facets equals d.

Definition 1. *A complex X of dimension n is an n-pseudomanifold if:*
i) X is pure;
ii) any $(n-1)$-face of X belongs to exactly two n-faces of X;
iii) for any two n-faces x, y of X, there exists a sequence $\langle x_0, \ldots, x_\ell \rangle$ of n-faces

of X such that, $x_0 = x$, $x_\ell = y$ and, for any i in $\{1, \ldots, \ell\}$, the intersection of x_{i-1} and x_i is an $(n-1)$-face of X.

The complex of Fig. 1f is pure, its dimension is 2, but it is not a 2-pseudomanifold. Indeed, it contains one 1-face that belongs to three 2-faces, and six 1-faces that belong to only one 2-face. Fig. 2a shows a subset of a 2-pseudomanifold.

Important notation. In this paper $\mathbb{M}$ stands for any n-pseudomanifold, where n is a positive integer.

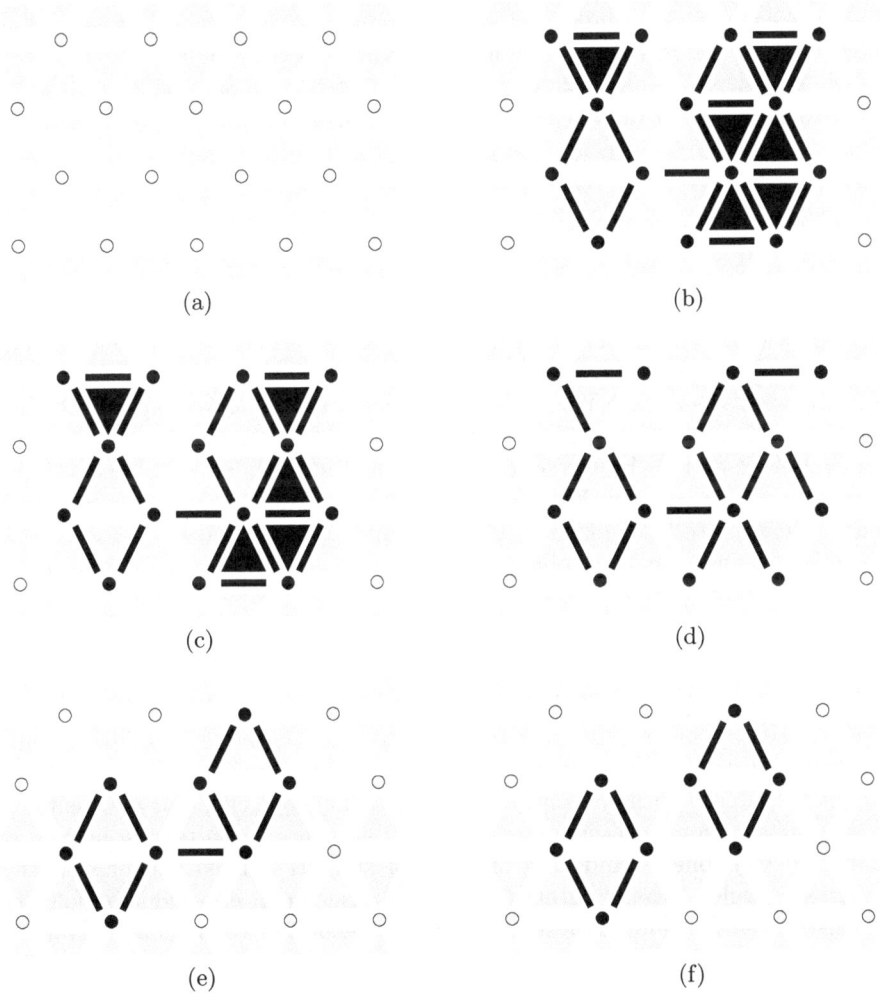

Fig. 2. (a) A subset of a 2-pseudomanifold $\mathbb{M}$. (b) A subcomplex X of $\mathbb{M}$ in black. (c) An elementary collapse Y of X. (d) An ultimate 2-collapse Z of both X and Y. (e) An ultimate 1-collapse W of Z. (f) A cut for $\overline{X}$.

Any subset of a complex X which is also a complex is called *subcomplex* of X. If Y is a subcomplex of X, we write $Y \preceq X$. If $X \subseteq \mathbb{M}$, we denote by $\overline{X}$ the complementary set of X in $\mathbb{M}$, *i.e.* $\overline{X} = \mathbb{M} \setminus X$. Note that if $X \preceq \mathbb{M}$, $\overline{X}$ is, in general, not a complex.

3   Collapse

In this section, we present the operation of collapse introduced by J.H.C. Whitehead [24], which is a discrete analogue of a retraction, that is, a continuous (homotopic) deformation of an object onto itself. This operation defined on the simplicial complexes is similar to the one of removal of a simple point defined on the subsets of the grid points $\mathbb{Z}^2$ and $\mathbb{Z}^3$ in the context of digital topology [18].

Let $X \preceq \mathbb{M}$, y be any face of X and $d - 1$ be the dimension of y. If there exists a unique face x of X which strictly contains y, we say that y is a *free face for* X and that the pair (x, y) is a *free pair* or a *free d-pair for* X.

Let $X \preceq \mathbb{M}$. If (x, y) is a free pair for X, the complex $X \setminus \{x, y\}$ is an *elementary collapse of* X or, more precisely, an *elementary d-collapse* if (x, y) is a free d-pair.

Let X and Y be two subcomplexes of $\mathbb{M}$. The complex Y is a *collapse of* X if there exists a collapse sequence from X to Y, *i.e.*, a sequence of complexes $\langle X_0, \ldots, X_\ell \rangle$ such that $X_0 = X$, $X_\ell = Y$ and X_i is an elementary collapse of X_{i-1}, for any i in $\{1, \ldots, \ell\}$. If each X_i is an elementary d-collapse of X_{i-1}, we also say that Y is a *d-collapse of* X. If Y is a d-collapse of X that does not contain any free d-pair for Y, Y is called an *ultimate d-collapse of* X.

Let X, Y, Z and W be the four subcomplexes in black in Figs. 2b,c,d and e respectively. The complex Y is an elementary 2-collapse of X. The complex Z is an ultimate 2-collapse of both X and Y, and W is a collapse of X, Y and Z which is also an ultimate 1-collapse of Z.

4   Cuts

Segmentation is the task of delineating objects of interest. In many cases, the result of such a process is a set of connected regions lying in a background which constitutes the separation between regions. Intuitively, a separation which cannot be reduced without connecting some regions is a cut. Our aim being to study segmentations in pseudomanifolds, we introduce hereafter the notions of connected components and cuts of a set of simplices. Then, we present two important results. First, we state (Property 3) that in an n-pseudomanifold the dimension of a nonempty cut is always $n - 1$. Secondly, we give an equivalence result (Theorem 5) between cuts and some subsets of the ultimate collapses. This last property leads to an efficient method to compute cuts in pseudomanifolds.

Let $A \subseteq \mathbb{M}$. Let $\pi = \langle x_0, \ldots, x_\ell \rangle$ be an ordered sequence of simplices in A, π is a *path* from x_0 to x_ℓ in A if, for any $i \in \{1, \ldots, \ell\}$, either $x_{i-1} \subseteq x_i$ or $x_i \subseteq x_{i-1}$. The set A is said to be *connected* if, for any two simplices x and y of A, there exists a path from x to y in A. Let $B \subseteq A$. We say that B is a *connected component* of A if B is a connected subset of A which is maximal for this property, *i.e.*, for any connected subset C of $\mathbb{M}$, $B \subseteq C \subseteq A$ implies $C = B$.

Definition 2. *Let A and B be two nonempty sets of simplices in $\mathbb{M}$.*
We say that B is an extension *of A if $A \subseteq B$, and if each connected component of B contains exactly one connected component of A. We also say that B is an extension of A if A and B are both empty.*
Let $X \subseteq \mathbb{M}$. We say that X is a cut *for A if $\overline{X}$ is an extension of A and if X is minimal for this property, i.e., for any $Y \subseteq X$, $\overline{Y}$ is an extension of A implies $Y = X$.*

For instance, the sets of gray simplices in Figs. 2c-f are extensions of the set A of gray simplices in Fig. 2b. The set of black simplices in Fig. 2f is a cut for A.

Intuitively, a "frontier" or a cut in an n-dimensional space should be an object of dimension $n - 1$. Nevertheless, we have shown [12] that the cuts (also called cleft in [12]) in the grids $\mathbb{Z}^2$, $\mathbb{Z}^3$, $\mathbb{Z}^n$ equipped with usual adjacency relations [18] cannot be considered as $(n-1)$-dimensional objects (the cuts are not necessarily thin). From this point of view, the next result shows that the framework of pseudomanifolds is an interesting alternative to usual adjacency relations.

Property 3. *Let A and X be two subsets of $\mathbb{M}$ such that $\overline{A}$ and X are two complexes. If X is a cut for A, then the complex X is either empty, or is a pure $(n - 1)$-complex. Furthermore, X does not contain any free n-pair.*

Let $X \subseteq \mathbb{M}$. We say that a face x of $\mathbb{M}$ is *adjacent to X* if there exists a simplex y in X such that either $x \subseteq y$ or $y \subseteq x$. A simplex x in X which is adjacent to more than one connected component of $\overline{X}$ is said to be *multi-connected for X*.

Definition 4. *Let $A \subseteq \mathbb{M}$ such that $\overline{A}$ is a complex. Let X be an ultimate n-collapse of $\overline{A}$ and let Y be the complex made of all faces which are multi-connected for X. We say that the complex Y is a* cut by collapse *for A.*

Following Section 3, in order to compute an ultimate n-collapse, we have to iterate steps of elementary collapse, until stability. Each step of elementary collapse requires only a local test. Moreover, the use of a breadth first strategy leads to a linear-time algorithm for computing ultimate collapses. However, the notion of a cut relies on a criterion, which is linked to connected components of the complement, and that requires a global computation (connected components labeling). Remark that the computation of a cut by collapse requires only one step of global computation: a labeling which can also be performed in linear-time. The following result establishes that any complex which is a cut, is a cut by collapse and that the converse is also true. Therefore, the above strategy constitutes a simple and linear-time algorithm to compute cuts.

Theorem 5. *Let $A \subseteq \mathbb{M}$ such that $\overline{A}$ is a complex. A complex X is a cut for A if and only if it is a cut by collapse for A.*

Remark 6. *Let $A \subseteq \mathbb{M}$. Note that if $\overline{A}$ is a complex, then there always exist a cut for A which is a complex. Indeed, in this case there exists by construction a* complex which is a cut by collapse for A and which is also, by Theorem 5, a cut for A. For instance, the complex in black in Fig. 2f is both a cut and a cut by collapse for A (in gray in Fig. 2b). On the other hand, if $\overline{A}$ is not a complex, there does not necessarily exist a cut for A which is a complex. For instance, the only cut (in black, Fig. 3b) for the set of gray simplices in Fig. 3a is not a complex. Observe also that this object contains a face of dimension 2, which would not be the case if $\overline{A}$ were a complex (see Property 3).

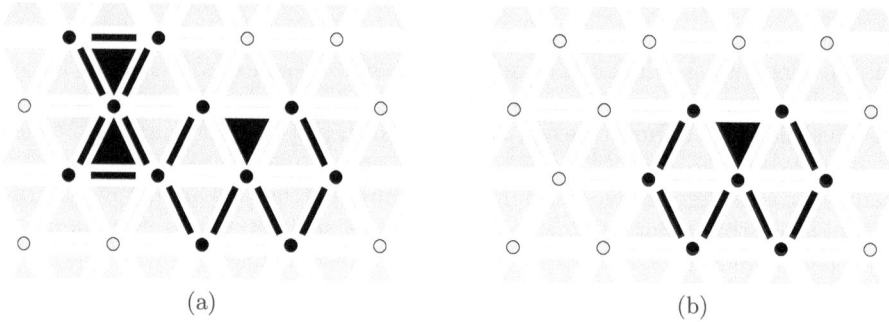

(a)   (b)

Fig. 3. Illustration of Remark 6. (a) A set A of simplices such that $\overline{A}$ (in black) is not a complex. (b) In black, the only cut for A.

5   Simplicial Stacks

This section presents some basic definitions relative to maps defined on a pseudomanifold. In particular, we introduce the simplicial stacks as the maps whose level sets are all simplicial complexes. This notion will be used in the next section to easily extend the operation of collapse from complexes to maps.

Here and subsequently $k_{\min}$ and $k_{\max}$ stand for two elements of $\mathbb{Z}$ such that $k_{\min} < k_{\max}$. We set $\mathbb{K} = \{k \in \mathbb{Z} \mid k_{\min} \leq k \leq k_{\max}\}$.

Let F be any map from $\mathbb{M}$ into $\mathbb{K}$. For any face x of $\mathbb{M}$, the value $F(x)$ is called the *altitude of x (for F)*. Let $k \in \mathbb{K}$. The *k-section of F*, denoted by $F[k]$, is the set of faces of $\mathbb{M}$ whose altitude is greater than or equal to k: $F[k] = \{x \in \mathbb{M} \mid F(x) \geq k\}$.

Let F be any map from $\mathbb{M}$ into $\mathbb{K}$, let $A \subseteq \mathbb{M}$ and let $k \in \mathbb{K}$. We say that A is a *minimum of F (at altitude k)* if A is a connected component of $\overline{F[k+1]}$ and if $A \cap \overline{F[k]} = \emptyset$. In the following, we denote by $M(F)$, the union of all minima of F and by $\overline{M}(F)$ its complementary set.

A desirable property is that any watershed of a stack F be both a complex and a cut for $M(F)$. However, as noted in Remark 6, if $\overline{M}(F)$ is not a complex, we cannot ensure that such a watershed exists. The notion of a simplicial stack, introduced hereafter, ensures that $\overline{M}(F)$ is a complex and thus that there exists a watershed of F which satisfies the above property.

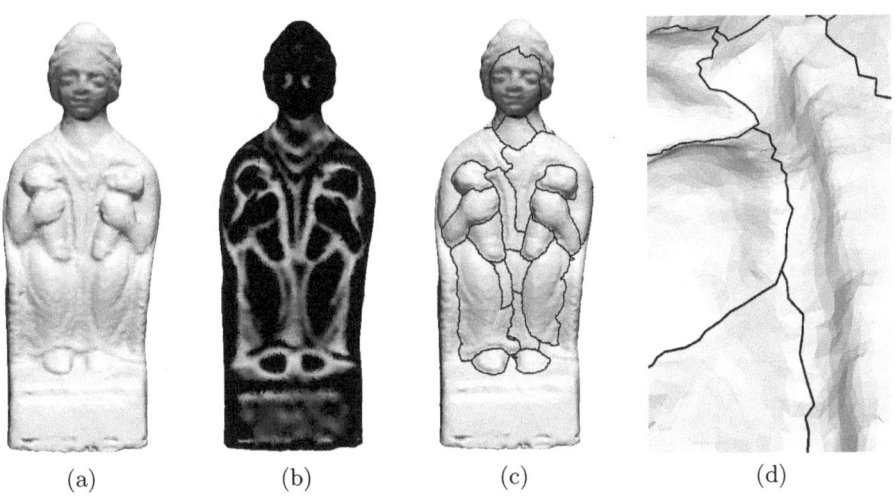

(a)   (b)   (c)   (d)

Fig. 4. (a) Rendering of a 2-pseudomanifold $\mathbb{M}$. (b) a simplicial stack F on $\mathbb{M}$ (which behaves like the inverse of the mean curvature of the surface, see [1]) (c) A watershed (in black) of F. (d) Zoom on a part of (c). The object $\mathbb{M}$ shown in (a) is provided by the French Museum Center for Research).

Definition 7. *A (simplicial) stack F (on $\mathbb{M}$) is a map from $\mathbb{M}$ into $\mathbb{K}$ such that, for any $k \in \mathbb{K}$, the k-section of F is a simplicial complex.*

Fig. 5a depicts a stack F and Figs. 5b, c and d depict in black the k-sections of F for respectively $k = 1, 2, 3$.

Remark 8. *Observe that we can obtain a stack H from any map F by considering the simplicial closure of F, i.e., H is the simplicial stack such that, for any $k \in \mathbb{K}$, the k-section of H is the simplicial closure of the k-section of F. Note that the simplicial closure H of a map F can be easily obtained by setting $H(x)$ to the maximum altitude for F of the simplices which contain x (i.e. $H(x) = \max\{F(y) \mid x \subseteq y \text{ and } y \in \mathbb{M}\}$), for any face x of $\mathbb{M}$.*

Important notation. In the sequel, F denotes a simplicial stack on $\mathbb{M}$.

6   Collapse of Simplicial Stacks

We propose an operation of a collapse of a simplicial stack based on the collapse operation in the sections of the stack. In this framework, an ultimate collapse of a stack can be seen as an analog of a homotopic grayscale skeleton in digital topology [5,21].

Let y be any face of $\mathbb{M}$, $d - 1$ be the dimension of y and $k = F(y)$. If y is a free face for $F[k]$, we say that y is a *free face for F*. If y is a free face for F, there exists a unique face x in $F[k]$ such that (x, y) is a free pair for $F[k]$ and we say that the pair (x, y) is a *free pair* or a *free d-pair for F*.

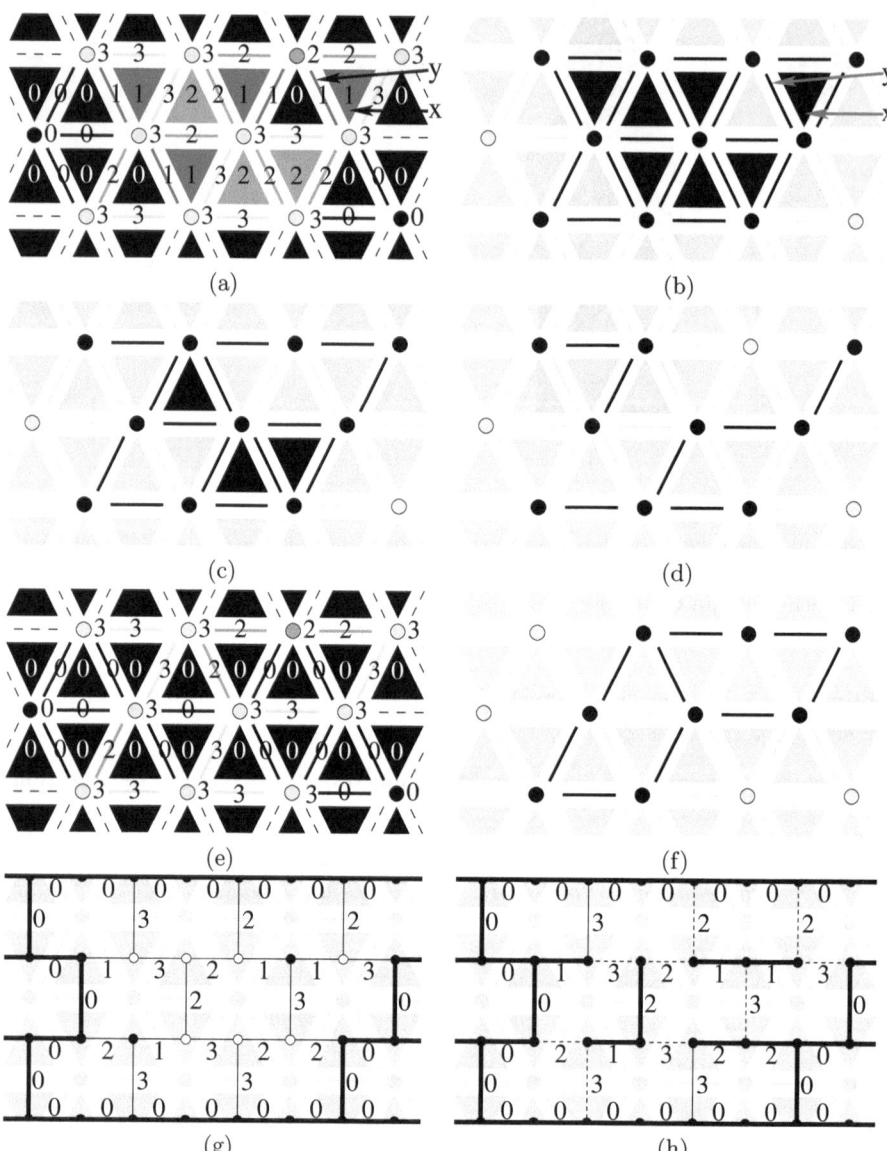

Fig. 5. (a) Representation of a simplicial stack F. The gray level of a face corresponds to its altitude which is also indicted by the superimposed number (the faces with no values are supposed to be at altitude 0). (b,c,d) The k-section of F for respectively $k = 1, 2, 3$. (e) An ultimate 2-collapse H of F. (f) A watershed X of F which is equal to the set of all multi-connected faces of $\overline{M}(H)$. (g) The graph G_M and the map F_G associated to the manifold $\mathbb{M}$ (light gray) and the stack F. In bold, a subgraph A of G_M. (h) An MSF relative to A (in bold) and an MSF cut for A (dashed edges).

In Fig. 5a, the 1-face y at altitude 1 is a free face for the depicted map F. Indeed, y is a free face for $F[1]$ (Fig. 5b). Thus, the pair (x, y) in Fig. 5a is a free pair for F. Note that, for any stack F, if (x, y) is a free pair for F, then $F(x) = F(y)$.

If (x, y) is a free pair for F, the simplicial stack H defined by $H(x) = H(y) = F(x) - 1$ and by $H(z) = F(z)$ for any $z \in \mathbb{M} \setminus \{x, y\}$ is called an *elementary collapse of F* or, more precisely, an *elementary d-collapse of F* if (x, y) is a free d-pair.

Let H be a simplicial stack on $\mathbb{M}$. We say H is a *collapse of F* if there exists a collapse sequence from F to H, i.e., a sequence of stacks $\langle F_0, \ldots, F_\ell \rangle$ such that $F_0 = F$, $F_\ell = H$ and F_i is an elementary collapse of F_{i-1}, $i \in \{1, \ldots, \ell\}$. If each F_i is an elementary d-collapse of F_{i-1}, we also say that H is a *d-collapse of F*. If H is a d-collapse of F which does not contain any free d-pair, H is called an *ultimate d-collapse of F*.

The stack H (Fig. 5e) depicts an ultimate 2-collapse of the map F (Fig. 5a).

Remark 9. *Note that a stack H is a collapse of F if and only if, for any $k \in \mathbb{K}$, the k-section of H is a collapse of the k-section of F. In this sense, we can say that the operation of collapse in simplicial stacks extends the one on simplicial complexes. In particular, if F is an* indicator map *of a complex X (i.e. a simplicial stack such that the altitude of the faces of X equals 1 and the altitude of the faces in $\overline{X}$ equals 0), it is equivalent to consider a collapse of F or to consider the indicator map of a collapse of the complex X.*

7   Watersheds

We now introduce the notion of a watershed in a pseudomanifold. Then, we present the main result (Theorem 12) of the section which establishes an equivalence between the watersheds of a map F and some sets of faces obtained from the ultimate collapses of F. In consequence, we can straightforwardly extract a watershed of a stack F from any ultimate collapse of F and, conversely, any watershed of F can be extracted from an ultimate collapse of F.

Intuitively, the "catchment basins" of a watershed constitute an extension of the minima and they are separated by a cut from which a drop of water can flow down towards distinct minima. Thus, before defining watersheds, we need the notion of a descending path.

Let $\pi = \langle x_0, \ldots, x_\ell \rangle$ be a path in $\mathbb{M}$. We say that the path π is *descending (for F)* if, for any $i \in \{1, \ldots, \ell\}$, $F(x_i) \leq F(x_{i-1})$.

Definition 10. *Let $X \preceq \mathbb{M}$ be a cut for $M(F)$. We say that X is a* watershed *of F if, for any $x \in X$, there exist two descending paths $\pi_1 = \langle x, x_0, \ldots, x_\ell \rangle$ and $\pi_2 = \langle x, y_0, \ldots, y_m \rangle$ such that:*
- x_ℓ and y_m are simplices of two distinct minima of F; and
- $x_i \notin X$, $y_j \notin X$, for any i in $\{0, \ldots, \ell\}$ and $j \in \{0, \ldots, m\}$.

For instance, the set of black faces in Fig. 5f is a watershed of the map F (Fig. 5a).

We call *divide* of a stack F the set $\overline{M}(F)$ of all faces of $\mathbb{M}$ which do not belong to any minimum of F. Observe that the divide of an ultimate collapse H of F is located on the "crests" of F (see, for instance, the non-null faces in Fig. 5e). On the other hand, we can say intuitively that a watershed of F corresponds to the "closed contours" located on the "crests" of F. Hence, a desirable property is that the watersheds of F correspond to the "closed contours" of the divide of the ultimate collapses of F. The following theorem asserts that this intuitive property is indeed true in the present framework.

Definition 11. *Let $X \subseteq \mathbb{M}$. We say that X is a* cut *by collapse for F if there exist an ultimate n-collapse H of F such that X is the set of all multi-connected elements for $\overline{M}(H)$.*

The next theorem, which is one of our main results, generalizes Theorem 5 to the case of stacks. It establishes a deep link between watersheds and homotopy.

Theorem 12. *Let $X \subseteq \mathbb{M}$. The set X is a watershed of F if and only if X is a cut by collapse for F.*

As far as we know, a similar property is not verified in other discrete frameworks. In particular, in digital topology, there is no such a straightforward relation between topological watersheds [3] and grayscale homotopic skeletons [5]. This point will be developed in a forthcoming extended version of the paper. On the other hand, a similar statement, linking watersheds and skeletons by influence zones, was presented by Najman and Schmitt [20] in the framework of continuous ($\mathcal{C}^2$) maps.

Remark that, there always exists a watershed of F. Indeed, since F is a stack, there exists by construction a cut by collapse for F which, by Theorem 12, is also a watershed of F. Furthermore, any watershed of F is by definition a complex and a cut for $M(F)$. Thus, we deduce the following result from Property 3.

Theorem 13. *Any watershed of F is either empty or is a pure $(n-1)$-complex.*

8   Minimum Spanning Forests

In this section, we establish a link between the framework studied in this paper and the one of watersheds and minimum spanning forests in edge-weighted graphs [14,15]. To this end, we first recall the definition of a minimum spanning forest relative to a graph. We will see that each of these forests induces a graph cut. Then, we propose a simple construction to derive a graph from $\mathbb{M}$ and we show the equivalence between the watersheds in $\mathbb{M}$ and the graph cuts induced by the minimum spanning forests relative to the minima in this graph. An important consequence is that the efficient algorithms to compute watersheds [14,15] and minimum spanning forests in edge-weighted graphs can be used to compute watersheds and collapses in simplicial complexes.

A *graph* is a simplicial complex of dimension 1. If G is a graph, we often consider the pair $(V(G), E(G))$ where $V(G)$ and $E(G)$ are respectively the set of 0-faces of G and the set of 1-faces of G. If G is a graph, any element of $V(G)$ is called a *vertex* or a *point of G* and any element of $E(G)$ is called an *edge of G*.

Let $A \subseteq \mathbb{M}$. We define the graph G_A such that the vertex set of G_A is composed of all n-simplices of A and such that two n-simplices x, y of A form an edge $\{x, y\}$ of G_A if $x \cap y$ is an $(n-1)$-simplex which belongs to A.

Important notation. In the sequel, we consider the graph $G_\mathbb{M}$ and the map F_G, from $E(G_\mathbb{M})$ into $\mathbb{K}$, defined by $F_G(\{x, y\}) = F(z)$ where $z = x \cap y$, for any $\{x, y\} \in E(G_\mathbb{M})$. The pair $(G_\mathbb{M}, F_G)$ forms an *edge-weighted graph*.

For instance, the edge-weighted graph $(G_\mathbb{M}, F_G)$ associated with the pseudo-manifold $\mathbb{M}$ and the stack F presented in Fig. 5a is shown in Fig. 5g.

Let us now introduce the notion of a spanning forest relative to a subgraph of $G_\mathbb{M}$. Generally, in graph theory, a forest is defined as a graph that does not contain any cycle. In this paper, the notion of forest is not sufficient since we want to deal with extensions of subgraphs that can contain cycles (such as the graph obtained from the minima of F). Therefore, we present hereafter the notion of a relative forest. Intuitively, a forest relative to a subgraph A of $G_\mathbb{M}$ is an extension B of A such that any cycle in B is also a cycle in A. In other words, to construct a forest relative to an arbitrary subgraph A of $G_\mathbb{M}$, one can add edges to A, provided that the added edges do not introduce new cycles and that the obtained graph remains an extension of A. Formally, the notion of cycle is not necessary to define a relative forest.

Let A and B be two nonempty subgraphs of $G_\mathbb{M}$. We say that B is a *forest relative to A* if:

i) B is an extension of A; and
ii) for any extension $C \subseteq B$ of A, we have $C = B$ whenever $V(C) = V(B)$.

We say that B is a *spanning forest relative to A (for $G_\mathbb{M}$)* if B is a forest relative to A and if $V(B) = V(G_\mathbb{M})$.

Informally speaking, condition ii) imposes that we cannot remove any edge from B while keeping an extension of A that has the same vertex set as B.

Remark that if the set A in the previous definition is a set of isolated vertices, we retrieve the usual notion of a forest.

Let A be a subgraph of $G_\mathbb{M}$, *the weight of A (for F_G)*, denoted by $F_G(A)$, is the sum of the weights of the edges in $E(A)$: $F_G(A) = \sum_{u \in E(A)} F_G(u)$.

Definition 14. *Let A and B be two subgraphs of $G_\mathbb{M}$. We say that B is a minimum spanning forest (MSF) relative to A (for F_G, in $G_\mathbb{M}$) if B is a spanning forest relative to A and if the weight of B is less than or equal to the weight of any other spanning forest relative to A.*

For instance, the graph made of bold edges and vertices in Fig. 5h is an MSF relative to the bold subgraph B of Fig. 5g.

The following theorem states that the graph associated with the minima of an ultimate n-collapse of F is an MSF relative to the graph associated with the minima of F. More remarkably, any MSF relative to the graph associated with the minima of F can be obtained thanks to an ultimate n-collapse of F. In this sense, the ultimate n-collapses on stacks can be said optimal.

Theorem 15. *Let A be a subgraph of G_M. The graph A is an MSF relative to the graph $G_{M(F)}$ if and only if there exists an ultimate n-collapse H of F such that $A = G_{M(H)}$.*

As a corollary of Theorem 15, we now establish the link between the watersheds in M and the minimum spanning forests in G_M.

Let A be a subgraph of G_M and let X be a set of edges of G_M. We say that X is an *MSF cut for A*, if there exists an MSF B relative to A such that X is the set of all edges of G_M adjacent to two distinct connected components of B.

If X is a set of $(n-1)$-faces of M, we set $Edge(X) = \{\{x, y\} \in E(G_M) \mid x \cap y \in X\}$.

Theorem 16. *Let X be a set of $(n-1)$-faces of M. The complex X^- is a watershed of F if and only if $Edge(X)$ is an MSF cut for $G_{M(F)}$.*

This theorem is illustrated in Fig. 5. Let Y be the set of 1-faces depicted in black in Fig. 5f. The complex $X = Y^-$ (Fig. 5f) is a watershed of the stack F (Fig. 5a) and the set $Edge(Y)$ (dashed edges in Fig. 5h) is an MSF cut for the graph $G_{M(F)}$ (shown in bold in Fig. 5g).

In consequence of Theorem 16, it should be noted that, to compute a watershed of a stack F, it is sufficient to compute, in the graph G_M, an MSF cut relative to the graph associated with the minima of F. An MSF cut can be computed using any minimum spanning tree algorithm (see [14]). The best complexity for solving this problem is reached by the quasi linear-time algorithm of Chazelle [8]. In fact, the problem of computing MSFs relative to the minima is simpler. In [14,15], two linear-time algorithms to compute MSF cuts relative to the minima are proposed. These two algorithms do not require any sorting step, or the use of a hierarchical queue or a representation to maintain unions of disjoint sets. Thanks to Theorems 15 and 16, these algorithms can be used to compute, in linear-time, ultimate n-collapses and watersheds in pseudomanifolds. The position of the contours produced by watershed algorithms on the plateaus is the subject of many discussions (see *e.g.* [20,22,2]). Depending on the implementation, the algorithm proposed in [15] allows for placing the watersheds either on the border of the plateaus or in the "middle". It is also possible to define explicitly the locus of the watershed on plateaus. For example, the algorithm presented in [11] allows for an optimal placement of the watershed on the plateaus with respect to random walkers criterion [17] or min cut criterion [7].

9   Conclusion

In this paper, we introduce the simplicial stacks as the maps on pseudomanifolds whose level sets are all simplicial complexes. Then, we propose definitions of watersheds and collapses for simplicial stacks. We establish a deep link between these two notions and also with the notion of a relative minimum spanning forest.

The proposed framework can be applied for segmenting the triangulated surface of 3D objects (see, *e.g.*, [1] and Fig. 4). It can also be easily transposed to cubical complexes which allows for handling digital images [4].

References

1. Alcoverro, M., Philipp-Foliguet, S., Jordan, M., Najman, L., Cousty, J.: Region-based artwork indexing and classification. In: Proc. 2-nd 3DTV-conference, pp. 393–396. IEEE, Los Alamitos (2008)
2. Audigier, R., Lotufo, R.: Uniquely-determined thinning of the tie-zone watershed based on label frequency. Journal of Mathematical Imaging and Vision 27(2), 157–173 (2007)
3. Bertrand, G.: On topological watersheds. Journal of Mathematical Imaging and Vision 22(2-3), 217–230 (2005)
4. Bertrand, G., Couprie, M., Cousty, J., Najman, L.: Chapitre 4: Ligne de partage des eaux dans les espaces discrets. In: Morphologie mathématique: approches déterministes, pp. 123–149. Hermes Sciences Publications (2008)
5. Bertrand, G., Everat, J.C., Couprie, M.: Image segmentation through operators based upon topology. Journal of Electronic Imaging 6(4), 395–405 (1997)
6. Beucher, S., Meyer, F.: The morphological approach to segmentation: the watershed transformation. In: Dougherty, E. (ed.) Mathematical Morphology in Image Processing, pp. 443–481. Marcel Decker, New York (1993)
7. Boykov, Y., Veksler, O., Zabih, R.: Fast approximate energy minimization via graph cuts. IEEE Transactions on Pattern Analysis and Machine Intelligence 23(11), 1222–1239 (2001)
8. Chazelle, B.: A minimum spanning tree algorithm with inverse-Ackermann type complexity. Journal of the ACM 47, 1028–1047 (2000)
9. Couprie, M., Bertrand, G.: New characterizations of simple points in 2D, 3D, and 4D discrete spaces. IEEE Transactions on Pattern Analysis and Machine Intelligence 31(4), 637–648 (2009)
10. Couprie, M., Bezerra, F.N., Bertrand, G.: Topological operators for grayscale image processing. Journal of Electronic Imaging 10(4), 1003–1015 (2001)
11. Couprie, C., Grady, L., Najman, L., Talbot, H.: Power watersheds: A new image segmentation framework extending graph cuts, random walker and optimal spanning forest. In: International Conference on Computer Vision. IEEE, Los Alamitos (to appear, 2009)
12. Cousty, J., Bertrand, G., Couprie, M., Najman, L.: Fusion graphs: merging properties and watersheds. Journal of Mathematical Imaging and Vision 30(1), 87–104 (2008)
13. Cousty, J., Bertrand, G., Couprie, M., Najman, L.: Watersheds and collapses in pseudomanifolds of arbitrary dimension (in preparation)
14. Cousty, J., Bertrand, G., Najman, L., Couprie, M.: Watershed cuts: Minimum spanning forests and the drop of water principle. IEEE Transactions on Pattern Analysis and Machine Intelligence 31(8), 1362–1374 (2009)
15. Cousty, J., Bertrand, G., Najman, L., Couprie, M.: Watershed cuts: thinnings, shortest-path forests and topological watersheds. IEEE Transactions on Pattern Analysis and Machine Intelligence (to appear, 2009)
16. Digabel, H., Lantuéjoul, C.: Iterative algorithms. In: 2nd European Symp. Quantitative Analysis of Microstructures in Material Science, Biology and Medicine, pp. 85–89 (1978)

17. Grady, L.: Random walks for image segmentation. IEEE Transactions on Pattern Analysis and Machine Intelligence 28(11), 1768–1783 (2006)
18. Kong, T., Rosenfeld, A.: Digital topology: Introduction and survey. Computer Vision, Graphics, and Image Processing 48(3), 357–393 (1989)
19. Maxwell, J.: On hills and dales. Philosophical Magazine 4/40, 421–427 (1870)
20. Najman, L., Schmitt, M.: Watershed of a continuous function. Signal Processing 38(1), 68–86 (1993)
21. Ranwez, V., Soille, P.: Order independent homotopic thinning for binary and grey tone anchored skeletons. Pattern Recognition Letters 23(6), 687–702 (2002)
22. Roerdink, J.B.T.M., Meijster, A.: The watershed transform: Definitions, algorithms and parallelization strategies. Fundamenta Informaticae 41(1-2), 187–228 (2001)
23. Vincent, L., Soille, P.: Watersheds in digital spaces: An efficient algorithm based on immersion simulations. IEEE Transactions on Pattern Analysis and Machine Intelligence 13(6), 583–598 (1991)
24. Whitehead, J.H.C.: Simplicial Spaces, Nuclei and m-Groups. Proc. London Math. Soc. s2-45(1), 243–327 (1939)

The Inscribed Square Conjecture in the Digital Plane

Feliú Sagols* and Raúl Marín

Departamento de Matemáticas
Centro de Investigación y de Estudios Avanzados del IPN
07000 México D.F., México
{sagols,rmarin}@math.cinvestav.mx

Abstract. The Inscribed Square Conjecture has been open since 1911. It states that any plane Jordan curve J contains four points on a non-degenerate square. In this article we prove that the conjecture holds for digital simple closed 4-curves, and that it is false for 8-curves. The given proof is based on a theorem due to Stromquist. We also discuss some properties of simple closed 4-curves in the digital plane containing a single non-degenerate inscribed square.

Keywords: digital topology, Inscribed Square Conjecture, Jordan Curve Theorem, simple closed digital curves.

1 Introduction

A *Jordan curve (simple closed curve)* is described by the set of points $\omega(t) = (x(t), y(t)) \in \mathbb{R}^2$, with $x, y : [0, 1] \to \mathbb{R}$ continuous, such that $\omega(0) = \omega(1)$, and if $0 \le t_1 < t_2 < 1$, then $\omega(t_1) \ne \omega(t_2)$. A polygon P is *inscribed* in a set S if all the vertices of P belong to S.

Toeplitz introduced in 1911 the following conjecture known as the "Inscribed Square Conjecture" (see [14] and [8]).

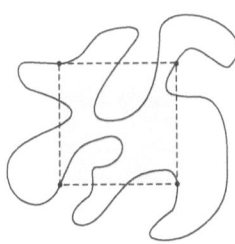

Fig. 1. A Jordan curve with a non-degenerate inscribed square

---

* Partially supported by the Mexican SNI (contract 7008).

P. Wiederhold and R.P. Barneva (Eds.): IWCIA 2009, LNCS 5852, pp. 411–424, 2009.
© Springer-Verlag Berlin Heidelberg 2009

Conjecture 1. Every Jordan curve (simple closed curve in the plane) contains four vertices of some non-degenerate square.

"Non-degenerate" means that the vertices of the square are distinct points. By "inscribed square" we denote in the following a "non-degenerate inscribed square".

It is quite simple to understand Conjecture 1, but it is hard to find solutions for particular Jordan curves, even for simple cases. Since 1911 several attempts to solve the conjecture have been made but the problem remains open. Nonetheless, some important results have been published since then.

Theorem 1 ([8]). *If the Jordan curve J is symmetric about a point c, and each ray issuing from c intersects J in a single point, then J admits an inscribed square.*

Theorem 2 ([8]). *Each convex Jordan curve admits an inscribed square.*

But the most remarkable result is due to Stromquist [13]. In order to introduce it we need a preliminary definition.

A simple closed curve ω in $\mathbb{R}^2$ is *locally monotone* if for every point p on the curve there are a real number $\epsilon > 0$ and a non-zero vector $n(p)$ such that for every pair p_1 and p_2 of distinct points in $B(p, \epsilon) \cap \omega$ (here $B(p, \epsilon)$ represents the open ball with center at p and radius ϵ), the equality $p_1 - p_2 = \lambda n(p)$ is not satisfied for any $\lambda \in \mathbb{R}$. In other words, no chord of ω contained in $B(p, \epsilon)$ is parallel to $n(p)$.

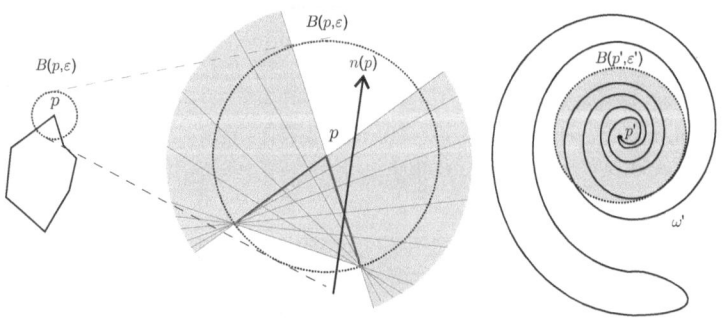

Fig. 2. Locally monotone curves

This definition was introduced by Stromquist in [13]. Actually it is quite general. Let us give an intuitive motivation using Figure 2. At the center of the figure we observe a shaded region representing the union of lines containing a chord in $\omega \cap B(p, \epsilon)$ (we term this the *span* of ω at p and ϵ) which separates the plane into two disjoint regions. Any vector crossing the span may play as $n(p)$. As long as ω is "smooth" in the neighborhood of p the span is close to a straight line and finding $n(p)$ is simple. But if ω has sudden changes within $B(p, \epsilon)$, then the span covers a greater area as shown in Figure 2 (at right), in this case the span of ω' at p' and ϵ' covers all the plane and there does not exist any possible assignment to $n(p')$. In this particular case, if we use a smaller value of ϵ' we can prove that the curve is locally monotone at p'. Now, if we replace the spiral around p' in Figure 2 (at right) by another similar spiral with infinitely many turns preserving the Jordan curve (see Figure 3 for an approximate graphic representation), the resulting curve will not be locally monotone. For Jordan curves containing these pathologies the Inscribed Square Conjecture remains open.

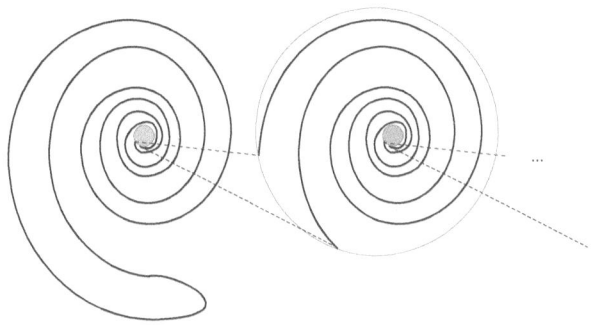

Fig. 3. Non-locally monotone curve

Theorem 3 (Stromquist [13]). *If ω is a locally monotone curve in $\mathbb{R}^2$, then ω admits an inscribed square.*

Smooth curves, convex curves, and most piecewise C^1 curves are locally monotone and satisfy Theorem 3. But in general, it is possible that most curves fail to admit inscribed squares (see [11]).

Other interesting articles on the Inscribed Square Conjecture include the following: [2], [6], [3], [7], [5], and [11].

This work addresses theoretical issues of Conjecture 1 on the Digital Plane. But it also opens the opportunity to devise new applications in Digital Topology. In the literature there are few reported applications to the Inscribed Square Conjecture; one application is provided by Fenn.

Theorem 4 (Table Theorem [3]). *Let D be a bounded, convex set in $\mathbb{R}^2$, and let $f : \mathbb{R}^2 \to \mathbb{R}$ be a continuous function which is non-negative in the interior of D and zero elsewhere. Let $d > 0$ be fixed. Then there is a square of side d whose center is in D, such that f takes the same value at all four vertices of the square.*

Fenn interprets this theorem in terms of making a square table stand level on a bumpy floor. This theorem is related to the Inscribed Square Conjecture when we consider the non-trivial level curves of f. In the digital plane, with the results given in this paper, we can prove a similar result which could be useful in digital terrain modeling applications, for example to find locations for facilities requiring square basements.

From the digital topology perspective introduced here, it is easy to devise other applications. For example (as pointed out by a reviewer), in image recognition we can use the number of squares contained in a digital curve as an invariant through rotations, translations and reflections on the digital plane. In [10] it is proved that if a digital 4-curve contains only one square then it is also invariant under scaling on the digital plane (see Figure 4). Other applications in areas such as robot motion planing and integrated circuit design seem plausible.

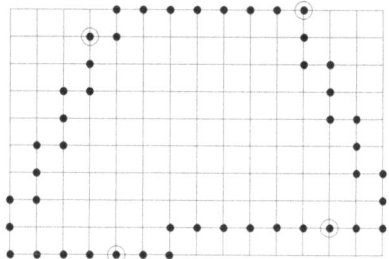

Fig. 4. Example of a 4-curve containing a single inscribed square (provided by Enrique García-Moreno [4])

In Section 2 we discuss a topological graph theory version of the Inscribed Square Conjecture and introduce some useful mathematical concepts. Section 3 introduces the digital plane as well as fundamental concepts from digital topology. In Section 4 we answer positively the digital version of Conjecture 1 under 4-connectivity and provide a counter example for 8-connectivity. Finally, in Section 5 we discuss some properties relevant to simple closed digital 4-curves with a single inscribed square.

2   The Inscribed Square Conjecture from a Topological Graph Theory Perspective

Before establishing and solving the digital topology version of the Inscribed Square Conjecture, we introduce in this section a Topological Graph Theory version of the problem. We use some conventional Graph Theory concepts (see [1]).

Given a simple graph G, a *planar graph embedding* of G is a function ψ_G transforming every vertex in $V(G)$ into a point on the plane and every edge in G into a homeomorphic copy of the closed real interval $[0, 1]$. The images under ψ_G of two distinct edges in $E(G)$ only meet at their endpoints and for each edge $vw \in E(G)$ the image $\psi_G(v)$ coincides with exactly one end point of $\psi_G(vw)$. A *path* in G of *length* n is a sequence of pairwise different vertices $v_0, \ldots, v_n$ (pairwise different with the possible exception of v_0 and v_n), where n is a non-negative integer, $v_0, v_1, \ldots, v_n \in V(G)$ and $(v_i, v_{i+1}) \in E(G)$ for $i = 0, \ldots, n-1$. If $v_0 = v_n$, then the path is a *cycle*. In a planar graph embedding ψ_G can be extended to paths making $\cup_{i=0,\ldots,n-1} \psi_G((v_i, v_{i+1}))$. In a planar graph embedding the image under ψ_G of a cycle is a Jordan curve.

The grid $\mathbb{Z}^2$ is the infinite graph $(V(\mathbb{Z}^2), E(\mathbb{Z}^2))$ where $V(\mathbb{Z}^2)$ are the ordered pairs (i, j) such that $i, j \in \mathbb{Z}$, and $((i_1, j_1), (i_2, j_2))$ are in $E(\mathbb{Z}^2)$ if and only if $(i_1, j_1) - (i_2, j_2) \in \{(1, 0), (0, 1), (-1, 0), (0, -1)\}$. We consider the grid $\mathbb{Z}^2$ as a planar graph embedding $\psi_{\mathbb{Z}^2}$ where $\psi_{\mathbb{Z}^2}((i, j)) = (i, j)$ for each $(i, j) \in V(\mathbb{Z}^2)$ and $\psi_{\mathbb{Z}^2}(((i_1, j_1), (i_2, j_2)))$ is the line segment $t(i_1, j_1) + (1-t)(i_2, j_2)$ with $t \in [0, 1]$.

We met the Inscribed Square Conjecture through the following variant introduced by Enrique García-Moreno (see [4]).

Conjecture 2. For every cycle C in $\mathbb{Z}^2$, $\psi_{\mathbb{Z}^2}(C)$ always contains an inscribed square with integer coordinates vertices.

The author of this conjecture believed that finding a positive answer could solve Conjecture 1. His idea was to refine infinitely the grid $\mathbb{Z}^2$ to approximate any continuous Jordan curve. Certainly, for each refinement a square exists, and a convergent subsequence of squares also exists due to compactness; however, we cannot ensure that the limit of such a subsequence is non-degenerate.

We do not know whether it is possible to overcome this difficulty and finally whether Conjecture 2 helps to solve Conjecture 1. But the condition on the integral coordinates of the inscribed square claimed in Conjecture 2 suggests an interesting question: Is there a way to state an equivalent digital topology conjecture?

The answer is positive. A key fact is that any inscribed square in a simple closed digital curve necessarily has vertices with integer coordinates.

Conjecture 2 is true; we can prove it using the proof of Theorem 5 detailed below. However, there is a technical detail which requires a routine but lengthy argument. For the digital version of Conjecture 1 this technical detail is solved easily. The solution uses the fact that simple closed digital 4-curves (to be defined in Section 3) have at least eight points, and each point is 4-adjacent to exactly two different points. A complete proof of Conjecture 2 is in [10].

3   The Digital Plane

Several digital space models exist (see [9]), we use a very simple one.

The *digital plane* is the set of points in $\mathbb{R}^2$ having integer coordinates (points in $\mathbb{Z}^2$). Two distinct points p and q of $\mathbb{Z}^2$ are *8-adjacent* if each coordinate of one differs from the corresponding coordinate of the other by at most 1. Points p and q are *4-adjacent* if they are 8-adjacent and differ in just one of their coordinates.

Let k be either of the numbers 4 or 8. A digital *k-path* is a finite sequence $p_0, p_1, \ldots, p_n$ of points in $\mathbb{Z}^2$ such that if $|i - j| = 1$, then p_i and p_j are

k-adjacent. A *closed k-path* is a k-path such that $p_0 = p_n$. This definition was introduced by Rosenfeld [12] in his study on arcs and curves in digital topology.

Definition 1. *A simple closed digital 8-curve J (or simply, a simple closed 8-curve) is a closed 8-path $p_0, p_1, \ldots, p_n$ with $n \geq 4$ and $p_i \neq p_j$ for all i, j such that $0 \leq i < j < n$, and each point is 8-adjacent to exactly two other points of J.*

Definition 2. *A simple closed digital 4-curve J (or simply, a simple closed 4-curve) is a closed 4-path $p_0, p_1, \ldots, p_n$ with $n \geq 8$ and $p_i \neq p_j$ for all i, j such that $0 \leq i < j < n$, and each point is 4-adjacent to exactly two other points of J.*

A digital set $C \subset \mathbb{Z}^2$ is k-*connected* ($k \in \{4, 8\}$) if for any two points $q_1, q_2 \in C$ there exists a k-path $p_0, p_1, \ldots, p_n$ in C with $p_0 = q_1$ and $p_n = q_2$.

The digital plane definition is given as a discrete approximation to the real plane. It is intended that basic properties of the real plane be fulfilled in the digital one. An example is the proof due to Rosenfeld [12] of a digital version of the Jordan Curve Theorem: every Jordan curve divides the plane into two components, the bounded one (the *inside* of the curve) and the unbounded (the *outside* of the curve). He proved that any simple closed 8-curve (or 4-curve) separates the digital plane into two disjoint 4-connected (resp. 8-connected) regions. Thus, the combined use of 4- and 8-connectivity reproduces in the digital plane a fundamental property of the real plane.

In Section 4 it is proved that the digital version of the Inscribed Square Conjecture is true for simple closed 4-curves and false for 8-curves. If the answer to the original conjecture on real plane were "yes", then for the Inscribed Square Conjecture, 4-connectivity should be a better way to model connectivity in the digital plane than 8-connectivity. Otherwise 8-connectivity should be better.

A *square* in the digital plane is a set of four distinct points $p_1 = (a_1, b_1)$, $p_2 = (a_2, b_2)$, $p_3 = (a_3, b_3)$, and $p_4 = (a_4, b_4)$, that are vertices of a square on the real plane (ordered clockwise around the square boundary). An alternate definition making no reference to the real plane establishes that if $(a_2, b_2) = (a_1 + b_1 - b_4, a_4 - a_1 + b_1)$ and $(a_3, b_3) = (a_2 + b_2 - b_1, a_1 - a_2 + b_2)$, then p_1, p_2, p_3 and p_4 are the vertices of a square in the digital plane. That is, a 90° rotation of p_4 around p_1 should be equal to p_2, and a 90° rotation of p_1 around p_2 should be equal to p_3. This implies the following:

Lemma 1. *If $p_1 p_2 p_3 p_4$ is a square in the real plane, and two consecutive points (clockwise or counter clockwise) have integer coordinates, then all four points have integer coordinates.*

4   The Digital Inscribed Square Conjecture

Let us introduce the digital version of Conjecture 1 for 8-connectivity.

Conjecture 3. Every simple closed 8-curve contains four points of some non-degenerate square.

Proposition 1. *Conjecture 3 is false.*

Proof. Let C be the simple closed 8-curve: $(0,1)$, $(1,2)$, $(2,3)$, $(3,3)$, $(4,2)$, $(5,1)$, $(4,0)$, $(3,0)$, $(2,0)$, $(1,0)$, $(0,1)$. No subset of four different points in C forms a square in the digital plane, hence the conjecture is false (Figure 5 gives a graphical representation of C in this proof). $\square$

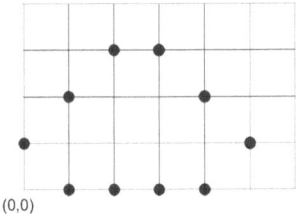

(0,0)

Fig. 5. A simple closed 8-curve without inscribed squares

In order to prove the main result in this paper, let us introduce some preliminary definitions. Let J be a simple closed 4-curve in the digital plane. As a sequence of points, it is a cycle in the infinite grid $\mathbb{Z}^2$ (see section 2) and the image of J under $\psi_{\mathbb{Z}^2}$ is by definition the *real plane embedding* of J.

Let J be a simple closed 4-curve and let p be a point in $\psi_{\mathbb{Z}^2}(J)$. Then p has *type h* with respect to $\psi_{\mathbb{Z}^2}(J)$ if p has integer coordinates or if p belongs to a line segment with end points (i,j), $(i+1,j)$; we use the notation $\text{type}(p) = h$. We say that p has *type v* with respect to $\psi_{\mathbb{Z}^2}(J)$ if p does not have type h and it belongs to a line segment with end points (i,j), $(i,j+1)$; we use the notation $\text{type}(p) = v$. If $p_1 p_2 p_3 p_4$ is a square, ordered clockwise, with vertices in $\psi_{\mathbb{Z}^2}(J)$, then the *type* of the square with respect to $\psi_{\mathbb{Z}^2}(J)$ is the sequence $\text{type}(p_1 p_2 p_3 p_4) := \text{type}(p_1)\text{type}(p_2)\text{type}(p_3)\text{type}(p_4)$. Note that the same square can have different types depending on the cyclic order of its vertices. When only one curve J is under discussion we usually omit the phrase "with respect to $\psi_{\mathbb{Z}^2}(J)$".

We develop some preliminary results to prove Theorem 5. First we prove that the plane embedding of any simple closed 4-curve is locally monotone (Lemma 2). Then, in Lemmas 3 and 6 we prove that if a non-degenerate square of type $hhhh$ and $hvhv$ is inscribed in the plane embedding of a simple closed 4-curve J, then J itself contains an inscribed square. We also prove in Lemmas 4 and 5 that squares of types $hhhv$ and $hhvv$ are never possible. Once we have these results the proof of Theorem 5 is straightforward.

Lemma 2. *If J is a simple closed 4-curve, then $\psi_{\mathbb{Z}^2}(J)$ is locally monotone.*

Proof. Let us prove that $\psi_{\mathbb{Z}^2}(J)$ is locally monotone. For every $p \in \psi_{\mathbb{Z}^2}(J)$ let us find a neighborhood $U(p)$ and a vector $n(p)$ such that no chord of J contained in $U(p)$ is parallel to $n(p)$.

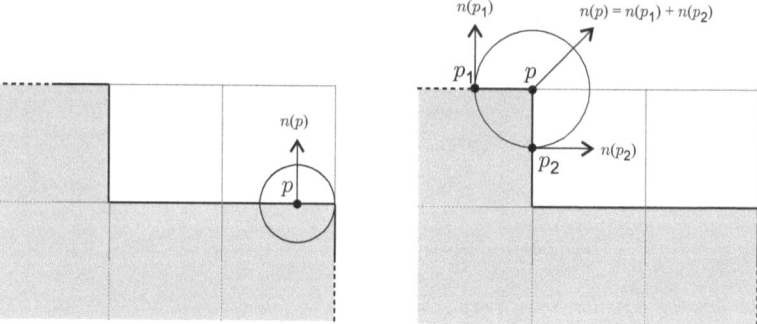

Fig. 6. Illustration of the cases contemplated by Lemma 2: case 1 (*left*) and case 2 (*right*)

Case 1. The first coordinate of p is not integer. Then $p = t(i, j) + (1-t)(i+1, j)$ for some $t \in (0, 1)$, and $i, j \in \mathbb{Z}$, then $U(p) = B(p, \min(\|p - (i, j)\|, \|(i+1, j) - p\|)$ and $n(p) = s(0, 0.5)$ where $s = 1$ or $s = -1$ depending upon whether $p + (0, 0.5)$ belongs to the outside of $\psi_{\mathbb{Z}^2}(J)$ or $p + (0, 0.5)$ belongs to the inside of $\psi_{\mathbb{Z}^2}(J)$, respectively. It is straightforward to prove that $U(p)$ and $n(p)$ satisfy the locally monotone condition. Analogous assignments to $U(p)$ and to $n(p)$ can be done if the second coordinate of p is not integer, that is $p = t(i, j) + (1-t)(i, j+1)$ for some $t \in (0, 1)$, and $i, j \in \mathbb{Z}$.

Case 2. The point p has integer coordinate entries, then let us denote by p_1 and p_2 the two meeting points between the circle with center at p and radius 0.5 and J. In this case $U(p) = B(p, 0.5)$ and $n(p) = n(p_1) + n(p_2)$ satisfy the locally monotone condition.   □

Henceforth, $p_1 p_2 p_3 p_4$ will denote a square in the real plane ordered clockwise. Let p and q be points in the real plane, $L(p, q)$ (resp. $R(p, q)$) denotes the point obtained when the point p is rotated $90°$ (resp. $-90°$) around q. Since the vertices of $p_1 p_2 p_3 p_4$ are ordered clockwise, the identities $R(p_2, p_1) = p_4$, $R(p_1, p_4) = p_3$, and $L(p_1, p_2) = p_3$ are satisfied.

Lemma 3. *Let J be a simple closed 4-curve. If $\psi_{\mathbb{Z}^2}(J)$ contains an inscribed square $p_1 p_2 p_3 p_4$ of type hhhh, then J also contains an inscribed square.*

Proof. From the definition of type, p_1, p_2, p_3 and p_4 have, respectively, coordinates $(a_1 + x_1, b_1)$, $(a_2 + x_2, b_2)$, $(a_3 + x_3, b_3)$ and $(a_4 + x_4, b_4)$, for some integer values $a_1, b_1, \ldots, a_4, b_4$ and some real values $x_1, x_2, x_3, x_4 \in [0, 1)$. Without losing generality we may assume that $a_1 = b_1 = 0$, and so $p_1 = (x_1, 0)$.

We know that $R(p_2, p_1) = p_4$, that is $(b_2 + x_1, -a_2 - x_2 + x_1) = (a_4 + x_4, b_4)$ (see Fig 7). This means that both $x_1 - x_4$ and $x_1 - x_2$ have integer values. But $x_1 - x_4$ cannot be greater than zero because otherwise x_1 would be greater than one and that is impossible. For a similar reason, $x_1 - x_4$ would not be less than

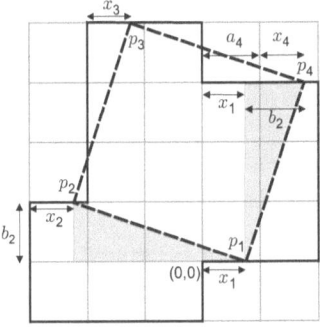 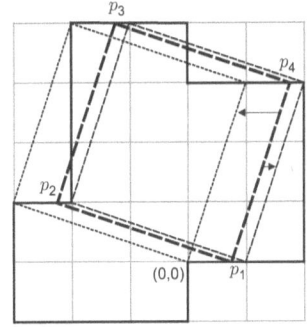

Fig. 7. Square of type $hhhh$ (*left*), and squares obtained from Lemma 3 (*right*)

zero. The only alternative for $x_1 - x_4$ to be an integer is $x_1 = x_4$. For the same reason $x_1 = x_2$.

Using the identity $R(p_1, p_4) = p_3$ we conclude that $x_1 = x_3$, and therefore $x_1 = x_2 = x_3 = x_4$. In this way, a translation of $p_1 p_2 p_3 p_4$ by x_1 units in the direction $(-1, 0)$ yields a non-degenerate square with vertices (a_1, b_1), (a_2, b_2), (a_3, b_3), and (a_4, b_4) which belong to $\psi_{\mathbb{Z}^2}(J)$ and have integer coordinates as required. This means that the vertices of this new square belong to J.

Note that if one of p_1, p_2, p_3, and p_4 is not in $\mathbb{Z}^2$, then there are two distinct non-degenerate squares inscribed in J. One is obtained from the translation given in the previous paragraph and the other is produced translating $p_1 p_2 p_3 p_4$ by $1 - x_1$ units along the direction $(1, 0)$ (see Fig 7). This observation will be useful to prove Proposition 2. □

Lemma 4. *Let J be a simple closed 4-curve. If $\psi_{\mathbb{Z}^2}(J)$ contains an inscribed square $p_1 p_2 p_3 p_4$, then type$(p_1 p_2 p_3 p_4)$ cannot be $hhhv$.*

Proof. Suppose that type$(p_1 p_2 p_3 p_4) = hhhv$. Then vertices p_1, p_2, p_3, and p_4 have coordinates $(a_1 + x_1, b_1)$, $(a_2 + x_2, b_2)$, $(a_3 + x_3, b_3)$, and $(a_4, b_4 + y_4)$, respectively. Here $x_1, x_2, x_3 \in [0, 1)$, $y_4 \in (0, 1)$, and $a_1, b_1, \ldots, a_4, b_4 \in \mathbb{Z}$. Without losing generality we may assume that $a_1 = b_1 = 0$, and so $p_1 = (x_1, 0)$.

We know that $R(p_2, p_1) = p_4$, so $(b_2 + x_1, -a_2 - x_2 + x_1) = (a_4, b_4 + y_4)$ and thus $x_1 = 0$ and $y_4 + x_2 - x_1$ is an integer. Similarly $L(p_1, p_2) = p_3$, that is $(b_2 + a_2 + x_2, x_1 - a_2 - x_2 + b_2) = (a_3 + x_3, b_3)$. Then $x_2 = x_3$ and $x_1 - x_2$ is an integer, as $x_1 = 0$ we have that $x_2 = 0$ as well. We conclude that $x_1 = x_2 = x_3 = 0$ and since $y_4 + x_2 - x_1$ must be an integer, y_4 must be an integer too, but the latter condition is impossible because $y_4 \in (0, 1)$. In other words, no inscribed square can have type $hhhv$. □

Lemma 5. *Let J be a simple closed 4-curve. If $\psi_{\mathbb{Z}^2}(J)$ contains an inscribed square $p_1 p_2 p_3 p_4$, then type$(p_1 p_2 p_3 p_4)$ cannot be $hhvv$.*

Proof. Suppose that type$(p_1 p_2 p_3 p_4) = hhvv$. Then p_1, p_2, p_3 and p_4 have, respectively, coordinates $(a_1 + x_1, b_1)$, $(a_2 + x_2, b_2)$, $(a_3, b_3 + y_3)$ and $(a_4, b_4 + y_4)$, for some integer values $a_1, b_1, \ldots, a_4, b_4$ and some real values $x_1, x_2 \in [0, 1)$ and $y_3, y_4 \in (0, 1)$. Without losing generality we may assume that $a_1 = b_1 = 0$, and so $p_1 = (x_1, 0)$.

Since $L(p_1, p_2) = (a_2 + b_2 + x_2, -a_2 - x_2 + b_2) = (a_3, b_3 + y_3)$, we have that $a_2 + b_2 + x_2 = a_3$, and thus, x_2 must be an integer, this entails $x_2 = 0$ and therefore y_3 is an integer. But this is impossible because $y_3 \in (0, 1)$. The conclusion is that $p_1 p_2 p_3 p_4$ cannot have type $hhvv$. □

Lemma 6. *Let J be a simple closed 4-curve. If $\psi_{\mathbb{Z}^2}(J)$ contains an inscribed square $p_1 p_2 p_3 p_4$ of type $hvhv$, then J contains an inscribed square.*

Proof. Here p_1, p_2, p_3 and p_4 have, respectively, coordinates $(a_1 + x_1, b_1)$, $(a_2, b_2 + y_2)$, $(a_3 + x_3, b_3)$ and $(a_4, b_4 + y_4)$, for some integer values $a_1, b_1, \ldots, a_4, b_4$ and some real values $x_1, x_3 \in [0, 1)$ and $y_2, y_4 \in (0, 1)$. Without losing generality we may assume that $a_1 = b_1 = 0$, and so $p_1 = (x_1, 0)$.

Since $R(p_2, p_1) = (b_2 + y_2 + x_1, x_1 - a_2) = (a_4, b_4 + y_4)$ (see Fig 8) we have that $y_4 - x_1$ and $y_2 + x_1$ are integers. Similarly from $L(p_1, p_2) = (b_2 + y_2 + a_2, x_1 - a_2 + y_2 + b_2) = (a_3 + x_3, b_3)$ we have that $y_2 - x_3$ is an integer.

Therefore $x_1 = y_4$, $y_2 = x_3$ and $y_2 + x_1 = 1$. Hence the additional identities $b_4 = -a_2$, $a_3 = b_2 + a_2$, $b_3 = b_2 - a_2 + 1$, and $a_4 = b_2 + 1$ are hold.

We claim that $p'_1 = (0, 0)$, $p'_2 = (a_2, b_2 + 1)$, $p'_3 = (a_3 + 1, b_3)$, and $p'_4 = (a_4, b_4)$ are vertices of an inscribed square of J. In fact, since the type of p_2 and p_4 is v, y_2 and y_4 are not integers, so x_1 and x_3 are not integers either because $x_1 = y_4$ and $y_2 = x_3$. From this and from the type of p_1, p_2, p_3 and p_4 we have that p'_1, p'_2, p'_3 and p'_4 are all points in J. We only have to verify that they form a square, that is, $L(p'_4, p'_1) = p'_2$, and $L(p'_1, p'_2) = p'_3$.

Developing both expressions we have that $L(p'_4, p'_1) = (-b_4, a_4) = (a_2, b_2 + 1) = p'_2$ and $L(p'_1, p'_2) = (a_2 + b_2 + 1, b_2 - a_2 + 1) = (a_3 + 1, b_3) = p'_3$, and thus $p'_1 p'_2 p'_3 p'_4$ is a square ordered clockwise.

Finally, it is proved that the square is non-degenerate; suppose contrarily that $p'_1 p'_2 p'_3 p'_4$ is a degenerate square. As $p'_1 = (0, 0)$, the identities $p'_2 = p'_3 = p'_4 =$

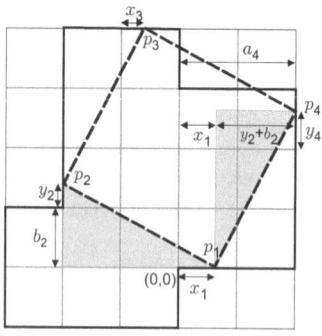
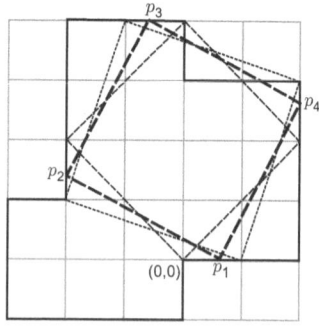

Fig. 8. Square of type $hvhv$ (*left*), and squares obtained from Lemma 6 (*right*)

$(0,0)$ must be satisfied, so $a_2 = b_3 = a_4 = b_4 = 0$ and $b_2 = a_3 = -1$. Now let us consider the original square vertices, $p_1 = (x_1, 0), p_2 = (0, -1 + y_2), p_3 = (-1 + x_3, 0)$, and $p_4 = (0, y_4)$. From the expression for p_1 and its type it follows that $(0, 0)$ belongs to J, and from the expressions and types of p_2, p_3 and p_4 it follows that $(0, -1), (-1, 0)$ and $(0, 1)$ belong to J. In other words, p_1' is 4-adjacent to 3 points in J: $(0, -1), (-1, 0)$ and $(0, 1)$, but that is impossible because J is a simple closed 4-curve (point p_1' must be 4-adjacent to exactly two other points in J). The conclusion is that $p_1' p_2' p_3' p_4'$ is a non-degenerate square.

In the same way, we can prove that $q_1 = (1, 0), q_2 = (a_2, b_2), q_3 = (a_3, b_3)$ and $q_4 = (a_4, b_4 + 1)$ form a second non-degenerate square (ordered clockwise) inscribed in J (see Fig 8). The proof that it is different from the previous one is given in Proposition 2. □

Theorem 5. *Every simple closed 4-curve J contains four vertices of some non-degenerate square.*

Proof. From Lemma 2 $\psi_{\mathbb{Z}^2}(J)$ is locally monotone, thus Theorem 3 guarantees the existence of an inscribed square S.

Let $p_1 p_2 p_3 p_4$ be vertices of S (recall that vertices are ordered clockwise). The sequence $\text{type}(p_1 p_2 p_3 p_4)$ belongs to $A := \{c_1 c_2 c_3 c_4 | c_1, c_2, c_3, c_4 \in \{h, v\}\}$. We can define an equivalence relation R on A: two elements in A are R related if and only if they represent the same cyclic sequence. Note that any cyclic order of $p_1 p_2 p_3 p_4$ represents S; in the same way, any cyclic order of $\text{type}(p_1)\text{type}(p_2)\text{type}(p_3)\text{type}(p_4)$ could be one type for S, and thus we only need to analyze a representative of each equivalence class of R to locate an inscribed square in $\psi_{\mathbb{Z}^2}(J)$ with integer coordinates. The equivalences classes of R are

1. $\{hhhh\}$
2. $\{hhhv, hvhh, vhhh, hhvh\}$
3. $\{hhvv, hvvh, vhhv, vvhh\}$
4. $\{hvhv, vhvh\}$
5. $\{vvvv\}$
6. $\{hvvv, vvhv, vhvv, vvvh\}$

We observe that a square of type $vvvv$ is transformed into a square of type $hhhh$ if we rotate the real plane $90°$. The same rotation transforms a square of type $hvvv$ into a square of type $vhhh$ if the first vertex does not have integer coordinates; otherwise it is transformed into an $hhhh$ type square. For this reason we only have to prove the existence of an inscribed square in J for types $hhhh, hhhv, hhvv, hvhv$. But from Lemmas 4 and 5 we know that the types $hhhv$ and $hhvv$ are impossible for $p_1 p_2 p_3 p_4$. For the types $hhhh$ and $hvhv$ the result follows from Lemmas 3, and 6. □

The next Proposition will be helpful in Section 5.

Proposition 2. *Let J be a simple closed 4-curve. If $p_1 p_2 p_3 p_4$ is an inscribed square in $\psi_{\mathbb{Z}^2}(J)$ and $p_i \notin \mathbb{Z}^2$ for some $i \in \{1, 2, 3, 4\}$, then J contains at least two different inscribed squares.*

Proof. Let $S = p_1p_2p_3p_4$ be an inscribed square in $\psi_{\mathbb{Z}^2}(J)$. For the same argument given in the proof of Theorem 5 we only have to prove the proposition when S has type $hhhh$ or $hvhv$.

If type$(p_1p_2p_3p_4)$ is $hhhh$ (resp. $hvhv$), the non-degenerate squares inscribed in J we are looking for are those found in the proof of Lemma 3 (resp. Lemma 6). For type $hhhh$ the squares are distinct, but for type $hvhv$ we have to prove that both squares are distinct.

Following the notation introduced in the proof of Lemma 6, the found vertices of the squares were $p_1' = (0,0)$, $p_2' = (a_2, b_2 + 1)$, $p_3' = (a_3 + 1, b_3)$, $p_4' = (a_4, b_4)$, and $q_1 = (1,0)$, $q_2 = (a_2, b_2)$, $q_3 = (a_3, b_3)$, $q_4 = (a_4, b_4 + 1)$. We should verify that $p_1'p_2'p_3'p_4'$ is not one of the four cyclic rotations of $q_1q_2q_3q_4$:

1. If $p_1'p_2'p_3'p_4' = q_1q_2q_3q_4$, then $(0,0) = p_1' = q_2 = (1,0)$, and so this case is impossible.

2. If $p_1'p_2'p_3'p_4' = q_4q_1q_2q_3$, then $p_1' = q_4$ and $p_2' = q_1$, and we conclude that $a_4 = 0$, $b_4 = -1$, $a_2 = 1$ and $b_2 = -1$, then $p_1' = (0,0)$, $p_2' = q_1 = (1,0)$, $p_3' = q_2 = (1,-1)$ and $p_4' = (0,-1)$. Points p_1', p_2', p_3' and p_4' are different and each one is 4-adjacent to exactly two points in J, but J is a simple closed 4-curve and by definition it contains at least eight different points. For this reason there is a point $p' \notin \{p_1', p_2', p_3', p_4'\}$ but 4-adjacent to p_j' for some $j \in \{1,2,3,4\}$. However, p_j' is 4-adjacent to three points in J: p', $p_{(j \bmod 4)+1}'$, and $p_{(j+2 \bmod 4)+1}'$. It is a contradiction, and then this case is impossible.

3. If $p_1'p_2'p_3'p_4' = q_3q_4q_1q_2$, then $p_3' = q_1 = (1,0)$, and since $p_1' = (0,0)$ and $p_1'p_2'p_3'p_4'$ is a square ordered clockwise, $p_2' = (0.5, 0.5)$ and $p_3' = (0.5, -0.5)$, but that is impossible because p_2' is in $\mathbb{Z}^2$.

4. If $p_1'p_2'p_3'p_4' = q_2q_3q_4q_1$, then $p_1' = q_2$ and $p_4' = q_1$, so it follows that $a_2 = 0$, $b_2 = 0$, $a_4 = 1$, and $b_4 = 0$. Thus $p_1' = (0,0)$, $p_2' = (0,1)$, $p_4' = (1,0)$, and then $p_3' = (1,1)$. Similarly to case 2 this is impossible.

We finally conclude that $p_1'p_2'p_3'p_4'$ and $q_1q_2q_3q_4$ are different inscribed squares in J of type $hvhv$.   □

5   Curves with a Single Inscribed Square

In the real plane there are infinite families of Jordan curves having more than one inscribed square; take for instance the family of circles. At the opposite extreme, there are infinite families of Jordan curves having a single inscribed square; ellipses with eccentricity greater than zero are examples of curves of this type. For digital topology, the number of inscribed squares contained in a simple closed 4-curve is always finite and it makes sense to ask whether some of these curves have exactly one inscribed square. The answer is yes; some examples of curves with this property appear in Figure 9. The curve at the upper-left corner of this figure is the minimal curve satisfying this property; it was found with a computer program.

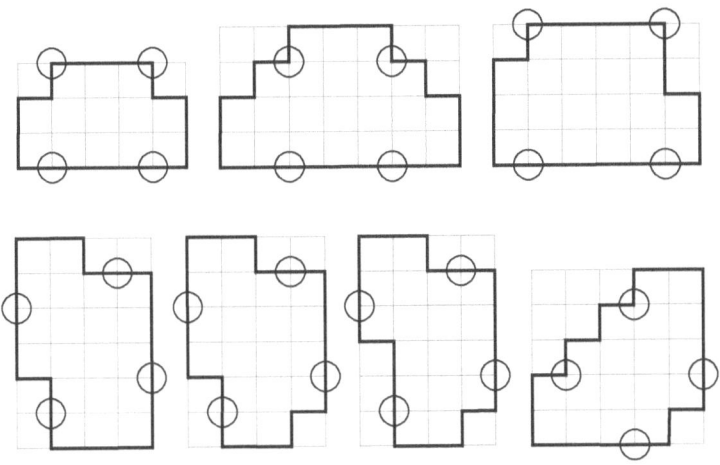

Fig. 9. Jordan curves with a single inscribed square

We conclude this paper with a theorem stating that simple closed 4-curves are a source for constructing Jordan curves in the real plane having a single inscribed square.

Theorem 6. *If J is a simple closed 4-curve with a single inscribed square, then $\psi_{\mathbb{Z}^2}(J)$ is a Jordan curve in the real plane with a single inscribed square.*

Proof. Let S be the single inscribed square in J. Then S is an inscribed square in $\psi_{\mathbb{Z}^2}(J)$ too. Let us assume that $\psi_{\mathbb{Z}^2}(J)$ contains a second inscribed square T different to S. Necessarily one of the vertices of T is not in $\mathbb{Z}^2$. From Proposition 2 J must contain two inscribed different squares, but that is impossible and so $\psi_{\mathbb{Z}^2}(J)$ has a single non-degenerate square inscribed.   □

We have found some infinite families of simple closed 4-curves having a single inscribed square. Unfortunately we do not have space enough to introduce them here. The problem of a full characterization of simple closed 4-curves with this property is left open.

Acknowledgments

The authors are grateful to the anonymous referees for their careful reading and valuable comments, to Shalom Eliahou for suggesting that the Inscribed Square Conjecture could be interpreted from a digital topology point of view, to Enrique García-Moreno for his comments about the Inscribed Square Conjecture and his example of a 4-curve with a single inscribed square, to Petra Wiederhold for her enthusiastic invitation to submit this work, and to Guillermo Morales and Michael Porter for the corrections to the final English presentation.

References

1. Diestel, R.: Graph Theory. Springer, Heidelberg (2005)
2. Emch, A.: Some Properties of Closed Convex Curves in a Plane. Amer. J. Math. 35, 407–412 (1913)
3. Fenn, R.: The Table Theorem. The Bulletin of the London Mathematical Society 2, 73–76 (1970)
4. García-Moreno, E.: On an Old Geometry Problem and Some of its Variants. In: International Workshop on Combinatorial and Computational Aspects of Optimization, Topology and Algebra (ACCOTA 2006), Mexico (2006)
5. Griffiths, H.B.: The topology of Square Pegs in Round Holes. Proc. London Math. Soc. 62, 647–672 (1991)
6. Guggenheimer, H.: Finite Sets on Curves and Surfaces. Israel J. Math. 3, 104–112 (1965)
7. Jerrard, R.P.: Inscribed Squares in Plane Curves. Transactions of the American Mathematical Society 98, 234–241 (1961)
8. Klee, V., Wagon, S.: Old and New Unsolved Problems in Plane Geometry and Number Theory. Dolciani Mathematical Expositions Series, vol. 11, pp. 58–65, 137–144 (1991)
9. Klette, R., Rosenfeld, A.: Digital Geometry. Morgan Kaufmann, San Francisco (2004)
10. Marín, R.: The Inscribed Square Conjecture from the Topological Graph Theory and Digital Topology Perspectives. Master Thesis, Departamento de Matemáticas, Centro de Investigación y de Estudios Avanzados del IPN (CINVESTAV-IPN), Mexico (to appear, 2009)
11. Nielsen, M.J., Wright, S.E.: Rectangles Inscribed in Symmetric Continua. Geometriae Dedicata 56, 285–297 (1995)
12. Rosenfeld, A.: Digital topology. American Mathematical Monthly 86, 621–630 (1979)
13. Stromquist, W.: Inscribed Squares and Square-like Quadrilaterals in Closed Curves. Mathematika 36, 187–197 (1989)
14. Toeplitz, O.: Verhandlungen der Schweizerischen Naturforschenden Gesellschaft in Solothurn. In: [§11], August 1, vol. 197 (1911)

Convenient Closure Operators on $\mathbb{Z}^2$

Josef Šlapal*

Brno University of Technology, Department of Mathematics,
616 69 Brno, Czech Republic
slapal@fme.vutbr.cz
http://at.yorku.ca/h/a/a/a/10.htm

Abstract. We discuss closure operators on $\mathbb{Z}^2$ with respect to which some cycles in a certain natural graph with the vertex set $\mathbb{Z}^2$ are Jordan curves. We deal with several Alexandroff T_0-pretopologies and topologies and also one closure operator that is not a pretopology.

Keywords: digital plane, closure operator, Alexandroff pretopology, connectedness graph, Jordan curve.

1 Introduction

To study geometric and topological properties of (two-dimensional) digital images, we need to provide the digital plane $\mathbb{Z}^2$ with a convenient structure. Here, convenience means that such a structure satisfies, in analogies, some basic geometric and topological properties of the Euclidean topology on the real plane $\mathbb{R}^2$. For example, it is usually required that an analogue of the Jordan curve theorem be valid. (Recall that the classical Jordan curve theorem states that any simple closed curve in the Euclidean plane divides this plane into exactly two components). In the classical approach to this problem (see e.g. [11] and [12]), graph theoretic tools are used for structuring $\mathbb{Z}^2$, namely the well-known binary relations of 4-adjacency and 8-adjacency. Unfortunately, neither 4-adjacency nor 8-adjacency itself allows an analogue of the Jordan curve theorem - cf. [8]. To eliminate this deficiency, a combination of the two binary relations has to be used. Despite this inconvenience, the graph-theoretic approach proved to be useful for solving many problems of digital image processing and for writing efficient graphic software. In [4], a new, purely topological approach to the problem is proposed which utilizes a convenient topology, called Khalimsky topology, for structuring the digital plane. The topological approach to digital topology was then developed by many authors - see, e.g., [3] and [6]-[9].

In [13] and [14] it was shown that closure operators more general than topologies, i.e., Kuratowski closure operators, may advantageously be used for structuring the digital plane $\mathbb{Z}^2$. In the present note, we continue the study of convenient closure operators on $\mathbb{Z}^2$. We introduce a certain natural graph on the vertex set

---
* The author acknowledges partial support from Ministry of Education of the Czech Republic, research plan no. MSM0021630518.

P. Wiederhold and R.P. Barneva (Eds.): IWCIA 2009, LNCS 5852, pp. 425–436, 2009.
© Springer-Verlag Berlin Heidelberg 2009

$\mathbb{Z}^2$ whose cycles are eligible for Jordan curves in $\mathbb{Z}^2$ and we solve the problem of finding closure operators on $\mathbb{Z}^2$ with respect to which some of these cycles are Jordan curves. Of these, a special attention will be paid to pretopologies and topologies. But a closure operator that is not a pretopology will be dealt with, too. The results obtained propose new structures on $\mathbb{Z}^2$ with natural Jordan curves which may, for example, be used in digital image processing for boundary detection.

2   Preliminaries

By a *closure operator* u on a set X we mean a map u: $\exp X \to \exp X$ (where $\exp X$ denotes the power set of X) fulfilling the following three axioms:

(i) $u\emptyset = \emptyset$,
(ii) $A \subseteq uA$ for all $A \subseteq X$,
(iii) $A \subseteq B \Rightarrow uA \subseteq uB$ whenever $A, B \subseteq X$.

The pair (X, u) is then called a *closure space* and, for every $A \subseteq X$, uA is called the *closure* of A. If axiom (iii) is replaced with the axiom

(iv) $u(A \cup B) = uA \cup uB$ whenever $A, B \subseteq X$,

which is stronger, then u is said to be a *pretopology* on X and (X, u) a *pretopological space*. Closure operators and pretopologies were studied, for example, by E. Čech in [1] and [2], respectively. A pretopology u on a set X that is idempotent (i.e., fulfils $uuA = uA$ whenever $A \subseteq X$) is nothing but the usual Kuratowski closure operator or, briefly, a *topology* on X (and the pair (X, u) is then a *topological space*).

If a closure operator u on a set X satisfies the axiom

(v) $uA = \bigcup_{x \in A} u\{x\}$ whenever $A \subseteq X$

(which is stronger than (iv)), then u and (X, u) are said to be *Alexandroff*. So, if u is an Alexandroff pretopology on a set X, then it is given by determining the closures of all points of X and there is an Alexandroff pretopology $\overline{u}$ on X given by $x \in \overline{u}\{y\} \Leftrightarrow y \in u\{x\}$ whenever $x, y \in X$. The pretopology $\overline{u}$ is said to be *dual* to u. Clearly, $\overline{\overline{u}} = u$ and a subset $A \subseteq X$ is closed (open) in (X, u) if and only it is open (closed) in $(X, \overline{u})$.

We will work with some basic topological concepts naturally extended from topological spaces to closure ones. Given a closure space space (X, u), a subset $A \subseteq X$ is called *closed* if $uA = A$, and it is called *open* if $X - A$ is closed. Observe that the closure of a subset of X need not be closed. A closure space (X, u) is said to be a *subspace* of a closure space (Y, v) if $X \subseteq Y$ and $uA = vA \cap X$ for each subset $A \subseteq X$. In this case we may simply say that X is a subspace of (Y, v) without explicitly mentioning the closure operator on X. A closure space (X, u) is said to be *connected* if $\emptyset$ and X are the only subsets of X which are both closed and open. A subset $X \subseteq Y$ is considered to be connected in a closure space (Y, v) if the subspace X of (Y, v) is connected. A maximal connected subset of a closure space is called a *component* of this space. Basic properties of connected sets and components in topological spaces (see e.g. [4]) are preserved also in closure spaces. A closure operator u on a set X is said to be a T_0-*closure operator* if, for arbitrary points $x, y \in X$, from $x \in u\{y\}$ and $y \in u\{x\}$ it follows that $x = y$, and it is called a $T_{\frac{1}{2}}$-*closure operator* if each singleton subset of X is closed or open (so that $T_{\frac{1}{2}}$ implies T_0).

As usual, given closure operators u and v on X, we put $u \leq v$ if $uA \subseteq vA$ for every $A \subseteq X$ (then u is said to be *finer* than v and v is said to be *coarser* than u).

By a *graph* on a set V we always mean an undirected simple graph without loops whose vertex set is V. Recall that a *path* in a graph is a finite (nonempty) sequence $x_0, x_1, ..., x_n$ of pairwise different vertices such that x_{i-1} and x_i are adjacent (i.e., joined by an edge) whenever $i \in \{1, 2, ...n\}$. By a *cycle* in a graph we understand any finite set of at least three vertices which can be ordered into a path whose first and last members are adjacent.

Recall also that the *connectedness graph* of a closure operator u on X is the graph on X in which a pair of vertices x, y is adjacent if and only if $x \neq y$ and $\{x, y\}$ is a connected subset of (X, u). If u is an Alexandroff pretopology on a set X, then a subset $A \subseteq X$ is connected in (X, u) if and only if each pair of points of A may be joined by a path in the connectedness graph of (X, u) contained in A. Clearly, u is then given by its connectedness graph provided that every edge of the graph is adjacent to a point which is known to be closed or to a point which is known to be open (in which case u is T_0). Indeed, the closure of a closed point consists of just this point, the closure of an open point consists of this point and all points adjacent to it and the closure of a mixed point (i.e., a point that is neither closed nor open) consists of this point and all closed points adjacent to it. In the sequel, only connectedness graphs of connected Alexandroff pretopological spaces (X, u) with $card\ X > 1$ will be displayed. In these graphs, the closed points will be ringed and the mixed ones boxed (so that the points neither ringed nor boxed will be open - note that no points of (X, u) may be both closed and open). Clearly, there are 2^n Alexandroff T_0-pretopologies having the same given connectedness graph with n edges (n a cardinal).

For every point $(x, y) \in \mathbb{Z}^2$, we denote by $A_4(x, y)$ or $A_8(x, y)$ the sets of all points that are 4-adjacent or 8-adjacent to (x, y), respectively. Thus, $A_4(x, y) = \{(x + i, y + j)\ i, j \in \{-1, 0, 1\},\ ij = 0,\ i + j \neq 0\}$ and $A_8(x, y) = A_4(x, y) \cup \{(x + i, y + j)\ i, j \in \{-1, 1\}\}$.

Definition 1. The *square-diagonal graph* is the graph on $\mathbb{Z}^2$ in which two points $z_1 = (x_1, y_1), z_2 = (x_2, y_2) \in \mathbb{Z}^2$ are adjacent if and only if one of the following four conditions is fulfilled:

1. $|y_1 - y_2| = 1$ and $x_1 = x_2 = 4k$ for some $k \in \mathbb{Z}$,
2. $|x_1 - x_2| = 1$ and $y_1 = y_2 = 4l$ for some $l \in \mathbb{Z}$;
3. $x_1 - x_2 = y_1 - y_2 = \pm 1$ and $x_1 - 4k = y_1$ for some $k \in \mathbb{Z}$,
4. $x_1 - x_2 = y_2 - y_1 = \pm 1$ and $x_1 = 4l - y_1$ for some $l \in \mathbb{Z}$.

A portion of the square-diagonal graph is shown in Figure 1.

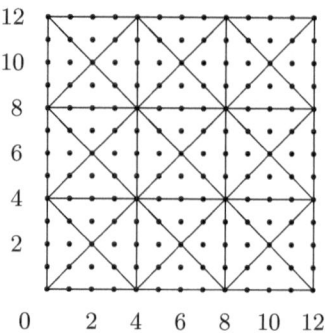

Fig. 1. A portion of the square-diagonal graph

When studying digital images, it may be advantageous to equip $\mathbb{Z}^2$ with a closure operator with respect to which some or even all cycles in the square-diagonal graph are Jordan curves. Such closure operators are considered to be convenient in this note.

When studying digital images, it may be advantageous to equip $\mathbb{Z}^2$ with a closure operator with respect to which some or even all cycles in the square-diagonal graph are Jordan curves. Such closure operators are considered to be convenient in this note.

3   Convenient Alexandroff Pretopologies on $\mathbb{Z}^2$

By a *(digital) simple closed curve* in an Alexandroff pretopological space (X, p) we mean, in accordance with [16], a nonempty, finite and connected subset $C \subseteq X$ such that, for each point $x \in C$, there are exactly two points of C adjacent to x in the connectedness graph of p. A simple closed curve C in (X, p) is said to be a *(digital) Jordan curve* if it separates (X, p) into precisely two components (i.e., if the subspace $X - C$ of (X, p) consists of precisely two components).

Recall [5] that the Khalimsky topology on $\mathbb{Z}^2$ is the Alexandroff topology t given as follows:

For any $z = (x, y) \in \mathbb{Z}^2$,

$$t\{z\} = \begin{cases} \{z\} \cup A_8(z) \text{ if } x, y \text{ are even,} \\ \{(x + i, y); i \in \{-1, 0, 1\}\} \text{ if } x \text{ is even and } y \text{ is odd,} \\ \{(x, y + j); j \in \{-1, 0, 1\}\} \text{ if } x \text{ is odd and } y \text{ is even,} \\ \{z\} \text{ otherwise.} \end{cases}$$

The Khalimsky topology is connected and T_0; a portion of its connectedness graph is shown in Figure 2.

Another well-known topology on $\mathbb{Z}^2$ is the Marcus-Wyse one (cf. [10]), i.e., the Alexandroff topology s on $\mathbb{Z}^2$ given as follows:

For any $z = (x, y) \in \mathbb{Z}^2$,

$$s\{z\} = \begin{cases} \{z\} \cup A_4(z) \text{ if } x + y \text{ is odd,} \\ \{z\} \text{ otherwise.} \end{cases}$$

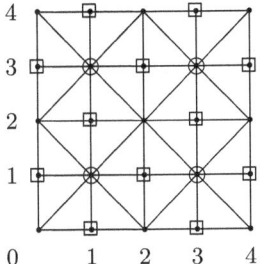

Fig. 2. A portion of the connectedness graph of the Khalimsky topology

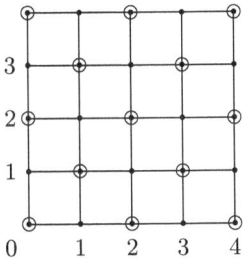

Fig. 3. A portion of the connectedness graph of the Marcus-Wyse topology

The Marcus-Wyse topology is connected and $T_{\frac{1}{2}}$. A portion of its connectedness graph is shown in the following Figure 3.

The topologies $\bar{t}$ and $\bar{s}$ dual to t and s are also called the Khalimsky and Marcus-Wyse topologies, respectively. It is readily confirmed that a cycle in the square-diagonal graph is a Jordan curve in the Marcus-Wyse topological space if and only if it does not employ diagonal edges. And a cycle in the square-diagonal graph is a Jordan curve in the Khalimsky topological space if and only if it does not turn, at any of its points, at the acute angle $\frac{\pi}{4}$ - cf. [5]. It could therefore be useful to replace the Khalimsky and Marcus-Wyse topologies with some more convenient connected topologies or pretopologies on $\mathbb{Z}^2$ that allow Jordan curves to turn at the acute angle $\frac{\pi}{4}$ at some points.

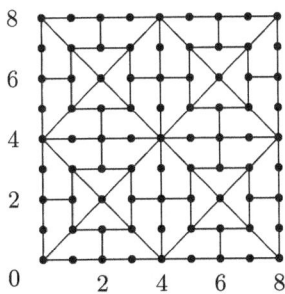 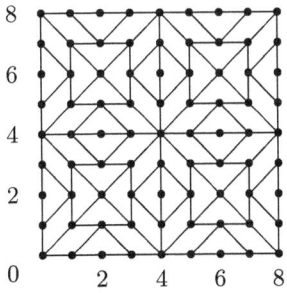

Fig. 4. Portions of two graphs on $\mathbb{Z}^2$

Theorem 1. *If u is an Alexandroff pretopology on $\mathbb{Z}^2$ whose connectedness graph coincides with one of the two graphs portions of which are shown in Figure 4, then every cycle in the square-diagonal graph is a Jordan curve in $(\mathbb{Z}^2, u)$.*

Proof. Clearly, any cycle in the square-diagonal graph is a simple closed curve in $(\mathbb{Z}^2, u)$. Let $z = (x, y) \in \mathbb{Z}^2$ be a point such that $x = 4k + p$ and $y = 4l + q$ for some $k, l, p, q \in \mathbb{Z}$ with $pq = \pm 2$. Then we define the *fundamental triangle* $T(z)$ to be the nine-point subset of $\mathbb{Z}^2$ given as follows:

$$
T(z) = \begin{cases}
\{(r, s) \in \mathbb{Z}^2;\ y - 1 \le s \le y + 1 - |r - x|\} \text{ if } x = 4k + 2 \text{ and} \\
\qquad y = 4l + 1 \text{ for some } k, l \in \mathbb{Z}, \\
\{(r, s) \in \mathbb{Z}^2;\ y - 1 + |r - x| \le s \le y + 1\} \text{ if } x = 4k + 2 \text{ and} \\
\qquad y = 4l - 1 \text{ for some } k, l \in \mathbb{Z}, \\
\{(r, s) \in \mathbb{Z}^2;\ x - 1 \le r \le x + 1 - |s - y|\} \text{ if } x = 4k + 1 \text{ and} \\
\qquad y = 4l + 2 \text{ for some } k, l \in \mathbb{Z}, \\
\{(r, s) \in \mathbb{Z}^2;\ x - 1 + |s - y| \le r \le x + 1\} \text{ if } x = 4k - 1 \text{ and} \\
\qquad y = 4l + 2 \text{ for some } k, l \in \mathbb{Z}.
\end{cases}
$$

Graphically, the fundamental triangle $T(z)$ consists of the point z and the eight points lying on the triangle surrounding z - the four types of fundamental triangles are represented in the following figure:

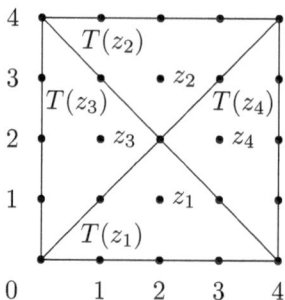

Given a fundamental triangle, we speak about its sides - it is clear from the above picture what sets are understood to be the sides (note that each side consists of five or three points and that two different fundamental triangles may have at most one common side).

Now, one can easily see that:

1. Every fundamental triangle is connected (so that the union of two fundamental triangles having a common side is connected) in $(\mathbb{Z}^2, u)$.
2. If we subtract from a fundamental triangle some of its sides, then the resulting set is still connected in $(\mathbb{Z}^2, u)$.
3. If S_1, S_2 are fundamental triangles having a common side D, then the set $(S_1 \cup S_2) - M$ is connected in $(\mathbb{Z}^2, u)$ whenever M is the union of some sides of S_1 or S_2 different from D.
4. Every connected subset of $(\mathbb{Z}^2, u)$ having at most two points is a subset of a fundamental triangle.

We will now show that the following is also true:

5. For every cycle C in the square-diagonal graph, there are sequences $\mathcal{S}_F, \mathcal{S}_I$ of fundamental triangles, $\mathcal{S}_F$ finite and $\mathcal{S}_I$ infinite, such that, whenever $\mathcal{S} \in \{\mathcal{S}_F, \mathcal{S}_I\}$, the following two conditions are satisfied:
   (a) Each member of $\mathcal{S}$, excluding the first one, has a common side with at least one of its predecessors.
   (b) C is the union of those sides of fundamental triangles from $\mathcal{S}$ that are not shared by two different fundamental triangles from $\mathcal{S}$.

Put $C_1 = C$ and let S_1^1 be an arbitrary fundamental triangle with $S_1^1 \cap C_1 \neq \emptyset$. For every $k \in \mathbb{Z}$, $1 \leq k$, if $S_1^1, S_2^1, ..., S_k^1$ are defined, let S_{k+1}^1 be a fundamental triangle with the following properties: $S_{k+1}^1 \cap C_1 \neq \emptyset$, S_{k+1}^1 has a side in common with S_k^1 which is not a subset of C_1 and $S_{k+1}^1 \neq S_i^1$ for all i, $1 \leq i \leq k$. Clearly, there will always be a (smallest) number $k \geq 1$ for which no such a fundamental triangle S_{k+1}^1 exists. We denote by k_1 this number so that we have defined a sequence $(S_1^1, S_2^1, ..., S_{k_1}^1)$ of fundamental triangles. Let C_2 be the union of those sides of fundamental triangles from $(S_1^1, S_2^1, ..., S_{k_1}^1)$ that are disjoint from C_1 and are not shared by two different fundamental triangles from $(S_1^1, S_2^1, ..., S_{k_1}^1)$. If $C_2 \neq \emptyset$, we construct a sequence $(S_1^2, S_2^2, ..., S_{k_2}^2)$ of fundamental triangles in an analogous way to $(S_1^1, S_2^1, ..., S_{k_1}^1)$ by taking C_2 instead of C_1 (and obtaining k_2 analogously to k_1). Repeating this construction, we get sequences $(S_1^3, S_2^3, ..., S_{k_3}^3)$, $(S_1^4, S_2^4, ..., S_{k_4}^1)$, etc. We put $\mathcal{S} = (S_1^1, S_2^1, ..., S_{k_1}^1, S_1^2, S_2^2, ..., S_{k_2}^2, S_1^3, S_2^3, ..., S_{k_3}^3, ...)$ if $C_i \neq \emptyset$ for all $i \geq 1$ and $\mathcal{S} = (S_1^1, S_2^1, ..., S_{k_1}^1, S_1^2, S_2^2, ..., S_{k_2}^2, ..., S_1^l, S_2^l, ..., S_{k_l}^l)$ if $C_i \neq \emptyset$ for all i with $1 \leq i \leq l$ and $C_i = \emptyset$ for $i = l+1$.

Further, let $S_1' = T(z)$ be a fundamental triangle such that $z \notin \mathcal{S}$ whenever S is a member of $\mathcal{S}$. Having defined S_1', let $\mathcal{S}' = (S_1', S_2', ...)$ be a sequence of fundamental triangles defined analogously to $\mathcal{S}$ (by taking S_1' in the role of S_1^1). Then one of the sequences $\mathcal{S}$, $\mathcal{S}'$ is finite and the other is infinite. (Indeed, $\mathcal{S}$ is finite or infinite, respectively, if and only if its first member equals such a fundamental triangle $T(z)$ for which $z = (k, l) \in \mathbb{Z}^2$ has the property that (1) k is even, l is odd and the cardinality of the set $\{(x, l) \in \mathcal{Z}^2; \; x > k\} \cap C$ is odd or even, respectively or (2) k is odd, l is even and the cardinality of the set $\{(k, y) \in \mathcal{Z}^2; \; y > l\} \cap C$ is odd or even, respectively. The same is true for $\mathcal{S}'$.) If we put $\{\mathcal{S}_F, \mathcal{S}_I\} = \{\mathcal{S}, \mathcal{S}'\}$ where $\mathcal{S}_F$ is finite and $\mathcal{S}_I$ is infinite, then the conditions (a) and (b) are clearly satisfied.

Given a cycle C in the square-diagonal graph, let S_F and S_I denote the union of all members of $\mathcal{S}_F$ and $\mathcal{S}_I$, respectively. Then $S_F \cup S_I = \mathbb{Z}^2$ and $S_F \cap S_I = C$. Let $\mathcal{S}_F^*$ and $\mathcal{S}_I^*$ be the sequences obtained from $\mathcal{S}_F$ and $\mathcal{S}_I$ by subtracting C from each member of $\mathcal{S}_F$ and $\mathcal{S}_I$, respectively. Let S_F^* and S_I^* denote the union of all members of $\mathcal{S}_F^*$ and $\mathcal{S}_I^*$, respectively. Then S_F^* and S_I^* are connected by (1), (2) and (3) and it is clear that $S_F^* = S_F - C$ and $S_I^* = S_I - C$. So, S_F^* and S_I^* are the two components of $\mathbb{Z}^2 - C$ by (4) ($S_F - C$ is the so-called *inside* component and $S_I - C$ is the so-called *outside* component). This proves the statement.

Observe that both connectedness graphs in Fig.4 are subgraphs of the 8-adjacency graph. (It is well known [3] that the Khalimsky and Marcus-Wyse topologies are the only topologies on $\mathbb{Z}^2$ whose connectedness graphs lie between the 4-adjacency graph and the 8-adjacency graph.) In the sequel, we will discuss some of the convenient Alexandroff pretopologies as specified in Theorem 1.

We denote by v the Alexandroff pretopology on $\mathbb{Z}^2$ given as follows:
For any point $z = (x, y) \in \mathbb{Z}^2$,

$$v\{z\} = \begin{cases} \{z\} \cup A_8(z) \text{ if } x = 4k, \ y = 4l, \ k, l \in \mathbb{Z}, \\ \{z\} \cup (A_8(z) - A_4(z)) \text{ if } x = 2 + 4k, \ y = 2 + 4l, \ k, l \in \mathbb{Z}, \\ \{z\} \cup \{(x-1, y), (x+1, y)\} \text{ if } x = 2 + 4k, \ y = 1 + 2l, \ k, l \in \mathbb{Z}, \\ \{z\} \cup \{(x, y-1), (x, y+1)\} \text{ if } x = 1 + 2k, \text{ and } y = 2 + 4l), \\ \{z\} \cup A_4(z) \text{ if either } x = 4k \text{ and } y = 2 + 4l \text{ or } x = 2 + 4k \text{ and } y = 4l, \ k, l \in \mathbb{Z}, \\ \{z\} \text{ otherwise.} \end{cases}$$

Clearly, v is connected and T_0. A portion of the connectedness graph of v is shown in Figure 5.

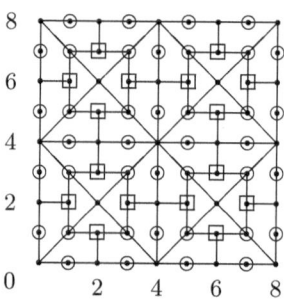

Fig. 5. A portion of the connectedness graph of v

By Theorem 1, we get

Corollary 1. *Every cycle in the square-diagonal graph is a Jordan curve in* $(\mathbb{Z}^2, v)$.

We denote by w the Alexandroff pretopology on $\mathbb{Z}^2$ given as follows:
For any point $z = (x, y) \in \mathbb{Z}^2$,

$$w\{z\} = \begin{cases} \{z\} \cup A_8(z) \text{ if } x = 4k, \ y = 4l, \ k, l \in \mathbb{Z}, \\ \{z\} \cup (A_8(z) - A_4(z)) \text{ if } x = 2 + 4k, \ y = 2 + 4l, \ k, l \in \mathbb{Z}, \\ \{z\} \cup (A_8(z) - (\{(x+i, y+1); i \in \{-1, 0, 1\}\} \cup \{(x, y-1)\})) \text{ if } x = 2 + 4k, \\ \quad y = 1 + 4l, \ k, l \in \mathbb{Z}, \\ \{z\} \cup (A_8(z) - (\{(x+i, y-1); i \in \{-1, 0, 1\}\} \cup \{(x, y+1)\})) \text{ if } x = 2 + 4k, \\ \quad y = 3 + 4l, \ k, l \in \mathbb{Z}, \\ \{z\} \cup (A_8(z) - (\{(x+1, y+j); j \in \{-1, 0, 1\}\} \cup \{(x-1, y)\})) \text{ if } x = 1 + 4k, \\ \quad y = 2 + 4l, \ k, l \in \mathbb{Z}, \\ \{z\} \cup (A_8(z) - (\{(x-1, y+j); j \in \{-1, 0, 1\}\} \cup \{(x+1, y)\})) \text{ if } x = 3 + 4k, \\ \quad y = 2 + 4l, \ k, l \in \mathbb{Z}, \\ \{(x+i, y); i \in \{-1, 0, 1\}\} \text{ if } x = 2 + 4k, \ y = 4l, \ k, l \in \mathbb{Z}, \\ \{(x, y+j); j \in \{-1, 0, 1\}\} \text{ if } x = 4k, \ y = 2 + 4l, \ k, l \in \mathbb{Z}, \\ \{z\} \text{ otherwise.} \end{cases}$$

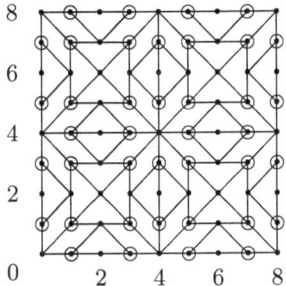

Fig. 6. A portion of the connectedness graph of w

Clearly, w is a connected $T_{\frac{1}{2}}$-topology. A portion of the connectedness graph of w is shown in Figure 6.

The topology w was introduced in [15] and then studied also in [16]. As a consequence of Theorem 1, we get

Corollary 2. *Every cycle in the square-diagonal graph is a Jordan curve in* $(\mathbb{Z}^2, w)$.

The previous statement is proved also in [15] (Theorem 11).

Let p, q and r be the Alexandroff pretopologies on $\mathbb{Z}^2$ with the connectedness graphs portions of which are shown in the following figures.

Pretopology p: Pretopology q:

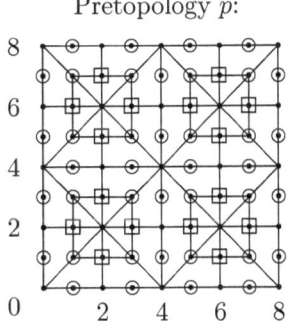   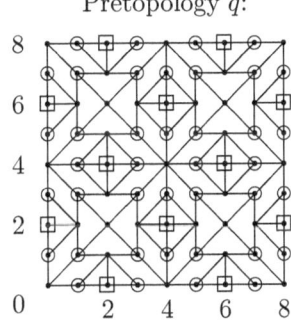

Pretopology r:

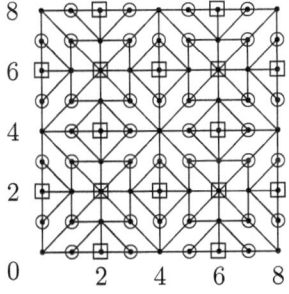

Clearly, p, q and r are T_0-pretopologies, q is even a topology and we have $v \leq p$ and $w \leq q \leq r$. Since Theorem 1 remains valid if we add, to either of the two graphs that occur in the Theorem, edges between arbitrary pairs of different vertices belonging to the same fundamental triangle (see the proof of the Theorem), we get

Corollary 3. *Every cycle in the square-diagonal graph is a Jordan curve in each of the pretopological spaces* $(\mathbb{Z}^2, p)$, $(\mathbb{Z}^2, q)$ *and* $(\mathbb{Z}^2, r)$.

Of course, Corollaries 1,2 and 3 remain valid when replacing Alexandroff pretopologies v, w and p, q, r by their duals, respectively (i.e., when interchanging open and closed points in these topologies).

4   A Convenient Closure Operator on $\mathbb{Z}^2$ That Is Not a Pretopology

Let $n > 1$ be an integer. An n-ary relation R on a set X is called *irreflexive* if $(x_1, ..., x_n) \in R$ implies that $x_i \neq x_j$ whenever $i, j \in \{1, ..., n\}$, $i \neq j$. A closure operator u on a set X is said to be a *relational closure operator of type n* and (X, u) is said to be a *relational closure space of type n* if there is an irreflexive n-ary relation R on X such that

$uA = A \cup \{x \in X;$ there exist $(x_1, ..., x_n) \in R$ and $i_0 \in \{2, ..., n\}$ such that $x = x_{i_0}$ and $x_i \in A$ for all i with $1 \leq i < i_0\}$

whenever $A \subseteq X$ (cf. [17]). We then put $u = u_R$. Clearly, relational closure operators u of type 2 are nothing but the Alexandroff pretopologies (the corresponding irreflexive binary relation R is given by $(x, y) \in R \Leftrightarrow y \in u\{x\} - \{x\}$). The definitions of a simple closed curve and a Jordan curve may be extended from Alexandroff pretopologies to relational closure operators as follows:

Definition 2. A *simple closed curve* in a relational closure space (X, u_R) of type n is a nonempty, finite and connected set $J \subseteq X$ such that every n-tuple $(z_1, ..., z_n) \in R$ with $\{z_1, z_2\} \subseteq J$ has the property $\{z_1, ..., z_n\} \subseteq J$ and every point $z \in J$ satisfies one of the following two conditions:

(1) There are precisely two n-tuples $(z_1, ..., z_n) \in R$ such that $\{z_1, ..., z_n\} \subseteq J$ and $z \in \{z_1, z_n\}$.
(2) There is precisely one n-tuple $(z_1, ..., z_n) \in R$ such that $\{z_1, ..., z_n\} \subseteq J$ and $z \in \{z_1, ..., z_n\} - \{z_1, z_n\}$.

A simple closed curve J in a relational closure space (X, u_R) is said to be a *(digital) Jordan curve* if it separates (X, u_R) into precisely two components.
   Let S be the ternary relation on $\mathbb{Z}^2$ given as follows:

$S = \{((x_i^1, x_i^2)| \ i = 1, 2, 3) \in (\mathbb{Z}^2)^3; \ \text{for every } j \in \{1, 2\}, \ x_1^j = x_2^j = x_3^j$
or there exists an odd number $k \in \mathbb{Z}$ fulfilling either $x_i^j = 2k + i - 1$ for all
$i = 1, 2, 3$ or $x_i^j = 2k - i + 1$ for all $i = 1, 2, 3\} - \{((x_i^1, x_i^2)| \ i = 1, 2, 3) \in (\mathbb{Z}^2)^3;$
$x_1^j = x_2^j = x_3^j$ for every $j \in \{1, 2\}\}$.

A portion of the relation S is demonstrated in the following figure. Each ordered triple $(a, b, c) \in S$ in this portion is represented as an arrow whose initial point, mid-point and terminal point are respectively a, b and c.

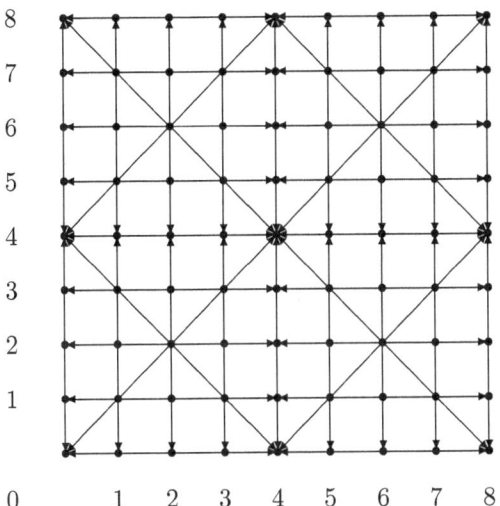

The following result may easily be obtained as a consequence of [14], Theorem 3.19:

Theorem 2. *Every cycle in the square diagonal graph that does not turn at any point $(4k + 2, 4l + 2)$, $k, l \in \mathbb{Z}$, is a Jordan curve in the relational closure space $(\mathbb{Z}^2, u_S)$ of type 3.*

It is obvious that u_S is not a pretopology. But it is a T_0-closure operator and has the property that, for every subset $A \subseteq \mathbb{Z}^2$, $u_S A = \bigcup \{u_S B;\ B \subseteq A,\ \text{card } B \leq 2\}$.

Example 1. Consider the following (digital picture of a) triangle:

E=(4,4)

(0,0)=A B   C D=(8,0)

While in $(\mathbb{Z}^2, u_S)$ the triangle ADE is a Jordan curve, in the Khalimsky space $(\mathbb{Z}^2, t)$ it is not. In order that this triangle be a Jordan curve in the Khalimsky space, we have to delete the points A,B,C and D. But this will lead to a considerable deformation of the triangle.

References

1. Čech, E.: Topological Spaces. In: Topological Papers of Eduard Čech, pp. 436–472. Academia, Prague (1968)
2. Čech, E.: Topological Spaces (Revised by Frolík, Z., Katětov, M.). Academia, Prague (1966)
3. Eckhardt, U., Latecki, L.J.: Topologies for the digital spaces $\mathbb{Z}^2$ and $\mathbb{Z}^3$. Comput. Vision Image Understanding 90, 295–312 (2003)
4. Engelking, R.: General Topology. Państwowe Wydawnictwo Naukowe, Warszawa (1977)
5. Khalimsky, E.D., Kopperman, R., Meyer, P.R.: Computer graphics and connected topologies on finite ordered sets. Topology Appl. 36, 1–17 (1990)
6. Khalimsky, E.D., Kopperman, R., Meyer, P.R.: Boundaries in digital planes. Jour. of Appl. Math. and Stoch. Anal. 3, 27–55 (1990)
7. Kiselman, C.O.: Digital Jordan curve theorems. In: Nyström, I., Sanniti di Baja, G., Borgefors, G. (eds.) DGCI 2000. LNCS, vol. 1953, pp. 46–56. Springer, Heidelberg (2000)
8. Kong, T.Y., Kopperman, R., Meyer, P.R.: A topological approach to digital topology. Amer. Math. Monthly 98, 902–917 (1991)
9. Kopperman, R., Meyer, P.R., Wilson, R.G.: A Jordan surface theorem for three-dimensional digital spaces. Discr. and Comput. Geom. 6, 155–161 (1991)
10. Marcus, D., et al.: A special topology for the integers (Problem 5712). Amer. Math. Monthly 77, 1119 (1970)
11. Rosenfeld, A.: Digital topology. Amer. Math. Monthly 86, 621–630 (1979)
12. Rosenfeld, A.: Picture Languages. Academic Press, New York (1979)
13. Šlapal, J.: A digital analogue of the Jordan curve theorem. Discr. Appl. Math. 139, 231–251 (2004)
14. Šlapal, J.: A quotient-universal digital topology. Theor. Comp. Sci. 405, 164–175 (2008)
15. Šlapal, J.: Closure operations for digital topology. Theor. Comp. Sci. 305, 457–471 (2003)
16. Šlapal, J.: Digital Jordan curves. Top. Appl. 153, 3255–3264 (2006)
17. Šlapal, J.: Relational closure operators. In: Contributions to General Algebra 16, pp. 251–259. Verlag Johannes Heyn, Klagenfurt (2005)

Author Index